NATURAL HISTORY
UNIVERSAL LIBRARY

西方博物学大系

主编：江晓原

AN INTRODUCTION TO BOTANY

植物学入门

［英］约翰·林德利 著

华东师范大学出版社

图书在版编目(CIP)数据

植物学入门 = An introduction to botany：英文 / (英)约翰·林德利(John Lindley)著. — 上海：华东师范大学出版社, 2018
（寰宇文献）
ISBN 978-7-5675-7995-8

Ⅰ.①植… Ⅱ.①约… Ⅲ.①植物学–普及读物–英文 Ⅳ.①Q94-49

中国版本图书馆CIP数据核字(2018)第181202号

植物学入门
An introduction to botany
(英)约翰·林德利(John Lindley)

特约策划	黄曙辉	徐 辰
责任编辑	庞 坚	
特约编辑	许 倩	
装帧设计	刘怡霖	

出版发行　华东师范大学出版社
社　　址　上海市中山北路3663号　邮编 200062
网　　址　www.ecnupress.com.cn
电　　话　021-60821666　行政传真　021-62572105
客服电话　021-62865537
门市（邮购）电话　021-62869887
地　　址　上海市中山北路3663号华东师范大学校内先锋路口
网　　店　http://hdsdcbs.tmall.com/

印 刷 者　虎彩印艺股份有限公司
开　　本　787×1092　16开
印　　张　38.5
版　　次　2018年8月第1版
印　　次　2018年8月第1次
书　　号　ISBN 978-7-5675-7995-8
定　　价　698.00元（精装全一册）

出 版 人　王　焰

（如发现本版图书有印订质量问题，请寄回本社客服中心调换或电话021-62865537联系）

《西方博物学大系》总序

江晓原

　　《西方博物学大系》收录博物学著作超过一百种，时间跨度为 15 世纪至 1919 年，作者分布于 16 个国家，写作语种有英语、法语、拉丁语、德语、弗莱芒语等，涉及对象包括植物、昆虫、软体动物、两栖动物、爬行动物、哺乳动物、鸟类和人类等，西方博物学史上的经典著作大备于此编。

中西方"博物"传统及观念之异同

　　今天中文里的"博物学"一词，学者们认为对应的英语词汇是 Natural History，考其本义，在中国传统文化中并无现成对应词汇。在中国传统文化中原有"博物"一词，与"自然史"当然并不精确相同，甚至还有着相当大的区别，但是在"搜集自然界的物品"这种最原始的意义上，两者确实也大有相通之处，故以"博物学"对译 Natural History 一词，大体仍属可取，而且已被广泛接受。

　　已故科学史前辈刘祖慰教授尝言：古代中国人处理知识，如开中药铺，有数十上百小抽屉，将百药分门别类放入其中，即心安矣。刘教授言此，其辞若有憾焉——认为中国人不致力于寻求世界"所以然之理"，故不如西方之分析传统优越。然而古代中国人这种处理知识的风格，正与西方的博物学相通。

　　与此相对，西方的分析传统致力于探求各种现象和物体之间的相互关系，试图以此解释宇宙运行的原因。自古希腊开始，西方哲人即孜孜不倦建构各种几何模型，欲用以说明宇宙如何运行，其中最典型的代表，即为托勒密（Ptolemy）的宇宙体系。

　　比较两者，差别即在于：古代中国人主要关心外部世界"如何"运行，而以希腊为源头的西方知识传统（西方并非没有别的知识传统，只是未能光大而已）更关心世界"为何"如此运行。在线性发展无限进步的科学主义观念体系中，我们习惯于认为"为何"是在解决了"如何"之后的更高境界，故西方的分析传统比中国的传统更高明。

然而考之古代实际情形，如此简单的优劣结论未必能够成立。例如以天文学言之，古代东西方世界天文学的终极问题是共同的：给定任意地点和时刻，计算出太阳、月亮和五大行星（七政）的位置。古代中国人虽不致力于建立几何模型去解释七政"为何"如此运行，但他们用抽象的周期叠加（古代巴比伦也使用类似方法），同样能在足够高的精度上计算并预报任意给定地点和时刻的七政位置。而通过持续观察天象变化以统计、收集各种天象周期，同样可视之为富有博物学色彩的活动。

还有一点需要注意：虽然我们已经接受了用"博物学"来对译 Natural History，但中国的博物传统，确实和西方的博物学有一个重大差别——即中国的博物传统是可以容纳怪力乱神的，而西方的博物学基本上没有怪力乱神的位置。

古代中国人的博物传统不限于"多识于鸟兽草木之名"。体现此种传统的典型著作，首推晋代张华《博物志》一书。书名"博物"，其义尽显。此书从内容到分类，无不充分体现它作为中国博物传统的代表资格。

《博物志》中内容，大致可分为五类：一、山川地理知识；二、奇禽异兽描述；三、古代神话材料；四、历史人物传说；五、神仙方伎故事。这五大类，完全符合中国文化中的博物传统，深合中国古代博物传统之旨。第一类，其中涉及宇宙学说，甚至还有"地动"思想，故为科学史家所重视。第二类，其中甚至出现了中国古代长期流传的"守宫砂"传说的早期文献：相传守宫砂点在处女胳膊上，永不褪色，只有性交之后才会自动消失。第三类，古代神话传说，其中甚至包括可猜想为现代"连体人"的记载。第四类，各种著名历史人物，比如三位著名刺客的传说，此三名刺客及所刺对象，历史上皆实有其人。第五类，包括各种古代方术传说，比如中国古代房中养生学说，房中术史上的传说人物之一"青牛道士封君达"等等。前两类与西方的博物学较为接近，但每一类都会带怪力乱神色彩。

"所有的科学不是物理学就是集邮"

在许多人心目中，画画花草图案，做做昆虫标本，拍拍植物照片，这类博物学活动，和精密的数理科学，比如天文学、物理学等等，那是无法同日而语的。博物学显得那么的初级、简单，甚至幼稚。这种观念，实际上是将"数理程度"作为唯一的标尺，用来衡量一切知识。但凡能够使用数学工具来描述的，或能够进行物理实验的，那就是"硬"科学。使用的数学工具越高深越复杂，似乎就越"硬"；物理实验设备越庞大，花费的金钱越多，似乎就越"高端"、越"先进"……

这样的观念，当然带着浓厚的"物理学沙文主义"色彩，在很多情况下是不正确的。而实际上，即使我们暂且同意上述"物理学沙文主义"的观念，博物学的"科学地位"也仍然可以保住。作为一个学天体物理专业出身，因而经常徜徉在"物理学沙文主义"幻影之下的人，我很乐意指出这样一个事实：现代天文学家们的研究工作中，仍然有绘制星图，编制星表，以及为此进行的巡天观测等等活动，这些活动和博物学家"寻花问柳"，绘制植物或昆虫图谱，本质上是完全一致的。

这里我们不妨重温物理学家卢瑟福（Ernest Rutherford）的金句："所有的科学不是物理学就是集邮（All science is either physics or stamp collecting）。"卢瑟福的这个金句堪称"物理学沙文主义"的极致，连天文学也没被他放在眼里。不过，按照中国传统的"博物"理念，集邮毫无疑问应该是博物学的一部分——尽管古代并没有邮票。卢瑟福的金句也可以从另一个角度来解读：既然在卢瑟福眼里天文学和博物学都只是"集邮"，那岂不就可以将博物学和天文学相提并论了？

如果我们摆脱了科学主义的语境，则西方模式的优越性将进一步被消解。例如，按照霍金（Stephen Hawking）在《大设计》（*The Grand Design*）中的意见，他所认同的是一种"依赖模型的实在论（model-dependent realism）"，即"不存在与图像或理论无关的实在性概念（There is no picture- or theory-independent concept of reality）"。在这样的认识中，我们以前所坚信的外部世界的客观性，已经不复存在。既然几何模型只不过是对外部世界图像的人为建构，则古代中国人干脆放弃这种建构直奔应用（毕竟在实际应用中我们只需要知道七政"如何"运行），又有何不可？

传说中的"神农尝百草"故事，也可以在类似意义下得到新的解读："尝百草"当然是富有博物学色彩的活动，神农通过这一活动，得知哪些草能够治病，哪些不能，然而在这个传说中，神农显然没有致力于解释"为何"某些草能够治病而另一些则不能，更不会去建立"模型"以说明之。

"帝国科学"的原罪

今日学者有倡言"博物学复兴"者，用意可有多种，诸如缓解压力、亲近自然、保护环境、绿色生活、可持续发展、科学主义解毒剂等等，皆属美善。编印《西方博物学大系》也是意欲为"博物学复兴"添一助力。

然而，对于这些博物学著作，有一点似乎从未见学者指出过，而鄙意以为，当我们披阅把玩欣赏这些著作时，意识到这一点是必须的。

这百余种著作的时间跨度为15世纪至1919年，注意这个时间跨度，正是西方列强"帝国科学"大行其道的时代。遥想当年，帝国的科学家们乘上帝国的军舰——达尔文在皇家海军"小猎犬号"上就是这样的场景之一，前往那些已经成为帝国的殖民地或还未成为殖民地的"未开化"的遥远地方，通常都是踌躇满志、充满优越感的。

作为一个典型的例子，英国学者法拉在（Patricia Fara）《性、植物学与帝国：林奈与班克斯》（*Sex, Botany and Empire, The Story of Carl Linnaeus and Joseph Banks*）一书中讲述了英国植物学家班克斯（Joseph Banks）的故事。1768年8月15日，班克斯告别未婚妻，登上了澳大利亚军舰"奋进号"。此次"奋进号"的远航是受英国海军部和皇家学会资助，目的是前往南太平洋的塔希提岛（Tahiti, 法属海外自治领，另一个常见的译名是"大溪地"）观测一次比较罕见的金星凌日。舰长库克（James Cook）是西方殖民史上最著名的舰长之一，多次远航探险，开拓海外殖民地。他还被认为是澳大利亚和夏威夷群岛的"发现"者，如今以他命名的群岛、海峡、山峰等不胜枚举。

当"奋进号"停靠塔希提岛时，班克斯一下就被当地美丽的土著女性迷昏了，他在她们的温柔乡里纵情狂欢，连库克舰长都看不下去了，"道德愤怒情绪偷偷溜进了他的日志当中，他发现自己根本不可能不去批评所见到的滥交行为"，而班克斯纵欲到了"连嫖妓都毫无激情"的地步——这是别人讽刺班克斯的说法，因为对于那时常年航行于茫茫大海上的男性来说，上岸嫖妓通常是一项能够唤起"激情"的活动。

而在"帝国科学"的宏大叙事中，科学家的私德是无关紧要的，人们关注的是科学家做出的科学发现。所以，尽管一面是班克斯在塔希提岛纵欲滥交，一面是他留在故乡的未婚妻正泪眼婆娑地"为远去的心上人绣织背心"，这样典型的"渣男"行径要是放在今天，非被互联网上的口水淹死不可，但是"班克斯很快从他们的分离之苦中走了出来，在外近三年，他活得倒十分滋润"。

法拉不无讽刺地指出了"帝国科学"的实质："班克斯接管了当地的女性和植物，而库克则保护了大英帝国在太平洋上的殖民地。"甚至对班克斯的植物学本身也调侃了一番："即使是植物学方面的科学术语也充满了性指涉。……这个体系主要依靠花朵之中雌雄生殖器官的数量来进行分类。"据说"要保护年轻妇女不受植物学教育的浸染，他们严令禁止各种各样的植物采集探险活动。"这简直就是将植物学看成一种"涉黄"的淫秽色情活动了。

在意识形态强烈影响着我们学术话语的时代，上面的故事通常是这样被描述的：库克舰长的"奋进号"军舰对殖民地和尚未成为殖民地的那些地方的所谓"访问"，其实是殖民者耀武扬威的侵略，搭载着达尔文的"小猎犬号"军舰也是同样行径；班克斯和当地女性的纵欲狂欢，当然是殖民者对土著妇女令人发指的蹂躏；即使是他采集当地植物标本的"科学考察"，也可以视为殖民者"窃取当地经济情报"的罪恶行为。

后来改革开放，上面那种意识形态话语被抛弃了，但似乎又走向了另一个极端，完全忘记或有意回避殖民者和帝国主义这个层面，只歌颂这些军舰上的科学家的伟大发现和成就，例如达尔文随着"小猎犬号"的航行，早已成为一曲祥和优美的科学颂歌。

其实达尔文也未能免俗，他在远航中也乐意与土著女性打打交道，当然他没有像班克斯那样滥情纵欲。在达尔文为"小猎犬号"远航写的《环球游记》中，我们读到："回程途中我们遇到一群黑人姑娘在聚会，……我们笑着看了很久，还给了她们一些钱，这着实令她们欣喜一番，拿着钱尖声大笑起来，很远还能听到那愉悦的笑声。"

有趣的是，在班克斯在塔希提岛纵欲六十多年后，达尔文随着"小猎犬号"也来到了塔希提岛，岛上的土著女性同样引起了达尔文的注意，在《环球游记》中他写道："我对这里妇女的外貌感到有些失望，然而她们却很爱美，把一朵白花或者红花戴在脑后的髮髻上……"接着他以居高临下的笔调描述了当地女性的几种发饰。

用今天的眼光来看，这些在别的民族土地上采集植物动物标本、测量地质水文数据等等的"科学考察"行为，有没有合法性问题？有没有侵犯主权的问题？这些行为得到当地人的同意了吗？当地人知道这些行为的性质和意义吗？他们有知情权吗？……这些问题，在今天的国际交往中，确实都是存在的。

也许有人会为这些帝国科学家辩解说：那时当地土著尚在未开化或半开化状态中，他们哪有"国家主权"的意识啊？他们也没有制止帝国科学家的考察活动啊？但是，这样的辩解是无法成立的。

姑不论当地土著当时究竟有没有试图制止帝国科学家的"科学考察"行为，现在早已不得而知，只要殖民者没有记录下来，我们通常就无法知道。况且殖民者有军舰有枪炮，土著就是想制止也无能为力。正如法拉所描述的："在几个塔希提人被杀之后，一套行之有效的易货贸易体制建立了起来。"

即使土著因为无知而没有制止帝国科学家的"科学考察"行为，这事也很像一个成年人闯进别人的家，难道因为那家只有不懂事的小孩子，闯入者就可以随便打探那家的隐私、拿走那家的东西、甚至将那家的房屋土地据为己有吗？事实上，很多情况下殖民者就是这样干的。所以，所谓的"帝国科学"，其实是有着原罪的。

如果沿用上述比喻，现在的局面是，家家户户都不会只有不懂事的孩子了，所以任何外来者要想进行"科学探索"，他也得和这家主人达成共识，得到这家主人的允许才能够进行。即使这种共识的达成依赖于利益的交换，至少也不能单方面强加于人。

博物学在今日中国

博物学在今日中国之复兴，北京大学刘华杰教授提倡之功殊不可没。自刘教授大力提倡之后，各界人士纷纷跟进，仿佛昔日蔡锷在云南起兵反袁之"滇黔首义，薄海同钦，一檄遥传，景从恐后"光景，这当然是和博物学本身特点密切相关的。

无论在西方还是在中国，无论在过去还是在当下，为何博物学在它繁荣时尚的阶段，就会应者云集？深究起来，恐怕和博物学本身的特点有关。博物学没有复杂的理论结构，它的专业训练也相对容易，至少没有天文学、物理学那样的数理"门槛"，所以和一些数理学科相比，博物学可以有更多的自学成才者。这次编印的《西方博物学大系》，卷帙浩繁，蔚为大观，同样说明了这一点。

最后，还有一点明显的差别必须在此处强调指出：用刘华杰教授喜欢的术语来说，《西方博物学大系》所收入的百余种著作，绝大部分属于"一阶"性质的工作，即直接对博物学作出了贡献的著作。事实上，这也是它们被收入《西方博物学大系》的主要理由之一。而在中国国内目前已经相当热的博物学时尚潮流中，绝大部分已经出版的书籍，不是属于"二阶"性质（比如介绍西方的博物学成就），就是文学性的吟风咏月野草闲花。

要寻找中国当代学者在博物学方面的"一阶"著作，如果有之，以笔者之孤陋寡闻，唯有刘华杰教授的《檀岛花事——夏威夷植物日记》三卷，可以当之。这是刘教授在夏威夷群岛实地考察当地植物的成果，不仅属于直接对博物学作出贡献之作，而且至少在形式上将昔日"帝国科学"的逻辑反其道而用之，岂不快哉！

2018年6月5日
于上海交通大学
科学史与科学文化研究院

植物学入门

出版说明

《植物学入门》是英国植物学家约翰·林德利（John Lindley，1799—1865）的一部博物学著作。林德利生于诺维奇，其父虽拥有一个果树园，但家庭经济状况不佳。为补贴家用，他幼时就帮父亲料理果园，到乡间采集花种，养成独到的观察自然的眼力。十六岁时，他离家到伦敦一家植物种子商铺做伙计，在那里结识了植物学家威廉·胡克，还被引荐给博物学家约瑟夫·班克斯，最后在班克斯的植物园找了份差事。他潜心钻研，二十一岁就出版了名作《蔷薇属》，并被选为伦敦林奈学会会员。1821年班克斯离世后，林德利被班克斯的朋友威廉·卡特雷聘用，负责在后者的植物园里记录、绘制新的植物品种。同年，卡特雷资助他出版了《毛地黄属》。此后，他担任了伦敦园艺协会理事和伦敦大学植物学讲座教授。1828年，他被选入皇家学会。除教职和本土研究工作外，林德利还曾于1838年远赴澳大利亚西部进行探险和植物样本采集工作，并长期担任园艺协会刊物的编辑，可谓精力旺盛的多面手。不过，长期辛劳也导致他晚年一直体弱，1863年赴法国温泉疗养也无起色，终于两年后逝世。

《植物学入门》是林德利的重要作品之一，是他在伦敦大学任教期间结合其园艺经验与学术研究方法撰著而成的，全书深入浅出地讲述了人应该如何培养对自然研究的兴趣、如何了解植物分类的基本方法、如何具体进行园艺配置和户外田野植物观察，并配以数百幅精美铜版插画。除专业学生和研究者外，此书也深受对自然科学兴趣盎然的普通市民的欢迎。

今据原版影印。

AN INTRODUCTION TO BOTANY.

LONDON:
Printed by A. & R. Spottiswoode,
New-Street-Square.

AN

INTRODUCTION

TO

BOTANY.

BY

JOHN LINDLEY, F.R.S. L.S. G.S.

MEMBER OF THE IMPERIAL ACADEMY NATURÆ CURIOSORUM; OF THE BOTANICAL SOCIETY OF RATISBON; OF THE PHYSIOGRAPHICAL SOCIETY OF LUND; OF THE HORTICULTURAL SOCIETY OF BERLIN; HONORARY MEMBER OF THE LYCEUM OF NATURAL HISTORY OF NEW YORK; ASSISTANT SECRETARY OF THE HORTICULTURAL SOCIETY OF LONDON, ETC. ETC.;

AND PROFESSOR OF BOTANY IN THE UNIVERSITY OF LONDON.

With Six Copper-Plates and numerous Wood-Engravings.

LONDON:
PRINTED FOR
LONGMAN, REES, ORME, BROWN, GREEN, & LONGMAN,
PATERNOSTER-ROW.
1832.

PREFACE.

Two hundred and ninety years have now elapsed since one of the earliest introductions to Botany upon record was published, in four pages folio, by Leonhart Fuchs, a learned physician of Tubingen. At that period Botany was nothing more than the art of distinguishing one plant from another, and of remembering the medical qualities, sometimes real, but more frequently imaginary, which experience, or error, or superstition, had ascribed to them. Little was known of Vegetable Physiology, nothing of Vegetable Anatomy, and even the art of arranging species systematically had still to be discovered; while scarcely a trace existed of those modern views which have raised the science from the mere business of the herb-gatherer to a station among the most intellectual branches of natural philosophy.

It now comprehends a knowledge not only of the names and uses of plants, but of their external and internal organisation, and of their anatomy and physiological phenomena; it embraces a consideration of the plan upon which those multitudes of vegetable forms that clothe the earth have been created, of the skilful combinations out of which so many various organs have emanated, of the laws that regulate the dispersion and location of species, and of the influence that climate exercises upon their developement;

and, lastly, from botany as now understood, in its most extensive signification, is inseparable the knowledge of the various ways in which the laws of vegetable life are applicable to the augmentation of the luxuries and comforts, or to the diminution of the wants and miseries of mankind. It is by no means, as some suppose, a science for the idle philosopher in his closet; neither is it merely an amusing accomplishment, as others appear to think; on the contrary, its field is in the midst of meadows, and gardens, and forests, on the sides of mountains, and in the depths of mines, — wherever vegetation still flourishes, or wherever it attests by its remains the existence of a former world. It is the science that converts the useless or noxious weed into the nutritious vegetable; which changes a bare volcanic rock, like Ascension, into a green and fertile island; and which enables the man of science, by the power it gives him of judging how far the productions of one climate are susceptible of cultivation in another, to guide the colonist in his enterprises, and to save him from those errors and losses into which all such persons unacquainted with Botany are liable to fall. This science, finally, it is which teaches the physician how to discover in every region the medicines that are best adapted for the maladies that prevail in it; and which, by furnishing him with a certain clue to the knowledge of the tribes in which particular properties are or are not to be found, renders him as much at ease, alone and seemingly without resources, in a land of unknown herbs, as if he were in the midst of a magazine of drugs in some civilised country.

The principles of such a science must necessarily be extremely complicated, and in certain branches, which have only for a short time occupied the attention of observers, or which depend upon obscure and ill-understood evidence, are by no means so clearly defined as could be wished. To explain those principles, to adduce the evidence by which their truth is supposed to be proved, or the reasoning upon which they are based in cases where direct proof is unattainable; to show the causes of errors that are now exploded, and the insufficiency of the arguments by which doubtful theories are still defended, — in fine, to draw a distinct line between what is certain and what is doubtful, — are some of the objects of this publication, which is intended for the use of those who, without being willing to occupy themselves with a detailed examination of the vast mass of evidence upon which the modern science of botany is founded, are, nevertheless, anxious to acquire a distinct idea of the nature of that evidence. Another and not less important purpose has been to demonstrate, by a series of well-connected proofs, that in no department of natural history are the simplicity and harmony that pervade the universe more strikingly manifest than in the vegetable kingdom, where the most varied forms are produced by the combination of a very small number of distinct organs, and the most important phenomena are distinctly explained by a few simple laws of life and structure.

In the execution of these objects, I have followed very nearly the method recommended by the celebrated Professor De Candolle, than whom no man is entitled to more deference, whether you consider the soundness of his judgment in all that relates to order and arrangement, or the great experience which a long and most successful career of public instruction has necessarily given him.

I have begun with what is called ORGANOGRAPHY

(Book I.); or an explanation of the exact structure of plants; a branch of the subject which comprehends all that relates either to the various forms of tissue of which vegetables are constructed, or to the external appearance their elementary organs assume in a state of combination. It is exceedingly desirable that these topics should be well understood, because they form the basis of all other parts of the science. In physiology, every function is executed through the agency of the organs: systematic arrangements depend upon characters arising out of their consideration; and descriptive Botany can have no logical precision without the principles of Organography being first exactly settled. Great difference of opinion exists among the most distinguished botanists, upon some points connected with this subject, so that it has been found expedient to enter occasionally into much detail, for the purpose of satisfying the student of the accuracy of the facts and reasonings upon which he is expected to rely.

To this succeeds VEGETABLE PHYSIOLOGY (Book II.); or the history of the vital phenonema that have been observed both in plants in general, and in particular species, and also in each of their organs taken separately. It is that part of the science which has the most direct bearing upon practical objects, and with which the enquirer who would occupy himself more with the *utile* than the *dulce* is most likely to be interested. Its laws, however, are either unintelligible, or susceptible of no exact appreciation, without a previous acquaintance with the more important details of Organography. Much of the subject is at present involved in mystery, and the accuracy of many of the conclusions of physiologists is inferred rather than demonstrated; so that it has been found essential that the grounds of the more popularly received opinions, whether admitted as true or rejected as erroneous, should be given at length. No particular chapter is assigned to the practical application of physiological principles, because this has been constantly taken as illustrative of the separate functions of individual parts.

Next follows TAXONOMY (Book III.); or some account of the Principles of Classification; — a very important subject, comprehending not only an account of the various methods of arrangement employed by botanists in their systematic works, but an explanation of the principles by which the limits of genera and species are determined. It also explains the mode of obtaining a correct view of vegetation, of conducting the examinations of unknown plants with precision, of avoiding errors in consequence of accidental aberrations from the ordinary structure, and of forming a just estimate of the mutual relation that one part of the vegetable kingdom bears to another.

After this I have taken GLOSSOLOGY (Book IV.); or, as it was formerly called, TERMINOLOGY; restricting it absolutely to the definition of the adjective terms, which are either used exclusively in Botany, or which are used in that science in some particular and unusual sense. The key to this book, and also to the substantive terms explained in Organography, will be found in a copious index at the end of the volume.

These four topics exhaust the science considered only with reference to first principles; there are, however, a few others which it has been thought advisable to append, on account of their practical value. These are, firstly, PHYTOGRAPHY (Book V.); or an exposition of the rules to be observed in describing and naming plants. As the great object of descriptions in natural history, is to enable every person to recognise a known species, after its station has been discovered by classification, and also to put those who have not had the opportunity of examining a plant themselves into possession of all the facts necessary to acquire a just notion of its structure and affinities; it is indispensable that the principles of making descriptions should be clearly understood, both to prevent their being too general to answer the intended purpose, or more prolix than is really requisite. It is the want of a knowledge of these rules that renders the short descriptions of the classical writers of antiquity, and the longer ones of many a modern traveller, equally vague and unintelligible. In this place are inserted a few notes upon the formation of an herbarium.

After this, has been introduced (Book VI.) a summary of the little which is known of the laws that regulate the distribution of plants upon the surface of the earth; a question which, however indefinite and unsatisfactory our information may at present be, has begun to assume such an appearance as to justify the expectation, that future discoveries will explain the causes of the characters of vegetation being determined, as they surely are, by climate.

Finally, the work is concluded by an exposition of what is called MORPHOLOGY; a subject which is in the vegetable what Comparative Anatomy is in the animal kingdom, and which is by far the most important branch of study after Elementary Anatomy and Vegetable Physiology. Organography itself is in all respects an exposition of the doctrines of Morphology; but the novelty of the subject, and a persuasion that it would be better understood if treated separately, has induced me to make it the subject of particular consideration. Unknown before the time of Linnæus, and first placed in its true light by the venerable poet Göthe, it lay neglected for nearly thirty years, until, having been revived by Du Petit Thouars, De Candolle, Brown, and others, it has come to be considered the basis of all scientific knowledge of vegetable structure.

It has been my wish to bring every subject that I have introduced down, as nearly as possible, to the state in which it is found at the present day; but, alas! tasks that are infinite can never be accomplished by finite means. While the MSS. has been going through the press, my table has been covered with works illustrating the fundamental principles of the science, of scarcely any of which has it been possible to make the slightest use. This I regret the more, as some of them are of high interest*, especially with regard to Vegetable Anatomy.

In the statements I have made, I have uniformly endeavoured to render due credit to all persons for the discoveries by which they may severally have contributed to the advancement of the science; and

---

* The more important of these works are,—
Agardh, C. A., Lehrbuch der Botanik. *Copenhagen*, 1831.
Kunth, K. S., Handbuch der Botanik. *Berlin*, 1831.
Meyer, F. J. F., Phytotomie. *Berlin*, 1830.
De Candolle, A. P., Physiologie Végétale, &c. *Paris*, 1832.
Martius, Palmæ. The part containing the Anatomy.
Arnott, G. W. The article Botany in the Edinburgh Encyclopædia.
Bischoff, G. W., Handbuch der Botanischen Terminologie und Systemkunde. *Nuremberg*, 1830.
Annales des Sciences Naturelles, par MM. Audouin, Ad. Brongniart, et Dumas. Several Numbers.
Mirbel, C. B., Mémoire sur le Marchantia. *Paris*, 1832.

if I have on any occasion either omitted to do so, or assumed to myself observations which belong to others, it has been unknowingly or inadvertently. It is, however, impracticable, and if practicable it would not be worth while, to remember upon all occasions from what particular sources information may have been derived. Discoveries, when once communicated to the world, become public property: they are thrown into the common stock for mutual benefit; and it is only in the case of debatable opinions, or of any recent and unconfirmed observations, that it really interests the world that authorities should be quoted at all. In the language of a highly valued friend, when writing upon another subject: — " The advanced state of a science is but the accumulation of the discoveries and inventions of many: to refer each of these to its author is the business of the history of science, but does not belong to a work which professes merely to give an account of the science as it is; all that is generally acknowledged must pass current from author to author."*

London, Sept. 10. 1832.

* Brett's Principles of Astronomy, p. v.

CONTENTS.

BOOK I.

	Page
ORGANOGRAPHY; OR, OF THE STRUCTURE OF PLANTS	1
CHAP. I. Of the Elementary Organs	ib.
1. Of Cellular Tissue	3
2. Of Woody Fibre	12
3. Of Vascular Tissue	17
4. Of Spurious Elementary Organs:	
1. Intercellular Passages	26
2. Receptacles of Secretion	27
3. Air Cells	ib.
4. Raphides	29
CHAP. II. Of the Compound Organs in Flowering Plants	31
1. Of the Cuticle and its Appendages:	
1. Cuticle	ib.
2. Stomata	33
3. Hairs	38
4. Scales	41
5. Glands	42
6. Prickles	44
2. Of the Stem or Ascending Axis:	
1. Of its Parts	45
2. Of its External Modifications	54
3. Of its Internal Modifications	58
1. Of the Exogenous Structure	59
2. Of the Endogenous Structure	72
3. Of the Root or Descending Axis	76
4. Of the Appendages of the Axis	78
1. Leaf	79
2. Stipulæ	98
3. Bracteæ	100

		Page
4. Flower	- - - - - -	105
5. Inflorescence	- - - - -	106
6. Calyx	- - - - - -	112
7. Corolla	- - - - - -	117
8. Stamens	- - - - - -	122
9. Disk	- - - - - -	137
10. Pistillum	- - - - -	138
11. Receptacle	- - - - -	152
12. Ovulum	- - - - -	153
13. Fruit	- - - - - -	160
14. Seed	- - - - - -	181
15. Naked Seeds	- - - - -	194

CHAP. III. Of the Compound Organs in Flowerless Plants - - 195

1. Ferns	- - - - - -	*ib.*
2. Equisetaceæ	- - - - -	197
3. Lycopodiaceæ	- - - - -	*ib.*
4. Marsileaceæ	- - - - -	198
5. Mosses	- - - - - -	200
6. Hepaticæ	- - - - -	202
7. Lichens	- - - - - -	204
8. Algæ	- - - - - -	206
9. Fungi	- - - - - -	208

BOOK II.

PHYSIOLOGY; OR, PLANTS CONSIDERED IN A STATE OF ACTION - 210

CHAP. I. Elementary Organs	- - - -	221
II. Root	- - - - -	227
III. Sap	- - - - -	230
IV. Pith, Wood, and Bark	- - - -	239
V. Leaves	- - - - -	247
VI. Bracteæ, Calyx, Corolla, and Disk	- - -	257
VII. Stamens and Pistillum	- - - -	262
VIII. Fruit	- - - - -	266
IX. Seeds	- - - - -	270
X. Colour, Smell, and Taste	- - - -	274
XI. Directions	- - - -	277
XII. Irritability	- - - -	288
XIII. Poisons	- - - -	293
XIV. Diseases	- - - -	297
XV. Hybrid Plants	- - - -	301
XVI. Flowerless Plants	- - - -	305

BOOK III.

TAXONOMY; OR, OF THE PRINCIPLES OF CLASSIFICATION.

	Page
CHAP. I. General Objects of Classification	306
II. Artificial Arrangements	309
III. Natural System	318
IV. Speculative Modes of Arrangement	324
V. Value of Characters	349
VI. Species, Varieties, Genera, Orders, and Classes	365

BOOK IV.

GLOSSOLOGY; OR, OF THE ADJECTIVE TERMS USED IN BOTANY — 369

CLASS I. Of Individual Terms — 371

   1. Of Individual Absolute Terms: — 372
      1. Figure — *ib.*
      2. Division — 386
      3. Surface — 393
      4. Texture or Substance — 397
      5. Size — 399
      6. Duration — 401
      7. Colour — 402
      8. Variegation or Marking — 407
      9. Veining — 408

   2. Of Individual Relative Terms:
      1. Æstivation — 409
      2. Direction — 411
      3. Insertion, or Origin — 415

CLASS II. Of Collective Terms:
      1. Of Arrangement — 417
      2. Of Number — 420

CLASS III. Of Terms of Qualification — 422
   Signs — *ib.*
   Abbreviations — 425

BOOK V.

Page

PHYTOGRAPHY; OR, OF THE RULES TO BE OBSERVED IN DESCRIBING AND NAMING PLANTS - - - - - - 432

    CHAP. I. Diagnoses; or, Generic and Specific Characters - 433
        II. Descriptions - - - - - 440
        III. Punctuation - - - - - 452
        IV. Nomenclature and Terminology - - - 454
        V. Synonyms - - - - - - 460
        VI. Herbaria - - - - - 463
        VII. Botanical Drawings - - - - 469

BOOK VI.

GEOGRAPHY; OR, THE DISTRIBUTION OF PLANTS UPON THE SURFACE OF THE GLOBE - - - - - - 472

BOOK VII.

MORPHOLOGY; OR, OF THE METAMORPHOSIS OF ORGANS - - 504

    CHAP. I. Regular Metamorphosis - - - - 509
        II. Irregular Metamorphosis - - - 521

EXPLANATION OF THE PLATES - - - - 527

INDEX:
    1. Substantives - - - - 537
    2. Adjectives - - - - 546

INTRODUCTION
TO
BOTANY.

BOOK I.

ORGANOGRAPHY; OR, OF THE STRUCTURE OF PLANTS.

CHAPTER I.

OF THE ELEMENTARY ORGANS.

IF plants are considered with reference to their internal organisation, they appear at first sight to consist of a vast multitude of exceedingly minute cavities, separated by a membranous substance; more exactly examined, it is found that these cavities have a variety of different figures, and that each is closed up from those that surround it; and if the enquiry is carried still farther, it will be discovered that the partitions between the cavities are all double, and that by maceration in water, or by other processes, the cavities with their enclosing membrane may be separated from each other into distinct bodies. These bodies constitute what is called Vegetable Tissue, or Elementary Organs: they are the *Similary parts* of Grew; the *Tissu organique* of Mirbel; and the *Parties élémentaires*, or *Parties similaires*, of De Candolle.

The chemical basis of the elementary organs has been found to be oxygen, hydrogen, and carbon, with occasionally a little nitrogen or azote, combined in various proportions: their organic basis is membrane and fibre. The latter only are here to be considered.

It is a common opinion that membrane only is the basis of the tissue of plants, and that fibre is itself a form of membrane. But as we find both developed in many of the most imperfectly organised plants, such as Scleroderma and other fungi; and as it is difficult to conceive how that can be a mere modification of membrane which is generated independently of it, which has no external resemblance to it, and which is obviously something superadded, it will be better to consider both membrane and fibre as the organic bases of vegetable tissue, rather than the former only.

The *membrane* varies in degree of transparency, being occasionally so exceedingly thin as to be scarcely discoverable, except by the little particles that stick to it, or by its refraction of light, and sometimes having a perceptible green colour, and a thickness which is considerable if compared with the diameter of the cavity it encloses. It generally tears readily, as if its component atoms do not cohere with greater force in one direction than another; but I have met with a remarkable instance to the contrary of this in Bromelia nudicaulis, in which the membrane of the cuticle breaks into little teeth of nearly equal width when torn. (Plate I. fig. 6.) It is in almost all cases destitute of visible pores; although as it is readily permeable by fluids, it must necessarily be furnished with invisible passages. An opinion to the contrary of this has been held by some botanists, who have described the existence of holes or pores in the membrane of tissue, and have even thought they saw a distinct rim to them; but this idea, which probably originated in imperfect observation with ill-constructed glasses, is now generally abandoned. The supposed pores, with their rim, have been ascertained to be nothing but grains of semi-transparent matter sticking to the membrane: this has been proved by Dutrochet, who found that boiling them in hot nitric acid rendered them opaque, and that dipping them in a solution of caustic potash restored their transparency, — a property incompatible with a perforation; and any one furnished with a good modern microscope may satisfy himself upon the point, without going through Dutrochet's process; by simple movement in water the grains may be often detached. It however occasionally happens that holes do exist in the membrane, of which mention will be made hereafter.

Fibre may be compared to hair of inconceivable fineness, its diameter often not exceeding the $\frac{1}{1200}$ of an inch. It has frequently a greenish colour, but is more commonly transparent and colourless. It appears to be sometimes capable of extension with the same rapidity as the membrane among which it lies, and to which it usually adheres; but occasionally elongates less rapidly, when it is broken into minute portions, and carried along by the growing membrane. In direction it is variable (Plates I. and II.); sometimes it is straight, and attains a considerable length, as in some fungi; sometimes it is short and straight, but hooked at the apex, as in the lining of the anther of Campanula; occasionally it is straight, and adheres to the side of membrane, as in the same part in Digitalis purpurea; but its most common direction is spiral. Whether it is solid or hollow has not been fully demonstrated; Dr. Purkinje asserts that it is hollow, as will be hereafter mentioned; but there can be no doubt that it is also, at least sometimes, solid, as in the fibrous cellules of the leaf of Oncidium altissimum; it is the opinion of many that it is hollow in the case of spiral vessels. Fibre has a constant tendency to anastomosing, in consequence of which reticulated appearances are occasionally found in tissue.

The forms under which the elementary organs are seen are,— 1. *Cellular tissue;* 2. *Woody fibre;* and, 3. *Vascular tissue.*

Sect. I. *Of Cellular Tissue.*

CELLULAR tissue (*Contextus cellulosus* or *Tela cellulosa*, Lat.; *Pulpa, Parenchyma*, or *pithy part*, of old writers; *Zellengewebe*, Germ.;) generally consists of little bladders or vesicles of various figures, adhering together in masses. Occasionally it is composed of fibre only, unconnected by membrane. It is transparent, and in all cases colourless: when it appears otherwise, its colour is always caused by matter contained within it.

If a thin slice of the pith of elder, or any other plant, be examined with a microscope, it will be found to have a sort of honeycomb appearance, as if there were a number of hexagonal

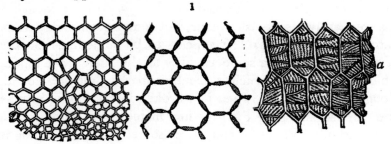

cavities, separated by partitions (*fig.* 1.). These little cavities are the inside of cellules of cellular tissue; and the partitions are caused by the cohesion of their sides, as may be easily proved by boiling the pith a short time, when the cellules readily separate from each other. In pulpy fruits, or in those which have their cellular tissue in a loose, dry state when ripe, the cellules may be readily separated from each other without boiling. It was formerly thought that cellular tissue might be compared to the air bubbles in a lather of soap and water, while by some it has been supposed to be formed by the doublings and foldings of a membrane in various directions: on both these suppositions, the partitions between the cells would be simple, and not composed of two membranes in a state of cohesion. But the facility with which, as has been just stated, the cellules may be separated, sufficiently disproves these opinions. It is probable, however, that although the double nature of the partitions in cellular tissue may be demonstrated, yet that the cellules usually grow so firmly together, that their sides really form in their union but one membrane.

Cellules are destitute of all perforation or visible pores, so that each is completely closed up from its neighbour, as far as we can see; although, as they have the power of filtering fluids with rapidity, it is certain that they must abound in invisible pores, and that they are not impermeable, as if they were made of glass. An opinion different from this has been and is still entertained by some observers, who have described and figured perforations of the membrane in various plants.

Mirbel states that "the sides of the cellules are sometimes riddled full of holes (*fig.* 2.), the aperture of which does not exceed the $\frac{1}{300}$ of a millimètre (or of half a line); or are less frequently pierced with transverse slits, which are occasionally so numerous as to transform the cellules into a real articulated tissue, as in the pith of the Nelumbium (*fig.* 3.)." This statement is now so well known to have been founded upon inaccurate observation, and such 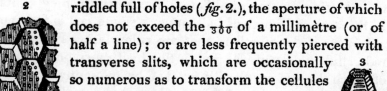 pores or slits are so universally admitted to be small portions of amylaceous matter sticking to the walls, that no additional disproof seems necessary. A good microscope is alone sufficient, generally, to show the real nature of these supposed pores, or if not, the test used by Dutrochet, and mentioned at page 2., is in all cases sufficient.

It may also be observed, that cellules often contain air-bubbles, which appear to have no direct means of escape, and that the limits of colour are often very accurately defined in petals, as, for instance, in the stripes of tulips and carnations, which could not be the case if cellular tissue were perforated by such holes as have been described; for in that case colours would necessarily run together.

One of the most striking instances with which I am acquainted, of cellular tissue having the appearance of pores is in Calycanthus, where it was pointed out to me by Mr. Varenne. (Plate I. fig. 1.) But even in this, a careful examination with glasses of different magnifying powers shows that the apparent pores are certainly not so, but composed of a solid subsance which may be distinctly seen by varying the direction of the rays of the transmitted light with which it is viewed. Sometimes they appear like luminous points; by a little alteration of light they acquire a brownish tint; and if seen with the highest powers of a compound microscope, where there is a great loss of light, they become perfectly opaque.

Cellular tissue is always transparent and colourless, or at most only slightly tinged with green. The brilliant colours of vegetable matter, the white, blue, yellow, scarlet, and other hues of the corolla, and the green of the bark and leaves, is not owing to any difference in the colour of the cellules, but to colouring matter of different kinds which they contain. In the stem of Impatiens balsamina, a single cell is frequently red in the midst of others that are colourless. Examine the red cellule, and you will find it filled with a colouring matter of which the rest are destitute. The bright satiny appearance of many richly coloured flowers depends upon the colourless quality of the tissue. Thus, in Thysanotus fascicularis, the flowers of which are of a deep brilliant violet, with a remarkable satiny lustre, that appearance will be found to arise from each particular cellule containing a single drop of coloured fluid which gleams through the white shining membrane of the cellules, and produces the flickering lustre that is perceived. The colouring matter of the cellular tissue is frequently fluid, but is in the leaves and other parts more commonly composed of granules of various sizes; this is particularly the case in all green parts; in which the granules lie amongst greenish liquid, the latter of which, as they grow older, dries up, while the granules themselves gradually change to olive green, and finally to brown.

Kieser distinguishes three sorts of globules among tissue: — 1. Round extremely transparent bodies, of a more or less regular figure, found principally in young plants and in cotyledons, and soluble in boiling water: it is these that constitute starch or fæcula. 2. Globules of a small size, a more irregular figure, and coloured either green or some other tint. They are not soluble in water, but are so in alcohol; but when dissolved, their matter is not precipitated by the addition of water, on which account they are distinguishable from resinous substances. 3. Extremely small round bodies, varying in colour, and found floating in the proper juices of vegetables.

The green granules are what M. Turpin calls *Globuline*. He believes them to be young cellules, and that it is from them that new tissue is developed. There does not, however, appear to be any evidence of this, which must be considered, at present, a gratuitous hypothesis, if, indeed, it were not rather said to be an untenable one. No one has ever seen the granules passing through the sides of the cellules; no rupture of the sides of the cellules, caused by such a passage, has ever been detected; and yet it seems inconceivable how the granules are to be constantly developing as new tissue, without some such trace of their passage being observable. Those who are curious to know the exact nature of this speculation, should consult the memoir of the author, in the *Mémoires du Muséum*, vol. 18. p. 212. The mode in which cellular or any other tissue is really formed, is buried in mystery. It has been suspected by Mr. Valentine, and I believe the same idea has also occurred to Mr. Bauer, that it may be caused by the extrication of gaseous matter among mucus; but it is obvious that there are many difficulties in the way of this supposition. Amici says the cellules pullulate.

The cellules develope, in some cases, with great rapidity. I have seen Lupinus polyphyllus grow in length, at the rate of an inch and a half a day. The leaf of Urania speciosa has been found by Mulder to lengthen at the rate of from one and a half to three and a half lines per hour, and even as much as from four to five inches per day. This may be computed to equal the developement of at least 4000 or 5000 cellules per hour. But the most remarkable instances of this sort are to be found in the mushroom tribe, which in all cases develope with surprising rapidity. It is stated by Junghuns, that he has known the Bovista giganteum, in damp warm weather, grow in a single night from the size of a mere point to that of a huge gourd. We are not further informed of the dimensions of this specimen; but supposing its cellules to be not less than the $\frac{1}{200}$ of an inch in diameter, and I suspect they are nearer the $\frac{1}{400}$, it may be fairly estimated to have consisted, when full grown, of about 47,000,000,000 cellules; so that, supposing it to have grown in the course of twelve hours, its cellules must have developed at the rate of near 4,000,000,000 per hour, or of more than sixty-six millions in a minute.

The cellules of cellular tissue are always very small, but are exceedingly variable in size. The largest are generally found in the gourd tribe (Cucurbitaceæ), or in pith, or in aquatic plants; and of these some are as much as the $\frac{1}{30}$ of an inch in diameter; the ordinary size is about the $\frac{1}{400}$ or the $\frac{1}{500}$, and they are sometimes not more than the $\frac{1}{1000}$. Kieser has computed that in the garden pink more than 5,100 are contained in half a cubic line.

Cellular tissue is found in two essentially different states, the *membranous* and the *fibrous*.

MEMBRANOUS CELLULAR TISSUE is that in which the sides consist of membrane only, without any trace of fibre; it is the most common, and was, till lately, supposed to be the only kind that exists. This sort of tissue is to be considered the basis of vegetable structure, and the only form indispensible to a plant. Many plants consist of nothing else; and while numberless vegetables are destitute of all other kinds of tissue, the membranous cellular tissue is never absent. It constitutes the whole of Mosses, Algæ, Fungi, Lichens, and the like; it forms all the pulpy parts, the parenchyma of leaves, the pith, medullary rays, and principal part of the bark, in the stem of exogenous plants, the soft substance of the stem of endogenous plants, the delicate membranes of flowers and their appendages, and both the hard and soft parts of fruits and seeds.

It appears that the spheroid is the figure which should be considered normal or typical in this kind of tissue; for that is the form in which cellules are always found when they are generated separately, without exercising any pressure upon each other; as, for example, is visible in the leaf of the white lily, and in the pulp of the strawberry or of other soft fruits, or in the dry berry of the jujube. All other forms of the cellules are considered to be caused by the compression or extension of such spheroids.

When a mass of spheroidal cellules is pressed together equally in all directions, rhomboidal dodecahedrons are produced, which, if cut across, exhibit the appearance of hexagons. (Plate I. fig. 12.) This is the state in which the tissue is found in the pith of all plants; and the rice paper, sold in the shops for making artificial flowers, and for drawing upon, which is really the pith of a Chinese plant, is an excellent illustration of it. If the force of extension or compression be greater in one direction than another, various other forms are produced, of which the following have been observed: —

1. The *oblong;* in the stem of Orchis latifolia, and in the inside of many leaves. (Plate I. fig. 9.)

2. The *lobed* (Plate I. fig. 2. *f*); in the inside of the leaf of Nuphar luteum, Lilium candidum, Vicia Faba, &c.: in this form of cellular tissue the vesicles are sometimes oblong with a sort of leg or projecting lobe towards one end; and sometimes irregularly triangular, with the sides pressed in and the angles truncated. They are well represented in the plates of M. Adolphe Brongniart's memoir upon the Organisation of Leaves, in the *Annales des Sciences*, vol. xxi.

3. The *square*; in the cuticle of some leaves, in the bark of many herbaceous plants, and frequently in pith. (Plate I. fig. 13.)

4. The *prismatical*; in some pith, in liber, and in the vicinity of vessels of any sort. (Plate 1. fig. 6.)

5. The *cylindrical* (Plate I. fig. 8. *a*); in Chara; this has been seen by Amici so large, that a single cellule measured four inches in length and one third of a line in diameter. (*Ann. des Sciences*, vol. ii. p. 246.)

6. The *fusiform* or the oblong pointed at each end; in wood, and in the membrane that surrounds the seed of a Gourd. These are what Dutrochet calls *clostres*. (Plate II. fig. 19. 8.; Plate I. fig. 5.)

7. The *muriform*; in the medullary rays. This consists of parallelopiped cellules compressed between woody fibre or vessels, with their principal diameter horizontal, and in the direction of the radii of the stem. It is so arranged that when viewed laterally it resembles the bricks in a wall; whence its name. (Plate I. fig. 7.)

8. The *compressed*; in the cuticle of all plants. The cellules are often so compressed as to appear to be only a single membrane. (Plate I. fig. 2. *a*; Plate III. fig. 3, 4, &c.)

9. The *irregular*; in the testa of many seeds, as Casuarina: here the form of the cellules is so very irregular that they can be reduced to no certain form.

10. The *sinuous*; in the cuticle, and also sometimes beneath it, as in the leaf of Lilium candidum. (Plate III. fig. 5.)

Cellular tissue is frequently called *Parenchyma*. Professor Link distinguishes both *Parenchyma* and *Prosenchyma*; referring to the former all tissue in which the cellules are applied by their plane faces (Plate I. fig. 1. 3. 6, 7, &c.), and which consequently have truncated extremities; and to the latter, forms of tissue in which the cellules taper to each end, and, consequently, overlap each other at their extremities. (Plate II. fig. 8. 19.)

FIBROUS CELLULAR or FIBRO-CELLULAR TISSUE is that in which the sides are composed either of both membrane and fibre together, or of fibre only.

It is only lately that this kind has been recognised. The first observation with which I am acquainted is that of Moldenhauer, who, in 1779, described the leaves of Sphagnum as marked by fibres twisted spirally. (Fig. 1. *a*, p. 4.) Link afterwards stated, that the supposed fibres were nothing but the lines where small cells contained in a larger one unite together; and his opinion was received. It is nevertheless certain, that the tissue of Sphagnum is as Moldenhauer described it. In November, 1827, I described the tissue of Maurandya Barclayana as consisting of cellules formed of spiral threads crossing each other, interlaced from the base to the apex, and connected by a membrane. A few other solitary cases of the observation of this kind of tissue had subsequently occurred, when the admirable investigation of a modern anatomist suddenly threw an entirely new light upon the subject.

Instead of being very rare, cellular tissue of this kind appears to be found in various parts; it has been already mentioned as existing in the leaves of Sphagnum; it is also found in the pith of Rubus odoratus. I originally discovered it in the parenchyma of the leaves of Oncidium altissimum, and in the testa of various seeds. Mr. Griffiths has detected it abundantly in the aerial roots of Orchideous plants, observations since confirmed by Mr. Brown; and Dr. Purkinje has shown, by a series of excellent observations and drawings, that it forms the lining of the valves of almost all anthers. The forms under which it exists in these parts are far more various than those of membranous cellular tissue. The principal varieties are these: —

A. *Membrane and Fibre combined.*

1. Fibres twisted spirally, adhering to a spheroidal or angular membrane, and often anastomosing irregularly, without the

spires touching each other. (Plate I. fig. 12.) This is what is found in Oncidium altissimum leaves, in the aerial roots of some Orchideous plants, in the lining of many anthers, and is what Mohl has figured (*Ueber die Poren, &c.* tab. i. fig. 9.), from the pith of Rubus odoratus. It approaches very nearly to the nature of spiral vessels, hereafter to be described, and appears only to be distinguishable by the spires of the fibres not being in contact, being incapable of unrolling, having no elasticity or tenacity; and by not being cylindrical and tapering to each end, but spheroidal.

2. Fibres crossing each other spirally, and forming a reticulated appearance by their anastomosing in oblong or botuliform cells. Of this nature are the reticulated cells of the testa of Maurandya Barclayana, Wightia gigantea, and the like. (Plate I. fig. 11.)

3. Fibres running straight along the sides of truncated cylindrical cells in the anthers of Calla æthiopica and many other plants. (Plate I. fig. 13.)

4. Fibres running transversely in parallel lines round three of the sides of prismatical right-angled cells, in the anthers of Nymphæaceæ, &c.

5. Fibres very short, attached to the sides of cells of various figures, to which they give a sort of toothed appearance, as in the anther of Phlomis fruticosa and other Labiatæ. (Plate I. fig. 15.)

The last three were first noticed by Dr. Purkinje.

6. The fibre twisted spirally, in the open membranous tubes that form the elaters of Jungermannia, apparently constitutes another form of tissue of this order. (Plate I. fig. 17.)

B. *Fibre without Membrane.*

7. Spiral fibres repressed by mucus, but having sufficient elasticity to uncoil when the mucus is dissolved, and then breaking up into rings. (Plate I. fig. 16.) These are what are found in the testa of Collomia linearis. They approach spiral vessels so very nearly, that when I originally discovered them I mistook them for such. They are known by their roundish or depressed figure when at rest, and by the want of an inclosing membrane, and by their brittleness when uncoiled.

8. Fibres short, straight, and radiating, so as to form little starlike appearances, found in the lining of the anthers of Polygala Chamæbuxus, &c. by Dr. Purkinje. (Plate I. fig. 19.)

9. Fibres originating in a circle, curving upwards into a sort of dome, and uniting at the summit, observed by the same anatomist in the anthers of Veronica perfoliata, &c.

10. Fibres standing in rows, each distinct from its neighbour, and having its point hooked, so that the whole has some resemblance to the teeth of a currycomb, in the anthers of Campanula; first noticed by Dr. Purkinje. (Plate I. fig 18.)

11. Fibres forming distinct arches, as seen in the anthers of Linaria cymbalaria, &c. by Dr. Purkinje. (Plate I. fig. 4.)*

In the centre of some of the cellules of the cellular tissue of many plants there is a roundish nucleus, apparently consisting of granular matter, the nature of which is unknown. It was originally remarked by Mr. Francis Bauer, in the cellules of the stigma of Phaius Tankervilliæ. A few other vegetable anatomists subsequently noticed its existence; and Mr. Brown, in his recent Memoir on the mode of impregnation in Orchideæ and Asclepiadeæ, has made it the subject of more extended observation. According to this gentleman, such nuclei not only occasionally appear on the cuticle of some plants (Plate III. fig. 9.), in the pubescence of Cypripedium and others, and in the internal tissue of the leaves, but also in the cells of the ovulum before impregnation. It would also seem that Mr. Brown considers stomata to be formed by the juxtaposition of two of these nuclei.

Sect. II. *Of Woody Fibre.*

This (*Vasa fibrosa*, Lat.; *Petits tubes*, Mirb.; *Tissu cellulaire allongé* or *ligneux*, Fr.; *Vaisseaux propres fasciculaires*, Mirb.; *Ligneæ fistulæ*, Malpighi; *Fasergefässe*, or *Baströhren*,

---

* According to the last mentioned author, the fibres themselves are generally tubular, and either perfectly round or somewhat compressed, or even three or four sided. He considers it proved, that they are hollow, by their appearance when compressed, by their occasionally containing bubbles of air, and by the difference between their state when dried and when recent.

Germ.; perhaps the *Vital vessels* of Schultz;) consists of very slender transparent membranous tubes, tapering acutely to each end, lying in bundles, and, like the cellular tissue, having no direct communication with each other, except by invisible pores. (Plate II. fig. 1. *a, b;* 2. 5. *a,* &c.)

Many vegetable anatomists consider it a mere form of cellular tissue, in an elongated state; but scarcely with justice; for if this mode of viewing the subject were pushed a little farther, it would be necessary to refer every modification of tissue to the cellular, which would be obviously improper.* Woody fibre may be at all times known by its elongated figure and extremely attenuated character; usually it has no sort of markings upon its surface, except occasionally a particle or two of greenish matter in its inside; but sometimes it is covered with spots that have been mistaken for pores, and that give it a peculiar character (Plate II. fig. 3. and 4.); and I have remarked an instance, in Oncidium altissimum, of its having tubercles on its surface. (Plate II. fig. 2.) Generally, while cellular tissue is brittle, and has little or no cohesion, woody fibre has great tenacity and strength; whence its capability of being manufactured into linen. Every thing prepared from flax, hemp, and the like, is composed of woody fibre.

That even the most delicate of it consists of tubes, may be readily seen by examining it with a high magnifying power, and also by the occasional detection of particles of greenish matter in its inside. (Plate II. fig. 2. *b.*) A very different opinion has nevertheless been held by some physiologists, who have thought that the woody fibre is capable of endless divisibility. "When," says Duhamel, "I have examined under the microscope one of the principal fibres of a pear tree, it seemed to me to consist of a bundle of yet finer fibres; and when I have detached one of those fibres, and submitted it to a more powerful magnifying power than the first, it has still

* The distinction between cellular tissue and woody fibre is more pronounced in the long club-shaped aerial radicle of Rhizophora Candelaria, than in any plant with which I am acquainted. It there consists of large, very long, transparent tubes, lying imbedded in fine brownish granular matter, which is minute cellular tissue.

appeared to be formed of a great number of yet more delicate fibres." (*Physique des Arbres*, i. 57.) To this opinion Du Petit Thouars assents, conceiving the tenuity of a fibre to be infinite, as well as its extensibility. (*Essais sur la Végétation*, p. 150.) These views have doubtless arisen from the use of very imperfect microscopes, under low powers of which such appearances as Duhamel describes are visible; but with modern glasses, and after maceration in nitric acid, or even in pure water, each particular fibre can be separated with the greatest facility. Their diameter is often very much less than that of the finest human hair; the tubes of hemp, for example, when completely separated, are nearly six times smaller. It must, however, be observed, that the fibres of this plant, as used in linen making, are by no means in a state of final separation, each of the finest fibres that meet the naked eye being in reality a bundle of tubes. While, however, some are of this extremely small size, others have a diameter as considerable as that of ordinary cellular tissue itself; in Coniferæ the tubes are often $\frac{1}{200}$ or $\frac{1}{300}$ of an inch in diameter, and in the Lime they average about $\frac{1}{150}$. Link states (*Elementa*, p. 85.) that they are very large in trees of hot countries, as, for instance, the Brazilian coffee.

It has been asserted by some writers, that the tubes of the woody fibre are occasionally divided internally by transverse septa or partitions; but the fact is denied by Link, who declares that " ejusmodi septa non existunt." It is no doubt true that, in general, there is no trace of such septa; but I think it is impossible to deny their existence in the tissue of the Lime tree, at least.

There are three distinct kinds of woody fibre:—

1. That in which the walls are not occupied with either granules or glands sticking to them, or in which the former are of very rare occurrence. (Plate II. fig. 1.) This is the finest and the commonest of all; and is also the most genuine state of woody fibre.

2. That in which the walls have uniformly considerable numbers of granules of regular size sticking to them in a scattered manner. (Plate II. fig. 3, 4, 5.) These granules have been, and are still considered by many anatomists as

pores in the sides of the tissue. They have been, in particular, so described and represented lately by Mons. Brongniart in Cycadeæ, in which the tubes are large, and the appearance very conspicuous. (*Annales des Sciences*, vol. xvi. tab. 21.) But I think it possible to demonstrate that this is an optical deception, and that the supposed perforations are of the same nature as the similar punctuations in cellular tissue, viz. semitransparent granules. In the first place, no colourless light passes through the supposed pores in any case; on the contrary, they are dark, and have a solid appearance at all times, except when, at a certain distance out of the focus of the microscope, they become luminous. Secondly, if they were holes, they would, at least, be seen open when the tissue is dry and contracted, although they might close up when it becomes swollen with moisture. That, however, they never are: on the contrary, they are more opaque when dry than when wet. Thirdly, they become more and more opaque as the magnifying power with which they are viewed is increased; a circumstance which seems incompatible with perforations. Finally, and it is this which will possibly be regarded most conclusive, if the tissue of Zamia be allowed to remain macerating for some time in dilute nitric acid, the apparent pores disappear: that is to say, the granules that cause the appearance of perforations are dissolved. It has been thought that such appearances as these were confined to Cycadeæ and Coniferæ; but I suspect that they are far from uncommon in other families. Such tissue constitutes a considerable part of the wood of Calycanthus (Plate II. fig. 4.), as has been already noticed; and it is abundant in an East Indian genus allied to Trichopus. This kind of tissue might be called *granular woody fibre:* it approaches very nearly to the character of ducts, into which, in Zamia, it seems to pass by almost insensible transitions. It may, however, be known from dotted ducts, either by its very acuminated extremities, or by its granules not being arranged in a spiral manner.

3. The third kind of woody fibre is the *glandular*. This has hitherto only been noticed in Coniferæ, in which it is uniformly found in every species. Its dimensions are more considerable than that of either of the last-mentioned forms;

and, like the second, it has been described as perforated with pores. It differs from granular woody fibre in the markings of the tube being vesicular, and usually transparent, with a darkened centre (Plate II. fig. 5, 6. 8.), which last is what has been described as a pore, the vesicle itself being considered a thickened rim. Kieser figures the glands as pores, both in Pine wood (*fig.* 4.), and in Ephedra (*fig.* 5.), and in other cases also. They may be most conveniently found by examining a thin shaving of Pinus Strobus with a microscope, when they will be seen in the form of transparent globules, having a dark centre, and placed upon the walls of the tissue. That these globules are not pores, seems to me to be proved thus: they are flaccid when dry, and distend when moistened, which is not the property of a pore; their centre is more generally opaque than transparent, which is also not the property of a pore; they may be torn through the middle without any hole becoming visible; and, finally, they may sometimes be detached from the tissue (Plate II. fig. 7.), or fall away spontaneously. In the latter case they leave a hole in the tissue at the place where they grew; and holes thus occasioned misled Kieser into the belief that the woody fibre of Ephedra was really pierced with pores of considerable magnitude. An illustration of the manner in which these perforations are caused will be found in Plate II. fig. 7.

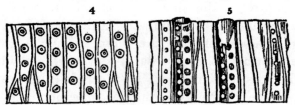

M. Adolphe Brongniart has rightly stated, that there exists in Gnetum Gnemon a form of tissue exactly the same as in Coniferæ. (*Voyage de Freycinet.*) In a species of that genus collected in Tavoy by Dr. Wallich, the glands are only different from what we commonly meet with in Coniferæ, in being arranged side by side, instead of being placed in single rows irregularly one above the other. (Plate II. fig. 5.)

Woody fibre constitutes a considerable proportion of the ligneous part of all plants; it is common in bark, and it forms the principal portion of the veins of leaves, to which it gives stiffness and tenacity.

Sect. III. *Of Vascular Tissue.*

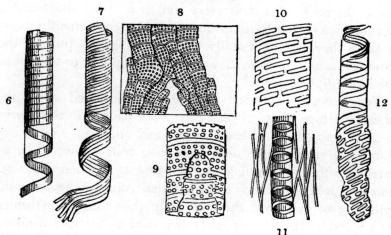

Vascular tissue consists of simple membranous tubes tapering to each end, but often ending abruptly, either having a fibre generated spirally in their inside, or having their walls marked by dots or transverse bars arranged in a spiral direction.

Such appears to me to be the most accurate mode of describing this kind of tissue, upon the exact nature of which anatomists are, however, much divided in opinion; some believing that the fibre coheres independently of any membrane, others doubting or denying the mode in which the vessels terminate; some describing the vessels as ramifying; and a fourth class ascribing to them pores and fissures, as we have already seen has been done in cellular tissue and woody fibre. It will be most convenient to consider all these points separately, along with the varieties into which vascular tissue passes.

There are two principal kinds of vascular tissue; viz. spiral vessels (Plate II. fig. 9. 11.), and ducts (Plate II. fig. 13. 15, 16. 18. 20.).

Spiral vessels (*fig.* 6, 7.) (*Vasa spiralia*, Lat.; *Tracheæ* of many; *Fistulæ spirales* of Malpighi; *Spiralgefässe* or *Schraubengefässe*, Germ.;) are membranous tubes with conical extremities; their inside being occupied by a fibre twisted spirally, and capable of unrolling with elasticity. To the eye they, when at rest, look like wire twisted round a cylinder that is afterwards removed. For the purpose of finding them for examination, the stalk of a strawberry leaf, or a young shoot of the Cornus alba (common dogwood) may be conveniently used; in these they may be readily detected by gently pulling the specimen asunder, when they unroll, and appear to the naked eye like a fine cobweb.

Very different opinions have been entertained as to the exact structure of spiral vessels. They have been considered to be composed of a fibre only, twisted spirally, without any connecting membrane; or to have their coils connected by an extremely thin membrane, which is destroyed when the vessel unrolls; or to consist of a fibre rolled round a membranous cylinder; or even, and this was Malpighi's idea, to be formed by a spiral fibre kept together as a tube by interlaced fibres. Again, the fibre itself has been by some thought to be a flat strap, by others a tube, and by a third class of observers a kind of gutter formed by a strap having its edges turned a little inwards. Finally, the mode in which they terminate, although formerly stated by Mirbel to be continuous with the cellular tissue, is so little known, that the learned M. De Candolle, in his *Organographie*, published in 1827, remarks, " Personne jusqu' ici n'a vu d'une manière claire, ni l'origine, ni la terminaison d'un vaisseau." (P. 58.) As doubts upon these points arise from the extreme minuteness of the vessels, and from the different degrees of skill that observers employ in the use of the microscope, I can scarcely hope that any observations of mine will have much weight. Nevertheless I may be permitted to state briefly what arguments occur to me in support of the definition of the spiral vessel as given above.

With regard to the presence of an external membrane within which the spiral fibre is developed, it must be confessed that direct observation is scarcely sufficient to settle that point. It is easy to prove the existence of a membrane, but it is difficult to demonstrate whether it is external or internal with respect to the fibre. The best mode of examination is to separate a vessel entire from the rest of the tissue, which may be done by boiling the subject, and then tearing it in pieces with the points of needles or any delicate sharp instrument; the real structure will then become much more apparent than if the vessel be viewed in connection with the surrounding tissue. From some beautiful preparations of this kind by Mr. Valentine and Mr. Griffiths, it appears that the membrane is external: in the root of the Hyacinth, for example, the coils of the spiral vessel touch each other, except towards its extremities; there they gradually separate, and it is then easy to see that the spiral fibre does not project beyond the membrane, but is bounded externally by the latter, which would not be the case if the membrane were internal: a representation of such a vessel is given at Plate II. fig. 9. Another argument as to the membrane being external may not unfairly be taken from the manifest analogy that a spiral vessel bears to that form of cellular tissue (p.11.), in which a spiral fibre is generated *within* a cellule: it is probable that the origin of the fibre is the same in both cases, and that its position with regard to the membrane is also the same.

It is much more difficult to determine whether the fibre is solid, or tubular, or flat like a strap; and Amici has even declared his belief that the question is not capable of solution with such optical instruments as are now in use. When magnified 500 times in diameter, a fibre appears to be transparent in the middle, and more or less opaque at the edges; a circumstance which has no doubt given rise to the idea that it is a strap or riband, with the edges either thickened, according to M. De Candolle, or rolled inwards according to Mirbel. But it is also the property of a transparent cylinder to exhibit this appearance when viewed by transmitted light, as any one may satisfy himself by examining a bit of a thermometer tube. A better mode of judging is, perhaps, to be found in the way in which the fibre bends when the vessel is flattened. If it were a flat thread, there would be no convexity at the angle of flexure, but the external edge of the bend would be straight. The fibre, however, always maintains its roundness, whatever the degree of pressure that I have been able to apply to it. (Plate II. fig. 10.) This I think conclusive as to the roundness of the fibre; but it does not determine the question of its being tubular or solid. I should have been induced to think, with Dr. Bischoff, who has investigated the nature of spiral vessels with singular skill (*De vera Vasorum Plantarum spiralium Structurâ et Functione Commentatio,* 1829) that it is solid, if it did not appear to have been ascertained by Hedwig that, when coloured fluids rise in spiral vessels, they follow the direction of the spires. This fact may, however, be explained upon the supposition that they rise in the channels formed by the approximation of cylindrical fibres, and not in the fibres themselves; in which case there could be little doubt that the fibres are really solid.

The nature of the termination of spiral vessels is now placed beyond all doubt, by the preparations of Mr. Valentine, above alluded to, and by some observations of my own. It is stated by Professor Nees von Esenbeck, in his *Handbuch der Botanik,* published in 1820, that they terminate in a conical manner; and in 1824 M. Dutrochet asserts, that they end in conical spires, the point of which becomes very acute; but one would not suppose, judging from the figure given by the latter writer, that he had seen the terminations very clearly. It is, however, certain that the statement of Nees von Esenbeck is correct, and that the spiral vessel generally terminates in a cone. If the point of such a vessel in the Hyacinth (Plate II. fig. 9.) be examined, it will be seen that the end of the spiral fibre lies just within the acute point of the vessel, and that the spires become gradually more and more relaxed as they approach the extremity, as if their power of extension gradually diminished, and the membrane acquired its pointed figure by the diminution of elasticity and extensibility in the fibre. It is not, however, always in a distinct membrane that the spiral vessel ends. In Nepenthes the fibres terminate in a blunt cone, in which no membrane is discoverable. (Plate II. fig. 11.) *

* A singular change occurs in the appearance of the spiral vessels of Nepenthes, after long maceration in dilute nitric acid, or caustic potash; the extremities cease to be conical and spirally fibrous, but become little transparent oblong sacs, in which the spires of the fibres gradually lose themselves. This alteration, which is a very likely cause of deception, is perhaps owing to the extremities of the vessels being more soluble than the other part, the sac being the confluent dissolved fibres. This is in some measure confirmed by the subsequent disappearance of all trace of fibres in any part of the vessels, under the influence of those powerful solvents.

A spiral vessel is formed by the convolutions either of a single spire, or of many. In the former case it is called *simple*, in the latter *compound*. The simple is the most common. (Plate II. fig. 9.) Kieser finds from two to nine fibres in the Banana; M. de la Chesnaye as many as twenty-two in the same plant. There are four in Nepenthes. (Plate II. fig. 11.) In general, compound spiral vessels are thought to be almost confined to Monocotyledonous plants, where they are very common in certain families, especially Marantaceæ, Scitamineæ, and Musaceæ; but their existence in Nepenthes, and, according to Rudolphi, in Heracleum speciosum, renders it probable that future observations will show them to be not uncommon among Dicotyledons also.

In Coniferæ the spiral vessels have in some cases their spires very remote, and even have glands upon their membrane between the spires. (Plate II. fig. 6.)

In size, spiral vessels, like other kinds of tissue, are variable; they are generally very small in the petals and filaments. Mirbel states them to be sometimes as much as the 288th of an inch in diameter; Hedwig finds them, in some cases, not exceeding the 3000th; a very common size is the 1000th.

An irritability of a curious kind has been noticed by Malpighi in the fibre of a spiral vessel. He says (*Anat.* p. 3.) that in herbaceous plants and some trees, especially in the winter, a beautiful sight may be observed, by tearing gently asunder a portion of a branch or stem still green, so as to separate the coils of the spires. The fibre will be found to have a peristaltic motion which lasts for a considerable time. An appearance of the same nature has been described by Mr. Don in the bark of Urtica nivea. These observations are, however, not conformable to the experience of others. M. De Candolle is of opinion that the motion seen by Malpighi is due to a hygroscopic quality combined with elasticity; and as spiral vessels do not exist in the bark of Urtica nivea, it seems that there is some inaccuracy in Mr. Don's remark.

The situation of spiral vessels is in that part of the axis of the stem surrounding the pith, and called the medullary sheath, and also in every part the tissue of which originates from it; such as the veins of leaves, and petals; and of all other modifications of leaves. It has been supposed that they are never found either in the bark, the wood, or the root; and this appears to be generally true. But there are exceptions to this: Mirbel and Amici have noticed their existence in roots; and Mr. Valentine and Mr. Griffiths have both extracted them from the root of the Hyacinth; they do not, however, appear to have been hitherto seen in the roots of Dicotyledonous plants. I know of no instance of their existence in bark, except in Nepenthes, where they are found in prodigious quantities, not only between the alburnum and the liber, embedded in cellular tissue, as was first pointed out to me by Mr. Valentine, but also sparingly both in the bark and wood. They have been described by myself as forming part of the testa of the seed of Collomia, and Mr. Brown has described them as existing abundantly in that of Casuarina. In the former case, the tissue was rather the fibro-cellular, as has been already explained (p. 11.); in the latter, they are apparently of an intermediate nature between the cellular-fibrous and the vascular; agreeing with the former in size, situation, and general appearance, but differing in being capable of unrolling. In the stem of Monocotyledonous plants, spiral vessels occur in the bundles of woody tissue that lie among its cellular substance; in the leaves of some plants of this description they are found in such abundance, that, according to M. de la Chesnaye, as quoted by De Candolle, they are collected in handfuls in some islands of the West Indies for Amadou. The same author informs us, that about a drachm and a half is yielded by every plantain, and that the fibres may be employed either in the manufacture of a sort of down, or may be spun into thread. In Coniferous plants they are few and very small, and in Flowerless plants they are for the most part altogether absent; the only exceptions being in Ferns and Lycopodiaceæ, orders occupying a sort of middle place between flowering and flowerless plants: in these they no doubt exist. My friend Mr. Griffiths has succeeded in unrolling them in the young shoots of Lycopodium denticulatum.

Some have thought that the spiral vessels terminate in those little openings of the cuticle called stomata; but there does not seem to be any foundation for this opinion.

DUCTS (*fig.* 8, 9, 10, 11, 12.) (*Fausses trachées*, Fr.; *Saft-röhren*, Germ; *Tubes corpusculifères of* Dutrochet, *Lymphæ-ducts*, or *Sap-vessels* of Grew and others; *Vaisseaux lymphatiques* of De Candolle, *Vaisseaux pneumatiques* of others;) are membranous tubes, with conical or rounded extremities; their sides being marked with transverse lines, or rings, or bars, or dots arranged spirally, and being incapable of unrolling.

In some states these approach so nearly to the spiral vessel, that it is impossible to doubt their being a mere modification of it, as is the case in the annular duct (Plate II. fig. 13.); but in other states, as in the dotted duct, it is impossible to trace the transition from the one form to the other. Some writers confound all the forms under the common name of spiral vessels, but it is more convenient to consider them as distinct, not only because of their peculiar appearances, but because they occupy a station in plants in which true spiral vessels are not found; and it is therefore probable that their functions are different.

All the forms of the duct seem reducible to the following varieties: —

1. The *Annular* (*fig.* 11., and Plate II. fig. 13.). These are well described by Bischoff as being formed of fibrous rings, placed at uncertain intervals; or, to speak more accurately, they, like spiral vessels, are formed of a spiral thread, but it is broken at every coil, so as to separate into a number of distinct rings. These rings are included within a membranous tube, by which they are held together. When the rings are distant from each other (Plate II. fig. 1. *b*), the duct has a very peculiar appearance; when the rings are packed together, so as to touch each other (Plate II. fig. 18.), the external appearance is exactly that of a spiral vessel, from which they are known by being incapable of unrolling. Both these forms are common in the soft parts of plants, particularly in the root, and also in Ferns and Lycopodiaceæ among flowerless plants.

2. The *Reticulated* (*fig.* 10. 12., and Plate II. fig. 13. *a*.). In these the spiral fibre, instead of separating into a number of distinct rings, is continuous in some places, anastomoses in others, so as to form a sort of netted appearance, or even

breaks into short lengths, which, adhering to the sides of the membrane, give the vessel the appearance of having transverse bars. It is these appearances that have given rise to the notion of cracked, or pierced ducts (*fig.* 10.) existing in plants; the membrane between the spires, or bars, having been mistaken for pores; hence the term *vaisseau fendu*, used by Mirbel and others. Vessels of this kind are found in the stem of some herbaceous plants; as, for example, the Impatiens Balsamina, in which they may be found in a great variety of states.

3. The *Dotted* (*fig.* 9.). Ducts of this kind are tubes having their sides marked with numerous dots, arranged in a more or less spiral manner, and being divided internally by transverse partitions. Usually, in addition to the dots, there is distinctly visible an oblique or annular transparent line upon the walls of the vessel. (Plate II. fig. 15. 17.) Hence Kieser considered them as spiral vessels, the spires of which, when old, elongate, and become connected by a dotted membrane. Bischoff, on the contrary, considers the dots to be caused by the separation of a spiral fibre into extremely minute portions; and he gives a figure (Plate II. fig 16.) of the manner in which he considers this change to occur.

It is certain, however, that the dotted duct is really an entirely distinct kind of vessel, or at least a modification of cellular rather than of vascular tissue, as has been asserted by Du Petit Thouars (*Ann. des Sciences*, vol. xxi. p. 224.); for the following reasons : — If it were such a modification of the spiral vessel as Kieser supposes, it would have none of those internal septa by which it is particularly known. The same remark applies to the theory of Bischoff, which is also imperfect, in not accounting for the nature of the transverse transparent lines that mark the sides of dotted ducts. Besides, the dotted ducts always terminate abruptly, not in acute cones, as has been seen by myself, and well represented by Mr. Griffiths, in his excellent illustrations of the anatomy of Phytocrene (Plate II. fig. 19. 20.), and they readily separate at the septa; none of which properties are those of a spiral vessel. That the partitions above alluded to really exist, as has been correctly stated by Dr. Dutrochet, there can be no doubt, notwithstanding the denial of the fact by Link and others. They may be seen with the naked eye in the ducts of the Cane, the Bamboo, and many other plants.

While, therefore, I conform to the general practice of classing this kind of duct among vascular tissue, I would suggest that it should rather be considered as made up of cylindrical cells, the *sides* of which are covered with oblong granules, arranged with their principal axis across the tube, and the *united ends* of which cause the partitions discoverable upon a longitudinal section. It is these partitions that cause externally the appearance of transverse transparent lines.

Dotted ducts are the largest of all kinds of tissue. The holes which are so evident to the naked eye, in a transverse section of the oak or the vine, are the mouths of dotted ducts; and the large openings in the ends of the woody bundles of Monocotyledonous stems, as in the Cane, are also almost always caused by the section of a dotted duct. The stem of Arundo Donax, or of any large grass, is an excellent subject for seeking them in; they can be readily extracted from it when boiled.

Vascular tissue always consists of tubes that are unbranched. They have been represented by Mirbel as ramifying in some cases; but this opinion has undoubtedly arisen from imperfect observation. When forming a series of vessels, the ends of the tubes overly each other, as represented in Plate II. fig. 18.

Some anatomists have added to the varieties above enumerated, what they call moniliform, or necklace-shaped, or strangulated vessels (*fig.* 8.) (*vaisseaux en chapelet* or *étranglés, vasa moniliformia, corpuscula vermiformia*). These are rightly determined by Bischoff to be mere accidental forms, caused by their irregular compression, when growing in knots or parts that are subject to an interrupted kind of developement. They may be found figured in Mirbel's *Elémens*, tab. x. fig. 15.; and in Kieser, fig. 56. and 57.

SECT. IV. *Of spurious elementary Organs; such as Air Cells, Receptacles of Secretion, Glands, &c. &c.*

The kinds of tissue now enumerated are all that have as yet been discovered in the fabric of a vegetable. There are, however, several other internal parts, which although not elementary, being themselves made up of some one or other of the forms of tissue already described, nevertheless have either been sometimes considered as elementary, or at least are not referable to the appendages of the axis, and can be treated of more conveniently in this place than elsewhere. These are, 1. *Intercellular passages*; 2. *Receptacles of secretion*; 3. *Air cells*; 4. *Raphides*.

1. Of Intercellular Passages.

As the elementary organs are all modifications of either the spherical or cylindrical figure, it must necessarily happen that when they are pressed together, spaces between them will remain, which will be more or less considerable in proportion as the tissue departs in a greater or less degree from the cylindrical or spherical form. When the pressure has been very uniform, as in the case of the tissue of the cuticle, and in many states of cellular substance, no spaces will exist. When they do exist, they are called *Intercellular passages* (*meatus* or *ductus intercellulares, canaux entrecellulaires*). They necessarily follow the course of the tissue, being horizontal, vertical, or oblique, according to the direction of the angles of the tissue by which they are formed. Their size varies according to the size of the tissue, and the quantity of sap. In plants of a dry character, they are frequently so small as to be scarcely discoverable; while in succulent plants they are so large as to approach the size of cellules, as in the stem of Tropæolum majus. (Plate II. fig. 14.) They are continually filled with fluid, so long as the part of the plant in which they are situated performs its vital functions, and only become dry when it has ceased to live, as in dry pith.

2. Of Receptacles of Secretion.

But it frequently occurs that the simple intercellular passages are dilated extremely by the secretions they receive, and either increase unusually in size, or rupture the coats of the neighbouring tissue; by which means cavities are formed replete with what is called the proper juice of the plant; that is to say, with the sap altered to the state which is peculiar to the particular species of tree producing it. Cavities of this nature are often called *vasa propria;* they are the *receptacula succi* of Link; the *vaisseaux propres* of Kieser and De Candolle; and the *réservoirs du suc propre* of the last author. To this class also are to be referred the *turpentine vessels,* and the *milk vessels* of Grew; the *réservoirs accidentels* of M. De Candolle; and also the *réservoirs en cæcum* of the latter, which latter are the clavate vessels filled with oily fluid that are found in the coat of the fruit of Umbelliferæ, and which are commonly called *vittæ.* Although the receptacles of secretion have no proper coat, yet they are so surrounded by cellular tissue, that a lining or wall is formed, of perfect regularity and symmetry. The tissue of this lining is generally much smaller than that of the neighbouring parts. In figure the receptacles are extremely variable, most commonly round, as in the leaves of the Orange and of all Myrtaceæ, where they are called *crypta,* or *glandulæ impressæ,* or *réservoirs vésiculaires,* or *glandes vésiculaires,* or *receptacles of oil.* In the Pistacia Terebinthus the receptacles are tubular; in Coniferæ they are very irregular in figure, and even position, chiefly forming large hollow cylindrical spaces in the bark. Those in the rind of the orange and lemon are little oblong or spherical cysts; their construction, which is very easily examined, gives an accurate idea of that of all the rest. (Plate II. fig. 21.)

3. Of Air Cells.

Besides the common intercellular passages, and the receptacles now described, there is another and a very remarkable sort of cavity among the tissue of plants. This is the *air cell;* the *lacuna* of Link, the *réservoir d'air* and *cellule d'air* of Kieser, and the *luftbehälter* of the Germans. Like the receptacles of secretion, the air cells are built up of tissue, but have no proper membrane of their own; and this sometimes takes place with a truly wonderful degree of uniformity and beauty. Each cell is often constructed so exactly like its neighbour, that it is impossible to regard them always as mere accidental distensions of the tissue; on the contrary, they are, in those plants to the existence of which they are necessary, evidently formed upon a plan which is uniform in the species, and which has been wisely contrived by Providence in that manner which is most suitable to the purpose for which they are destined.

They differ from receptacles of secretion in containing air only, and not the proper juice of the plant; a peculiarity which is provided for by a curious contrivance of Nature. In receptacles, the orifices of the intercellular passages through which the fluid that is to be deposited drains, are all open; but, to prevent any discharge of fluid into the air cells, the orifices of all the intercellular passages that would otherwise open into them are closed up.

Air cells are very variable in size, figure, and arrangement. In the stems of fistular plants, as Allium, they from a cavity from the base to the summit; in the stem of the Rush (Juncus articulatus), they consist of a number of tubular cavities placed one above the other, and separated by membranous partitions composed of a combination of minute cellules; in some aquatic plants they are very small, as in Butomus umbellatus. In form they are either cylindrical, or they assume the figure of the cellules by which they are formed, as in Limnocharis Plumieri (Plate III. fig. 1. and 2.), in which the structure of the air cells and their coats forms one of the most beautiful of microscopical objects.

The inner surface of the air cells, when they are essential to the life of a plant, is smooth and uniform; but in grasses, umbelliferous plants, and others where they are not essential, they seem to be caused by the growth of the stem being more rapid than the formation of the air cells; so that the tissue is torn asunder into cavities of an irregular figure and surface. Kieser was the first to observe that in many plants in which the air cells of the stem are regularly separated by partitions, the intercellulary passages of the cellules forming the partitions are sometimes left open, so that a free communication is maintained between all the tiers of air cells. (Plate II. fig. 2.)

4. Of Raphides.

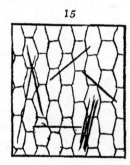

Among the tissue, and particularly in the intercellular passages of Monocotyledonous plants, are found certain needle-shaped transparent bodies, lying either singly or in bundles, and called *raphides*. They were first discovered by Rafn, who found them in the milky juice of Euphorbiæ; afterwards they were met with by M. Jurine, in the leaves of Leucojum vernum, and elsewhere; and they are now well known to all vegetable anatomists. It is probable that the first discoverers considered them a kind of special organ; but they have subsequently been recognised to be crystals of extreme minuteness, and, according to M. Raspail, of oxalate of lime. If a common Hyacinth is wounded, a considerable discharge of fluid takes place, and in this myriads of raphides are found floating; or if the cuticle of the leaf of Mirabilis Jalapa is lifted up, little whitish spots are observable, which are composed of them; all these are acicular in form, whence their name. But in the Cactus peruvianus they are, according to M. Turpin, found in the inside of the vesicles of cellular tissue, and, instead of being needle-shaped, have the form of extremely minute conglomerated crystals, which are rectangular prisms with tetraedral summits, some with a square, others with an oblong base. Crystals of a similar figure have been remarked by the same observer in Rheum palmatum; and their presence, according to him, is sufficient to distinguish samples really from China and Turkey, from those produced in Europe. The former abound in these crystals, the latter have hardly any.

In the above figure, 15 represents the raphides of Aloe verrucosa (from Kieser); 14, those of Cactus peruvianus; 13, those of Rheum palmatum: the two latter from Turpin.

CHAPTER II.

OF THE COMPOUND ORGANS IN FLOWERING PLANTS.

HAVING now explained the more important circumstances connected with modifications in the elementary organs of vegetation, the next subject of enquiry will be the manner in which they are combined into those masses which constitute the external or compound organs, or in other words the parts that present themselves to us under the form of roots, stems, leaves, flowers, and fruit, and that constitute the apparatus through which all the actions of vegetable life are performed. In doing this, I shall limit myself in the first place to Flowering Plants (*Introduction to the Natural System*, p. 1.); reserving for the subject of a separate chapter the explanation of some of the compound organs of Flowerless plants (*Ibid.* p. 307.), which differ so much in structure from all others, as to require in most cases a special and distinct notice.

SECT. I.   *Of the Cuticle and its Appendages.*

1. Of the Cuticle.

VEGETABLES, like animals, are covered externally by a thin membrane or cuticle, which usually adheres firmly to the cellular substance beneath it. To the naked eye it appears like a transparent homogeneous pellicle, but under the microscope it is found to be traversed in various directions by lines, which, by constantly anastomosing, give it a reticulated character. In some of the lower tribes of plants, consisting entirely of cellular tissue, it is not distinguishable, but in all others it is to be found upon every part, except the stigma and the spongioles of the roots. Its usual character is that of a delicate membrane, but in some plants it is so hard as almost to resist the blade of a knife, as in the pseudo-bulbs of certain Orchideous plants. The most usual form of the reticulations is the hexagonal (Plate III. fig. 11.): sometimes they are exceedingly irregular in figure; often prismatical; and not unfrequently bounded by sinuous lines, so irregular in their direction as to give the meshes no determinate figure (fig. 5.).

Botanists have not agreed as to the exact nature of the cuticle; while the greater number incline to the opinion that it is an external layer of cellular tissue in a dry and compressed state; others, among whom are included both Kieser and Amici, consider it a membrane of a peculiar nature, transversed by veins, or vasa lymphatica.

By the latter it is contended, that the sinuous direction of the lines in many cuticles is incompatible with the idea of any thing formed by the adhesion of cellular tissue; that when it is once removed, the subjacent tissue dies, and does not become cuticle in its turn, and that it may often be torn up readily without laceration.

On the other hand, it is contended, that the reticulations of the cuticle are mostly of some figure analogous to that of cellular tissue, and that the sinuous meshes themselves are not so different as to be incompatible with the idea of a membrane formed of adhering cellules. We are accustomed to see so much variety in the mere form of all parts of plants, that an anomalous configuration in cellular tissue should not surprise us. The lines, or supposed vasa lymphatica, are nothing more than the united sides of the cellules, and are altogether the same as are presented to the eye by any section of a mass of cellular substance. It is certain that the cuticle cannot be removed without lacerating the subjacent tissue, with however much facility it may be sometimes separable: on the under surface of the leaf of the Box, for instance, there has plainly been some tearing of the tissue, before the cuticle acquired the loose state in which it is finally found. If the subjacent epidermis never becomes cuticle when the latter is removed, this is no reason why the cuticle itself should not be composed of cellular tissue; for it is an axiom in vegetable physiology, that *a part once fully formed is incapable of any subsequent change.* Thus, pith never alters its dimensions, after the medullary sheath that encloses it has been once completed, and a zone of wood never contracts or expands after it has been deposited: new matter may be added to any part, but the arrangement of the tissue, once fixed, remains unchangeable.

The principal argument, however, in favour of cuticle being compressed cellular tissue, is, that in the cuticle of many plants the cellular state is distinctly visible upon a section (Plate I. fig. 2. *a*); that it even consists occasionally of several layers of cellules, as in many epiphytes of the Orchis tribe; and that, as there is no reason to doubt that Nature is as uniform in the plan upon which cuticle is constructed as in all her other works, in those cases in which the cellular structure is less distinctly visible, we are nevertheless justified by sound philosophy in recognising it; while, on the other hand, it would be highly unphilosophical to suppose that the cuticle is formed in some plants upon one plan, and in others upon a totally different one. It may be farther remarked, that separable cuticle may often be traced into that which, being younger, is both inseparable and undistinguishable from the other cellular substance with which it is in contact, and from which it possesses no organic difference.

There is some reason to suppose that there is occasionally present, on the outside of the cuticle, a transparent, very delicate membrane, having no organic structure, as far as can be discovered with the most powerful microscopes. Something of this kind has been noticed by M. Adolphe Brongniart in the Cabbage leaf, and an analogous structure has been remarked by Professor Henslow in the Digitalis.

2. Of Stomata.

In many plants the cuticle has certain openings of a very peculiar character, which appear connected with respiration, and which are called *Stomata*. (Plate III. *passim.*)

STOMATA (*Pores of the epidermis; Pores corticaux, allongés, évaporatoires, or grands pores; Glands corticales, miliaires, or epidermoidales; Glandulæ cutaneæ; Oeffnungen; Stomatia;*) are passages through the cuticle, having the appearance of areolæ, in the centre of which is a slit that opens or closes according to circumstances, and lies over a cavity in the subjacent tissue.

There is, perhaps, nothing in the structure of plants upon which it is more difficult to form any satisfactory opinion than these stomata. Malpighi, and Grew, who seems first to have figured them (*see* his plate xlviii. fig. 4.), call them openings or apertures, but had no exact idea of their structure. Mirbel also, for a long time, considered them pores, and figured them as such; admitting, however, that he suspected the openings to be an optical deception. M. De Candolle entertains no doubt of their being passages through the cuticle. He says their edge has the appearance of a kind of oval sphincter, capable of opening and shutting. The membrane that surrounds this sphincter is always continuous with those which constitute the network of the cuticle: under the latter, and in the interval between the pore and the edge of the sphincter, are often found molecules of adhesive green matter (*Organogr.* i. 80.); and recently M. Adolphe Brongniart, in his beautiful figures of the anatomy of leaves, would seem to have settled the question beyond all dispute. (*Annales des Sciences*, vol. xxi.) Nevertheless, there are anatomists of high reputation who entertain a directly opposite opinion; denying the existence of passages, and considering the stomata rather in the light of glands. Nees von Esenbeck and Link deny the existence of any perforation in the stomata, and consider that the supposed opening is a space more pellucid than the surrounding tissue, and that what seems a closed up slit is the thickened border of the space. Link further adds, that the obscuration of the centre of the stomata is caused by a peculiar secretion of matter, as is plainly visible in Baryosma serratum. (*Elementa*, p. 225.) To the views of these writers is to be added the testimony of Mr. Brown (*Suppl. prim. Prodr.* p. 1.), who describes the stomata as glands which are really almost always imperforate, with a disk formed by a membrane of greater or less opaqueness, and even occasionally coloured; at the same time he speaks of this disk being, *perhaps*, sometimes perforated.

In the midst of such conflicting testimony, an observer

necessarily finds much difficulty in fixing his opinion. I shall take the liberty of stating what I myself have seen, without, however, supposing that my humble authority can in any degree contribute to the determination of the question. In no plants are stomata larger than in some Monocotyledons; they are, therefore, the best subjects for examination for general purposes. In Crinum amabile they evidently consist of two kidney-shaped bodies filled with green matter, lying upon an area of the cuticle smaller than those that surround it, and having their incurved sides next each other. In some, at the part where the kidney-shaped bodies come in contact, there is an elevated ridge, dark, as if filled with air, and having its principal diameter distinctly divided by a line. (Plate III. fig. 11.) In this state the stoma is at rest: but in others the kidney-shaped bodies are much more curved; their sides are more separated from each other; and there is no elevated ridge: at their former line of contact there is an opening so distinct and wide as to be equal to half the diameter of one of the kidney-shaped bodies; this, I presume, is the stoma open. That what is described to be an opening, is really so, seems to be demonstrated by the following tests:—
1. It is more transparent than any part of the most transparent portion of the cuticle; 2. It admits transmitted light without interruption; as is seen by gradually augmenting the magnifying power by which it is viewed, when the opening continues transparent, notwithstanding the great loss of light that attends the use of very high powers in compound microscopes; and, 3. None of those arts which the microscopic observer knows so well how to employ, such as shifting, augmenting or decreasing the light, interposing moveable shadows between the mirror and the object, and the like, give the least indication of the presence of any membrane across the orifice of the stoma. I therefore conclude, that, in the Crinum amabile, the stomata are formed by two elastic reniform cellules, lying over an opening in the centre of a contracted area of cuticle; that these cellules, when expanded, meet, and press powerfully against each other, like two opposing springs; thus causing the elevated ridge-like appearance visible in the axis of the stoma in the figure above

referred to; and that, when contracted, they curve in an opposite direction, separating from each other, and ceasing to close up the aperture over which they lie. If it were possible to be absolutely certain of the accuracy of this description, the structure of the stoma in Crinum amabile might be safely taken as the type of all others; for, no doubt, they are all constructed upon a similar plan. Without actually asserting so much as this, I may venture to state, that, of many hundreds of observations I have made upon this subject, I have not met with any thing that has led me to doubt the uniformity of their nature, or their general accordance with what is found in Crinum amabile, whatever that may be. Or at least, the only difference is this, that while the two cellules that form the edges of the aperture are distinctly separated at their extremities in this plant, they are often confluent in others, as in Caladium esculentum. (Plate III. fig. 9.) Several varieties are represented at Plate III.; besides which, they have been noticed by Link to be occasionally quadrangular, as in Yucca gloriosa (Plate III. fig. 10.), and Agave americana, and by Mr. Brown to be very rarely angular, of which, however, no instance is cited by that botanist. The former case is one in which the quadrangular figure is caused by the cellules being straight; I am not aware if Mr. Brown means the same thing.

I have never been so fortunate as to discover the membrane which Mr. Brown describes as generally overlying the apertures; nor do I know of any other botanist having confirmed that observation.

Stomata are not found in Mosses, Hepaticæ, Fungi, Algæ, or Lichens (see *Introduction to the Natural System*); in no submersed plants, or submersed parts of amphibious plants; it is also said, not in Monotropa hypopithys, Neottia nidus avis, and Cuscuta europæa. They are not formed in the cuticle of plants growing in darkness; are very small in trees and shrubs, particularly evergreens, and more especially in such as have coriaceous leaves, and acrid or aromatic juices. (*Rudolphi.*) They are not present upon roots, or the ribs of leaves. It also frequently happens that they are found upon one surface of a leaf, but not on another, and generally in most abundance on the under side. In succulent plants, or in the succulent parts of other plants, they are either rare, or wholly wanting. They may be generally seen upon the calyx; often on the corolla; and rarely, but sometimes, upon the filaments, anthers, and styles. In fruit, they have only been noticed upon such as are membranous, and never upon the coat of the seed; they exist, however, upon the surface of cotyledons.

Mr. Brown thinks that the uniformity of the stomata, in figure, position, and size, with respect to the meshes of the cuticle, is often such as to indicate the limits, and sometimes the affinities, of genera, and of their natural sections. He has shown, with his usual skill, that this is the case in Proteaceæ. He also remarks, that on the microscopic character of the equal existence of stomata on both surfaces of the leaf depends that want of lustre which is so remarkable in the forests of New Holland. (*Journal of the Royal Geogr. Society*, i. 21.)

The same botanist is of opinion, that the two glands, or cellules, of which a stoma is composed, are each analogous to the single cellules found occupying the inner face of the meshes of the cuticle. (Plate iii. fig. 9.) See the *Memoir on the Impregnation of Orchideæ.*)

The surface of the cuticle is either perfectly smooth, or furnished with numerous processes, consisting of cellular tissue in different states of combination, which may be arranged under the heads of *hairs, scales, glands,* and *prickles.* All these originate either directly from the cuticle, or from the cellular substance beneath it; never having any communication with the vascular or ligneous system.

In Nepenthes the cuticle in the inside of the pitchers is pierced by a great number of holes, each of which is closed up by a firm thick disk of small cellular tissue, deep brown in colour, and connected with the cavernous parenchyma of the pitcher. Besides these, Nepenthes has also stomata.

Nerium oleander and some other plants have, in lieu of stomata, cavities in the cuticle, curiously filled up or protected by hairs. (See *Annales des Sciences,* xxi. 438.)

3. Of Hairs.

These (*fig.* 16.) are minute, transparent, filiform, acute processes, composed of cellular tissue more or less elongated, and arranged in a single row. They are found occasionally upon every part of a plant, even in the cavities of the and stem, as in Nymphæa and other aquatic plants. In the Cotton Plant (Gossypium herbaceum, &c.) they form the substance which envelopes the seeds, and is wrought into linen; in the Cowhage (Mucuna urens and pruriens), it is they that produce the itching; and in the Palm tribe they are the long, entangled, soft, strangulated filaments that are used for Amadou. They vary extremely in length, density, rigidity, and other particulars; on which account they have been distinguished by the following names: —

Down or *Pubescence* (*pubes*, adj. *pubescens*), when they form a short soft stratum, which only partially covers the cuticle, as in Geranium molle.

Hairiness (*hirsuties*, adj. *hirsutus*), when they are rather longer and more rigid, as in Galeopsis Tetrahit.

Pili (*pilus*, adj. *pilosus*), when they are long, soft, and erect, as in Daucus Carota.

Villus (adj. *villosus*), when they are very long, very soft, erect, and straight, as in Epilobium hirsutum. *Crini* (adj. *crinitus*) are this variety in excess.

Velvet (*velumen*, adj. *velutinus*), when they are short, very dense and soft, but rather rigid, and forming a surface like velvet, as in many Lasiandras.

Tomentum (adj. *tomentosus*), when they are entangled, and close pressed to the stem, as in Geranium rotundifolium.

Ciliæ (adj. *ciliatus*), when long, and forming a fringe to a margin, like an eyelash, as in Sempervivum tectorum.

Bristles (*setæ*, adj. *setosus*), when short and stiff, as on the stems of Echium.

Stings (*stimuli*, adj. *stimulans; pili subulati* of De Candolle), when stiff and pungent, giving out an acrid juice if touched, as in the Nettle.

Glandular hairs (*pili capitati*), when they are tipped with a glandular exudation, as in Primula sinensis. These must not be confounded with stalked glands.

Hooks (*hami, unci, rostella*), when curved back at the point, as in the nuts of Myosotis Lappula.

Barbs (*glochis*, adj. *glochidatus*), if forked at the apex, both divisions of the fork being hooked, as in the nuts of the same plant.

Hairs also give the following names to the surface of any thing: —

1. *Silky* (*sericeus*), when they are long, very fine, and pressed closely to the surface, so as to present a sublucid silky appearance: *ex.* Protea argentea.

2. *Arachnoid*, when very long, and loosely entangled, so as to resemble cobweb: *ex.* Calceolaria arachnoidea.

3. *Manicate*, when interwoven into a mass that can be easily separated from the surface: *ex.* Cacalia canescens, Bupleurum giganteum.

4. *Bearded* (*barbatus*), when the hairs are long, and placed in tufts: *ex.* the lip of Chelone barbata.

5. *Rough* (*asper*), when the surface is clothed with hairs, the lower joint of which resembles a little bulb, and the upper a short rigid bristle: *ex.* Borago officinalis.

Hairs are either formed of a single cell of cellular tissue (Plate I. fig. 8. *b*), or of several placed end to end in a single series (Plate I. fig. A, B.), whence, if viewed externally, they have the appearance of being divided internally by transverse partitions. They are sometimes divided into two or three forks at the extremity, as in Alyssum, some species of Apargia, &c. Occasionally they emit little branches along their whole length: when such branches are very short, the hairs are said to be *toothed* or *toothleted*, as in the fruit of Torilis Anthriscus; when they are something longer, the hairs are called *branched*, as in the petioles of the gooseberry; if longer and finer still, the term is *pinnate*, as in Hieracium Pilosella; if the branches are themselves pinnate, as in Hieracium undulatum, the hairs are then said to be *plumose*. It sometimes happens that little branchlets are produced on one side only of a hair, as on the leaves of Siegesbeckia orientalis, in which case the hair is called *one-sided (secundatus)*; very rarely they appear upon the articulations of the hair, which in that case is called *ganglioneous*. (Plate I. fig. 9. Verbascum Lychnitis): the *poils en goupillon* of De Candolle are referable to this form. Besides these, there are many other modifications. Hairs are conical, cylindrical, or moniliform, thickened slightly at the articulations (*torulose*), as in Lamium album, or much enlarged at the same point (*nodulose*), as in the calyx of Achyranthes lappacea.

Hairs are sometimes said to be *fixed by their middle* (Plate I. fig. 10. c); a remarkable structure, common to many different genera; as Capsella, Malpighia, Indigofera, &c. This expression, however, like many others commonly used in botany, conveys a false idea of the real structure of such hairs. They are in reality formed by an elevation of one cellule of the cuticle above the level of the rest, and by the developement of a simple hair from its two opposite sides. Such would be more correctly named *divaricating* hairs. When the central cellule has an unusual size, as in Malpighia, these hairs are called *poils en navette* (*pili Malpighiacei*) by M. De Candolle; and when the central cellule is not very apparent, *poils en fausse navette* (*pili pseudo-Malpighiacei, biacuminati*), as in Indigofera, Astragalus, Asper, &c. In many plants the hairs grow in clusters, as in Malvaceæ, and are occasionally united at their base: such are called *stellate*, and are frequently peculiar to certain natural orders. (Plate I. fig. 10. a.)

All these varieties belong to one or other of the two principal kinds of hairs; viz. the Lymphatic and the Secreting. Of these, *lymphatic* hairs consist of tissue tapering gradually from the base to the apex; and *secreting*, of cellules visibly distended either at the apex or base into receptacles of fluid. Malpighiaceous and glandular hairs, stings, and those which cause asperity on the surface of any thing, belong to the latter; almost all the other varieties to the former.

When hairs arise from one surface only of any of the

appendages of the axis, it is almost always from the under surface; but the seed leaves of the nettle, and the common leaves of Passerina hirsuta, are mentioned by M. De Candolle as exceptions to this rule: certain states of Rosa canina might also be mentioned as exhibiting a similar phenomenon. When a portion only of the surface of any thing is covered by hairs, that portion is uniformly the ribs or veins. According to M. De Candolle, hairs are not found either upon true roots, except at the moment of germination, nor upon any part of the stem that is formed under ground, nor upon any parts that grow under water.

4. Of Scales.

SCALES are thin flat membranous processes, formed of cellular tissue springing from the cuticle. They may be considered as hairs of a higher order, — as organs of the same nature, but more developed; for they differ from hairs only in their degree of composition. They are of two kinds, *Scales* properly so called, and *Ramenta*. Care must be taken not to confound scales of this description with scales of the stem, to be described hereafter: those now under consideration being mere processes of the cuticle; those to be noticed hereafter being peculiar modifications of leaves.

Scales, properly so called, are the small, roundish, flattened particles which give a leprous appearance to the surface of certain plants, as the Elæagnus and the Ananassa. (Plate I. fig. 10. *b.*) They consist of a thin transparent membrane, attached by its middle, and, owing to the imperfect union towards its circumference, of the cellular tissue of which it is composed, having a lacerated irregular margin. A scale of this nature is called in Latin composition *lepis*, and a surface covered by such scales *lepidotus*, and not *squamosus*, which is only applied to a surface covered with the rudiments of leaves. Scales are the *poils en écusson* (*pili scutati*) of De Candolle.

Ramenta (*Vaginellæ*) are thin, brown, foliaceous scales, appearing sometimes in great abundance upon young shoots. They are particularly numerous, and highly developed, upon the petioles and the backs of the leaves of Ferns. They consist of cellular tissue alone, without any vascular bundles, and are known from leaves not only by their anatomical structure, but also by their irregular position, and by the absence of buds from their axillæ. The student must particularly remark this, or he will confound with them leaves having a ramentaceous appearance, such as are produced upon the young shoots of Pinus. Link remarks, that they are very similar in structure to the leaves of mosses. The term *striga* has occasionally been applied to them (*Dec. Théor. Elém.* ed. 2. 376. *Link*, *Elem.* 240.); but that word was employed by Linnæus to designate any stiff bristle-like process, as the *spines* of the Cactus, the divaricating *hairs* of Malpighia, and the stiff stellated *hairs* of Hibiscus. So vague an application of the term is very properly avoided at the present day, and the substantive is rejected from modern glossology; the adjective term *strigose* is, however, occasionally still employed to express a surface covered with stiff hairs.

5. Of Glands.

GLANDS are elevated spaces in the stratum of parenchyma lying immediately below the cuticle, in which they cause projections. They are of several kinds.

Stalked glands are elevated on a stalk which is either simple or branched: they secrete some peculiar matter at their extremities, and are often confounded with the glandular hairs above described, from which they have been well distinguished by Link. According to that botanist, they are either simple or compound; the former consisting of a single cell, and placed upon a hair acting as a simple conduit, occasionally interrupted by divisions; the latter consisting of several cells, and seated upon a stalk containing several conduits, formed by rows of cellular tissue. They are common upon the rose and the bramble, in which they become very rigid, and assume the nature of aculei. For the sake of distinguishing them from the latter, they have been called *setæ* by Woods and myself, but improperly; they are also the *aiguillons* of the French. In Hypericum they abound on the calyx and corolla of some species, but do not give out any exudation; they contain, however, a deep red juice within their cells. In some Jatrophas they are much branched; in many Diosmeæ they form a curious humid appendage at the apex of the stamens.

Sessile glands are produced upon various parts, and are extremely variable in figure. In Cassias, they are seated upon the upper edge of the petiole, and are usually cylindrical or conical; in Cruciferous plants they are little roundish shining bodies, arising from just below the base of the ovarium; in the leafless Acacias, they are a little depressed, with a thickened rim, and placed on the upper edge of the phyllodium; they are little kidney-shaped bodies upon the petiole of the Peach and other drupaceous plants; and they assume many more appearances.

Verrucæ, or warts, are roundish excrescences, formed of cellular tissue filled with opaque matter, and are situated upon various parts. They are common upon the surface of the leaves of the Aloe, where they are very large; upon the stem, as in Euonymus verrucosus; upon the petiole, as in Passiflora; they are also found upon the calyx, as in some species of Campanula, and at the serratures of the leaves, when they are considered by M. Röper (*De Floribus Balsaminearum*, p. 15.) to be abortive ovula. They also appear upon the pericarpium and the testa of the seed; in the latter case they are called *spongiolæ seminales* by De Candolle. They are round, oblong, or reniform, and occasionally cupulate, when they receive the name of *glandes à godet* (*glandulæ urceolares*) from some French writers. Verrucæ are the *glandes cellulaires* of Mirbel; but they must not be confounded with the *glandes vasculaires* of the same writer, which are not mere excrescences of the epidermis, but modifications of well known organs. (See *Discus*, further on.) The presence of minute verrucæ upon the surface of a leaf gives rise to a peculiar kind of roughness which is called *scabrities*, and such a surface is then said to be *scabrous* (scaber): this must not be confounded with *asperity*.

Papillæ (*Glandulæ utriculaires* of Guettard) are minute transparent elevated points of the cuticle, filled with fluid, and covering closely the whole surface upon which they appear. In other words, they are elevated, distended cellules of the cuticle. The presence of papillæ upon the leaves of the ice plant gives rise to the peculiar crystalline nature of its surface; they also cause the satiny appearance of the petals, upon which they almost always exist in great quantities. Link remarks, that the petals of Plantago, which are destitute of papillæ, are also without the usual satiny lustre of those organs. When the papillæ are much elongated beyond the surface, as in many stigmas, of which they form the collecting fringes, they receive sometimes the name of *papulæ*. It should be observed, that in M. De Candolle's *Théorie Elémentaire*, these two terms are transposed, each having received the definition belonging to the other.

Lenticular glands (*Lenticelles* of De Candolle; *Glandes lenticulaires* of Guettard;) are brown oval spots found upon the bark of many plants, especially willows: they indicate the points from which roots will appear if the branch be placed in circumstances favourable to their production. They are considered by M. De Candolle to bear the same relation to the roots that buds bear to young branches. (*De Candolle, Premier Mém. sur les Lentic.*, in the *Ann. des Sciences Naturelles.*)

6. Of Prickles.

PRICKLES (*aculei*) are rigid, opaque, conical processes, formed of masses of cellular tissue, and terminating in an acute point. They may be, not improperly, considered as very compound indurated hairs. They have no connection with the woody fibre, by which character they are obviously distinguished from spines, of which mention will be made under the head of branches, of which spines are an abortion. Prickles are found upon all parts of a plant, except the stipulæ and stamens. They are very rarely found upon the corolla, as in Solanum Hystrix; their most usual place is upon the stem, as in Rosa, Rubus, &c.

SECT. II. *Of the Stem or Ascending Axis.*

WHEN a plant first begins to grow from the seed, it is a little body called an embryo, with two opposite extremities, of which the one elongates in the direction of the earth's centre, and the other, taking a direction exactly the contrary, extends upwards into the air. This disposition to develope in two diametrically opposite directions is found in all seeds, properly so called, there being no known exception to it; and the tendency is moreover so powerful, that, as we shall hereafter see (Book II.), no external influence is sufficient to overcome it. The result of this developement is the axis, or centre, round which the leaves and other appendages are arranged. That part of the axis which forces its way downwards, constantly avoiding light, and withdrawing from the influence of the air, is the descending axis, or the root; and that which seeks the light, always striving to expose itself to the air, and expanding itself, to the utmost extent of its nature, to the solar rays, is the ascending axis, or the stem. As the double elongation just mentioned exists in all plants, it follows that all plants must necessarily have, at an early period of their existence at least, both stem and root; and that, consequently, when plants are said to be rootless, or stemless, such expressions are not to be considered physiologically correct.

The STEM has received many names; such as *caudex ascendens, caudex intermedius, culmus, stipes, truncus,* and *truncus ascendens*. It always consists of bundles of vascular and woody tissue, embedded in cellular substance in various ways, and the whole enclosed within a cuticle. The manner in which these parts are arranged with respect to each other will be explained hereafter. The more immediate subject of consideration must be the parts that are common to all stems.

1. Of its Parts.

Where the stem and root, or the ascending and descending axes, diverge, there commences in many plants a difference of anatomical structure, and in all a very essential physiological dissimilarity; as will be hereafter seen. This portion of the axis is called the neck or collum, (*coarcture* of Grew, *nœud vital* of Lamarck, *limes communis*, or *fundus plantæ*, of Jungius,) and has been thought by some to be the seat of vegetable vitality; an erroneous idea, of which more will be said in the next book. At first it is a space that we have no difficulty in distinguishing, so long as the embryo, or young plant, has not undergone any considerable change; but in process of time it is externally obliterated; so that in trees of a few years' growth its existence becomes a matter of theory, instead of being actually evident to our senses.

Immediately consequent upon the growth of a plant is the formation of leaves. The point of the stem whence these arise is called the *nodus* (*geniculum*, Jungius; *nœud*, Fr.;) and the space between two nodi is called an *internodium* (*entrenœud* Fr.; *merithallus*, Du Petit Thouars). In internodia the arrangement of the vascular and fibrous tissue, of whatever nature it may be, of which they are composed, is nearly parallel, or, at least, experiences no horizontal interruption. At the nodi, on the contrary, vessels are sent off horizontally into the leaf; the general developement of the axis is momentarily arrested while this horizontal communication is effecting, and all the tissue is more or less contracted. In many plants this contraction, although it always exists, is scarcely appreciable; but in others it takes place in so remarkable a degree as to give their stems a peculiar character; as, for instance, in the Bamboo, in which it causes diaphragms that continue to grow and harden, notwithstanding the powerfully rapid horizontal distension to which the stems of that plant are subject. In all cases, without exception, a leaf-bud or buds is formed at a nodus immediately above the base of the leaf; generally such a bud is either sufficiently apparent to be readily recognised by the naked eye, or, at least, it becomes apparent at some time or other: but in certain plants, as Heaths, the buds are often never discoverable; nevertheless, they always exist, in however rudimentary a state, as is proved by their occasional developement under favourable or uncommon circumstances. By some writers nodi, upon which buds are obviously formed, are called *compound*, or *artiphyllous;* and those in which no apparent buds are discoverable, are named *simple*, or *pleiophyllous:* they are also said to be *divided*, when they do not surround the stem, as in the apple and other alternate-leaved genera; or *entire*, when they do surround it, as in grapes and umbelliferous plants: they are further said to be *pervious*, when the pith passes through them without interruption, or *closed*, when the canal of the pith is interrupted, as if by a partition. Pervious and divided, and closed and entire nodi usually accompany each other. For other remarks upon this subject, see Link's *Elementa*.

All the divisions of a stem are in general terms called *branches* (*rami*); but it is occasionally found convenient to express particular kinds of branches by special names. Thus, the twigs, or youngest shoots, are called *ramuli* or branchlets (*brindilles* or *ramilles*, Fr.), and by the older botanists *flagella;* the assemblage of branches which forms the head of a forest tree is called the *coma: cyma* is sometimes used to express the same thing, but improperly.

Shoots which have not completed their growth have received the name of *innovations*, a term usually applied to mosses. When such a shoot is covered with scales upon its first appearance, as the Asparagus, it is called *turio:* by the old botanists all such shoots were named *asparagi*. When a shoot is long and flexible, it receives the name of *vimen*. This word, however, is seldom used; its adjective being employed instead: thus, we say, *rami viminei*, or *caulis vimineus;* and not *vimen*. From this kind of branch, that called a *virgate* stem, *caulis virgatus*, differs only in being less flexible and more rigid. A young slender branch of a tree or shrub is sometimes named *virgultum*. When the branches diverge nearly at right angles from the stem, they are said to be *brachiate*. Small stems, which proceed from buds formed at the neck of a plant without the previous production of a leaf, are called *cauliculi*.

Besides these terms, Du Petit Thouars employed certain French words in a way peculiar to himself. The first young shoot produced during the year by a tree, he named *scion;* any subsequent shoots formed by the scion, he termed *ramilles;* the shoot that supports the scion was a *rameau;* that which supports the *rameau* a *branche;* and the trunk which bears the whole the *tronc*. Professor Link calls a stem which proceeds straight from the earth to the summit, bearing its branches on its sides, as Pinus, a *caulis excurrens*, and a stem which at a certain distance above the earth breaks out into irregular ramifications, a *caulis deliquescens*.

From the constitution and ramifications of their branches, plants are divided into trees, shrubs, and herbs. When the branches are perennial, and supported upon a trunk, a *tree* (*arbor*) is said to be formed; for a small tree, the term *arbusculus* is sometimes employed. When the branches are perennial, proceeding directly from the surface of the earth without any supporting trunk, we have a shrub (*frutex* or *arbustum*, Lat., and *arbrisseau*, Fr.), which occasionally, when very small, receives the diminutive name of *fruticulus*. If a shrub is low, and very much branched, it is often called *dumosus* (subst. *dumus*): this kind of shrub is what the French understand by their word *buisson*. The *suffrutex*, *under-shrub*, or *sous-arbrisseau*, differs from the shrub, in perishing annually, either wholly or in part; and from the herb, in having branches of a woody texture, which frequently exist more than one year: such is the Mignonette (Reseda odorata) in its native country, or in the state in which it is known in gardens as the Tree Mignonette. The under-shrub is exactly intermediate between the shrub and the herb. All plants producing shoots of annual duration from the surface of the earth are called herbs.

Some botanists distinguish two sorts of stems, the characters of which are derived from their mode of growth. When a stem is never terminated by a flower-bud, nor has its growth stopped by any other organic cause, as in Veronica arvensis, and all perennial and arborescent plants, it is said to be *indeterminatus*; but when a stem has its growth uniformly stopped at a particular period of its existence by the production of a terminal bud, or by some such cause, it is called *determinatus*. The capitate and verticillate species of Mint owe their differences to causes of this nature; the stem of the former being determinate, the latter indeterminate.

Some branches are imperfectly formed, lose their power of extension, become unusually hard, and acquire a sharp point. They are then called *spines* (*spinæ*), and must not be confounded with prickles, already described, from which they are distinguishable by their woody vascular centre.

Occasionally, as in the Whitethorn, they bear leaves. In domesticated plants they often entirely disappear, as in the Apple and Pear, the wild varieties of which are spiny, and the cultivated ones spineless.

The point whence two branches diverge is called the *axilla*, or, in old botanical language, the *ala*.

Leaf-buds (*Gemma*, Linn. *Bourgeon*, Fr.), being the rudiments of young branches, are of great importance in regard to the general structure of a plant. They consist of scales

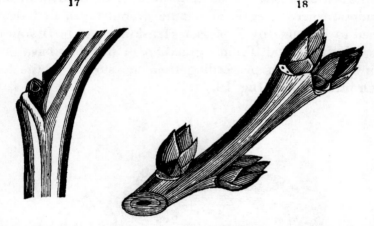

imbricated over each other, the outermost being the hardest and thickest, and surrounding a minute axis which is in direct communication with the woody and cellular tissue of the stem. Linnæus called this bud *Hybernaculum*, because it serves for the winter protection of the young and tender parts; and distinguished it into the *Gemma*, or leaf-bud of the stem, and the *Bulb*, or leaf-bud of the root.

The leaf-bud has been compared by Du Petit Thouars and some other botanists to the embryo, and has even been denominated a *fixed embryo*. This comparison must not, however, be understood to indicate any positive identity between these two parts in structure, but merely an analogous function, both being formed for the purpose of reproduction; both in origin and structure they are entirely different. The leaf-bud consists of both vascular and cellular tissue, the embryo of cellular tissue only: the leaf-bud is produced without sexual intercourse, to the embryo this is essential:

finally, the leaf-bud perpetuates the individual, the embryo continues the species.

The usual, or normal, situation of leaf-buds is in the axillæ of leaves; and all departure from this position is either irregular or accidental. Botanists give them the name of *regular* when they are placed in their normal station, and they call all others *latent* or *adventitious*. The latter have been found in almost every part of plants; the roots, the internodia, the petiole, the leaf itself, have all been remarked producing them. On the leaf they usually proceed from the margin, as in Malaxis paludosa, where they form minute granulations, first determined to be buds by Professor Henslow, or as in Bryophyllum calycinum and Tellima grandiflora; but they have been seen by M. Turpin proceeding from the surface of the leaf of Ornithogalum. (fig. 19.)

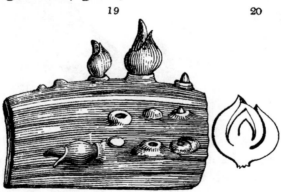

We are wholly unacquainted with the cause of the formation of leaf-buds; all we know is, that they appear to proceed from woody or vascular, and not from cellular tissue. There is, indeed, an opinion, which I believe is that of Mr. Knight, that the sap itself can at any time generate buds without any previously formed rudiment; and that they depend, not upon a specific alteration of the arrangement of the vascular system, called into action by particular circumstances, but upon a state of the sap favourable to their creation. In proof of this it has been said, that if a bud of the Prunus Pseudo-cerasus, or Chinese Cherry, be inserted upon a cherry stock it will grow freely, and after a time will emit small roots from just above its union with the stock;

at the time when these little roots are formed, let the shoot be cut back to within a short distance of the stock, and the little roots will then, in consequence of the great impulsion of sap into them, become branches emitting leaves.

The leaf-buds of the deciduous trees of cold climates are covered by scales, which are also called *tegmenta ;* these afford protection against cold and external accidents, and vary much in texture, thickness, and other characters. Thus, in the *Beech*, the tegmenta are thin, smooth, and dry; in many *Willows* they are covered with a thick down; in *Populus balsamifera* they exude a tenacious viscid juice. In herbaceous plants and trees of climates in which vegetation is not exposed to severe cold, the leaf-buds have no tegmenta; which is also, but very rarely, the case in some northern shrubs, as Rhamnus Frangula.

The scales of the bud, however dissimilar they may be to leaves in their ordinary appearance, are nevertheless, in reality, leaves in an imperfectly formed state. They are the last leaves of the season, developed at a period when the current of vegetation is stopping, and when the vital powers have become almost torpid. That such is really their nature is apparent from the gradual transition from scales to perfect leaves, that occurs in such plants as Virburnum prunifolium, Magnolia acuminata, Liriodendron tulipifera, and Æsculus Pavia; in the latter the transition is, perhaps, most satisfactorily manifested. In this plant the scales on the outside are short, hard, dry, and brown; those next them are longer, and greenish, and delicate; within these they become dilated, are slightly coloured pink, and occasionally bear a few imperfect leaflets at their apex; next to them are developed leaves of the ordinary character, except that their petiole is dilated and membranous like the inner scales of the bud; and, finally, perfectly formed leaves complete the series of transitions.

Among the varieties of root is sometimes classed what botanists call a *bulb ;* a scaly body, formed at or beneath the surface of the ground, emitting roots from its base, and producing a stem from its centre. Linnæus considered it the leaf-bud of a root; but in this he was partly mistaken, roots being essentially characterised by the absence of buds. He was, however, perfectly correct in identifying it with a leaf-bud. A bulb has the power of propagating itself by developing in the axillæ of its scales new bulbs, or what gardeners call *cloves*, (*Cayeu*, French; *Nucleus* and *Adnascens* of the older botanists; *Adnatum* of Richard;) which grow at the expense of their parent bulb, and eventually destroy it. Every true bulb is, therefore, necessarily formed of imbricated scales, and a solid bulb has no existence. The *bulbi solidi*, as they have been called, of the Crocus, the Colchicum, and others, are, as we shall hereafter see (see *Cormus*), a kind of ubterranean stem: they are distinct from the bulb in being not an imbricated scaly bud, but a solid fleshy stem, itself emitting buds. It has been supposed that they were buds, the scales of which had become consolidated; but this hypothesis leads to this very inadmissible conclusion,—that as the cormus or bulbus solidus of a Crocus is essentially the same, except in size and situation, as the stem of a Palm, the stem of a Palm must be a bulbus solidus also, which is absurd. In truth, the bulb is analogous to the bud that is seated upon the cormus, and not to the cormus itself; a bulb being an enlarged subterranean bud without a stem, the cormus a subterranean stem without one enlarged bud.

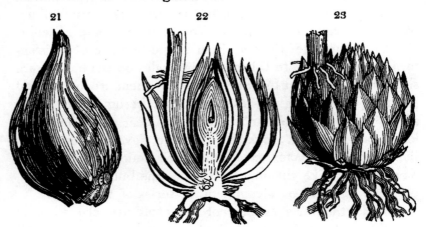

Of the bulb, properly so called, there are two kinds.

1. The *tunicated bulb* (fig. 21.), of which the outer scales are thin and membranous, and cohere in the form of a distinct covering, as in the onion; and, 2. the *naked bulb* (*Bulbus*

squamosus) (fig. 23. 22.), in which the outer scales are not membranous and coherent, but distinct and fleshy like the inner scales, as in *Lilium.* The outer covering of a bulb of the first kind is called the *tunic.*

Besides the bulbs properly so called, there are certain leaf-buds, developed upon stems in the air, and separating spontaneously from the part that bears them, which are altogether of the nature of bulbs. Such are found in Lilium tigrinum, some Alliums, &c. They have been called *bulbilli, propagines, sautilles, bacilli,* &c. Care must be taken not to follow some botanists, in confounding with them the seeds of certain Amaryllideæ, which have a fleshy testa; but which, with a vague external resemblance to bulbs, have in every respect the structure of genuine seeds.

The tegmenta, or scales of the bud, have received the following names, according to the part of the leaf of which they appear to be a transformation; such terms are, however, but seldom employed: —

1. *Foliacea*, when they are abortive leaves, as in Daphne Mezereum.

2. *Petiolacea*, when they are formed by the persistent base of the petiole, as in Juglans regia.

3. *Stipulacea*, when they arise from the union of stipulæ, which roll together and envelope the young shoot, as in Carpinus, Ostrya, Magnolia, &c.

4. *Fulcracea*, when they are formed of petioles and stipules combined, as in Prunus domestica, &c. — (*Rich. Nouv. Elem.* 134. ed. 3.)

The manner in which the nascent leaves are arranged within the leaf-bud is called *foliation* or *vernation.* The names applied to the various modifications of this will be explained in Glossology. They are of great practical importance both for distinguishing species, genera, and even natural orders; but have, nevertheless, received very little general attention. The vernation of Prunus Cerasus is *conduplicate;* of Prunus domestica, *convolute;* of Filices and Cycadeæ, *circinate,* and so on.

2. Of its External Modifications.

It has already been stated, that the first direction taken by the stem immediately upon its developement is upward into the air. While this ascending tendency is by many plants maintained during the whole period of their existence, by others it is departed from at an early age, and a horizontal course is taken instead; while also free communication with light and air is essential to most stems, others remain during all their lives buried under ground, and shun rather than seek the light. From these and other causes, the stems of plants assume a number of different states, to which botanists attach particular terms. It will be most convenient to divide the subject into the consideration of the varieties of —

1. The *subterranean* stem; and,
2. The *aerial* stem.

The SUBTERRANEAN stem, often called *souche* by the French, was confounded by all the older botanists, as it still is by the vulgar, with the root, to which it bears an external resemblance, but from which it is positively distinguished both by its ascending direction, and by its anatomical structure. (*See* Root.)

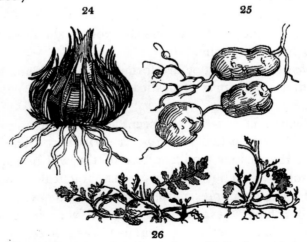

The following are the varieties which have been distinguished: —

The *Cormus*, fig. 24. (*Lecus* of Du Petit Thouars, *Plateau* of De Candolle), is the dilated base of the stem of Monocotyledonous plants, intervening between the roots and the first buds; and forming the reproductive portion of the stem of such plants when they are not caulescent. It is composed of cellular tissue, traversed by bundles of vessels and woody fibre, and has the form of a flattened disk. The fleshy root of the Arum, that of the Crocus and the Colchicum, are all different forms of the Cormus.

It has been called *bulbo-tuber* by Mr. Ker, and *bulbus solidus* by many others; the last is a contradiction in terms. (*See* Bulb.)

The stems of Palms have by some writers been considered as an extended cormus, and not a true stem, but this seems an extravagant application of the term; or rather an application which reduces the signification of the term to nothing. A cormus is a depressed subterranean stem of a particular kind; the trunk of a Palm is, as far as its external character is concerned, as much a stem as that of an oak. M. De Candolle applies the name cormus only to the stems of Cryptogamous plants, and refers to it the *Anabices* of Necker.

The *Tuber*, fig. 25. (*Tuberculum* if very small), is an annual thickened subterranean stem, provided at the sides with latent buds, from which new plants are produced the succeeding year, as in the Potato and Arrow root. A tuber is, in reality, a part of a subterranean stem, excessively enlarged by the developement to an unusual degree of cellular tissue. The consequences frequently attendant upon a state of anamorphosis, as this is called, take place; the regular and symmetrical arrangement of the buds is disturbed, the buds themselves are sunk beneath the surface, or half obliterated, and the whole becomes a shapeless mass. Such is not, however, always the case; the enlargement sometimes occurs without being accompanied by much distortion. In most, perhaps all, tubers, a great quantity of amylaceous matter is deposited, on which account they are frequently found to possess highly nutritive properties.

The *Creeping stem*, fig. 26. (*soboles*), is a slender stem, which creeps along horizontally below the surface of the earth, emitting roots and new plants at intervals, as in the Triticum repens.

This is what many botanists call a creeping *root*. It is one of those provisions of nature, by which the barren sands that bound the sea are confined within their limits; most of the plants which cover such soils being provided with subterranean stems of this kind. It is also extremely tenacious of life, the buds at every nodus being capable of renewing the existence of the individual; hence the almost indestructible properties of the Couch grass, Triticum repens, by the ordinary operations of husbandry; divisions of its creeping stem by cutting and tearing, producing no other effect than that of calling new individuals into existence as fast as others are destroyed. The term soboles is applied by Link and De Candolle to the sucker of trees and shrubs. (See *Surculus*.)

Of the AERIAL stem, the most remarkable forms are the following:—

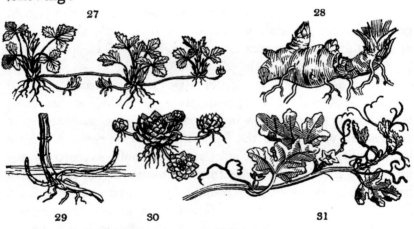

The term *stem* (*caulis*) is generally applied to the ascending caudex of herbaceous plants or shrubs, and not to trees, in which the word *truncus* is employed to indicate their main stem; sometimes, however, this is called *caulis arboreus*. From the *caulis*, Linnæus, following the older botanists, distinguished the *culmus* or *straw* (*Chaume*, Fr.), which is the stem of Grasses; and M. De Candolle has further adopted the name *Calamus* (*Chalumeau*, Fr.) for all fistulous simple stems without articulations, as those of Rushes; but neither of these differ in any material degree from common stems, and the employment of either term is superfluous.

This has been already remarked with respect to Culmus by Link, who very justly enquires (Linnæa, ii. 235.) "cur Graminibus caulem denegares et culmum diceres?"

The *Runner*, fig. 27. (*sarmentum* of Fuchsius and Linnæus, *coulant* of the French,) is a prostrate filiform stem, forming at its extremity roots and a young plant, which itself gives birth to new runners, as in the Strawberry. Rightly considered it is a prostrate viviparous scape; that is to say, a scape which produces roots and leaves instead of flowers. It has been called *flagellum* by some modern botanists, but that term properly applies to the trailing shoots of the vine.

The *Sucker*, fig. 29. (*surculus*), called by the French *Dragon* or *Surgeon*, is a branch which proceeds from the neck of a plant beneath the surface, and becomes erect as soon as it emerges from the earth, immediately producing leaves and branches, and subsequently roots from its base, as in Rosa spinosissima, and many other plants. Link applies the term *soboles* to this form of stem. From this has been distinguished by some botanists the *Stole*, (*Stolo*, Lat.; and *Jet*, French;) which may be considered the reverse of the sucker, from which it differs in proceeding from the stem above the surface of the earth, into which it afterwards descends and takes root, as in *Aster junceus ;* but there does not appear to be any material distinction between them. Willdenow confines the term *surculus* to the creeping stems of Mosses. By the older botanists a sucker was always understood by the word *Stolo*, and *Surculus* indicated a vigorous young shoot without branches.

The shoots thrown up from the buried stems of Monocotyledonous plants, as the Pineapple for example, (the *Adnata*, *Adnascentia*, or *Appendices* of Fuchsius,) are of the nature of suckers.

It may be here remarked, that *Stolo* has given rise to the name *Stool*, which is applied to the parent plant, whence young individuals are propagated by the process of *laying*, as it is technically called by gardeners. The branch laid down was termed *Propago* by the older botanists, and the layer was called *Malleolus*, which literally signifies a hammer, and which was thus applied, because when the layer is separated from its parent, its lower end resembles a hammer head, of which the new plant represents the handle.

The *Offset*, fig. 30. (*propaculum*, Link), is a short lateral branch in some herbaceous plants, terminated by a cluster of leaves, and capable of taking root when separated from the mother plant, as in Sempervivum. It differs very little from the runner.

The *Rootstock*, fig. 28. (*rhizoma*), is a prostrate thickened rooting stem, which yearly produces young branches or plants. It is chiefly found in Irideæ and epiphytous Orchideæ, and is often called *Caudex repens*. Link considers it scarcely distinct from the *sucker* or *stole*. The old botanists called it *Cervix*,—a name now forgotten.

The *Vine*, fig. 31. (*viticula*, Fuchs.), is a stem which trails along the ground without rooting, or entangles itself with other plants, to which it adheres by means of its tendrils, as the Cucumber and the Vine. This term is now rarely employed. M. De Candolle refers it to the runner or sarmentum; but it is essentially distinct from that form of stem.

If a plant is apparently destitute of an aerial stem, it is technically called *stemless* (*acaulis*); a term which must not however be understood to be exact, because it is, from the nature of things, impossible that any plant can exist without a stem in a greater or less degree of developement. All that the term *acaulis* really means, is that the stem is very short.

The *Pseudobulb* is an enlarged aerial stem, resembling a tuber, from which it scarcely differs, except in its being formed above ground, in having a cuticle that is often extremely hard, and in retaining upon its surface the scars of leaves that it once bore. This is only known in Orchideous plants, in which it is very common: the tuber of Arrow root is intermediate between the Pseudobulb and the genuine tuber.

3. Of its Internal Modifications.

THE internal structure of the stems of Flowering plants, is subject to two principal and to several subordinate modifications. The former are well illustrated by such plants as the Oak and the Cane, specimens of which can be easily obtained for comparison. A transverse slice of the former exhibits a central cellular substance or *pith*, an external cellular and fibrous ring or *bark*, an intermediate *woody* mass, and certain fine lines radiating from the pith to the bark, through the wood, and called *medullary rays*; this is called *Exogenous* structure. In the Cane, on the contrary, neither bark, nor pith, nor wood, nor medullary rays, are distinguishable; the transverse section exhibits, on the contrary, a large number of hard spots caused by the section of bundles of woody tissue, and a mass of cellular substance in which they lie imbedded. This kind of structure is named ENDOGENOUS.

In both cases there is a *cellular* and *vascular system* distinct from each other; by a diversity in the respective arrangement of which the differences above described are caused. In describing in detail the peculiar structure of Exogenous and Endogenous stems, it will be more convenient to consider them with reference to those two systems, than to follow the usual method of leaving the fact of there being two distinct systems out of consideration.

§ 1. *Of the* Exogenous *Structure.*

32

The Cellular system in an Exogenous stem chiefly occupies the centre and the circumference, which are connected by thin vertical plates of the same nature as themselves. The central part (*a* fig. 33.) is the *pith*, that of the circumference (*b*) is the *bark*, and the connecting vertical plates (*c*) are *medullary rays*.

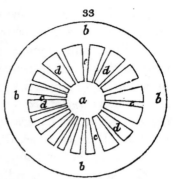

The *pith* is a cylindrical or angular column of cellular tissue, arising at the neck of the stem and terminating at the leaf-buds, with all of which, whether they are lateral or terminal, it is in direct communication. Its tissue, when cut through, almost always exhibits an hexagonal character, and is frequently larger than in any other part. When newly formed, it is green, and filled with fluid; but its colour gradually disappears as it dries up, and it finally becomes colourless. After this it undergoes no further change, unless by the deposition in it, in course of time, of some of the peculiar secretions of the species to which it belongs. It has been contended, indeed, by some physiologists, that it is gradually pressed upon by the surrounding part of the vascular system, until it is either much reduced in diameter or wholly disappears; and in proof of this assertion, the Elder has been referred to, in which the pith is very large in the young shoots, and very small in the old trunks. Those, however, who entertain this opinion, seem not to consider that the diameter of the pith of all trees is different in different shoots, according to the age of those shoots;—that in the first that arises after germination, the pith is a mere thread, or at least of very small dimensions — that in the shoots of the succeeding year it becomes larger — and that its dimensions increase in proportion to the general rapidity of developement of the vegetable system: the pith, therefore, in the first-formed shoots, in which it is so small compared with that in the branches of subsequent years, is not so because of the pressure of surrounding parts; it never was any bigger.

The pith is always, when first forming, a uniform compact mass, connected without interruption in any part; but the vascular system sometimes developing more rapidly than itself, it occasionally happens that it is either torn or divided into irregular cavities, as in the Horse Chestnut, the Rice-paper plant, and many others; or that it is so much lacerated as to lose all resemblance to its original state, and to remain in the shape of ragged fragments adhering to the inside of the vascular system: this is what happens in Umbelliferous and other fistular-stemmed plants.

Sometimes the pith is much more compact at the nodi than in the internodia, as in the Ash; whence an idea has arisen that it is actually interrupted at those places: this is, however, an obvious mistake; there is no interruption of continuity, but a mere alteration in compactness.

It very seldom happens that any part of the vascular system intermixes with the pith, which is almost always composed of cellular tissue exclusively; but in Ferula and the Marvel of Peru, it has been proved by Messrs. Mirbel and De Candolle, that bundles of woody fibre are intermixed; and in the Nepenthes there is a considerable quantity of spiral vessels scattered among the cellular tissue of the pith.

The *Bark* is the external coating of the stem, lying immediately over the wood, to which it forms a sort of sheath, and from which it is always distinctly separable. When but one year old, it consists of an exterior coating of cellular substance, called the *cellular integument* or the *epidermis*, and of an interior lining of woody fibre, called the *liber* or *inner bark:* if more than one year old, then it is composed of as many layers of cellular integument and woody fibre as it is years old, the former being invariably external, and the latter internal, in each layer; and every layer being formed beneath the previous one, and therefore next the wood. In consequence of the new bark being continually generated within that of the previous year, it is necessary that the latter, which is pushed outwards, should be extensible; and in many plants this extensibility takes place to a considerable degree. In the Apple, several successive zones of bark are formed without any appearance of a dislocation or disruption of the tissue of the outside; and in the Daphne Lagetto, the fibres of the liber are so tenacious that, instead of being ruptured by the force of the inward growth, they are separated into lozenge-shaped meshes, arranged in such beautiful order, as to have acquired for the plant itself the name of the Lace Bark Tree. There exists, however, in all cases, a limit to the extensibility of the old layers of bark; and when this ceases, the outer bark either splits into deep fissures, as in the Oak, the Elm, the Cork, and most of our European trees, or it falls away in broad plates, as in the Plane, or it peels off in long thin ribands, as in the Birch.

As there is a double layer of cellular integument and woody fibre formed every year, it follows that the age of a tree ought to be indicated by the number of such deposits contained in its bark. But the arrangement of the zones is so very soon disturbed, and the distinction between them becomes so imperfect, that even when the outermost coating is still entire, it is scarcely practicable to count the zones; and as soon as the outside begins to split or peel off, all traces of their full number necessarily disappear.

That the bark really increases by constant deposits of new matter between it and the wood, is demonstrated by introducing a piece of metal into the liber of a tree, and watching it subsequently: in process of time it will be protruded to the outside, and will finally fall away.

Notwithstanding the fibrous character of a certain portion of the bark, it is generally so brittle as to be capable of breaking in all directions with a clean fracture, as soon as it becomes dry and ceases to live; but in many plants, when young, it is so tough as to be applied to different economical purposes. For instance, the Russia mats of commerce are prepared from the liber of two or three species of Tilia, that of many Malvaceæ is manufactured into cordage, and similar properties are found in that of many other plants.

When stems are old, the bark usually bears but a small proportion in thickness to the wood; yet in some plants its dimensions are of a magnitude that is very remarkable. For instance, Pinus Douglasii specimens have been brought to Europe twelve inches thick, and these are said not to be of the largest size.

Lacunæ and Vasa propria are exceedingly common in the

bark, but there is no well authenticated instance of any spiral vessels having been found in it; except in Nepenthes, in which they are found in almost every part, and exist in no inconsiderable numbers in the bark. Mr. Don states that spiral vessels abound in the bark of Urtica nivea, but I have not succeeded in discovering them there.

Beneath the bark and above the wood is interposed in the spring a mucous viscid layer, which, when highly magnified, is found to consist of numerous minute transparent granules, and to exhibit faint traces of a delicate cellular organization. This secretion is named the CAMBIUM, and appears to be exuded both by the bark and wood, certainly by the latter.

The cellular system of the pith and that of the bark are, in the embryo, and youngest shoots, in contact; but the vascular system, as it forms, gradually interposes between them, till after a few weeks they are distinctly separated, and in very aged trunks are sometimes divided by a space of several feet; that is to say, by half the diameter of the wood. But whatever may be the distance between them, a horizontal communication of the most perfect kind continues to be maintained. When the vascular system is first insinuated into the cellular system, dividing the pith and bark, it does not completely separate them, but pushes aside a quantity of cellular tissue, pressing it tightly into thin vertical radiating plates; as the vascular system extends, these plates increase outwardly, continuing to maintain the connection between the centre and the circumference. Botanists call them *medullary rays* (or *plates*); and carpenters, the *silver grain*. They are composed of muriform cellular tissue (Plate I. fig. 7.), often not consisting of more than a single layer of cellules; but sometimes, as in Aristolochias, the number of layers is very considerable (Plate II. fig. 12. *a*). In horizontal sections of an Exogenous stem, they are seen as fine lines radiating from the centre to the circumference; in longitudinal sections they give that glancing satiny lustre which is in all discoverable, and which gives to some, such as the Plane and the Sycamore, a character of remarkable beauty.

No vascular tissue is ever found in the medullary rays,

unless those curious plates described by Mr. Griffith in the wood of Phytocrene gigantea, in which vessels exist, should prove to belong to the medullary system.

The vascular system in an Exogenous stem is confined to the space between the pith and the bark, where it chiefly consists of ducts and woody fibre collected into compact wedge-shaped vertical plates (fig. 33. *d*), the edges of which rest on the pith and bark, and the sides of which are in contact with the medullary rays.

That portion of the vascular system which is first generated is in immediate contact with the pith, to which it forms a complete sheath, interrupted only by the passage of the medullary rays through it. It consists of spiral vessels and woody fibre intermixed, and forms an exceedingly thin layer, called the *medullary sheath*. This is the only part of the vascular system of the stem in which spiral vessels are ordinarily found; the whole of the vessels subsequently deposited over the medullary sheath being ducts, and mostly dotted ones, with a few exceptions. The medullary sheath establishes a connection between the axis and all its appendages, the veins of leaves, flowers, and fruits, being in all cases prolongations of it. It has been remarked by Senebier, and since by M. De Candolle, that it preserves a green colour even in old trunks, which proves that it still continues to retain its vitality when that of the surrounding parts has ceased.

The vascular system of a stem one year old consists of a zone of wood lying between the pith and the bark, lined in the inside by the medullary sheath, and separated into wedge-shaped vertical plates by the medullary rays that pass through it. All that part of the first zone which is on the outside of the medullary sheath is composed of woody fibre and ducts intermixed in no apparent order; but the ducts are generally either in greater abundance next the medullary sheath, or confined to that side of the zone, and the woody fibre alone forms a compact mass on the outside. The second year another zone is formed on the outside of the first, with which it agrees exactly in structure, except that there is no medullary sheath; the third year a third zone is formed on the outside the second, in all respects like it; and so on, one zone being deposited every year as long as the plant continues to live. As each new zone is formed over that of the previous year, the latter undergoes no alteration of structure when once formed: wood is not subject to distension by a force beneath it, as the bark is, but, whatever the first arrangement or direction of its tissue may be, such they remain to the end of its life. The formation of the wood is, therefore, the reverse of that of the bark; the latter increasing by addition to its inside, the former by successive deposits upon its outside. It is for this reason that stems of this kind are called Exogenous (from two Greek words, signifying to grow outwardly). According to M. Dutrochet, each zone of wood is in these plants separated from its neighbour by a layer of cellular tissue, forming part of the system of the pith and bark.

After wood has arrived at the age of a few years, or sometimes even sooner, it acquires a colour different from that which it possessed when first deposited, becoming what is called *heart-wood*, or *duramen*. For instance, in the beech it becomes light brown, in the oak deep brown, in Brazil wood and Guaiacum green, and in ebony black. In all these it was originally colourless, and owes its different tints to matter deposited at first in the ducts, and subsequently in all parts of the tissue; as may be easily proved by throwing a piece of heart-wood into nitric acid, or some other solvent, when the colouring matter is discharged, and the tissue recovers its original colourless character. That part of the wood in which no colouring matter is yet deposited, and consequently that which, being last formed, is interposed between the bark and duramen, is called alburnum. The distinction between these is physiologically important, as will hereafter be explained.

Each zone of the vascular system of an Exogenous stem being the result of a single year's growth, it should follow that to count the zones apparent in a transverse section is sufficient to determine the age of the individual under examination; and further, that, as there is not much difference in the average depth of the zones in very old trees, a certain rate of growth being ascertained to be peculiar to particular species, the examination of a mere fragment of a tree, the diameter of which is known, should suffice to enable the botanist to judge with considerable accuracy of the age of the individual to which it belonged. It is true, indeed, that the zones become less and less deep as a tree advances in age; that in cold seasons, or after transplantation, or in consequence of any causes that may have impeded its growth, the formation of wood is so imperfect as scarcely to form a perceptible zone: yet the learned M. De Candolle has endeavoured to show, in a very able paper *Sur la Longévité des Arbres*, that the general accuracy of calculations is not much affected by such accidents; occasional interruptions to growth being scarcely appreciable in the average of many years. This is possibly true in European trees, and in those of other cold or temperate regions in which the seasons are distinctly marked; in such the zones are not only separated with tolerable distinctness, but do not vary much in annual dimensions. But in many hot countries the difference between the growing season and that of rest, if any occur, is so small, that the zones are as it were confounded, and the observer finds himself incapable of distinguishing with exactness the formation of one year from that of another. In the wood of Guaiacum, Phlomis fruticosa, Metrosideros polymorpha, and many other Myrtaceæ, for instance, the zones are extremely indistinct; in some Bauhinias they are formed with great irregularity; and in Holböllia latifolia, some kinds of Ficus, certain species of Aristolochia, as A. labiosa, and many other plants, they are so confounded, that there is not the slightest trace of annual separation.

With regard to judging of the age of a tree by the inspection of a fragment, the diameter of the stem being known, a little reflection will show that this is to be done with great caution, and that it is liable to excessive error. If, indeed, the zones upon both sides of a tree were always of the same, or nearly the same, thickness, much error would, perhaps, not attend such an investigation; but it happens that, from various causes, there is often a great difference between the growth of the two sides, and consequently, that a fragment taken from either side must necessarily lead to the falsest inferences. For example, I have now before me four specimens of wood, taken almost at hazard from among a fine collection, for which I am indebted to the munificence of the East India Company. The measurements of either side, and their age, as indicated by the number of zones they comprehend, are as follows: —

	Diameter of		Total.	Real Age, or No. of Zones.
	Side A.	Side B.		
Cornus capitata	9 lines.	36 lines.	45 lines.	40
Pyrus foliolosa	8 lines.	22 lines.	30 lines.	36
Magnolia insignis	11 lines.	20 lines.	31 lines.	17
Alnus napalensis	11 lines.	23 lines.	34 lines.	8

Now, in the first of these cases, suppose that a portion of the side A. were examined, and the observer were told that the diameter of the whole stem was 45 lines, he would find that each zone was 0.45 of a line deep, and that, consequently, the tree had been twenty years forming a radius of 9 lines, or increasing 18 lines in diameter: the total diameter being 45 lines, he would necessarily infer that the age of the tree was one hundred years; its real age being only forty, as indicated by its zones. And so of the rest.

When we hear of the Baobab trees of Senegal being 5150 years old, as computed by Adanson, and the Taxodium distichum still more aged, according to the calculations of the ingenious M. Alphonse De Candolle, it is impossible to avoid suspecting that some such error as that just explained has vitiated their conclusions.

To the characters above assigned to the stem of Exogenous plants there are several remarkable exceptions, some of which have been described by botanists; others are mentioned now for the first time. M. Mirbel has noticed the unusual structure of CALYCANTHUS, (*Annales des Sciences*, vol. xiv.) in the bark of which, at equal distances, are found four minute extremely eccentrical woody axes, the principal diameter of which is inwards; that is to say, next the wood. The existence of this structure, noticed by the discoverer only in C. floridus, I have since ascertained in all the other species, and also in Chimonanthus.

In Coniferous wood (*fig.* 34.) there is scarcely any mixture of ducts among woody fibre, as in other exogenous plants; in consequence of which a cross section exhibits none of those open mouths which are caused by the division of ducts, and which give what is vulgarly called porosity to wood. Instead of this, the vascular system generally consists exclusively of that kind of woody fibre which has been described at p. 15., under the name of glandular, with the exception of  the medullary sheath, in which spiral vessels are present in small numbers. The Yew is the principal exception: in this plant the woody fibre is the same as that of other Coniferæ; but many tubes have a great quantity of little fibres lying obliquely across them at nearly equal distances, sometimes arranged with considerable regularity, — sometimes disturbed as it were, so that the transverse fibres, although they retain their obliquity, are not parallel, — and sometimes, but more rarely, so regular as to give to the tubes of woody fibre the appearance of spiral vessels, the coils of which are separated by considerable intervals. The latter only is represented by Kieser, at his tab. xxi. fig. 103, 104.; but the former is by far the most common appearance.

In Cycadeæ the vascular system is destitute of ducts, as in Coniferæ; their place being supplied by such woody fibre as has been already described at p. 14. But the zones of wood are separated by a layer of cellular substance resembling that of the pith, and often as thick as the zones themselves. This structure is represented by M. Adolphe Brongniart, in the 16th volume of the Annales des Sciences.

My friend Mr. Griffith has beautifully illustrated the structure of a plant called Phytocrene (*fig.* 35.), in Dr. Wallich's Plantæ Asiaticæ, vol. iii. t. 216. In this curious production, the wood consists of vessels encompassed by woody fibre; and in the place of medullary rays are thick plates, connected neither with the medulla nor with the bark, nor even with each other in different zones. When the wood is dry, these plates separate from the wood, in which they finally lie loose; and, what is very remarkable, *they contain vessels.*

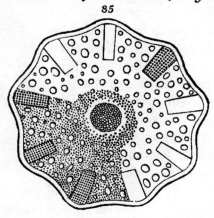
85

In Nepenthes distillatoria the pith contains a great quantity of spiral vessels; the place of the medullary sheath is occupied by a deep and very dense layer of woody fibre, in which no vessels, or scarcely any, are discoverable; there are no medullary rays; the wood has no concentric zones; between the bark and the wood is interposed a thick layer of cellular tissue, in which an immense quantity of very large spiral vessels is formed; on the outside of this layer is a thinner coating of woody fibre, containing some very minute spiral vessels; and, finally, the whole is enclosed in a cellular integument, also containing spiral vessels of small size. In this singular plant the outer layers are, it is to be presumed, liber and epidermis; and the cellular deposit between the former and the wood is analogous to cambium in an organised state, belonging equally to the wood and the bark. What is so exceedingly remarkable is the complete intermixture of the vascular and cellular systems, so that limits no longer exist between the two.

I have a specimen of the twisted compressed stem of a Bauhinia from Colombia (*fig.* 36.), in which there are no concentric circles, properly so called; but in which there are certain irregular flexuous zones, consisting of a layer of cellular tissue coated by a stratum of woody fibre, enclosing, at irregular distances from the centre, very unequal portions of the vascular system. The pith is exceedingly eccentrical; and the medullary rays, which are very imperfectly formed, do not all radiate from the pith, but on the thickest side form curves passing from one side of the stem to the other, their convexities turned towards the pith.

In the stem of an unknown climber in my possession from Colombia (*fig.* 37.), the vascular system is divided into four nearly equal parts, by four short thick plates radiating from the pith, and consisting of woody fibre with a very few ducts. These plates are not more than one third the depth of the wood; so that between their back and the bark there is a considerable vacancy, by which the four divisions of the vascular system are separated. This vacancy is nearly filled with bark, which projects into the cavity.

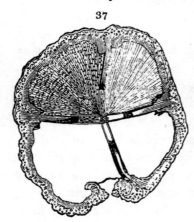

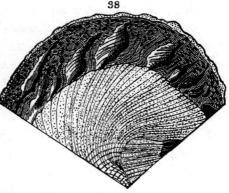

In Hollböllia latifolia (*fig.* 38.), which has a twining stem, there are no concentric circles, and the medullary rays are curved, part from right to left, and part from left to right, diverging at one point and converging at another; the bark is pierced with extensive longitudinal perforations.

CHAP. II.   COMPOUND ORGANS IN FLOWERING PLANTS.   71

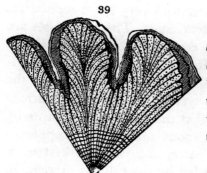

In Euonymus tingens (*fig.* 39.) the vessels near the centre of the stem are arranged in concentric interrupted circles, but towards the bark there is no trace of such circles; and the vessels are all confounded in an uniform mass.

In Menispermum laurifolium (*fig.* 40.) the concentric lines evidently belong to the medullary system; they are extremely interrupted and unequal, often only half encircling the stem, or even less, and they anastomose in various ways; the medullary rays are unusually large, and lie across the wood like parallel bars; and, finally, the plates of which the wood consists each contain but one vessel, which is situated at the external edge of the plate.

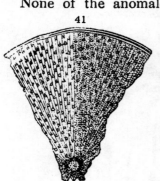

None of the anomalous forms of Exogenous stems are, however, so remarkable as an unknown Burmese tree (*fig.* 41.), for a specimen of which I am indebted to my friend Dr. Wallich. In a section of this the general appearance is so much that of an Endogenous stem, that without an attentive examination it might be actually mistaken for one. The diameter of this stem is two inches seven lines; it is nearly perfectly circular, and has a very thin but distinct bark, with a central pith surrounded by very compact woody fibre. There are neither zones nor medullary rays; but the vascular system consists of an uniform mass of ducts and woody fibre, disposed with great symmetry, and of the same degree of compactness at the circumference as well as in the centre. Amongst his wood are interspersed, at the distance of about half a line, with very great regularity, passages containing loose cellular tissue. These passages are convex at the back and rather concave in front, run parallel with the ducts, and do not seem to have any kind of communication with each other. They, no doubt, represent the medullary rays of the cellular system of this highly curious plant. It must be remarked, that the resemblance borne by this stem to that of an Endogenous plant is more apparent than real; for whilst, in the latter, the vascular system is separated into bundles surrounded by the cellular system, in this, on the contrary, the cellular system consists of tubular passages surrounded by masses of the vascular system.

These examples of anomalous structure will show the student that it is neither medullary rays nor concentric zones in the wood that are the certain indications of Exogenous growth; both the one and the other being sometimes absent; but that the presence of a central pith, and a greater degree of hardness in the centre than in the circumference, are the signs from which alone any absolute evidence can be derived.

§ 2. *Of the* Endogenous *Structure.*

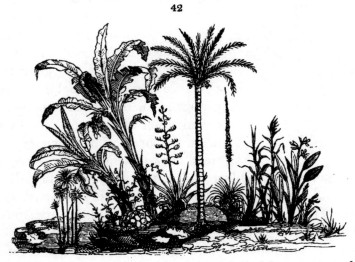

Plants of an arborescent habit having this structure being almost exclusively extra-European, and most of them being natives only of the tropics, botanists have had much fewer occasions of examining them, and, consequently, their knowledge concerning them is far more limited. It is, therefore, probable that in this department of the subject there will be hereafter much to add and much to correct. In the mean while it must suffice, that what has been published concerning Endogenous plants is given with such corrections and additions as my own experience may have suggested.

In Endogenous plants the vascular and cellular systems are as distinct as in Exogenous, but they are differently arranged. The cellular system, instead of being distinguishable into pith, bark, and medullary rays, is a uniform mass, in which the vascular system lies imbedded in the form of thick fibres; 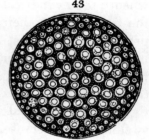 and the vascular system itself has no tendency to collect into zones or wedges resembling wood, but in all cases retains the form of bundles resembling fibres. These bundles consist of woody fibre, enclosing spiral vessels or ducts; most commonly the latter.

The diameter of an Endogenous stem is increased by the constant addition of fibrous bundles to the centre, whence the name. Those bundles displace such as are previously formed, pushing them outwards; so that the centre, being always most newly formed, is the softest; and the outside, being older, and being gradually rendered more and more compact by the pressure exercised upon the bundles lying next it by those forming in the centre, is the hardest. In Endogenous plants that attain a considerable age, such as many Palms, this operation goes on till the outside becomes sometimes hard enough to resist the blow of a hatchet. It does not, however, appear that each successive bundle of fibres passes exactly down the centre, or that there is even much regularity in the manner in which they are arranged in that part: it is only certain that it is about the centre that they descend, and that on the outside no new formation takes place. This appears from the manner in which the bundles cross and interlace one another, as is shown in the figure of

Pandanus odoratissimus given by M. De Candolle in his *Organographie* (tab. vi.), or still more clearly in the lax tissue of the inside of the stems of Dracæna Draco.

The epidermis of an Endogenous stem seems capable of very little distension. In many plants of this kind the diameter of the stem is the same, or not very widely different, at the period when it is first formed, and when it has arrived at its greatest age: Palms are, in particular, an instance of this; whence the cylindrical form that is so common in them. That the increase in their diameter is really inconsiderable, is proved in a curious, and at the same time very conclusive, manner by the circumstance of gigantic woody climbing plants sometimes coiling round such stems, and retaining them in their embrace for many years, without the stem thus tightly wound round indicating in the slightest manner, by swelling or otherwise, that such ligatures inconvenience it. A specimen illustrative of this is preserved in the Museum of Natural History at Paris, and has been figured both by M. Mirbel in his *Elémens* (tab. xix.), and M. De Candolle in his *Organographie* (tab. iv.). We know, from the effect of the common Bindweed upon the Exogenous plants of our hedges, that the embrace of a twining plant is, in a single year, destructive of the life of every thing that increases in diameter; or at least produces, above the strangled part, extensive swellings that always end in death.

It is, however, certain that other Exogenous plants do increase extensively in diameter up to a certain point; but this is effected with great rapidity; and the horizontal growth once stopped appears never to be renewed. Thus, in the Bamboo, stems are sometimes found as much as two feet in circumference, which were originally not more than half an inch in diameter. Others would seem to have an unlimited power of distension. In the Dracænas, called in French colonies in Africa Bois-chandelles, the first shoot from the ground is a Turio (sucker), an inch in diameter, and perhaps fifteen feet high; but in time it distends so much that sometimes two men can scarcely embrace it in their extended arms. (*Thouars, Essais*, p. 3.)

As Endogenous stems contain no concentric zones, there is

nothing in their internal structure to indicate age; but, in the opinion of some botanists, there are sometimes external characters that will afford sufficient evidence. It is said that the number of external rings that indicate the fall of leaves from the trunk of the Palm tribe coincide with the number of years that the individual has lived. There is, however, nothing like proof of this at present before the public; such statements must therefore be received with great caution. It may further be remarked, with reference to this subject, that in many Palms these rings disappear after a certain number of years.

In arborescent Endogenous plants, it usually happens that only one terminal leaf-bud developes; and in such cases the stem is cylindrical, or very nearly so, as in Palms. If two terminal leaf-buds constantly develope, the stem becomes dichotomous, but the branches are all cylindrical, as in Pandanus and the Doom Palms of Egypt; but if axillary leaf-buds are regularly developed, as in the Asparagus, Dracæna Draco, or in arborescent grasses, then the conical form that prevails in Exogenous plants uniformly exists in Endogenous ones also.

Besides the difference now mentioned, there is one other form of the Endogenous stem that it is necessary to describe; viz. that of Grasses. In those plants the stem is hollow except at the nodi, where transverse partitions intercept the cavity, dividing it into many cells. In the Bamboo these cells and partitions are so large that, as is well known, lengths of that plant are used as cases to contain papers. In consequence of this great apparent deviation from the usual structure, a celebrated Swedish botanist has remarked, that Grasses are the least Endogenous of all Endogenous plants.

But if the gradual developement of a grass be attentively observed, it will be found that the stem is originally solid; that it then becomes hollow in consequence of its increasing in diameter more rapidly than new tissue can be formed; and that, finally, in old arborescent stems, it again becomes solid by the constant addition of matter to its inside; so that its deviation from the ordinary characters of Endogenous structure is much less considerable than it seems to be at first sight.

Sect. III. *Of the Root, or descending Axis.*

At or about the same time that the ascending axis seeks the light and becomes a stem, does the opposite extremity of the seed or bud bury itself in the earth and become a root, with a tendency downwards so powerful, that no known force is sufficient to overcome it. Correctly speaking, nothing can be considered a root except what has such an origin; for those roots which are emitted by the stems of plants, are in reality the roots of the buds above them, as will be hereafter explained. Nevertheless, nothing is more common than even for botanists to confound subterranean stems or buds with roots, as has been already seen. (See Bulb, Tuber, Soboles, &c. &c.)

Independently of its origin, the root is to be distinguished from the stem by many absolute characters. In the first place, its ramifications occur irregularly, without any symmetrical arrangement: they do not, like branches, proceed from certain fixed points (buds), but are produced from all and any points of the root. Secondly, a root has no leaf-buds, unless indeed, as is sometimes the case, it has the power of forming adventitious ones; but, in such a case, the irregular manner in which such are produced is sufficient evidence of their nature. Thirdly, roots have no scales, leaves, or other appendages; neither do they ever indicate upon their surface, by means of scars, any trace of such: all underground bodies upon which scales have been found are stems, whatever they may have been called; the only appendages they ever have are such things as the little hollow floating bladders found in Utricularia. A fourth distinction between roots and stems is, that the former have never any stomata upon their cuticle; and, finally, in Exogenous plants, the root has never any pith. It has been also said that roots are always colourless, while stems are always coloured; but aerial roots are often green, and all underground stems are colourless.

The body of the root is sometimes called the *caudex;* the minute subdivisions have been sometimes called *radiculæ,* — a term that should be confined to the root in the embryo; others name them *fibrillæ,* a term more generally adopted; while the terms *rhizina* and *rhizula* have been given by Professor Link to the young roots of mosses and lichens.

A *fibrilla* is a little bundle of annular ducts, or sometimes of spiral vessels, encased in woody fibre, and covered by a lax cellular integument: it is in direct communication with the vascular system of the root, of which it is, in fact, only a subdivision; and its apex consists of extremely lax cellular tissue and mucus. This apex has the property of absorbing fluid with great rapidity, and has been called by M. De Candolle the *Spongiole*. It must not be considered a particular organ; it is only the newly formed and forming tender tissue. In Pandanus the spongioles of the aerial roots consist of numerous very thin exfoliations of the epidermis, which form a sort of cup fit for holding water in.

The proportion borne by the root to the branches is extremely variable: in some plants it is nearly equal to them, in others, as in Lucerne (Medicago sativa), the roots are many times larger and longer than the stems; in all succulent plants and Cucurbitaceæ they are much smaller. When the root is divided into a multitude of branches and fibres, it is called *fibrous*: if the fibres have occasionally dilatations at short intervals, they are called *nodulose*. When the main root perishes at the extremity, it receives the name of *præmorse*, or *bitten* off: frequently it consists of one fleshy elongated centre tapering to the extremity, when it is termed *fusiform* (or *tap-rooted* by the English, and *pivotante* by the French); or it dilates immediately below the surface of the earth into a globose form, when it is named *turnip-shaped*, as in the common turnip; if it is terminated by several distinct buds, as in some herbaceous plants, it is called *many-headed* (*multiceps*).

The roots of many plants are often fleshy, and composed of lobes, which appear to serve as reservoirs of nutriment to the fibrillæ that accompany them; as in many terrestrial Orchideous plants, Dahlias, &c. These must not be confounded either with tubers or bulbs, as they have been by some writers, but are rather to be considered a special form of the root, to which the name of *Pseudo-tuber* (*fig.* 44.) would not be inapplicable. In Orchis the pseudo-tubers are often palmated or lobed; in the

Dahlia, and many Asphodeleæ, they hang in clusters, or are *fasciculated*.

In internal structure the root differs little from the stem, except in being often extremely fleshy; the cellular system being subject to an unusually high degree of developement in a great many plants, as the Turnip, the Parsnep, and other edible roots. In Endogenous plants, the mutual arrangement of the cellular and vascular systems of the root and stem is absolutely the same; but in Exogenous plants there is never any trace of pith in the root.

Sect. IV. *Of the Appendages of the Axis.*

From the outside of the stem, but connected immediately with its vascular system, arises a variety of thin flat expansions, arranged with great symmetry, and usually falling off after having existed for a few months. These are called, collectively, appendages of the axis; and, individually, scales, leaves, bracteæ, flowers, sexes, and fruit. They must not be confounded with mere expansions of the cuticle, such as ramenta, already described (p. 41.), from which they are known by having a connection with the vascular system of the axis. Till lately, botanists were accustomed to consider all these as essentially distinct organs; but, since the appearance of an admirable treatise by Goethe in 1790, *On the metamorphosis of plants*, proofs of their being merely modifications of one common type, the leaf, have been gradually discovered; so that that which, forty years ago, was considered as the romance of a poet, is now universally acknowledged to be an indisputable truth. It is not my intention to enter into much separate discussion of this doctrine; proof of it will be more conveniently adduced as the different modifications of the appendages of the axis come separately under consideration. The leaf, as the first that is formed, the most perfect of them all, and that which is most constantly present, is properly considered the type from which all the others are deviations, and is that with the structure of which it is first necessary to become acquainted.

CHAP. II. COMPOUND ORGANS IN FLOWERING PLANTS.

§ 1. *Of the* Leaf.

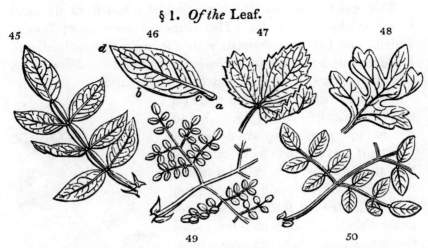

The leaf is an expansion of the bark at the base of a leaf-bud, prior to which it is developed. In most plants it consists of cellular tissue filling up the interstices of a net-work of fibres that proceed from the stem, and ultimately separating from the bark by an articulation; in many Monocotyledonous plants, Ferns, and Mosses no articulation exists, and the base of the leaf only separates from its parent stem by rotting away.

This difference of organisation has given rise to a distinction, on the part of Oken, between the articulated leaves of Dicotyledones and the inarticulated leaves of Monocotyledones and Acotyledones: the former he calls true leaves, and distinguishes by the name of *Laub;* the latter he considers foliaceous dilatations of the stem, analogous to leaves, and calls *Blatt.*

A leaf consists of two parts; namely, its stalk, which is called the *petiole* (*fig.* 46. *a*), and its expanded surface, which is called the *lamina* (*fig.* 46. *c, b, d*): in ordinary language the latter term is not employed, but in very precise descriptions it is indispensable.

The point where the base of the upper side of a leaf joins the stem is called the *axilla;* any thing which arises out of that point is said to be *axillary.* If a branch or other process proceeds from above the axilla, it is called *supra-axillary;* if from below it, *infra-axillary.*

The scar formed by the separation of a leaf from its stem is called the *cicatricula*. The withered remains of leaves, which, not being articulated with the stem, cannot fall off, but decay upon it, are called *reliquiæ* or *induviæ* (*débris*, Fr.), and the part so covered is said to be *induviatus*.

When leaves are placed in pairs on opposite sides of a stem (*fig.* 51.), and on the same plane, they are called *opposite*: if more than two are opposite, they then form what is called a *whorl*, or *verticillus*, and are said to be *whorled*, or *verticillate*: but if they arise at regular distances from each other round the stem, and not from the same plane, they are then called alternate.

In plants having Exogenous stems, the first leaves, — namely, those which are present in the embryo itself (*cotyledons*), — are uniformly opposite; but those subsequently developed are either opposite, verticillate, or alternate in different species: on the contrary, in Endogenous plants, the embryo leaf is either solitary, or, if there are two, they are alternate; and those subsequently developed are usually alternate also, but few cases occurring in which they are opposite.

Hence some have formed an opinion that the normal position of the leaves of Exogenous plants is opposite, or verticillate; and that when they are alternate, this arises from the extension of a nodus; while that of Endogenæ is alternate, the verticilli being the result of the contraction of internodia.

But it seems more probable that the normal position of all leaves is alternate, and their position upon the stem an elongated spiral, as is in many cases exceedingly apparent, as, for instance, in the genus Pinus, in Pandanus, which is actually named Screw-pine, in consequence of the resemblance its leaves bear to a screw, and in the Pine apple; the Apple, the Pear, the Willow, the Oak, will also be found to indicate the same arrangement, which is only less apparent because of the distance between the leaves, and the irregularity of their direction. If, in the Apple tree, for instance, a line be drawn from the base of one leaf to the base of another, and the leaves be then broken off, it will be found that a perfectly spiral line will have been formed. Upon this supposition, opposite or verticillate leaves are to be considered the result of a peculiar contraction or non-developement of internodia, and the consequent confluence of as many nodi as there are leaves in the whorl. The Rhododendron ponticum will furnish the student with an illustration of this: on many of its branches the leaves are some alternate and some opposite; and many intermediate states between these two will be perceivable. In many plants, the leaves of which are usually alternate, there is a manifest tendency to the approximation of the nodi, and consequently to an opposite arrangement of the leaves, as in Solanum nigrum, and many other Solaneæ * ; while, on the other hand, leaves that are usually opposite separate their nodi and become alternate, as in Erica mediterranea : but this is more rare.

The best argument in support of the hypothesis that all verticilli arise from the contraction of internodia and confluence of nodi, is, however, to be derived from flowers, which are several series of verticilli, as will be seen hereafter. In plants with alternate leaves, the flowers often change into young branches, and then the verticilli of which they consist are broken, the nodi separate, and those parts that were before opposite become alternate; while, in monstrous Tulips, the verticilli of which the flower consists are plainly shown to arise from the gradual approximation of leaves, that in their unchanged state are alternate.

In this normal state leaves are obviously distinct, both from each other and from the stem. But, in some cases, adhesions of various kinds occur, and give them a new character. Thus, in Cardui, and many other thistle-like plants, the elongated bases of the leaves adhere to the stem, and become what is called *decurrent*. In Bupleurum perfoliatum the lobes of the base of the leaf not only cohere with the stem, but, projecting beyond it, grow together, so as to resemble a leaf through

* Introduction to the Natural System of Botany, p. 231.

which the stem has pierced: this is called being *perfoliate*. Frequently two opposite leaves grow together at the base, as in Caprifolium perfoliatum; to this modification the latter term is often also applied, but that of *connate* is what more properly belongs to it.

The anatomical structure of the leaf is this:—From the medullary sheath diverges a bundle of woody tissue, accompanied by spiral vessels: this passes through the bark, and proceeds, at an angle more or less acute, to a determinate distance from the stem, branching off at intervals, and, by numerous ramifications, forming a kind of net-work. At the point of the stem whence the bundle of fibrovascular tissue issues, the cellular tissue of the bark also diverges, accompanying the fibrovascular tissue, expanding with its ramifications, and filling up their interstices. The tissue that proceeds from the medullary sheath, after having passed from the origin of the leaf to its extremity, doubles back upon itself, forming underneath the first a new layer of fibre, which, upon its return, converges just as the first layer diverged, at length combines into a single bundle, corresponding in bulk and position to that which first emerged, and finally discharging itself into the liber. If, therefore, a section of the leaf and stem be carefully made at a nodus, it will be found that the bundle of woody tissue which forms the frame-work of the leaf communicates above with the medullary sheath, and below with the liber. This is easily seen in the spring, when the leaves are young; but is not so visible in the autumn, when their existence is drawing to a close. The double layer of fibrovascular tissue is also perceptible in a leaf which has laid during the winter in some damp ditch, where its cellular substance has decayed, so that the cohesion between the upper and lower layers is destroyed: they can then be easily separated. The curious Indian leaves which have the property of opening, upon slight violence, like the leg of a silk stocking, so that the hand may be thrust between their upper and lower surfaces, derive that singular separability from an imperfect union between the layer of excurrent and recurrent fibre. M. De Candolle remarks, that when the fibres expand to form the limb of a leaf, they may (whether this phenomenon occurs at the extremity of a petiole, or at the point of separation from the stem,) do so after two different systems: they may either constantly preserve the same plane when the common flat leaves are formed; or they may expand in any direction, when cylindrical, or swollen, or triangular leaves are the result. (*Organogr.* p. 270.)

The cellular tissue of which the rest of the leaf is composed is parenchyma, which Link then calls *diachyma*, or that immediately beneath the two surfaces *cortex*, and the intermediate substance *diploe*. M. De Candolle calls these two, taken together, the *mesophyllum*. The whole is protected, in leaves exposed to air, by a coating of cuticle, furnished with stomata; but in submersed leaves the parenchyma is naked, no cuticle overlaying it.

The general nature of the parenchymatous part of leaves has been very well explained, both by Link and others, and figured by Dr. Mohl, in 1828. (*Uber die Poren des Pflanzenzellgewebes*, tab. i. fig. 4, &c.) But the most complete account is that of M. Adolphe Brongniart, in 1830 (*Annales des Sc.* vol. xxi. p. 420.), of which the principal part of what follows is an abstract.

The cuticle is a layer of cellules adhering firmly to each other, and sometimes but slightly to the subjacent tissue, from which they are entirely different in form and nature: in form, for the cellules are depressed, and, in consequence of the variety of outline that they present, form meshes either regular or irregular; and in nature, because these cellules are perfectly transparent, colourless, and probably filled with air, — for the manner in which light passes through them proves that they do not contain dense fluid. They scarcely ever contain any organic particles, and are probably but little permeable either to fluids or gaseous matter; while, on the other hand, the cellules of the subjacent parenchyma are filled with the green substance that determines the colour of the leaf. The cuticle is not always formed of a single layer of cellules, but in some cases consists of two, or even three. No trace whatever is discoverable of vessels either terminating in or beneath the cuticle; M. Brongniart states this most explicitly, and my own observations are entirely in accordance with his: an opinion, therefore, which some distinguished botanists have entertained, that spiral vessels terminate in the stomata (D. C. *Organogr.* p. 272, &c.), must hereafter be abandoned. At the margin of a leaf the cuticle is generally harder than elsewhere, and sometimes becomes so indurated as to assume a flinty texture, as in the Aloe, and many other plants.

Stomata (p. 33.) are found upon various parts of the cuticle: in some plants only on that of the under side of leaves, in others on the upper also; in floating leaves upon the latter only. When leaves are so turned that their margins are directed towards the earth and the heavens, the two faces are then alike in appearance, and are both equally furnished with stomata. In succulent leaves they are said to be either altogether absent or very rare; but this is a statement that requires confirmation. According to the observations of M. De Candolle (*Organogr.* p. 272.), they are, in the Orange and the Mesembryanthemum, as ten in the former to one in the latter.

The parenchyma is, if casually examined, or even if viewed in slices of too great thickness, apparently composed of heaps of little green cells, arranged with little order or regularity; but, if very thin slices are taken and viewed with a high magnifying power, it will be seen that nothing can be more perfect than the plan upon which the whole structure is contrived, and that, instead of disorder, the most wise order pervades the whole. Upon this subject I extract the words of M. Adolphe Brongniart: — " There exists beneath the upper cuticle two or three layers of oblong blunt vesicles, placed perpendicular to the surface of the leaf, and generally much less in diameter than the cells of the cuticle; so that they are easily seen through it. These vesicles, which appear specially destined to give solidity to the parenchyma of the leaf, have no other intervals than the little spaces that result from the contact of this sort of cylinder: nevertheless, in plants that have stomata on the upper surface of their leaves, as is the case in most herbaceous plants, and in such as float on the surface of water, there exist here and there among the vesicles some large spaces, through which the stomata communicate with the interior of the leaf.

This parenchyma is entirely different from what is found beneath the cuticle of the lower side. There, instead of consisting of regular cylindrical vesicles, it is composed of irregular ones, often having two or three branches, which unite with the limbs of the vesicles next them, and so form a reticulated parenchyma; the spaces between whose vesicles are much larger than the vesicles themselves.

It is this reticulated tissue, with large spaces in it (to which the name of cavernous or spongy parenchyma might not improperly be applied), that, in most cases, occupies at least half the thickness of the leaves between the veins. The arrangement of the vesicles is very obvious if the lower cuticle of certain leaves be lifted up with the layer of parenchyma that is applied against it; it may then be seen that these anastomosing vesicles form a net with large meshes,—a sort of grating inside the cuticle. It must not, however, be supposed that this structure, which I have remarked in several ferns, and in a great many dicotyledonous plants, is without exception. In many monocotyledonous and succulent plants we have some remarkable modifications of this structure. Thus, in the Lily, and several plants of the same family, the vesicles of parenchyma that are in contact with the lower cuticle are lengthened out, sinuous, and toothed, as it were, at the sides: these projections join those of the contiguous vesicle; and a number of cavities is the consequence, which render this sort of parenchyma permeable to air. An analogous arrangement exists in the lower parenchyma of Galega. In the Iris, there is scarcely any space between the oblong and polyedral vesicles which form the parenchyma; but it is remarked, that the subjacent parenchyma is wanting at every point where the cuticle is pierced by a stoma. In such succulent plants as I have examined, the spaces between the cellules of parenchyma are very small; but, nevertheless, here and there, there are often larger cavities, which either correspond directly with the stomata, or are in communication with them. The same thing happens in plants with floating leaves, where the stomata placed on the upper surface correspond with the layer of cylindrical and parallel vesicles; in such case there are, here and there, between these vesicles, empty spaces which almost always correspond to the points where the stomata exist, and which permit the air to penetrate between the vesicles as far as the middle of the parenchyma of the leaf.

Thus much M. Brongniart; who adds, that in submersed leaves there is no cuticle, but the whole consists of solid parenchyma alone, in which there are no other cavities than such as are necessary to float the leaves.

The veins, being elongations of the medullary sheath, necessarily consist of woody fibre and spiral vessels, to which are sometimes added annular ducts. In submersed leaves spiral vessels are often wanting, the veins consisting of nothing but woody fibre. In these veins M. Schultz finds what he calls *vessels of the latex,* or of the nutritive fluid; but it is difficult to understand, either from his figures or descriptions, which kind of tissue in particular he means to designate by that name. M. Adolphe Brongniart says, the latex vessels are the vasa propria; but what are the vasa propria of leaves, in which there is nothing but woody fibre, spiral vessels, and ducts?

Such are the general anatomical characters of leaves; but it must be borne in mind, that, in different species, they undergo a variety of remarkable modifications. These arise either from the addition of parenchyma when leaves become *succulent,* or from the non-developement of it when they become *membranous,* or from the total suppression of it, and even of the veins also in great part, as in those which are called *ramentaceous,* such as the primordial leaves of the genus Pinus.

I have dwelt thus much at length upon the structure of the leaf, because it is by far the most important part of a plant, and that of which the functions are the best ascertained. Let us now turn our attention to the modifications of the leaf. It has already been seen that a leaf may consist of two distinct parts; the *petiole,* or stalk, and *lamina,* or leaf itself: both of these demand separate consideration.

The *lamina,* or *limbus,* as it is called by some, is subject to many diversities of figure and division; most commonly it forms an approach to oval, being longer than broad. When speaking of the leaf, it is usual to take the opportunity of explaining the terms employed by botanists to distinguish varieties of figure; but, as those terms are equally applicable to any other part with a similar dilated surface, it has appeared to me expedient to include them in Glossology, where they will accordingly be found.

That extremity of the lamina which is next the stem is called its *base;* the opposite extremity, its *apex;* and the line representing its two edges, the *margin* or *circumscription.*

If the lamina consists of one piece only, the leaf is said to be *simple*, whatever may be the depth of its divisions: thus, the entire lamina of Box, the serrated lamina of the Apple, the toothed lamina of Coltsfoot, the runcinate lamina of Taraxacum, the pinnatifid lamina of Hawthorn (which is often divided almost to its very midrib), are all considered to belong to the class of simple leaves. But if the petiole branches out, separating the cellular tissue into more than one distinct portion, each forming a perfect lamina by itself, such a leaf is often said to be *compound*, whether the divisions be two, as in the conjugate leaf of Zygophyllum, or indefinite in number, as in the many varieties of pinnated leaves. Nevertheless, a more accurate notion of a compound leaf is found to consist in its divisions being articulated with the petiole, by which it is much better distinguished from the simple leaf than by the number of its divisions. Thus, the pinnated leaf of a Zamia, and the pedate leaf of an Arum, both in this sense belong to the class of simple leaves; while the solitary lamina of the Orange, the common Berberry, &c. are referable to the class of compound leaves. This distinction is of some importance to the student of natural affinities; for, while division, of whatever degree it may be, may be expected to occur in different species of the same genus or order (provided there is no articulation), it rarely happens that truly compound leaves,—that is to say, such as are articulated with their petiole,—are found in the same natural assemblage with those in which no articulation exists.

In speaking of the *surface* of a leaf it is usual to make use of the word *pagina*. Thus, the upper surface is called *pagina superior;* the lower surface, *pagina inferior.* The upper surface is more shining and compact than the under, and less generally clothed with hairs; its veins are sunken; while those of the lower surface are usually prominent. The cuticle readily separates from the lower surface, but with difficulty from the upper. There are frequently hairs upon the under surface while the upper is perfectly smooth; but there is scarcely any instance of the upper surface being hairy while the lower is smooth.

The ramifications of the petiole among the cellular tissue of the leaf are called *veins*, and the manner of their distribution is termed *venation*. This influences in a great degree the figure and general appearance of the foliage, and requires a more careful consideration than it generally receives in elementary works.

The vein which forms a continuation of the petiole and the axis of the leaf is called the midrib or *costa*: from this all the rest diverge, either from its sides or base. If other veins similar to the midrib pass from the base to the apex of a leaf, such veins have been named *nerves*; and a leaf with such an arrangement of its veins has been called a *nerved* leaf. If the veins diverge from the midrib towards the margin, ramifying as they proceed, such a leaf has been called a *venous* or *reticulated* leaf. This is the sense in which these terms were used by Linnæus; but Link and some others depart from so strict an application of them, calling all the veins of a plant nerves, whatever may be their origin or direction.

Till within a few years the distribution of veins in the leaf had not received much attention; the terms just mentioned had been contrived to express certain of the most striking forms of venation; but the application of these was far from being sufficiently precise. Many improvements have been proposed by modern botanists; it however appears to me that the whole nomenclature of venation is essentially defective, and requires complete revision. My ideas upon this subject have been already laid before the public in the Botanical Register for Sept. 1826, page 1004.; and, as I am not aware that any objection to them has yet been taken, I shall repeat them here, in a form better adapted to an elementary work than that under which they first appeared.

The objections that I take to the present modes of distinguishing veins are these: — 1st, That the veins are very improperly, as I think, called nerves, either in all cases, as by Link, which is bad, or in certain cases only, when they have a particular size or direction, as by Linnæus and his followers, which is worse. Nothing is more destructive of accurate ideas in natural history than giving names well understood in one kingdom of nature to organs in another kingdom of an entirely different kind, unless it is the, perhaps, more reprehensible practice of giving two names conveying totally different ideas to the same organ in the same kingdom of nature. Thus, when the veins of a plant are termed nerves, it is necessarily understood that they exercise functions of a similar nature to those of the nerves of animals: if otherwise, why are they so called? But they exercise no such functions, being, beyond all doubt, mere channels for the transmission of fluid. Again, if one portion of the skeleton of a leaf is called a vein, and another portion a nerve, this apparently precise mode of speaking leads yet more strongly to the belief (especially when such a distinction is seen admitted into works which are said to be of the highest authority in science), that the structure and function of those two parts are as widely different as the structure and function of a vein and a nerve in the animal economy; else why should such nice caution be taken to distinguish them? But it must be confessed that there is no difference whatever, except in size, between the veins and nerves of a leaf. Let us, then, abandon a term which is one of those relics of a barbarous age, which it is the duty of modern science to expel. My second objection is caused by the vague manner in which the veins of leaves are at present described; whence it happens that no precise idea can be attached to the different terms that have been contrived to designate particular forms of venation. A third objection is this,—that, while slight modifications in the arrangement of the veins have received distinctive names, others of much greater importance, and of a more decided character, have received no distinctive appellation whatever. For these reasons, the practical weight of which I have long experienced, it has occurred to me that the following changes in the language used in speaking of venation will be found better, at least, than that for which they are substituted, if they are not entirely what could be desired.

It has been usual to call that bundle of vessels only which passes directly from the base to the apex of a leaf the *costa*, or midrib. This term I would extend to all main veins which proceed directly from the base to the apex, or to the points of the lobes. There is no difference in size in these costæ; and in lobed leaves, which may be understood as simple leaves, approaching composition, each costa has its own particular set of veins.

The costa (*fig. 52, 7.*) sends forth, alternately right and left along its whole length, ramifications of less dimensions than itself, but more nearly approaching it than any other veins:

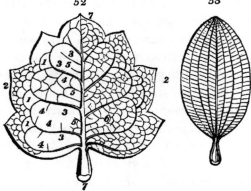

these I would call *venæ primariæ* (*fig. 52, 3.*). They diverge from the costa at various angles, and pass to the margin of the leaf, curving towards the apex in their course, and finally, at some distance within the margin, forming what is called an *anastomosis*, or junction, with the back of the vena primaria, which lies next them. That part of the vena primaria which is between the anastomoses thus described, having a curved direction, may be called the *vena arcuata*. Between this latter and the margin, other veins, proceeding from the venæ arcuatæ, with the same curved direction, and of the same magnitude, occasionally intervene: they may be distinguished by the name of *venæ externæ* (*fig. 52, 1.*). The margin itself and these last are connected by a fine net-work of minute veins, which I would distinguish by the name of *venulæ marginales*. From the costa are generally produced, at right angles with it, and alternate with the venæ primariæ, smaller veins; which may be considered imperfect venæ primariæ, and may not improperly be named *venæ costales* (*fig. 52, 5.*). The venæ primariæ are themselves connected by fine veins, which anastomose in the area between them. These veins, when they immediately leave the venæ primariæ, I call *venulæ propriæ* (*fig.* 52, 4.); and where they anastomose, *venulæ communes*. The area of parenchyma, lying between two or more veins or veinlets, I name with the old botanists *intervenium*.

These distinctions may to some appear over-refined; but I am convinced that no one can accurately describe a leaf without the use of them, or of equivalent terms yet to be invented. Upon these principles leaves may be conveniently divided into the following kinds: —

1. *Veinless* (*avenium*), when no veins at all are formed, except a slight approach to a costa, as in Mosses, Fuci, &c. Leaves of this description exist only in the lowest tribes of foliaceous plants, and must not be confounded with the fleshy or thickened leaves common among the higher orders of vegetation, in which the veins are by no means absent, but only concealed within the substance of the parenchyma. (*See* No. 10.) Of this M. De Candolle has two forms, — first, his *folia nullinervia*, in which there is not even a trace of a costa, as in *Ulva*; and second, his *folia falsinervia*, in which a trace of a costa is perceptible. These terms appear to me unnecessary; but, if they be employed, the termination *nervia* must be changed to *venia*.

2. *Equal-veined* (*æqualivenium*), when the costa is perfectly formed, and the veins are all of equal size, as in Ferns. This kind of leaf has not been before distinguished: it may be considered intermediate between those without veins and those in which venæ primariæ are first apparent. The veins are equal in power to the venulæ propriæ of leaves of a higher class.

3. *Straight-veined* (*rectivenium*). In this the veins consist only of venæ primariæ, generally very much attenuated, and arising from towards the base of the costa, with which they lie nearly parallel: they are connected by venulæ propriæ; but there are no venulæ communes. The leaves of Grasses and of Palms and Orchideous plants are of this nature. This form has been called by Link *paralleli* and *convergenti-nervosum*, according to the degree of parallelism of the venæ primariæ; and to these two he has added what he calls *venuloso-*

nervosum, when the venæ primariæ are connected by venulæ propriæ: but as this is always so, although it is not in all cases equally apparent, the term is superfluous. Ach. Richard calls this form *laterinervium*, and De Candolle *rectinervium;* from which I do not find it advisable to distinguish his *ruptinervium*, which indicates the straight-veined leaf, when the veins are thickened and indurated, as in the Palm tribe.

4. *Curve-veined* (*curvivenium*). This is a particular modification of the last form, in which the venæ primariæ are also parallel, simple, and connected by unbranched venulæ propriæ; do not pass from near the base to the apex of the leaf, but diverge from the costa along its whole length, and lose themselves in the margin. This is the *folium hinoideum* and *venuloso-hinoideum* of Link, the *f. penninervium* of A. Richard, and the *f. curvinervium* of De Candolle. It is common in Scitamineæ. It is not improbable that both this and the last ought to be regarded as peculiar modifications of petiole (a kind of phyllodia), rather than as true leaves analogous to those next to be described.

5. *Netted* (*reticulatum*). Here the whole of the veins that constitute a completely developed leaf are present, arranged as I have above described them, there being no peculiar combination of any class of veins. This is the common form of the leaves of Dicotyledones, as of the Lilac, the Rose, &c. It is the *folium venosum* of Linnæus, the *f. indirectè venosum* of Link, the *f. mixtinervium* of A. Richard, and the *f. retinervium* of De Candolle. If the venæ externæ and venulæ marginales are conspicuous, Link calls this form *combinatè venosum;* but if they are indistinct, he calls it *evanescentè venosum*.

6. *Ribbed* (*costatum*). In this three or more costæ proceed from the base to the apex of the leaf, and are connected by branching venæ primariæ of the form and magnitude of venulæ propriæ, as in Melastoma. This must not be confounded with the *straight-veined* leaf, from which it may in all cases of doubt be distinguished by the ramified veins that connect the costæ. This is a very material difference, which has never been properly explained. Linnæus and his followers confound the two forms; but modern writers separate them: although it must be confessed that it is difficult to discover

their distinctions from the characters hitherto assigned to them. Link calls these leaves *f. nervata*, A. Richard *f. basinervia*, and De Candolle *f. triplinervia* and *f. quintuplinervia*. If a ribbed leaf has three costæ springing from the base, it is said to be *three-ribbed* (*tri-costatum*, *trinerve* of authors); if five, *five-ribbed*, and so on. But if the ribs do not proceed exactly from the base, but from a little above it, the leaf is then said to be *triply-ribbed* (*triplicostatum*), as in the Helianthus.

7. *Falsely ribbed* (*pseudocostatum*), is when the venæ arcuatæ and venæ externæ, both or either, in a reticulated leaf, become confluent into a line parallel with the margin, as in all Myrtaceæ. This has not been before distinguished.

8. *Radiating* (*radiatum*), when several costæ radiate from the base of a reticulated leaf, to its circumference, as in lobed leaves. This and the following form the *f. directè venosum* of Link: it is the *f. digitinervium* of A. Richard. Hither I refer without distinguishing them the *f. pedalinervia, palminervia*, and *peltinervia* of M. De Candolle; the differences of which do not arise out of any peculiarity in the venation, but from the particular form of the leaves themselves.

9. *Feather-veined* (*pennivenium*), when the venæ primariæ of a reticulated leaf pass in a right line from the costa to the margin, as in Castanea. This has the same relation to the radiating leaf that the curve-veined bears to the straight-veined; it is the *folium pennivenium* of M. De Candolle.

10. *Hidden-veined* (*introvenium*). To this I refer all leaves the veins of which are hidden from view by the parenchyma being in excess, as in the *Hoya*, and many others. Such a leaf is often inaccurately called veinless. M. De Candolle calls a leaf of this nature, in which the veins are dispersed through a large mass of parenchyma, as in Mesembryanthemum, *vaginervium*.

It is often necessary to explain the direction that the venæ primariæ take when they diverge from the costa: this may be denoted by measuring the angle which is formed by the costa and the diverging vein, and can either be stated in distinct words, or by applying the following terms thus:— if the angle formed by the divergence is between 10° and 20°, the vein may be said to be *nearly parallel (subparallela)*; if between 20° and 40°, *diverging*; between 40° and 60°, *spreading*; between 60° and 80°, *divaricating*; between 80° and 90°, *right-angled*; between 90° and 120°, *oblique*; beyond 120°, *reflexed (retroflexa)*.

The *petiole* (*fig.* 55 *a—b.*) is the part which connects the lamina with the stem, of which it was considered by Linnæus as a part. It consists of one or more bundles of fibrovascular tissue surrounded by cellular tissue. Its figure is generally half cylindrical, frequently channelled on the surface presented to the heavens; but in most monocotyledonous plants it is perfectly cylindrical. If the petiole is entirely absent, which is often the case, the leaf is then said to be *sessile*. Generally the petiole is simple, and continuous with the axis of the leaf; sometimes it is divided into several parts, each bearing a separate leaf or *leaflet (foliolus)*: in such case it is by some said to be compound; each of the stalks of the leaflets being called *petiolules (ramastra,* Jungius). In all simple leaves the petiole is continuous with the axis of the lamina, from which it never separates; in all truly compound leaves the petiole is articulated with each petiolule; so that when the leaf perishes, it separates into as many portions as there are leaflets, as in the Sensitive plant: hence, whenever an apparently simple leaf is found to be articulated with its petiole, as in the Orange, such a leaf is not to be considered a simple leaf, but the terminal leaflet of a pinnated leaf, of which the lateral leaflets are not developed. This is a most important difference, and must be borne constantly in mind by all persons who are engaged in the investigation of natural affinities. It is an occult sign which must never be neglected.

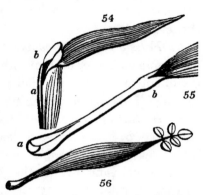

At the base of the petiole, where it joins the stem, and upon its lower surface, the cellular tissue increases in quantity, and produces a protuberance or gibbosity, which Ruellius, and after him Link, called the *pulvinus*, and M. De Candolle *coussinet* (*fig.* 55, *a.*). At the opposite extremity of the petiole, where it is connected with the lamina, a similar swelling is often remarkable, as in *Sterculia, Mimosa sensitiva,* and others: this is called the *struma*, or, by the French, *bourrelet* (*fig.* 55, *b.*).

Occasionally the petiole embraces the branch from which it springs, and in such case is said to be *sheathing* ; and is even called a *sheath* or *vagina*, as in grasses (*fig.* 54, *a.*). When the lower part only of the petiole is sheathing, as in Umbelliferæ, that part is sometimes called the *pericladium*. In grasses there is a peculiar membranous process at the top of the vagina, between it and the lamina, which has received the name of *ligula* (*fig.* 54, *b.*) (*languette*, Fr. ; *collare*, Rich.); the nature of this process has not yet been determined. In the Asparagus, the petiole has the form of a small sheath, is destitute of lamina, and surrounds the base of certain small branches having the appearance of leaves : such a petiole has been named *hypophyllium* by Link. In Trapa natans, Pontedera crassipes, and other plants, the petiole is excessively dilated by air, and acts as a bladder to float the leaves : except in this state of dilatation, it differs in no wise from common petioles : it has, nevertheless, received the name of *vesicula* from M. De Candolle, who considers it the same as the bladdery expansions of Fuci. The petiole is generally straight : occasionally it becomes rigid and twisted, so that the plant can climb by it.

It has been said that the figure of the petiole usually approaches more or less closely to the cylindrical : this, however, is not always the case. In many plants, especially of an herbaceous habit, it is very thin, with foliaceous margins ; it is then called *winged*. There are, moreover, certain leafless plants, as the greater number of species of Acacia, in which the petiole becomes so much developed as to assume the appearance of a leaf, all the functions of which it performs. Petioles of this nature have received the name of *Phyllodia* (*fig.* 56.). They may always be distinguished from true leaves by the following characters : — 1. If observed when the plant is very young, they will be found to bear leaflets. 2. Both their surfaces are alike. 3. They very generally present their margins to the earth and heavens, — not their surfaces. 4. They are always straight-veined; and, as they only occur among dicotyledonous plants which have reticulated leaves, this peculiarity alone will characterise them.

But, besides the curious transformation undergone by the petiole when it becomes a phyllodium, there are several others still more remarkable: among these the first to be noticed is the *tendril* (*Vrille*, Fr.; *Cirrhus*, Linn.; *Capreolus* and *Clavicula* of the old botanists). It is one of the contrivances employed by nature to enable plants to support themselves upon others that are stronger than themselves. It was included by Linnæus among what he called *fulcra;* and has generally, even by very recent writers, been spoken of as a peculiar organ. But, as it is manifestly in most cases a particular form of the petiole, I see no reason for regarding it in any other light. It may, indeed, be a modification of the inflorescence, as in the Vine; but this, I conceive, is an exception, showing, not that the cirrhus is not a modification of the petiole, but that any part may become cirrhose.

In some cases, the petiole of a compound leaf is elongated, branched, and endowed with the power of twisting round any small body that is near it, as in the Pea: it then becomes what is called a *cirrhus petiolaris*. At other times, it branches off on each side at its base below the lamina into a twisting ramification, as in *Smilax horrida;* when it is called a *cirrhus peduncularis*. At other times it passes, in the form of midrib, beyond the apex of a simple leaf, twisting and carrying with it a portion of the parenchyma, as in *Gloriosa superba;* when it is said to be a *cirrhus foliaris*. M. De Candolle refers to tendrils the acuminate, or rather caudate, divisions of the corolla of Strophanthus, under the name of *cirrhus corollaris;* but these do not appear to me to possess any of the requisites of a tendril.

As another modification of the petiole, I am disposed to consider, with Link (Elem. 202.), the singular form of leaf in Sarracenia and Nepenthes, which has been called *Ascidium* or *Vasculum* (*outre* De Candolle). This consists of a fistular green body, occupying the place and performing the functions of a leaf, and closed at its extremity by a lid termed the *operculum*. To me it appears that the ascidium itself, or fistular part, is

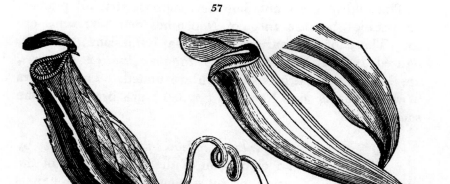

the petiole, and the operculum the lamina of a leaf in an extraordinary state of transformation. Look, for example, at Dionæa muscipula; in this plant the leaf consists of a broad winged petiole, articulated with a collapsing lamina, the margins of which are pectinate and inflexed. Only suppose the broad winged petiole to collapse also, and that its margins, when they meet, as they would in consequence of collapsion, cohere; a fistular body would then be formed, just like the ascidium of Sarracenia; and there would be no difficulty in identifying the acknowledged lamina of Dionæa with the operculum of Sarracenia also. From Sarracenia the transition to Nepenthes would perhaps not be considered improbable.

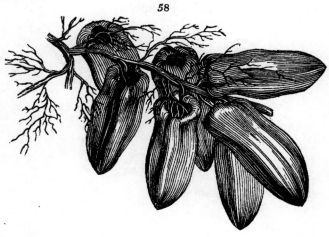

The student must not, however, suppose that all pitchers are petioles, because those of Nepenthes and Sarracenia are so. Those of the curious Dischidia Rafflesiana (*fig.* 58.), figured by Dr. Wallich in his *Plantæ Asiaticæ Rariores*, are leaves, the margins of which are united. The pitchers of Marcgraavia and Norantea (*fig.* 59.) are bracteæ in the same state.

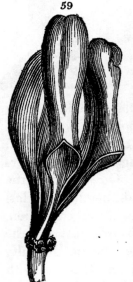

Spines of the leaves are formed either by an elongation of the woody tissue of the veins, or by a contraction of the parenchyma of the leaves: in the former case they project beyond the surface or margin of the leaf, as in the Holly (*Ilex aquifolium*): in the latter case they are the veins themselves become indurated, as in the palmated spines of *Berberis vulgaris*. The spiny petiole of many Leguminous plants is of the same nature as the latter. So strong is the tendency in some plants to assume a spiny state, that in a species of Prosopis from Chili, of which I have a living specimen now before me, half the leaflets of its bipinnate leaves have the upper half converted into spines.

2. *Of* Stipulæ.

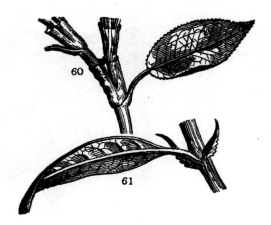

At the base of the petiole, on each side, is frequently seated a small appendage, most commonly of a texture less firm than the petiole, and having a subulate termination. These two appendages are called *stipulæ*. They either adhere to the base of the petiole or are separate; — they either endure as long as the leaf, or fall off before it; — they are membranous, leathery, or spiny; — finally, they are entire or lacineated. By Link they have been called *Paraphyllia*; an unnecessary term. When they are membranous, and surround the stem like a vagina, cohering by their anterior margins, as in Polygonum (*fig.* 60.), they have been termed *ochrea* by Willdenow. Of this the fibrous sheath at the base of the leaves of Palms, called *reticulum* by some, may possibly be a modification. In pinnated leaves there are often two stipulæ at the base of each leaflet as well as at the base of the common petiole: stipulæ, under such circumstances, are called *stipellæ*.

The exact analogy of stipulæ is not well made out. M. De Candolle seems, from some expressions in his *Organographie*, to suspect their analogy with leaves; while, in other places in the same work, it may be collected that he rather considers them special organs. I am clearly of opinion that, notwithstanding the difference in their appearance, they are really accessory leaves: first, because occasionally they are transformed into leaves, as in Rosa bracteata, in which I have seen them converted into pinnated leaves; secondly, because they often are undistinguishable from leaves, of which they obviously perform all the functions, as in Lathyrus, Lotus, and many other Leguminosæ: and, finally, because there are cases in which buds develope in their axilla, as in Salix; a property peculiar to leaves and their modifications. M. De Candolle, in suggesting, after Seringe, that the tendrils of Cucurbitaceæ are modified stipulæ, assigns the latter a tendency to a transformation exclusively confined either to the midrib of a leaf, or to a branch; and they cannot be the latter.

It is sometimes difficult to distinguish from true stipulæ, certain membranous expansions, or ciliæ, or glandular appendages of the margin of the base of the petiole, such as are found in Ranunculaceæ, Apocyneæ, Umbelliferæ, and many other plants. In these cases the real nature of the parts is only to be collected from analogy, and a comparison of them with the same part differently modified in neighbouring species.

M. De Candolle remarks, that no Monocotyledonous plants have stipulæ; but they certainly exist, at least in Fluviales and Aroideæ. The *ligula* of grasses, a membranous appendage at the apex of their sheathing petiole, which some have considered stipulæ, should rather be understood as a membranous expansion analogous to the corona of some Caryophylleæ, such as Silene.

It has been already noted, that when they surround the stem of a plant they become an *ochrea*; in this case their anterior and posterior margins are united by cohesion; a property that they possess in common with all modifications of leaves, and of which different instances may be pointed out in Magnoliaceæ, where the back margins only cohere, in certain Cinchonaceæ, in which the anterior margins of the stipulæ of opposite leaves are united, and in a multitude of other plants.

3. *Of* Bracteæ.

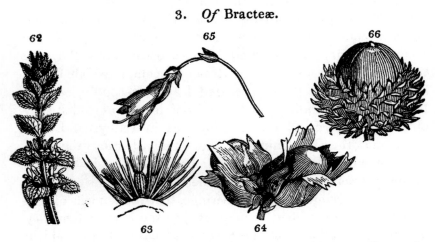

All the parts that have hitherto been subjects of enquiry are called *organs of vegetation;* their duty being exclusively to perform the nutritive parts of the vegetable economy. Those which are about to be mentioned are called *organs of fructification;* their office being to reproduce the species by a process in some respects analogous to that which takes place in the animal kingdom. The latter are, however, all modifications of the former, as will hereafter be seen: and as the subject of this division is in itself a kind of proof; bracteæ not being exactly organs either of vegetation or reproduction, but between the two.

Botanists call *Bracteæ* either the leaf from the axilla of which a flower is developed, such as we find in Veronica agrestis; or else all those leaves that are found upon the inflorescence, and are situated between the true leaves and the calyx. There are, in reality, no exact limits between bracteæ and common leaves; but in general the former may be known by their situation immediately below the calyx, by their smaller size, difference of outline, colour, and other marks. They are generally entire, however much the leaves may be divided; frequently scariose, either wholly or in part; often deciduous before the flowers expand; but rarely very much dilated, as in Origanum, Dictamnus, and a few other plants. It is often more difficult to distinguish bracteæ from the sepals of a polyphyllous calyx than even from the leaves of the stem. In fact, there is in many cases no other mode than ascertaining the usual number of sepals in other plants of the same natural order, and considering every leaf-like appendage on the outside of the usual number of sepals as bracteæ. In Camellia, for example, if it were not known that the normal number of sepals of kindred genera is five, it would be impossible to determine the number of its sepals. When the bracteæ are very small, they are called *bracteolæ*; or if they are of different sizes upon the same inflorescence, the smallest receive that name. It rarely occurs that an inflorescence is destitute of bracteæ. In Cruciferæ this is a frequent character, and is observed by Link to indicate an extremely irregular structure. When bracteæ do not immediately support a flower or its stalk, they are called *empty* (*vacuæ*). As a general rule, it is to be understood, that whatever intervenes between the true leaves and the calyx, whatever be their form, colour, size, or other peculiarity, comes within the meaning of the term.

Under particular circumstances bracteæ have received the following peculiar names: —

When they are empty, and terminate the inflorescence, they form a *coma*, as in Salvia Horminum. In this case they are generally enlarged and coloured.

If they are verticillate, and surround several flowers, they constitute an *involucrum*. In Umbelliferous plants, the bracteæ which surround the general umbel are called an *universal involucrum;* and those which surround the umbellules a *partial involucrum*, or *involucellum*. In Compositæ, the involucrum often consists of several rows of imbricated bracteæ, and has received a variety of names, for none of which does there appear to be the least occasion. Linnæus called it *calyx communis*, Necker *perigynandra communis*, Richard *periphoranthium*, Cassini *periclinium*. There is often found at the base of the involucrum of Compositæ an exterior rank of bracteæ, which Linnæus called *calyculus;* and such involucra as were so circumstanced *calyx calyculatus*. M. Cassini restricts the term *involucrum* to this; but it seems most convenient to call these exterior bracteæ *bracteolæ*, and to say that an involucrum in which they are present is *basi bracteolatus*, bracteolate at the base.

Another and very remarkable form of the involucrum is the *cupula* (*fig.* 66.). It consists of bracteæ not developed till after flowering, when they cohere by their bases, and form a kind of cup. In the Oak the cupula is woody, entire, and scaly, with indurated bracteæ: in Fagus it forms a sort of coriaceous valvular spurious pericarpium: in Corylus (*fig.* 64.) it is foliaceous and lacerated: in Taxus it is fleshy and entire, with no appearance of bracteæ.

The name *squama* or *scale* is usually applied to the bracteæ of the amentum; it is also occasionally used to indicate any kind of bracteæ which has a scaly appearance.

The bracteæ which are stationed upon the receptacle of Compositæ, between the florets, have generally a membranous texture and no colour, and are called *paleæ*, Englished by some botanists *chaff of the receptacle*. The French call this sort of bracteæ *paillette*, Cassini *squamelles* (*fig.* 63.).

In Palms and Aroideæ there are seated, at the base of the spadix, large, coloured bracteæ, in which the spadix is in æstivation wholly enwrapped, and which may perhaps perform

CHAP. II.   COMPOUND ORGANS IN FLOWERING PLANTS.   103

in those plants the office of corolla. This is called the *spatha* (*fig.* 84.). Link considers it a modification of the petiole! (*Elementa*, p. 253.)

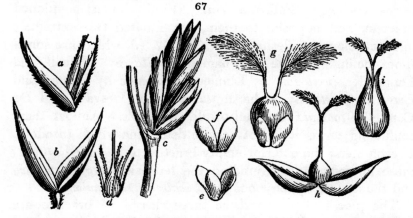

The most remarkable arrangement of bracteæ takes place in Grasses, in which they occupy the place of calyx and corolla, and have received a great variety of names from different systematic writers. In order to explain distinctly the application of these terms, I must describe with some minuteness the structure of a locusta or spicula, as the partial inflorescence of Grasses is denominated. Take, for example, any common Bromus; each locusta will be seen to have at its base two opposite empty bracteæ (*fig.* 67, *b*.), one of which is attached to the rachis a little above the base of the other: these are the *gluma* of Linnæus and most botanists, the *gluma exterior* or *calycinalis* of some writers, the *tegmen* of Palisot de Beauvois, the *lepicena* of Richard, the *cætonium* of Trinius, and, finally, the *peristachyum* of Panzer. Above the gluma are several florets sitting in denticulations of the rachis (*fig.* 67, *c*.): each of these consists of one bractea, with the midrib quitting the lamina a little below the apex, and elongated into a bristle called the *awn*, *beard*, or *arista*, and of another bractea facing the first, with its back to the rachis, bifid at the apex, with no dorsal vein, but with its edges inflexed, and a rib on each side at the line of inflexion (*fig.* 67, *a*.). These bracteæ are the *corolla* of Linnæus, the *calyx* of Jussieu, the *perianthium* of Mr. Brown, the *gluma interior* or *corollina* and *perigonium* of some, the *stragulum* of Palisot de Beauvois, the *gluma* of

H 4

Richard, the *bale* or *Glumella* of De Candolle and Desvaux, the *paleæ* of others. When the *arista* proceeds from the very apex of the bracteæ, and not from below it, it is denominated in the writings of Palisot a *seta*. Within the last-mentioned bracteæ, and opposite to them, are situated two extremely minute colourless fleshy scales (*fig.* 67, *e.*), which are sometimes connate: these are named *corolla* by Micheli and Dumortier, *nectarium* by Linnæus, *squamulæ* by Jussieu and Brown, *glumella* by Richard, *glumellula* by Desvaux and De Candolle, *lodicula* by Palisot de Beauvois. Amidst these conflicting terms it is not easy to determine which to adopt. I recommend the exterior empty bracteæ to be called *glumæ;* those immediately surrounding the fertilising organs *paleæ;* and the minute hypogynous ones, *scales* or *squamulæ*.

The pieces of which these three classes of bracteæ are composed are called *valves* or *valvulæ* by the greater part of botanists; but as that term has been thought not to convey an accurate idea of their nature, Desvaux has proposed to substitute that of *spathella*, which is adopted by M. De Candolle. Palisot proposed to restrict the term *gluma* to the pieces of the gluma, and to call the pieces of the perianthium *paleæ*. Richard called the pieces of both gluma and perianthium *paleæ*, and the squamulæ *paleolæ*. It seems to me most convenient to use the term *valvula*, because it is more familiar to botanists than any other, and because I do not see the force of the objection which is taken to it.

In the genus Carex two bracteæ (*fig.* 67, *e, h.*) become confluent at the edges, and enclose the pistillum, leaving a passage for the stigmata at their apex. They thus form a single urceolate body named *urceolus* or *perigynium*. M. De Candolle justly observes, in his *Théorie*, that some botanists call this *nectarium*, although it does not produce honey; others *capsula*, although it has nothing to do with the fruit; but he does not seem to me more correct than those he criticises in arranging the urceolus among his miscellaneous appendages of the floral organs, which are " ni organes génitaux ni tégumens." I believe I was the first who explained the true nature of the urceolus, in my translation of Richards's *Analyse du Fruit,* printed in 1819. (p. 13.)

At the base of the ovarium of Cyperaceæ are often found little filiform appendages, called *hypogynous setæ* by most botanists. These are probably of the nature of the squamulæ of Grasses, and have been named *perisporum* by some French writers.

Bracteæ are generally distinct from each other, and imbricated or alternate. Nevertheless, there are some striking exceptions to this; as remarkable instances of which may be cited Althæa and Lavatera among Malvaceæ, all Dipsaceæ, and some Trifolia, particularly my Tr. cyathiferum (*Hooker, Fl. Boreali-Amer.*), in all which the bracteæ are accurately verticillate, and their margins confluent, as in a true calyx.

4. *Of the* Flower.

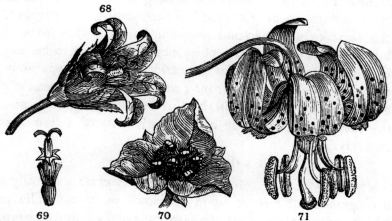

The Flower is a terminal bud enclosing the organs of reproduction by seed. By the ancients the term flower was restricted to what is now called the corolla; but Linnæus wisely extended its application to the union of all the organs which contribute to the process of fecundation. The flower, therefore, as now understood, comprehends the *calyx*, the *corolla*, the *stamens*, and the *pistillum*, of which the two last only are indispensable. The calyx and corolla may be wanting, and a flower will nevertheless exist; but if neither stamens nor pistillum nor their rudiments are to be found, no assemblage of leaves, whatever may be their form or colour, or how much soever they may resemble the calyx and corolla, can constitute a flower.

The flower, when in the state of a bud, is called the *alabastrus* (*bouton* of the French), a name used by Pliny for the rose-bud. Some writers say *alabastrum*, forgetting, as it would seem, that that term was used by the Romans for a scent-box, and not for the bud of a flower. Link calls the parts of a flower generally, whether united or connate, *moria*, whence a flower is *bi-polymorious* (*Elem.*, 243.); but I know of no writer who employs these terms, which indeed are quite superfluous.

The flowers of an anthodium, which are small, and somewhat different in structure from ordinary flowers, are called *florets* (*flosculi; elytriculi* of Necker; *fleuron* of the French).

The period of opening of a flower is called its *anthesis;* the manner in which its parts are arranged with respect to each other before opening is called the *æstivation*. Æstivation is the same to a flower-bud as vernation (p. 53.) is to a leaf-bud: the terms expressive of its modifications are to be sought in Glossology. This term æstivation is applied separately to the parts of which a flower may consist; thus, we speak of the æstivation of the calyx, of the corolla, of the stamens, and of the pistillum; but never of the æstivation of a flower, collectively.

5. *Of the* Inflorescence.

Inflorescence is a term contrived to express generally the arrangement of flowers upon a branch or stem. The part which immediately bears the flowers is called the *pedunculus* or peduncle, and is to be distinguished from any portion of a branch by not producing perfect leaves; those which are found upon it called *bracteæ* being much reduced in size and figure from what are borne by the rest of the plant.

The term *peduncle*, although it may be understood to apply to all the parts of the inflorescence that bear the flowers, is only made use of practically, to denote the immediate support of a single solitary flower, and is therefore confined to that part of the inflorescence which first proceeds from the stem. If it is divided, its principal divisions are called branches; and its ultimate ramifications, which bear the flowers, are named *pedicels*. There are also other names which are applied to modifications of the peduncle.

CHAP. II.   COMPOUND ORGANS IN FLOWERING PLANTS.   107

In plants which are destitute of stem, it often rises above the ground, supporting the flowers on its apex, as in the Cowslip. Such a peduncle is named a *scape* (*hampe*, Fr.). Some botanists distinguish from the scape the *pedunculus radicalis*, confining the former term to the peduncle which arises from the central bud of the plant, as in the Hyacinth; and applying the latter to a peduncle proceeding from a lateral bud, as in Plantago media.

When a peduncle proceeds in a nearly right line from the base to the apex of the inflorescence, it is called the *rachis*, or the *axis* of the inflorescence. This latter term was used by Palisot de Beauvois to express the rachis of Grasses, and is perhaps the better term of the two, especially as the term rachis is applied by Willdenow and others, without much necessity it must be confessed, to the petiole and costa of Ferns. In the locustæ of Grasses the rachis has an unusual toothed flexuose appearance, and has received the name of *scobina* from M. Dumortier. If it is reduced to a mere bristle, as in some of the single-flowered locustæ, the same writer then distinguishes it by the name of *acicula*. I mention these and similar terms, in order that nothing which can even remotely lead to information may be omitted; but I cannot recommend their adoption.

When the part which bears the flowers is repressed in its developement, so that, instead of being elongated into a rachis, it forms a flattened area on which the flowers are arranged, as in Compositæ, it becomes what is called a *receptacle;* or, in the language of some botanists, the *receptacle of the flower* (*fig.* 72.).

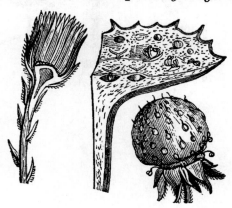

72   73   74

When the receptacle is not fleshy, but is surrounded by an involucrum, it is called the *clinanthium* (the *thalamus* of Tournefort), as in Compositæ, or, in the language of M. Richard, *phoranthium;* the former term is that generally adopted. But if the receptacle is fleshy, and is not enclosed within an involucrum, as in Dorstenia and Ficus (*fig.* 73.), it is then called by Link *Hypanthodium;* the same writer formerly named it *Amphanthium*, a term now abandoned.

According to the different modes in which the inflorescence is arranged, it has received different names, the right application of which is of the first importance in descriptive botany. If flowers are sessile along a common axis, as in Plantago, the inflorescence is called a *spike* (*épi*, Fr.), (*fig.* 76.); if they are pedicellate, under the same circumstances, they form a *raceme* (*grappe*, Fr.), (*fig.* 77.) as in the Hyacinth: the raceme and the spike differ, therefore, in nothing, except that the flowers of the latter are sessile, of the former pedicellate. These are the true characters of the raceme and spike, which have been confused and misunderstood in a most extraordinary manner by some French writers.

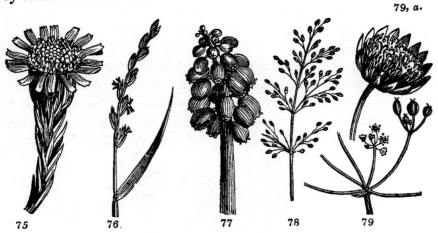

79, a.

75   76.   77   78   79

CHAP. II. COMPOUND ORGANS IN FLOWERING PLANTS. 109

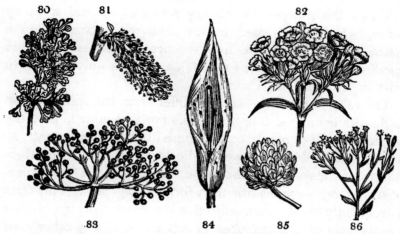

When the flowers of a spike are destitute of calyx and corolla, the place of which is taken by bracteæ, and when with such a formation the whole inflorescence falls off in a single piece, either after flowering or ripening the fruit, as in Corylus, Salix, &c., such an inflorescence is called an *amentum* or *catkin* (*chaton*, Fr.; *Catulus, Iulus, nucamentum*, of old writers), (*fig.* 81.) Linnæus considered the catkin to be an elongated filiform receptacle, analogous to that of Compositæ, in which he is followed by Sir James Edward Smith, Link, and others. This opinion arises from a distinction being drawn between the axis of a spike and the receptacle of Compositæ; but, as I have already stated, the latter can be considered in no other light than that of a depressed axis or rachis: so that, when the amentum is said not to be a true axis, but an elongated receptacle, a difference is drawn between words rather than things; for if a receptacle is only a depressed axis, an elongated receptacle is necessarily a return to the common form of the axis.

If a spike consists of flowers destitute of calyx and corolla, the place of which is occupied by bractæa, supported by other bracteæ which enclose no flowers, and when with such a formation the rachis, which is flexuose and toothed, does not fall off with the flowers, as in Grasses, each part of the inflorescence so arranged is called a *spicula* or *locusta* (*épillet*, Dec.; *paquet*, Tournefort). Link is of opinion that the rachis of a spicula, as well as that of the amentum, is a kind of receptacle.

When the flowers are closely arranged around a fleshy rachis, which is enclosed in the kind of bracteæ called a spatha (see p. 103.), the inflorescence is termed a *spadix* (*spadice* or *poinçon*, Fr.), (*fig.* 84.). This is only known to exist in Aroideæ and Palms.

The raceme has been said to differ from the spike only in its flowers being pedicellate: to this must be added, that the pedicels are all of nearly equal length; but in many plants, as Alyssum saxatile, the lower pedicels are so long that their flowers are elevated to the same level as that of the uppermost flowers; a *corymbus* is then formed (*fig.* 86.). This term is frequently used in an adjective sense, to express a similar arrangement of the branches of a plant or of any other kind of inflorescence: thus, in Stevia, the branches are said to be corymbose; in others, the panicle is said to be corymbose; and so on. When corymbose branches are very loose and irregular, they have given rise to the term *muscarium ;* a name formerly used by Tournefort, but not now employed.

If the expansion of an apparent corymb is centrifugal, instead of centripetal; that is to say, commences at the centre, and not at the circumference, as in Dianthus Carthusianorum, we then have the *fasciculus* (*fig.* 82.); a term which may not incorrectly be understood as synonymous with *compound corymbus*. The modern *corymbus* must not be confounded with that of Pliny, which was analogous to our *capitulum*.

When the pedicels all proceed from a single point, as in *Astrantia*, and are of equal length, or corymbose, we have what is called an *umbel* (*fig.* 79.). If each of the pedicels bears a single flower, as in *Eryngium*, the umbel is said to be *simple* (*fig.* 79, *a.*); but if they divide and bear other umbels, as in Heracleum, the umbel is called compound; and then the assemblage of umbels is called the *umbella universalis*, while each of the secondary umbels, or the umbellules, is named an *umbella partialis*. The peduncles which support the partial umbels are named *radii*. The late M. Richard confined the word umbel to the compound umbel, and named the simple umbel *sertulum* (*bouquet*); but this was an unnecessary change.

Suppose the flowers of a simple umbel to be deprived of their pedicels, and to be seated on a receptacle or enlarged axis, and we have a *capitulum* or head, named *glomus* by some, *glomerulus* by others. If this is flat, and surrounded by an involucrum, the compound flower, as it is inaccurately called by the school of Linnæus, of Compositæ, is produced; which is often named by modern botanists *anthodium;* a term invented by Ehrhart, and to which there seems to be no objection. It was called *cephalanthium* by Richard, and *calathidium* by Mirbel. The flowers or florets borne by the anthodium in its circumference are usually ligulate, and different from those produced within the circumference. Those in the former station are called *florets of the ray*, and those in the latter *florets of the disk*.

I have said that the school of Linnæus inaccurately calls the anthodium a compound flower, from which opinion I should think that few persons would at the present day dissent, unless they applied the same term to the umbella, the spica, and all other forms of inflorescence, of which the anthodium is palpably a mere modification. Professor Link, however, has in a late work defended the nomenclature of Linnæus; urging, that the rays of an anthodium may be considered a sort of corolla, the florets of the disk a representation of the stamens and pistillum; and that this mode of viewing the subject is much confirmed by the property possessed by the rays of many Compositæ, of closing at night, or in cloudy or rainy weather. What this sort of argument may be worth, I profess not to understand; but it seems to me that we may as well call a branch clothed with leaves a compound leaf, or a flock of sheep a compound sheep, as the cluster of flowers of an anthodium a compound flower.

All the forms of inflorescence which have been yet mentioned are to be considered as reductions of the spike or raceme. Those which are now to be described are decompositions, more or less irregular, of the raceme.

The first of these is the *panicle* and its varieties. The *simple panicle* differs from the raceme in bearing branches of flowers where the raceme bears single flowers, as in Poa (*fig.* 80.); but it often happens that the rachis itself separates into irregular branches, so that it ceases to exist as an axis, as in some Oncidiums. This is called by Willdenow a *deliquescent panicle.* When the panicle was very loose and diffuse, the older botanists named it a *juba;* but this is obsolete. If the lower branches of a panicle are shorter than those of the middle, and the panicle itself is very compact, as in Syringa, it then receives the name of *thyrsus.*

Suppose the branches of a deliquescent panicle to become short and corymbose, with a centrifugal expansion indicated by the presence of a solitary flower seated in the axillæ of the dichotomous ramifications, and a clear conception is formed of what is called a *cyme.* This kind of inflorescence is found in Sambucus, Viburnum, and other plants (*fig.* 83.).

If the cyme is reduced to a very few flowers, and those few become corymbose, such a disposition has been called a *verticillaster* by Hoffmansegg. (*Verzeichniss z. Pflanz. Cult.*, ii. 203.) It constitutes the normal form of inflorescence in Labiatæ, in which two verticillastri are situated opposite to each other in the axillæ of the opposite leaves. By Linnæus, the union of two such verticillastri was called a *verticillus* or whorl; and by others, with more accuracy, a *verticillus spurius* or false whorl. Link terms this inflorescence a *thyrsula;* but Hoffmansegg's name seems preferable.

It occasionally happens, as in the Vine, that the rachis of some of the masses of inflorescence lose their flowers, but at the same time acquire the property of twining round any body within their reach, and so of supporting the stem, which is too feeble to support itself. Such rachises form what is called a spurious cirrhus, or a *cirrhus peduncularis,* and are a striking exception to the general law that the cirrhus takes its rise from the petiole or costa.

6. *Of the* Calyx.

The *calyx* is the most exterior integument of the Flower, consisting of several verticillate leaves, either united by their margins or distinct, usually of a green colour, and of a ruder and less delicate texture than the corolla.

Authors have long disputed about the definition of a calyx, and the limits which really exist between it and the corolla: the above, which is copied from Link, seems to be the only one that can be considered accurate. The fact is, that in many cases they pass so insensibly into each other, as in Calycanthus and Nymphæa, that no one can say where the calyx ends and the corolla begins, although it is evident that both are present. Linnæus, indeed, thought that it was possible to distinguish them by their position with regard to the stamens, asserting that the divisions of the calyx are opposite those organs, of the corolla alternate with them; but, if this distinction were admitted, the corolla of the Primrose would be an inner calyx, which is manifestly an absurdity. Jussieu defines a calyx by its being continuous with the peduncle, which the corolla never is; and this may seem in some cases a good distinction; but there are plenty of true calyxes, of all Papaveraceous and Cruciferous plants, for instance, in which the calyx is deciduous, and not more continuous with the peduncle than the corolla itself. The only just mode of distinguishing the calyx seems to me to be to consider it in all cases the most exterior verticillate series of the integuments of the flower within the bracteæ, whether it be half-coloured, deciduous, and of many pieces, as in Cruciferæ; membranous and wholly-coloured, as in Mirabilis; green and campanulate, or tubular, as in Laurus and Lythrum. Upon this principle, whenever there is only one series of floral integuments, that series is the calyx. A calyx, therefore, can exist without a corolla; but a corolla cannot exist without a calyx.

In some elementary works the term Perianthium is given as synonymous with that of calyx; but this is an error.

The word Perianthium signifies the calyx and corolla combined, and is therefore strictly a collective term. It should only be employed to designate a calyx and corolla, the limits of which are undefined, so that they cannot be satisfactorily distinguished from each other, as in most Monocotyledonous plants, the Tulip and the Orchis for example. But since, even in such plants as these, there can be no reasonable doubt that the three outer floral leaves are the calyx, and the three inner the corolla (as is shewn both by Tradescantia and its allies, in which the usual limits between calyx and corolla exist, and by the usual origin of those parts in two distinct whorls), the utility of the term Perianthium is rendered extremely doubtful. It is, in reality, an evasion of the task of ascertaining the exact nature of the floral envelopes in doubtful cases. Some writers, among whom are Link and De Candolle, have substituted *Perigonium* for Perianthium; but the latter is in most common use, its application is perfectly well understood, and there is no good reason for its being changed. Ehrhart, with whom the name Perigonium originated, called it double when the calyx and corolla are evidently distinct, and single if they are not distinguishable; but this use of terms is obsolete.

The divisions of a calyx are called its *sepals* (*sepala*); a term first invented by Necker, and recently revived by M. De Candolle. Botanists of the school of Linnæus call them the leaflets or foliola. Link says the word sepalum is barbarous, and proposes to substitute *phyllum*. The sepals are generally longer than the corolla in æstivation, and during that period act as its protectors: during flowering they are mostly shorter.

The calyx, if deciduous, falls off from the peduncle by its base. In many cases the sepals drop off separately, as leaves fall from the stem; but occasionally they cohere firmly into a sort of cap or lid, which is pushed off entire by the increase of the corolla and stamens: in these cases the calyx is said to be *operculate*, if it falls off without any lateral rupture of its cap, as in Eucalyptus; and *calyptrate*, if at the period of falling it bursts on one side, as in Eschscholtzia. In the former of these two cases, the cohesion between the sepals is complete and never destroyed; in the latter, two of the sepals separate, the cohesion between the remainder continuing complete.

The calyx of Compositæ is so very different in appearance from the calyx of other plants, that it is known by the particular name of *pappus*. It usually consists of hair-like processes proceeding from the apex of the ovarium, in which case it is said to be *pilose:* if those hairs are themselves divided it is *plumose;* if they are very unusually stiff, it is *setose*, in which case the setæ are often reduced in number to two, or even one; if the divisions of the pappus are broad and membranous it is said to be *paleaceous:* finally, it is sometimes reduced to a mere rim; in which case it is usually said either to be marginate, or to be *none*, or to have no existence. If the pappus is in two rows, which it occasionally is, the inner circle only is to be understood as calyx: the exterior must then be accounted bracteæ or paleæ of the receptacle confluent with the ovarium.

In such cases as those above mentioned, when the calyx is altogether obsolete, the definition of that organ, as the most external of the floral envelopes, appears to be destroyed; but there can be no doubt that it is present in the form of a membrane adhering to the side of the ovarium, although it is not visible to our eyes. The same may be said of such plants as those Acanthaceæ (*Introduction to the Nat. Syst.*, p. 233.), in which, although the calyx is reduced to a mere ring, yet it does exist in the shape of that ring.

The Calyx being composed of leaves analogous to those of the stem, but reduced in size and altered in appearance, it will follow that it is subject to the same laws of developement as stem-leaves; and, as the latter, in all cases, originate immediately from the axis, *below* those that succeed them in the order of developement, so the calyx must always have an origin beneath those other organs which succeed it in the form of corolla, stamen, and pistillum or ovarium. Hence has arisen the axiom in botany, that whatever the apparent station of the calyx may be, it always derives its origin from below the ovarium: nevertheless, it is often said to be superior.

If it is distinct from the ovarium, as in Silene, it is said to be *inferior* (*calyx inferus*, or *liberus*); and the ovarium is then called *superior* (*ovarium superum*, or *liberum*) (Plate V. fig. 3.); but if it is firmly attached to the sides of the ovarium, so that it cannot be separated, as in Myriophyllum, it is then called *superior* (*calyx superus*), and the ovarium *inferior* (*ovarium inferum*) (Plate V. fig. 7. 9.). From what has been said of pappus it will be obvious that it is a *superior calyx*.

The general opinion of botanists, in regard to the real nature of the superior calyx, is such as I have stated; and the accuracy of it in the majority of cases is indisputable. But it is by no means certain that, in some instances, what is called the tube of the calyx is not, as I have elsewhere stated

(*Introduction to the Natural System*, p. 26.), " sometimes a peculiar extension or hollowing out of the apex of the pedicel, of which we see an example in Eschscholtzia, and of which Rosa and Calycanthus, and, perhaps, all supposed tubes without apparent veins, may also be instances." And if this be so, the superior calyx may be so in consequence of the cohesion of the ovarium with the inside of an excavated pedicel, and not with the calyx itself.

When the sepals cohere by their contiguous edges into a kind of tube or cup, the calyx is said to be *monophyllous ;* an inaccurate term, which originated in what may be called the dark age of botany, when the real nature of organs was unknown, and when a monophyllous calyx was thought to consist really of a single leaf, clipped into teeth at its margin. To avoid this inaccuracy, the word *gamosepalous* has been proposed. But as the real nature of a monophyllous calyx is now understood, changing the term is more embarrassing to the student than profitable to science.

Various terms are employed to express the degree in which the sepals of a monophyllous calyx cohere: they will be explained in Glossology. When no cohesion whatever takes place between the leaves of a calyx, the term *sepalous* is employed with that Greek numeral prefixed, which is equivalent to the number of pieces; as, for example, if they are two, the calyx is disepalous; if three, trisepalous; if four, tetrasepalous, and so on.

Sometimes the calyx has certain expansions or dilatations, as in Scutellaria and Salsola. These are generally named *appendages*, and such a calyx is said to be *appendiculate ;* but Mœnch has proposed a particular term for them, *peraphyllum*, which is, however, never used.

7. *Of the* Corolla.

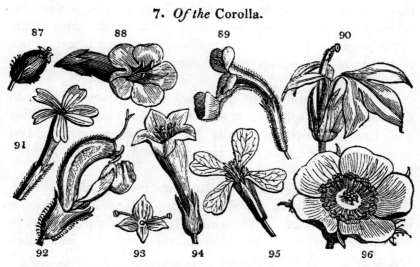

That envelope of the flower which forms a second whorl within the calyx, and between it and the stamens, is called the *corolla*. Its divisions always, without exception, alternate with those of the calyx, and are called *petals*. Like the sepals, they are either united by their margins, or distinct; but, unlike the calyx, they are rarely green, being for the most part either white, or of some colour, such as red, blue, or yellow, or of any of the hues produced by their intermixture. The corolla is generally also much larger than the calyx.

Necker called the corolla *perigynandra interior*, and Linnæus occasionally gave it the name of *Aulæum*, which literally signifies the drapery of a room.

The alternation of the segments of the corolla with those of the calyx is a necessary consequence of their both being modifications of verticilli of leaves, and therefore subject to the same laws of arrangement. If two verticilli of leaves are examined, those of Galium, for example, they will always be found to be mutually so arranged, that if the internodium that separates them were removed, they would exactly alternate with each other; and as there are no known exceptions to this law in real leaves, it is natural that it should not be departed from in any modifications of them.

When the petals of a corolla are all distinct, then the corolla is said to be *polypetalous;* but if they cohere at all by their contiguous margins, so as to form a tube, it then becomes what is called *monopetalous;* an inaccurate term of the same origin as that of monophyllous, in regard to calyx (*see* p. 116.), and for which that of *gamopetalous* has been sometimes substituted.

If the petals adhere to the bases of the stamens, so as to form a sort of spurious monopetalous corolla, as in Malva and Camellia, such a corolla has been occasionally called *catapetalous;* but this term is never used, all such corollas being considered polypetalous.

When the petals are confluent into a monopetalous corolla, the unguis form what is called a *tube;* the orifice of which is the *faux* or *throat.* The principal forms of such a corolla are rotate (*fig.* 93.), hypocrateriform (*fig.* 91.), infundibuliform (*fig.* 94.), campanulate (*fig.* 68.), and labiate (*fig.* 92.). When the divisions of a monopetalous corolla do not spread regularly round their centre, as in Campanula, but part take a direction upwards, and the remainder a direction downwards, as in Labiatæ, the upper form what is called the *upper lip,* and the lower, the *lower lip,* or *labellum;* the latter term is chiefly applied to the lower lip of Orchideous plants. If the upper lip is arched, as in Lamium album, it is termed the *galea* or *helmet.* When the two lips are separated from each other by a wide regular orifice, as in Lamium, the corolla is said to be *labiate* or *ringent;* if the upper and lower sides of the orifice are pressed together, as in Antirrhinum, it is *personate* or *masked,* resembling the face of some grinning animal. In the latter, the lower side of the orifice is elevated into two longitudinal ridges, divided by a depression corresponding to the sinus of the lip; this part of the orifice is called the *palate.* In ringent and personate corollas the orifice is sometimes named the *rictus;* but this term is superfluous and little used.

A petal consists of the following parts: — the *limbus* or *lamina* (lame, *Fr.*); and the *unguis* or *claw* (onglet, *Fr.*). The unguis is the narrow part at the base which takes the place of the foot-stalk of a leaf, of which it is a modification; the limbus is the dilated part supported upon the unguis, and is a modification of the lamina of a leaf. In many petals there is no unguis, as in Rosa; in many it is very long, as in

Dianthus. When the unguis is present, the petal is said to be *unguiculate*. In some unnaturally deformed flowers the limbus is absent, as in the garden variety of Rosa, called R. Œillet, in which the petals consist wholly of unguis.

According to the manner in which the petals of a polypetalous corolla are arranged, they have received different names, which are thus defined by Link:—the *rosaceous* corolla (*fig.* 96.) has no unguis, or it is very small; the *liliaceous* (*fig.* 71.) has its ungues gradually dilating into a lamina, and standing side by side; a *caryophyllaceous* has long, narrow, distant ungues; the *alsinaceous* has short distant ones; the *cruciate* flower has four valvaceous sepals, four petals, and six stamens, of which two are shorter than the rest, and placed singly in front of the lateral sepals, and four longer, and standing in pairs opposite the two other sepals. If the corolla is very irregular, with one petal very large and helmet-shaped, or hooded, as in Aconitum, it is sometimes called *cassideous*; if it resembles what is called labiate in gamopetalous corollas, it is termed *labiose*. The corolla of the Pea, and most Leguminous plants, has received the fanciful name of *papilionaceous* or *butterfly-shaped* (*fig.* 97, 98.); in this there are five petals, of which the upper is erect and more expanded than the rest, and is named the *vexillum* or *standard*, (étendard, *Fr.*); the two lateral are oblong, at right angles with the vexillum, and parallel with each other, and are called the *alæ* or *wings* (ailes, *Fr.*); and the two lower, shaped like the alæ and parallel with them, cohere by their lower margin, and form the *carina* or *keel* (carène or nacelle, *Fr.*). The alæ were formerly called *talaræ* by Link, and the carina *scaphium* by the same author.

When the corolla is very small, or when it forms a part of an anthodium, it is called *corollula*: that of a floret is so called.

If the flower has no corolla, it is said to be *apetalous*.

Sometimes a petal is lengthened at the base into a hollow

tube, as in Orchis, &c.: this is called the spur or *calcar*, and by some *nectarotheca*.

In Umbelliferæ the petal is abruptly acuminate; and the acumen is inflexed. The latter is named the *lacinula*.

A corolla is said to be *regular* when its segments form equal rays of a circle supposed to be described, with the axis of the flower for a centre. If they are unequal, the corolla is called *irregular*. *Equal* and *unequal* are occasionally substituted for regular and irregular.

In anatomical structure, the petal should agree with a leaf, of which it is a mere modification; and, in fact, it does so in all that it is important, its differences consisting chiefly in an attenuation and coloration of the tissue, with a suppression of woody fibre. Like a leaf, they consist of a flat plate of parenchyma, articulated with the stem, traversed by veins, and frequently having stomata upon its surface. Their veins consist almost entirely of delicate spiral vessels, upon which the parenchyma is immediately placed. It is therefore by mistake that the learned M. De Candolle has stated (*Organogr.* p. 454.) that stomata and spiral vessels are usually absent. The latter may be very readily seen in the corolla of Anagallis, where they form a beautiful microscopical object, as I first learned from Mr. Solly.

The petals are usually deciduous soon after flowering, or even at the instant of expansion; a very rare instance of their persistence and change from minute colourless bodies into leafy, richly coloured expansions, occurs in Dr. Wallich's Melanorrhæa usitatissima.

Their colours are due to the secretion within the cellules of their parenchyma of a peculiar substance: even white petals are so in consequence of the deposit of an opaque white substance, and not because of the absence of colouring matter.

In most corollas the petals, in their natural state, form but one verticillus within that of the calyx: but instances exist in which they naturally are found in several whorls, as in Nymphæa, Nuphar, Magnolia, &c. It sometimes happens that, if there is more than one row of petals, all within the first row assume a different appearance from the first; the filamentous processes of the crown of Passiflora are also apparently of this nature.

CHAP. II.   COMPOUND ORGANS IN FLOWERING PLANTS.   121

The petals are often furnished with little appendages (*fig.* 104.), which are either inner rows of petals in a state of adhesion to the first row, or modified stamens; which it is sometimes difficult to ascertain, but always certainly one of the two. Many of these enter into Linnæus's notion of *nectarium,* although nearly the whole of them are destitute of any power of secreting *nectar* or honey.

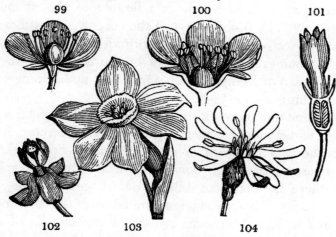

The most common form of appendage is the *corona,* which proceeds from the base of the limb, forming sometimes an undivided cup, as in Narcissus (*fig.* 103.), when it becomes the *scyphus* of Haller; sometimes dividing into several foliaceous erect scales, as in Silene and Brodiæa, when it forms the *lamella* of some writers; occasionally appearing as cylindrical or clavate processes, as in Schwenckia and Tricoryne, where they are manifestly modified stamens: and even in some instances forming a thick solid mass covering over the ovarium, and adhering to the stamens, as in Stapelia; when it is called the *orbiculus.* Parts of this last form of corona bear several names, which are found useful in avoiding repetition in describing the complicated structure of this kind of appendage. The whole mass of the corona is the *orbiculus,* or *saccus,* or *stylotegium;* certain horn-like processes are *cornua,* or horns; the upper end of these is the beak, or *rostrum,* and their back, if it is dilated and compressed, is the *ala,* or *appendix;* occasionally there is an additional set of horns proceeding from the base of the orbiculus, and alternate with the *horns,* these are *ligulæ;* the circular space in the middle of the top of the orbiculus is the *scutum*. Mr. Brown names the orbiculus *corona staminea*, and its divisions *foliola*, or leaflets.

In some plants, as Cynoglossum, the lamellæ are very small, scale-like, and overarch the orifice of the tube; such have received the name of *fornix*.

Link calls every appendage which is referable to the corolla a *paracorolla;* or, if consisting of several pieces, *parapetalum;* and every appendage which is referable to the stamens a *parastemon.* The filiform rays of the corona of Passiflora the same author calls *paraphyses* or *parastades.*

Mœnch names such appendages of the corolla as the filamentous beard of Menyanthes *perapetalum*, and Sprengel calls the same thing *nectarilyma*.

In Ranunculus there exists at the base of each petal a little shining, sometimes elevated, space which secretes honey. This is the true *nectarium* or *nectarostigma* of Sprengel. By some writers it has been considered a kind of reservoir, in which there is much plausibility; but it seems to me, from analogy, to be a barren stamen, united with the base of the petal, and to be of the same nature as the lamella of other plants.

8. *Of the* Stamens.

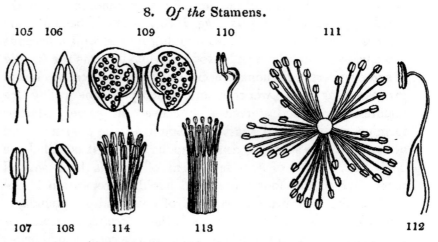

Next the petals, in the inside, are seated the organs called *Stamens* — the *Apices* of old botanists. These constitute the male apparatus of the flower, like the calyx and corolla are modifications of leaves, and consist of the *filament*, the *anthera*, and the *pollen*, of which the two latter are essential: the first is

CHAP. II.   COMPOUND ORGANS IN FLOWERING PLANTS.   123

not essential; that is to say, a stamen may exist without a filament, but it cannot exist without an anther and pollen. All bodies, therefore, which resemble stamens, or which occupy their place, but which are destitute of anthera, are either petals, or appendages of the petals, or abortive stamens.

As the petals are naturally alternate with the sepals, so the natural station of the stamens, if of equal number with the petals, is alternately with them; and all deviations from this law are to be understood as irregularities arising from the snppression or addition of parts. Thus, when in Primula we find the stamens opposite the segments of the corolla, and equal to them in number, it is to be supposed that those stamens which are present constitute the second of two rows of which the exterior is not developed; and when in Silene we find the stamens ten, while the petals are five, the former are to be considered to consist of two rows, although appearing to consist of one. This may be understood by examining Oxalis, in which the stamens are all apparently in one row, but are alternately of different lengths. When the number of stamens exceeds twice that of the petals, they will still be divisible by the number of which they were at first a multiple, until their number is excessively increased, when they seem to cease to bear any kind of proportion to the petals.

The stamens always originate from the space between the base of the petals and the base of the ovarium. But botanists are nevertheless in the habit of saying that they are inserted into the calyx or corolla (*fig.* 119.) (*perigynous*), or under the pistillum (*fig.* 117.) (*hypogynous*), or into the pistillum (*fig.* 118.)

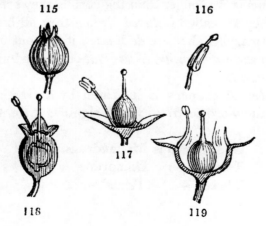

(*epigynous*), all expressions inaccurate and leading to erroneous notions of structure. The student, therefore, must understand, that when in Primula the stamens are said to be inserted into the faux of the corolla, it is meant that they cohere with the corolla as far as the faux, where they first separate from it; when in Rosa they are said to be inserted into the calyx, it is meant that they cohere with the calyx up to a certain point, where they separate from it; when in Arabis they are said to be inserted under the pistillum, it is meant that they cohere with neither calyx nor corolla, but stand erect from the point which immediately produces them; and finally, when in Orchis or Heracleum they are said to be inserted into the pistillum, such an expression is to be taken as meaning that they cohere with the pistillum more or less perfectly. For excellent arguments in support of this hypothesis, see *Considérations sur la Nature et les Rapports de quelques uns des Organes de la Fleur;* by Mons. DUNAL. Montpellier, 1829, 4to. I do not use them, or any such, here, because it seems to be so self evident a fact, when once pointed out, as to require no demonstration.

When their filaments are combined into a single mass, the mass is said to be a brotherhood or an *adelphia:* if there is one combination, as in Malva, they are *monadelphous* (*fig.*113.); if two, as in Fumaria or Pisum, *diadelphous;* if three, as in some Hypericums, *triadelphous;* if several, as in Melaleuca, *polyadelphous* (*fig.* 111.). The tube formed by the union of the filaments in a monadelphous combination is called, by Mirbel, *androphorum*.

If the stamens are longer than the corolla they are *exserted;* if shorter, they are called *included;* when they all bend to one side, as in Amaryllis, they are *declinate;* if two out of four are shorter, they are *didynamous;* if two out of six are longest, they are *tetradynamous*.

The *number* of stamens is indicated by a Greek numeral prefixed to the word *androus*, which signifies male, thus:—

        One stamen is Monandrous.
        Two   —   Diandrous.
        Three —   Triandrous.

CHAP. II. COMPOUND ORGANS IN FLOWERING PLANTS. 125

     Four stamens, Tetrandrous.
     Five — Pentandrous.
     Six — Hexandrous.
     Seven — Heptandrous.
     Eight — Octandrous.
     Nine — Enneandrous.
     Ten — Decandrous.
 Eleven or twelve stamens, Dodecandrous.
 Twelve to twenty — Icosandrous.
 Above twenty — Polyandrous, or Indefinite.

 The *filament* (Plate III.) (*capillamentum*, or *pediculus* of some) is the part that supports the anther. It consists of a bundle of delicate woody fibre and spiral vessels, surrounded by cellular tissue, and is in all respects the same as the petiole of a leaf, of which it is a modification, except that its parts are more delicate. As the petiole is unessential to the leaf, so is the filament to the anther, it being frequently absent, or at least so strictly united to the sides of the calyx or corolla as to be undistinguishable. Its most common figure is filiform or cylindrical (Plate III. fig. 12, 13. 20, 21.), and it is almost always destitute of colour; but there are exceptions to both these characters. In Fuchsia, for instance, the filaments are red like the petals; in Adamia they are blue; in Œnothera they are yellow; and a return to the foliaceous state of which they usually are a distinct modification is by no means rare. (Plate IV. fig. 6. 8.) Thus the filament in Canna is undistinguishable from petals except by its having an anther; in the same genus and its allies, and in all Scitamineæ, the inner series of what seem to be petals are modifications of filaments (See *Introduction to the Nat. Syst.* p. 265.): and this is a very common circumstance in sterile stamens.

 The filament also varies in other respects: in Thalictrum it is thickest at the upper end, or *clavate* (Plate III. fig. 23.); in Mahernia *geniculate* (Plate III. fig. 25.), in Hirtella *spiral,* in Crambe *bifurcate,* in Anthericum *bearded* or *stupose.* In some plants the filaments are combined into a solid body called the *columna,* as in Stapelia, Stylidium (Plate 4. fig. 1,

2, 3.), Rafflesia, and others: this has in Orchideæ received from M. Richard the name of *gynostemium*.

Care must be taken not to confound the pedicel and single stamen of the naked male flowers of Euphorbia with a filament, as was done by all writers, until Mr. Brown detected the error; and as is still done by some botanists from whom better things might be expected. For modifications of filaments see Plates III. and IV.

The Anther (*Theca* of Grew; *Capsula*, Malpighi; *Apex*, Ray; *Testiculus* or *Testis*, Vaillant; *Capitulum*, Jungius; *Spermatocystidium*, Hedwig) is a body generally attached to the apex of the filament, composed of two parallel lobes or cells (*thecæ*, or *coniothecæ*, or *loculi*), containing *pollen*, and united by the *connectivum*. It consists entirely of cellular tissue, with the exception sometimes of a bundle of very minute vascular tissue, which diverges on each side from the filament, and passes through that part of the anther from which the pollen has been incorrectly supposed to separate, and which is called the *receptacle of the pollen* by some, the *trophopollen* by Turpin, and the *raphe* by Link, but with greater propriety the *septum* of the anther. Its coat is called by Purkinje *exothecium*.

In the most common state of the anther the cells are parallel with each other (Plate III. fig. 14.), and open with two *valves* (Plate III. fig. 13. *a*), by a longitudinal fissure from the base to the apex; in Labiatæ and Scrophularineæ the cells diverge more or less at the base (Plate III. fig. 15. 18.), so as in some cases to assume the appearance of a one-celled horizontal anther, especially after they have burst. In Cucurbitaceæ the lobes are very long and narrow, sinuous and folded back upon themselves (Plate III. fig. 24.). In Salvia the connectivum divides into two unequal portions, one of which supports a cell and the other is cell-less; in this case the connectivum has been called by Richard, *distractile*. Lacistema (Plate IV. fig. 7.), affords another instance of a divided connectivum. In many of the cases of excessive divergence of the cells the line of dehiscence of the anther is changed from longitudinal to vertical (Plate III. fig. 20. 17.), and has actually been supposed to be really transverse; an error which in most cases has arisen from not understanding the real structure of the anther. Some anthers, however, no doubt have cells that burst transversely, as Lemna, Alchemilla arvensis, Securinega, &c. (See Plate III., fig. 12. 16. 30.)

All anthers are not two-celled, their internal structure being subject to several modifications. It sometimes happens that the septum, instead of being very obscurely formed, projects forward into the cavity of the anther, till it meets the inflexed lips of the fissure : in such a case the anther is spuriously four-celled, as in Tetratheca. In Epacris the two parallel cells become confluent into one, and the anther is therefore one-celled. In Maranta and Canna only one cell is produced, the other being entirely suppressed. In most Amarantaceæ, and some other plants, the anther seems to be absolutely one-celled. (Plate IV. fig. 8.)

Other deviations from the normal form of anther occur, which are less easy to reconcile with the idea of a two-celled type. In some Laurineæ the anther is divided into four cells, one placed above the other in pairs; in Ægiceras it consists of numerous little cavities; and in the singular genus Rafflesia the interior is separated into many cellules of irregular figure and position, described by Mr. Brown as " somewhat concentrical, longitudinal, the exterior ones becoming connivent towards the apex, sometimes confluent, and occasionally interrupted by transverse partitions." In these instances the septa may be understood to arise from portions of the cellular tissue of the anther remaining unconverted into pollen, and may be considered of a nature analogous to that of the fruit of Diplophractum.

With regard to the deviations from the usual mode of dehiscence, Mr. Brown observes (*Linn. Trans.* xiii. 214.), " that they are numerous: in some cases consisting either in the aperture being confined to a definite portion, — generally the upper extremity of the longitudinal furrow, — as in Dillenia and Solanum; in the apex of each theca being produced beyond the receptacle of the pollen into a tube opening at top, as in several Ericineæ (Plate III. fig. 22.); or in the two thecæ being confluent at the apex, and bursting by a common foramen or tube, as in Tetratheca (See Plate IV. fig. 4.). In other cases a separation of determinate portions of the membrane takes place, either the whole length of the theca, as in Hamamelideæ and Berberideæ, or corresponding with its subdivisions, as in several Laurineæ, or lastly, having no obvious relation to internal structure as in certain species of Rhizophora. In Laurineæ and Berberideæ the anthers are technically said to burst by valves (Plate IV. fig. 10, 11.), that is to say, the dehiscence does not take place by a central line, but the whole face of the cell separates from the anther, and curls backwards, adhering to it only at the apex to which it is, as it were, hinged.

The cells of the anther have frequently little appendages, as in different species of Erica, where they resemble setæ, aristæ, or crests. (Plate III. fig. 29.).

The anthers are attached to the filament either by their base, when they are called *innate* (Plate III. fig. 27. 21. 23.), or by their back, when they are *adnate* (Plate III. fig. 13.), or by a single point of the connectivum from which they lightly swing: in the latter case they are said to be *versatile*. This form is common to all true Grasses.

When the line of dehiscence is towards the pistillum, the anthers are called by Mr. Brown *anticæ*, but by other botanists *introrsæ*, or turned inwards: when the line is towards the petals they are said by Mr. Brown to be *posticæ*, and by other botanists to be *extrorsæ*, or turned outwards.

The *connectivum* is usually continuous with the filament, and terminates just at the apex of the anther; but in some plants, as Compositæ, it is articulated with the filament (Plate IV. fig. 5.) In others it is elongated far beyond the apex (Plate IV. fig. 6. 9.), now into a kind of crest, as in many Scitamineæ; now into a sort of horn, as in Asclepiadeæ; now into a kind of secreting cup-like body articulated with the apex, as in Adenostemon. Very frequently it is enlarged in various ways. For cases of this kind see Plates III. and IV. Its being sometimes two-lobed, or forked, has been already noticed (Plate IV. fig. 7.). The lining of the anther has received particular illustration from M. Purkinje, who calls it *endothecium,* and who has found that it consists of that very remarkable kind of tissue which has been already described under the name of fibrous cellular (p. 10.). According to that botanist the forms of this tissue are extremely variable, the cells being sometimes oblong, sometimes round, frequently cylindrical, usually fully developed, or, in some cases, merely rudimentary; the cellules are in some species erect, in others decumbent; but in all cases more or less fibrous. (See Plate I. figs. 4. 13, 14, 15. 18, 19, 20.) For an elaborate treatise on the subject see JOH. EV. PURKINJE *de Cellulis Antherarum Fibrosis*. Vratislaviæ, 1830. 4to. with 18 plates.

The *pollen* is the pulverulent substance which fills the cells of the anthers: it consists of a multitude of little grains, most commonly called *granules*, or sometimes *utriculi*.

The origin of the granules is still involved in some mystery. The best account of it has been given by M. Adolphe Brongniart, in the *Annales des Sciences*, vol. 12. Gleichen considered it to take its origin in the midst of a mucilaginous mass, occupying the cells of the anther, and merely becoming indurated and solidified towards maturity. Mr. Brown, in the year 1820, without entering into any details on the subject, described it (*Linn. Trans.* xiii. 211.) as produced on the surface or in the cells of a pulpy substance with which the thecæ are filled. But this hypothesis is objected to by Link (*Elem.* 294). M. Guillemin (*Recherches* p. 5.) declares that the granules are always arranged in regular rows, and generally in the direction of the valves, and that they are always distinct, at first floating in a viscid liquid, but finally quite separate from it. M. Brongniart concludes, from a series of very interesting observations, " that the pollen is formed in the interior of the cellules of a single and distinct cellular mass, which fills each cell of the anther without adhering to its walls, and consequently without being a continuation of the parenchyma of that organ, from which it also differs in the size and form of the cellules that compose it; that sometimes these cellules, which are at first in close cohesion, separate from each other, when each becomes a *grain of pollen*; and that sometimes the cellules contain an uncertain number of grains of pollen, which, at the time of their perfect developement, rupture and almost entirely destroy their membrane, some remains of which may sometimes be found among the grains of pollen.

In 1831, Mr. Brown speaks thus of the evolution of the pollen of Tradescantia virginica. " In the very early stage of the flower bud, while the antheræ are yet colourless, their loculi are filled with minute lenticular grains, having a transparent flat limb, with a slightly convex and minutely granular semi-opake disk. This disk is the nucleus of the cell, which probably loses its membrane or limb, and, gradually enlarging, forms in the next stage a grain also lenticular, and which is marked either with only one transparent line, dividing it into two equal parts, or with two lines crossing at right angles, and dividing it into four equal parts. In each of the quadrants a small nucleus is visible: and even where one transparent line only is distinguishable, two nuclei may often be found in each semicircular division. These nuclei may be readily extracted from the containing grain by pressure, and, after separation, retain their original form. In the next stage examined, the greater number of grains consisted of the semicircular divisions already noticed, which had naturally separated, and now contained only one nucleus, which had greatly increased in size. In the succeeding state the grain apparently consisted of the nucleus of the former stage, considerably enlarged, having a regular oval form, a somewhat granular surface, and originally a small nucleus. This oval grain continuing to increase in size, and in the thickness and opacity of its membrane, acquires a pale yellow colour, and is now the perfect grain of pollen." (*On Orchid. and Asclep.* p. 21.)

The granules of pollen are commonly distinct from each other. They are, nevertheless, in certain cases, found in various states of cohesion. In many plants they cohere in threes or fours, as in many Orchideæ; or in clusters of many grains, as in Acacia (Plate IV. fig. 28.). In some, as the Fuchsia, Œnothera, &c., they hang together by a sort of cobweb substance, which is the remains of the cellular matter in which they were engendered. In other cases they coalesce in masses, having a waxy texture and colour, and occupying the whole cavity of a cell of the anther, as in Asclepiadeæ. But the most curious instances of the cohesion of the grains of pollen is to be found in Orchideæ; in which some genera have the pollen in its common pulverulent state, with no remains of the cellular substance in which it was developed; others have the granules held together by some of this cellular substance in an elastic state, and forming a distinct appendage to the pollen called the *caudicula;* while others have the grains united either by threes or fours, or in wedge-shaped masses, or in a hard, dry, solid body. It appears from Mr. Francis Bauer's observations, that the masses of pollen of both Asclepiadeæ and Orchideæ, in the most solid state, are really cellular, the grains of pollen being contained in cavities, the walls of which are either separable from each other as in some Orchideæ, or are ruptured without a separation of the cavities as in Asclepiadeæ. (See the Observations on Orchideæ and Asclepiadeæ before referred to.)

The granules are generally discharged at once, upon the dehiscence of the anther, or at least are at that time wholly formed. But in some Aroideæ, which emit their pollen by a hole in the apex of their anther, the formation or developement of pollen must be going on for a considerable time after the first emission. A single anther continues to secrete and discharge pollen, till, as Mr. Brown remarks, the whole quantity produced greatly exceeds the size of the secreting organ.

The *surface* of the pollen is commonly smooth. In some plants it is hispid, as in the Gourd and Ipomæa purpurea; in others it is covered with strong points, as Hibiscus syriacus; and in all cases, when there are asperities of the surface or angles in its outline, it is asserted by M. Guillemin to have a mucous surface, which was first observed in Proteaceæ by Mr. Brown.

The *figure* of the granules is very various; most frequently it is spherical or slightly oblong. Many other forms have, however, been described. The cylindrical exists in Anethum segetum, and in a very remarkable degree in Tradescantia virginica, where the grains become curved. In Colutea arborescens, it was observed by M. Guillemin to be nearly square; in Lavatera acerifolia to be oval, much attenuated to each end. In Œnothera it is triangular, with the angles so much dilated as to give the sides a curved form. In Jacaranda tomentosa I have remarked it to be spherical, with three projecting ribs tapering to either apex. In the Chicoraceæ of Jussieu the granules are spherical with facettes; in Dipsaceæ they are a depressed polyedron; in Scabiosa caucasica patelliform and angular. (For other Modifications see Plate IV. fig. 12. to 37.)

In most spherical or elliptical pollen, with a smooth surface, a line is observable along the axis of the granules, when they are dry, which disappears upon the application of moisture. This was long ago remarked by Malpighi, who compared granules of pollen of this kind to grains of wheat, one side of which is convex and the other furrowed. M. Guillemin is of opinion that this supposed furrow exists on both sides of a grain of pollen, because, let there be never so many of this description examined at the same instant, the appearance will be visible in all. But it is probable that the strong transmitted light which is used in microscopical examinations of minute objects would render the furrow visible on both sides of a grain although it really existed only in one.

As to the nature of this supposed furrow, nothing positive is known. M. Guillemin supposes it to be a slit intended to facilitate the admission of water into the interior of the granules and the emission of their fovilla, and he further compares it to the line of dehiscence of each lobe of the Anthera. In Passiflora a curious contrivance exists for the emission of the contents of the pollen. Each spherical grain has on its surface three equidistant circles, which indicate the lines of dehiscence: at the proper time those parts of the coat of the grain contained within the circles separate from the rest like little lids, and allow the contents of the pollen to escape. This economy is well represented by Purkinje.

Many botanists are of opinion that the coat of the pollen is a simple cellular substance; others think it a solid membrane; and a third class of writers insist upon its consisting of two integuments, the outer of which is cellular, the inner membranous and extensible: the last of these opinions is entertained by M. Adolphe Brongniart and Amici. Mr. Brown says that the existence of an inner membrane is manifest in several Coniferæ, in which the outer coat regularly bursts and is deciduous; and further, he considers that the structure in Asclepiadeæ, as discovered by Mr. Francis Bauer, furnishes the strongest argument in support of the opinion of the existence of two membranes. In parts of such extreme minuteness and delicacy of structure, a point of this kind cannot be determined by sections; for the sharpest knives in the most skilful hands will only crush the grain of pollen into a shapeless mass. The evidence of the existence of an internal membrane is derived from the appearance of a thin transparent coating round the fovilla when it is emitted upon the stigma, and which is sometimes extended to a considerable length. Its existence having been called in question, M. Adolphe Brongniart was induced, in his examination of the anther, to pay particular attention to the circumstance; and he declares that, " in all the pollen that he has examined with care, after it had been a greater or less space of time upon the stigma, he has found a tubular appendage, of variable length, formed of an extremely thin and transparent membrane, which evidently proceeded from the interior of the grain of pollen, either through an accidental opening, or through a special passage formed in the external membrane." (See Plate IV. figs. 34. to 38.) He calls this appendage the *boyau* or intestine. Notwithstanding the precise manner in which this is stated, it has nevertheless been doubted by some whether the boyau or pollen-tube is any thing more than mucus surrounding the fovilla when emitted. Mr. Brown, in 1828, declared his difference in opinion from M. Brongniart as to the existence of a membrane forming the coat of the pollen-tube; but, in 1831, he states, in another place, that several arguments may be adduced in favour of M. Brongniart's opinion that the pollen-tubes belong to the inner membrane of the grain; and he particularly cites the structure in Asclepiadeæ as favourable to the opinion. For my own part I can only state that I have not succeeded in discovering two membranes in any really *simple* pollen that I have examined; and that in some, particularly in Gesneria bulbosa (Plate IV. fig. 32.) there is incontestibly but one membrane, the extension of which forms the pollen-tube. That the pollen-tube itself has not been found by some observers, has probably arisen from its having been looked for in pollen made to burst on the field of the microscope, immersed in water, when the pollen-tube is scarcely ever emitted. The vital action that causes the emission seems to depend upon contact with the secretion of the stigmatic surface. The pollen-tube is, therefore, only to be sought in pollen that has been some time upon the stigma.

The *colour* of pollen is chiefly yellow. In Epilobium angustifolium it is blue, in Verbascum it is red, and it occasionally assumes almost every other colour, except green, According to Messrs. Fourcroy and Vauquelin, the pollen of the Date tree consists of malic acid, phosphate of magnesia, and lime, and also an insoluble animal matter intermediate between gluten and albumen. M. Macaire Prinsep has ascertained that the pollen of the Cedar contains acid, malate of potass, sulphate of potass, phosphate of lime, silica, sugar, gum, yellow resin, and a substance which by its characters approximated to starch. Being analysed as a whole, it gave, per cent., 40 carbon, 11.7 hydrogen, and 48.3 oxygen, but no azote. — *Bibl. Univers.* 1830. 45.

The matter contained in the granules is called the *fovilla*. Under common magnifiers it appears like a turbid fluid; under glasses of greater power it has been found to consist of a multitude of particles moving on their axes with activity, of such excessive minuteness as to be invisible, unless viewed with a magnifying power equal to 300 diameters, and measuring from the 4000th or 5000th to the 20,000th or 30,000th of an inch in length. This motion was first distinctly noticed by Gleichen; but it seems to have escaped the recollection of succeeding botanists until the fact was confirmed by Amici, who some time before 1824 saw and described a distinct, active, molecular motion in the pollen of Portulaca oleracea. In 1825 the existence of this motion was confirmed by M. Guillemin, who ascertained its presence in other species. In June 1827 I was shown the motion by Mr. Brown, who subsequently published some valuable observations upon the subject, without however noticing those of either Amici or Guillemin. The most important addition that was made by

Mr. Brown to the knowledge that previously existed, consisted in the discovery of the presence of two kinds of active particles in pollen, of which one is spheroidal, extremely minute, and not distinguishable from the moving ultimate organic particles common to all parts of a vegetable, the other much larger, often oblong, and unlike any other kind of particle hitherto detected in plants.

The supposed functions of these particles will be explained hereafter. For the present it will be sufficient to remark, that some of the best subjects in which to witness their motions are Clarkia pulchella, Mirabilis jalapa, and Lolium perenne.

The stamen deviates in a greater degree than any other organ from the structure of the leaf, from a modification of which it is produced; and, at first sight, in many cases, it appears impossible to discover any analogy between the type and its modification; as, for instance, between the stamen and leaf of a Rose. Nevertheless, if we watch the transitions that take place between the several organs in certain species, what was before mysterious or even inscrutable, becomes clear and intelligible. In Nymphæa alba the petals so gradually change into stamens, that the process may be distinctly seen to depend upon a contraction of the lower half of a petal into the filament, and by a development of yellow matter within the substance of the upper end of the same petal on each side into pollen. A similar kind of passage from petals to stamens may be found in Calycanthus, Illicium, and many other plants. Now, as no one can doubt that a petal is a modified leaf, it will necessarily follow, from what has been stated, that a stamen is one also. But it is not from parts in their normal state that the best ideas of the real nature of the stamen may be formed; it is rather by parts in a monstrous state, when reverting to the form of that organ from which they were transformed, that we can most correctly judge of the exact nature of the modification. Take for example that well-known double Rose, called by the French R. Œillet. In that very remarkable variety, the unguis of the petals may at all times be found in every degree of gradation from its common state to that of a filament, and the lamina sometimes almost of its usual degree of developement, — sometimes contracting into a lobe of the anther on one side, or perhaps on both sides, — now having the part assuming the character of the anther merely yellow, — now polliniferous, — and finally acquiring, in many instances, all the characters of an undoubted though somewhat distorted stamen. Double Pæonies, Double Tulips, and many other monstrous flowers, particularly of an icosandrous or polyandrous structure, afford equally instructive specimens. It is for these reasons that it is stated in the *Outlines of the first Principles of Botany*, 307., that " the anther is a modification of the lamina, and the filament of the petiole."

Such is the structure of the stamens in their perfect state. It often, however, happens that, owing to causes with which we are unacquainted, some of the stamens are developed imperfectly, without the anther and pollen. In such cases they are called *sterile stamens*, and are frequently only to be recognised by the position they bear with respect to the other parts of the flower. Botanists consider every appendage, or process, or organ, that forms part of the same verticillus of organs as the true stamens, or that originates between them and the pistillum, as stamens, or as belonging to what Röper calls the *androcæum*, namely, to the *male system;* and every thing on the outside of the fertile stamens is in like manner usually referred to modifications of petals; a remarkable instance of which is exhibited by Passiflora. The appearances assumed by these sterile stamens are often exceedingly curious, and generally extremely unlike those of the fertile stamens: thus in Canna they are exactly like the petals; in Hamamelis they are oblong fleshy bodies, alternating with the fertile stamens; in Pentapetes they are filiform, and placed between every three fertile ones; in Scitamineæ they are minute gland-like corpuscles, a very common form (Plate IV. fig. 10. *c*); in Brodiæa they are bifid petaloid scales; and in Asclepiadeæ they undergo yet more remarkable transformations. M. Dunal calls these sterile stamens *lepals* (*lepala*); a term which has not yet been adopted.

9. *Of the* Disk.

By this term are meant certain bodies or projections, situated between the base of the stamens and the base of the ovarium, but forming part with neither; they are referred by the school of Linnæus, along with other things, to nectarium: Link calls them *sarcoma* and *perigynium;* and Turpin, *phycostemones.* The most common form is that of a fleshy ring, either entire or variously lobed, surrounding the base of the ovarium (Plate V. fig. 4. *e*, 8. *d*), as in Lamium, Cobæa, Gratiola, Orobanche, &c.; in Gesnerieæ and Proteaceæ the disk consists of fleshy bodies of a conical figure, which are usually called *glandulæ hypogynæ.* It occasionally assumes the appearance of a cup, named by M. De Candolle in Pæonias and Aconites *lepisma*, a bad term, for which it is better to say *discus cyathiformis*. In flowers with an ovarium inferum (Plate 5. fig. 9. *c*. 7. *c*) the discus necessarily ceases to be hypogynous, and generally also to appear in the form of scales. In Compositæ it is a fleshy solid body, interposed between the top of the ovarium and the base of the style; and has given rise, when much enlarged, to the unfounded belief in the existence of an ovarium superum in that order, as in Tarchonanthus. In Umbelliferæ it is dilated and covers the whole summit of the ovarium, adhering firmly to the base of the styles; by Hoffman it is then called *stylopodium*, a word which is seldom used.

Besides these forms of discus there is still another, called the *gynobasis*, a term which is used when the discus is enlarged, and as it were inserted under the ovarium to which it forms a sort of receptacle (Plate V. fig. 3. *a*). It occurs in Labiatæ, and Boragineæ, and especially in Ochnaceæ; and in many other orders. It ought, perhaps, to be considered a combination of the discus and receptacle.

It is an opinion that daily gains ground, that the discus is really only a rudimentary state of the stamens; and it is thought that proofs of the correctness of this hypothesis are to be found in the frequent separation of the cyathiform disk into bodies alternating with the true stamens, as in Gesneria; in its resemblance in Parnassia to bundles of polyadelphous stamens, and particularly in the fact noticed by Mr. Brown, that an anther is occasionally produced upon the highly developed disk of Pæonia Moutan. To which may be added the observation of Dunal, that half the disk of Cistus vaginatus occasionally turns into stamens. (*Considerations*, &c. p. 44.)

Like the petals, sepals, and stamens, the disk always originates from below the pistillum; but it often contracts an adhesion with the sides of the calyx, when it becomes *perigynous*, as in Amygdalus; or with both the calyx and the sides of an inferior ovarium, when it becomes *epigynous*, as in umbelliferous plants.

10. *Of the* Pistillum.

The last organ to enumerate in the flower is that which constitutes the *female system*, or *gynæceum* of Röper, and which is usually called the *pistillum*. In all cases it occupies the centre of the flower, terminating the axis of growth of the peduncle; and is consequently the part around which every other organ without exception is arranged.

It is distinguished into three parts; viz. the ovarium (Plate V. fig. 7. *a*), the *style* (fig. 7. *f*), and the *stigma* (fig. 7. *g*).

The ovarium, called germen by Linnæus, is a hollow case placed at the base of the pistillum, enclosing the *ovula*, and often containing two or more *cells* or cavities. It is the part which ultimately becomes the fruit; and consequently, whatever may be the structure of the ovarium such must necessarily be that of the fruit; allowance being made, as will hereafter be explained, for changes that may occur during the progress of the ovarium to maturity.

Notwithstanding what has been stated of the pistillum constantly occupying the centre of the flower, and being the part *around* which all the other parts are arranged, an apparent exception exists in those flowers, the calyx of which is said to be superior (Plate V. fig. 7. & 9.), as the Apple blossom. In this instance the ovarium seems to originate *below* the calyx, corolla, and male system; on which account it is said to be *inferior* in such cases, while in the opposite state it is called *superior*. But in reality, the inferior ovarium is only so in consequence of the tube of the calyx *contracting an adhesion with its sides;* and such being the case, the exactness of the description of the constant place of the pistillum as above, is unshaken. This is proved in many ways. In Saxifrageæ, the genus Leiogyne has the ovarium superior; in Saxifraga itself the calyx partially adheres to the sides of the ovarium, which then becomes half inferior, while in Chrysosplenium the union between the calyx and ovarium is complete, and the lattter is wholly inferior. Again, in Pomaceæ, the ovaria partially cohere with the calyx in Photinia, completely in Pyrus, and by their backs only in Cotoneaster; whence the ovarium is half superior in the first instance, quite inferior in the second, and what is called *parietal* in the third. Botanists call any thing parietal which arises from the inner lining or wall of an organ; thus in Cotoneaster the ovaria are parietal, because they adhere to the inner lining of the calyx, and in Papaver the placentæ are parietal, because they originate in the inner lining of the fruit.

Sometimes the ovarium, instead of being sessile, as is usually the case, is seated upon a long stalk; as in the Passion flower and the genus Cleome. This stalk is often called the *thecaphore* or *gynophore;* but it is obviously analogous to the petiole of a leaf, as will hereafter be seen, and the application of a special term to it appears unnecessary. M. Cassini calls the elongated apex of the ovarium of some Compositæ *le plateau*.

That part of the ovarium from which the ovula arise is called the *placenta* (*Trophospermium*, Richard; *Spermaphorum, Colum, Receptacle of the Seeds*). It generally occupies the whole or a portion of one angle of each cell (Plate V. fig. 1. *e*, 2. *c*, &c.), and will be spoken of more particularly hereafter. It is sometimes elongated in the form of a little cord, as in the Hazel nut, and many Cruciferæ: such a part is called the *umbilical cord* (*funiculus umbilicalis, podospermium*).

The swelling of the ovarium after fertilisation is termed *grossification*.

The *style* (*tuba* of old authors) is that elongation of the ovarium which supports the stigma (Plate V. fig. 7.*f*). It is frequently absent, and then the stigma is sessile : it is not more essential to a pistillum than the petiole to a leaf, or the unguis to a petal, or the filament to a stamen. Anatomically considered, it consists of a column of one or more bundles of vascular tissue, surrounded by cellular tissue; the former communicating on the one hand with the stigma, and on the other with the vascular tissue of the ovarium. It is usually taper, often filiform, sometimes very thick, and occasionally angular: rarely thin, flat, and coloured, as in Iris and in Canna. In some plants it is continuous with the ovarium, the one passing insensibly into the other, as in Digitalis; in others it is articulated with the ovarium, and falls off, by a clean scar, immediately after fertilisation has been accomplished, as in the Scirpus. Its usual point of origin is from the apex of the ovarium; nevertheless, cases occur, in which it proceeds from the side, as in Alchemilla, or even from the base, as in Labiatæ and Boragineæ. In these cases, however, it is to be understood that the geometrical and organic apices are different, the latter being determined by the origin of the style. For this reason, when the style is said to proceed from the side or base of the ovarium, it would be more correct to say that the ovarium is obliquely inflated or dilated, or gibbous at the base of the style.

The surface of the style is commonly smooth; but in Compositæ, Campanulaceæ, and others, it is often densely covered with hairs, called *collectors*, which seem intended as brushes to clear the pollen out of the cells of the anthers. In Lobelia these hairs are collected in a whorl below the stigma; in Goodenoviæ they are united into a cup, in which the stigma is enclosed, and which is called the *indusium* (Plate V. fig. 13. *b.*). Many styles which appear to be perfectly simple, as for instance those of the Primrose, the Lamium, the Lily, or the Borage, are in reality composed of several grown together; as is indicated by the lobes of their stigma, or by the number of cells or divisions of their ovarium. In Malva an example may be seen of a partial union only of the styles, which are distinct upwards, but united below. In speaking of styles in this latter state, botanists are apt to describe them as *divided* in different ways, which is manifestly an inaccurate mode of expression.

The *stigma* is the upper extremity of the style, without a cuticle; in consequence of which it has almost uniformly either a humid or papillose surface. In the first case it is so in consequence of the fluids of the style being allowed to flow up through the intercellular passages of the tissue, there being no cuticle to repress and conceal them; in the latter case the papillæ are really the rounded sides of vesicles of cellular tissue. When perfectly simple, it is usually notched on one side, the notch corresponding with the side from which the placenta arises: see the stigma of Rosa, Prunus, Pyrus, and others. If it belongs to a single carpellum (p. 143.), it is either undivided, or its divisions, if any, are all placed side by side, as in some Euphorbiaceæ, Crocus, &c.; but if it is formed by the union of the stigmas of several carpella, its lobes are either opposite each other, as in Mimulus, or placed in a verticillus, as in Geranium. Such being the case, it is always to be understood that an apparently simple ovarium, to which two or more opposite stigmata belong, is really of a compound nature, some of its parts being abortive, as in Compositæ.

Nothing is, properly speaking, stigma, except the secreting surface of the style: it very often, however, happens, that the term is carelessly applied to certain portions of the style. For example, in the genus Iris the three petaloid lobed styles in the centre are called stigmata; while the stigma is in reality confined to a narrow humid space at the back of each style; in Labiatæ, my friend Mr. Bentham has shown that what is called a two-lobed stigma is a two-lobed style, the points only of the lobes of which are stigmatic: and in Lathyrus, and many other Papilionaceous plants, Linnæan botanists call the hairy back of the style the stigma; while, in fact, the latter is confined to the mere point of the style.

Nevertheless, there are certain stigmata in which no denuded or secreting surface can be detected. Of this nature is that of Tupistra, in which the apparent stigma is a fungous mass with a surface of the same nature as that of the style;

in such a stigma the mode of fertilisation forms a very interesting problem, which botanists have yet to solve.

In almost all cases the stigmata are perfectly distinct from all surrounding bodies, and are freely exposed to the influence of the pollen; but in Asclepiadeæ they adhere to the anthers in a solid mass, of which the angles only that are in contact with the cells of the anther are free and susceptible of fertilisation.

The centre of a stigma consists of tissue of a peculiar character, which communicates directly with the placenta, and which is called the *stigmatic tissue*. It is more lax than that which surrounds it, and serves for the conveyance of the fertilising matter of the pollen into the ovula.

Such is a general view of the more remarkable peculiarities of the female system. This part, however, bears so important an office in the functions of vegetation, is so valuable as a means of scientific arrangement, and is liable to such a great variety of modifications, that it will be necessary now to consider it in another and more philosophical point of view. For we have yet to consider the structure of the compound pistillum, and to learn to understand the exact nature of its cells, and dissepiments, and placentæ, and the precise relation that these parts bear to each other; and also to prove that the necessary consequence of the laws under which pistilla are constructed is, that they can be subject to only a particular course of modification, within which every form must absolutely, and without exception, fall. This enquiry would, perhaps, be less important if none but structure of a very regular and uniform kind were to exist; but, considering the numberless anomalies that the pistillum exhibits, it becomes at once one of the most difficult and most essential parts of a student's investigation.

In the days of Linnæus and Gærtner, and even in those of the celebrated L. C. Richard, nothing whatever was known of this matter, and consequently the writings of those carpologists are a mere tissue of ingenious misconceptions. Nor did the subject become at all intelligible until the admirable Treatise upon Vegetable Metamorphosis, which had been published by Goethe in 1790, but which had long been neglected, was again brought into notice, and illustrated by the skilful demonstrations of De Candolle, Turpin, Du Petit Thouars, and others.

120   121

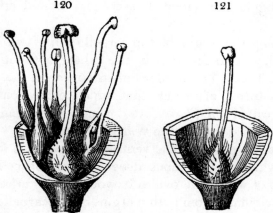

According to these writers, the pistillum is either the modification of a single leaf, or of one or more whorls of such leaves, which are technically called *carpella*. Each carpellum has its own ovarium, style, and stigma, and is formed by a folded leaf, the upper surface of which is turned inwards, the lower outwards, and the two margins of which develope one or a greater number of buds, which are in a rudimentary state, and are called the ovules.

A very distinct idea of the manner in which this occurs may be obtained from the carpellum of a double cherry, in which the pistillum loses its normal carpellary character, and reverts to the structure of the leaf. In this plant the pistillum is a little contracted leaf the sides of which are pressed face to face, the midrib elongated, and its apex discoloured, or a little distended. If we compare this with the pistillum of a single cherry, the margins of the leaf with the ventral suture, the elongated midrib with the style, the discoloured distended apex with the stigma, they will be found to correspond exactly.

In this case there is an indisputable identity of origin and nature between the ovarium and the lamina of a leaf, — between the little suture that occupies one angle of the carpellum of a cherry, and the line of union of the two edges of the leaf, — and between the elongated midrib, with its distended apex, and the style and stigma. There can be no doubt that the plan of all carpella is the same; so that the ovarium is the lamina of a leaf, the style an elongated midrib, and the stigma the denuded, secreting, humid apex of the latter.

Such being the origin of the carpellum, its two edges will correspond, one to the midrib, the other to the united margins of the leaf. These edges often appear in the carpellum like two sutures, of which that which corresponds to the midrib is called the *dorsal*, that which corresponds to the united margins is named the *ventral suture*.

It is at some point of the ventral suture that is formed the *placenta*, which is a copious developement of cellular substance, out of which the ovules or young seeds arise. It, the placenta, originates from both margins of the carpellary leaf,— but as they are generally in a state of cohesion, there appears to be but one placenta,—nevertheless if, as sometimes happens, the margins of the carpellary leaf do not unite, there will be two obvious placentæ to each carpellum. Now, as the stigma is the termination of the dorsal suture, it occupies the same position as that suture with regard to the two placentæ; consequently the normal position of the two placentæ of a single carpellum will, if they are separate, be right and left of the stigma. This is a fact very important to bear in mind.

Pistilla consisting of but one carpellum are simple; of several, are compound. If the carpella of a compound pistillum are distinct entirely or in part, they are *apocarpous*, as in Caltha; if they are completely united into an undivided body, as in Pyrus, they are *syncarpous*. That syncarpous pistilla are really made up of a number of united carpella is easily shown, as Goethe has well remarked, in the genus Nigella, in which N. orientalis has the carpella partially united, while N. damascena has them completely so. In the latter case, however, the styles are distinct; they and the stigmata are all consolidated in a single body, when the pistillum acquires its most complete state of complication, as in the Tulip; which is, however, if carefully examined, nothing but an obvious modification of such a pistillum as that of Nigella damascena.

There is this important conclusion that is deducible from the foregoing considerations: viz., that, as the carpella are modified leaves, they are necessarily subject to the same laws of arrangement, *and to no others*, as leaves developed round a common axis upon one or several planes. For no axiom appears more incontestible in botany, than that all modifications of a given organ are controlled essentially in the same way, and by the same influences, as the organ itself in an unmodified state: and hence every theory of the structure of fruit which is not reducible to that which would be applicable to the structure of whorls of leaves is vicious of necessity. I shall proceed to demonstrate the perfect accordance of the carpellary theory of structure in every point with these principles.

The placenta arises from the two margins, either distinct or more usually combined, of a leaf folded inwards. When a leaf is folded inwards, its margins will point towards the stem or axis around which it is developed; and in a whorl of leaves such inflected margins would all be collected round a common centre; or, if the axis were imaginary, in consequence of the whorl being terminal, would be placed next each other, in a circle of which the back of the leaves would represent the circumference. Therefore the placentæ will always be turned towards the axis, or will actually meet there, forming a common centre; and, which is a very important consequence of this law, if one carpellum only, with its single placenta, be formed in a flower, the true centre of that flower will be indicated by the side of the carpellum occupied by the placenta. Proofs of this may be found in every blossom: but particularly in such as habitually having but one carpellum occasionally form two, as the Wisteria sinensis, Alchemilla arvensis, Cerasus acida, &c.; in these the second carpellum, when added, does not arise by the side of the first, but opposite to it, the face of its placenta being in front of that of the habitual carpellum. A fourth proof of this uniform direction of the placentæ towards the axis, is afforded by those pistilla in which a great number of carpella is developed in several rows, as in the Strawberry and the Ranunculus: in all these the placentæ will be, without exception, found directed towards the axis, and conse-

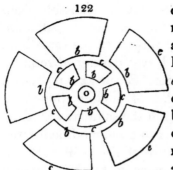

quently towards the back of every row, except the inner. For example, in the following diagram let O be the axis, *b b* placentæ, *c c* the backs of carpella; the placentæ, *b b*, of the inner row will be next the centre O; the placentæ, *b b*, of the second row will be next the backs, *c c*, of the first row; and so on.

If the order of developement of leaves were exactly followed in that of the stamens and carpella, it would happen that the latter would be invariably alternate with the inner row of stamens; for if *a a* (*fig.* 123.) is the station of five stamens, *b b* would be the situations of the carpella: this relative position is therefore considered the normal one, and is in fact that which usually exists in perfectly regular flowers; but as all the parts of a flower are subject to deviations, either real or apparent, from what is considered their normal state, in consequence of the non-developement of some parts, or the excessive developement of others, it frequently happens that the carpella either bear no apparent relation to the stamens or are opposite them. In Papilionaceous plants, for example, where only one carpellum is present, it is difficult to say that it bears any exact relation to the stamens, although it is probable that its position is really normal with regard to them; and so also in Rosaceous plants, with numerous carpella, no exact relation can be proved to exist between the latter and the stamens, unless it may be said to be indicated by those genera, such as Spiræa, in which the carpella are reduced to five; and, finally, in such plants as Delphinium, in which the carpella are three, while the floral envelopes and male system are divided upon a quinary plan, it is manifest that no alternation can exist between the stamens and carpella.

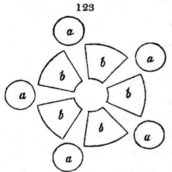

As the vessels and petals most commonly consist each of a

CHAP. II. COMPOUND ORGANS IN FLOWERING PLANTS. 147

single whorl of parts, so the pistillum is more frequently composed of one whorl of carpella than of more. There are, however, certain families in which several whorls are produced one within the other, as in Fragaria, Ranunculus, Magnolia, Annona, and the like. In these cases it mostly happens that the carpella are either entirely separate or nearly so; but it sometimes happens that syncarpous pistilla are habitually produced with more than one whorl of carpella, and consequently of cells, as Nicotiana multivalvis, and some varieties of the genus Citrus. In such instances the placentæ of the outer series will necessarily be applied to the backs of the inner series, as has been just demonstrated.

This mutual relation of the different rows of carpella is sometimes observed when the receptacle from which they arise is either convex or concave: in the former state the outer series will obviously be lowermost, and in the latter uppermost; a circumstance that leads to no intricacy of structure when the carpella are distinct, but which may cause an exceedingly anomalous structure in syncarpous pistilla, especially when accompanied by other unusual modifications of structure. There can be no doubt that the true nature of the composition of the pomegranate is to be explained upon this principle. In order to make these considerations more clear, let *figs*. 124, 125, and 126. represent — *fig*. 124. a convex receptacle, with distinct carpella; *fig*. 125. a concave one, with the same; and *fig*. 126. a concave one, with the carpella con-

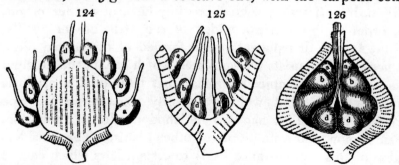

solidated. In these, *a a* are the outer row of carpella, *b b* the next, and *d d* the central row. The relative position of these, as the receptacle is convex or concave, will now be apparent.

I have stated that the placenta, however simple it may appear to be, is really the result of the union of two united margins of a carpellary leaf: it is, therefore, essentially double; and, accordingly, we find that in polyspermous ovaria the ovula are almost always arranged in two rows, as in the Pea and Bean, the Quince, the Pæony, &c.; nevertheless there are instances in which the placentæ occupy a considerable portion of the wall of the ovarium, and bear the ovula in a great many rows, but in no certain order, as in Nymphæa; and, on the other hand, some plants have the placentæ so little developed, that not more than one ovulum is generated between the two placentæ, as in Boragineæ, Labiatæ, Umbelliferæ, Stellatæ, Compositæ, and many others. There can be no doubt, however, that all the latter cases are mere instances of suppressed structure, in consequence of the general incompleteness of developement.

When two leaves are developed upon a stem, they are always opposite, and never side by side. As carpella are modified leaves they necessarily obey this law; and, consequently, when a pair of carpella form a bilocular ovarium, the separation of the two cells is directly across the axis of the flower.

The partitions that are formed in ovaria, by the united sides of cohering carpella, and which separate the inside into cells, are called *dissepiments* or *septa*. It is extremely important to bear in mind, not only that such is really their origin, but that they cannot possibly have any other origin, in order to form an exact idea of the structure of pistilla. Now, as each dissepiment is thus formed of two united sides, it necessarily consists of two plates, which are, in the ovarium state, often so completely united, that their double origin is undiscoverable, but which frequently separate in the ripe Pericarpium. This happens in Rhododendron, Euphorbia, Pentstemon, and a multitude of other plants. The consideration of this circumstance leads to certain laws which cannot be subject to exception, but which are of great importance; the principal of which are these: —

1. *All dissepiments are vertical and never horizontal.* — For

CHAP. II.   COMPOUND ORGANS IN FLOWERING PLANTS.   149

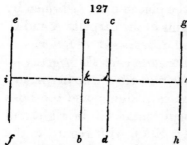

if *a, b* in *fig.* 127. represents the side of one carpellum and *c, d* that of another, the dissepiment *a, c, b, d* formed by this union will have precisely the same direction as that of the carpella, and can never acquire any other; and the same would be true of the sides *e, f* and *g, h,* if they formed themselves into dissepiments by uniting with other carpella: consequently a partition in any cell in the direction of *i, k* could not be a dissepiment, but would be of a different nature.

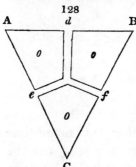

2. *They are uniformly equal in number to the carpella out of which the pistillum is formed.* — Suppose the triangle A, B, C represented a transverse section of an ovarium formed by the union of three carpella *o, o, o;* then *d, e, f* would be the dissepiments, and could not be either more or less.

3. *They proceed directly from the placentæ.* — As the placenta is the margin of the carpellary leaf, and as the dissepiment is the side of the carpellary leaf, it is evident that a dissepiment cannot exist apart from the placenta. Hence, when any partition exists in an ovarium which is not connected with the placenta, it follows that such a partition is not a dissepiment, however much it may otherwise resemble one.

4. *They are alternate with placentæ, formed by the cohesion of the margins of the same carpellum, and opposite to placentæ, formed by the cohesion of the contiguous margins of different carpella.* — Let the triangle A, B, C represent a transverse section of a three-celled ovarium of which *d, e, f* are the dissepiments: the dissepiments *d* and *e* will alternate with the placentæ *m, g,* both belonging to the carpellum A; but

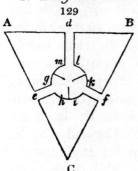

the dissepiment *d* will be opposite the placentæ *m, l*, formed by the cohesion of the contiguous margins of the carpella A and B.

5. *A single carpellum can have no dissepiment whatever.*

6. *The dissepiment will always alternate with the stigma;* — for the stigma is the extremity of the mid-rib of the carpellary leaf, or of the dorsal suture of the carpellum; and the sides of either of these (which form dissepiments) will be right and left of the stigma, or in the same position with regard to the latter organ as the sides of the lamina of a leaf to its apex.

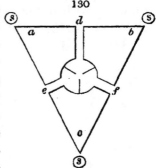

Let the triangle *a, b, c* represent a transverse section of a three-celled ovarium, of which *d, e, f* are the dissepiments. The stigmata would occupy a position equal to that of the spaces *s, s, s*, and would consequently be alternate with *d, e, f*, the dissepiments: they could not possibly be placed opposite *d, e, f*, upon any principle of structure with which we are acquainted. This law proves, that neither the membrane which separates the two cells of a Cruciferous siliqua, nor the vertical plate that divides the ovarium of Astragalus into two equal portions, are dissepiments; both are expansions of the placenta, or some other part, in different degrees.

Such is the structure of an ovarium in its most common state; certain deviations from it remain to be explained. We have seen that when carpella become syncarpous, they form a pistillum, the ovarium of which has as many cells and dissepiments as there are carpella employed in its construction. But sometimes the united sides of the carpella do not project so far into the cavity of the ovarium as to meet in the axis, as in the Poppy; and then an ovarium is the result, which, although composed of many carpella, is nevertheless one-celled (*fig.* 133.) In such case the dissepiments project a short distance only beyond the inner lining, or *paries*, of the ovarium, and, bearing on their edges the placentæ, the latter are said to be parietal. In other plants, such as Corydalis, Viola, and Orchis, the carpella are not folded together at all, but are spread open and united by their edges (*fig.* 132.): in that case

CHAP. II.   COMPOUND ORGANS IN FLOWERING PLANTS.   151

the placentæ do not project at all into the cavity of the ovarium, but are still more strictly parietal than the last.

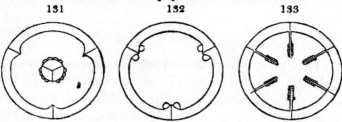

Another class of anomalies of a still more remarkable character, is that in which the dissepiments are obliterated, while the placentæ remain a distinct mass in the centre of the ovarium, as in Lychnis; forming what is called a *free central placenta* (*fig.* 131.). But, if we examine these plants at a very early period of their formation, long before the flowers expand, the explanation of the anomaly will be obvious. Such plants are, at that time, constructed upon the ordinary plan, with their dissepiments meeting in the centre and forming there a fungous placenta; but subsequently the shell of the ovarium grows more rapidly than the dissepiments, and breaks away from them; while the excessive growth of the placenta afterwards destroys almost all trace of them: their previous presence is only to be detected by lines upon the shell of the ovarium, or by a separation of the mass of ovula into distinct parcels upon the placenta.

All partitions whose position is at variance with the foregoing laws are spurious. Such *spurious dissepiments* are caused by many circumstances, the chief of which are the following:— they are caused by expansions of the placenta, as in Cruciferæ, when they form a partition stretching from one side to the other of the fruit; or they are mere dilatations of the lining of the pericarpium, as in Cathartocarpus Fistula, in which they are horizontal; or they are internal expansions of the dorsal or ventral suture, as in Amelanchier, Astragalus, and Thespesia, in which they are distinguishable from their dissepiments by not bearing the placentæ, and by being opposite the stigma, or by projecting beyond the placentæ; or, finally, they are caused by the sides of the ovarium projecting into the cavity, uniting and forming many supernumerary cells, as in Diplophractum.

11. *Of the* Receptacle.

The part upon which the carpella are seated is the apex of the peduncle, or the summit of the floral branch, of which the carpella are the termination. Usually this part, which is called the receptacle, is flat, or merely a vanishing point; but in other cases it is very much dilated, and then assumes a variety of curious appearances. This receptacle is called *torus*, or *thalamus* as well as *receptaculum*, and, in Greek compounds, has the name of Clinium.

In Annonaceæ and Magnoliaceæ it elevates itself from the base of the calyx, and bears the numerous stamens peculiar to these orders: here it is called *Gonophore* (Gonophorum) by De Candolle. In Caryophylleæ the receptacle is elongated, and bears on its summit the petals and stamens: M. De Candolle calls this form *Anthophore* (Anthophorum). When the receptacle bears only the ovarium, and is not a support to either corolla or stamens (Plate V. fig. 1. *a*.) it is called *Carpophore*, or *Gynophore* (Carpophorum, Gynophorum): this may either be a simple rounded stalk to the fruit, and of the same texture with it, as in Capparis, Phaca, and others,—when Richard calls it *Basigynium* or *Podogynium*, and Ehrhart *Thecaphore*, — or it may be succulent and much dilated, so as to resemble the receptacle of a Composita, bearing at the same time many ovaria, as in the Strawberry and Raspberry, when Richard calls it *Polyphorum:* most commonly such a receptacle is sufficiently described by the adjective fleshy. In Geranium it is remarkable for being lengthened into a tapering cone, to which the styles adhere, in the form of a beak; and in Nelumbium it is excavated into a number of cavities, in which the ovaria are half hidden.

12. *Of the* Ovulum.

The *ovulum* (Plate V. fig. 16. to 26.) is a small, semipellucid, pulpy body, borne by the placenta, and gradually changing into a seed. Its internal structure is exceedingly difficult to determine, either in consequence of its minuteness, or of the extreme delicacy of its parts, which are easily torn and crushed by the dissecting knife. It is doubtless to this circumstance chiefly, that the anatomy of the ovulum was almost unknown to botanists of the last century, and that it has only begun to be understood within ten or twelve years, during which it has received ample illustration from several skilful observers. Mr. Brown, indeed, claims to have pointed out its real nature so long ago as 1814; but the brief and incomplete terms then used by that gentleman, in the midst of a long description of a single species, in the Appendix to Captain Flinders' Voyage, unaccompanied as they were by any explanatory remarks, prove indeed that he knew something of the subject, but by no means entitle him to the credit of having, at that time, made the world acquainted with it. The late Mr. Thomas Smith seems to deserve the credit of having first made any *general* remarks upon the subject: of what extent they exactly were is not known, as his discoveries, in 1818, were communicated, as it would seem, in conversation only; but it is to be collected from Mr. Brown's statement that they were of a highly important nature. Since that period the structure of the ovulum has received much attention from Messrs. Brown, in England; Turpin and Adolphe Brongniart, in France; and Treviranus, in Germany; by all of whom the subject has been greatly illustrated. It is, however, to the learned M. Mirbel,—who, by collecting the discoveries of others, examining their accuracy, snd combining them with numerous admirable observations of his own, has given a full account of the gradual developement and the different modifications of the ovulum—that we are indebted for by far the best description of that important organ. His two papers read before the Academy of Sciences at Paris, in 1828 and 1829, are a perfect model of candour and patient investigation, and form the basis of what is here about to be recorded on the subject. I regret, however that the space which can now be devoted to the explanation of the structure of the ovulum is by no means such as its intricacy and interest demand.

As the ovula are the production of the placentæ, they necessarily originate in the margins of the carpellary leaf; and hence they have not only been compared to the buds found upon the margins of some true leaves, in a theoretical point of view, but it also has been attempted to be shown that they are analogous to them in structure. Of the truth of the former there can be little doubt; for, to say nothing of such plants as Bryophyllum, which habitually form buds on the margins of the leaves, the case of Malaxis paludosa, first recorded by Professor Henslow, in which the edge of the leaf is frosted by little microscopical points, which are neither exactly ovula nor exactly buds, seems a sufficient proof: but that in structure they are analogous to buds is by no means well made out. M. Turpin indeed has attempted, with his usual ingenuity, to demonstrate an analogy between the bracteæ of Marcgraavia and the outer integument or primine of the ovulum; but his statement cannot be regarded, as a sufficient evidence, of the fact. It is, therefore, only safe, at present, to regard the ovulum as analogous to the marginal buds of leaves in nature and position, but not in structure.

In almost all cases the ovulum is enclosed within an ovarium, as would necessarily happen in consequence of the convolute nature of the carpellary leaves; but if the convolution is imperfect, as in Reseda, the ovula are partially naked; and if it does not exist at all, as in Cycadeæ and Coniferæ, the ovula are then entirely naked; and, instead of being fertilised by molecules conveyed through the stigma and the style, as in other plants, are exposed to the direct influence of the pollen. This was first noticed by Mr. Brown; and, although since contradicted, is no doubt perfectly true.

When the ovula are attached to the placenta by a kind of cord, that cord is called the *funiculus* (Plate 5. fig. 26, *a.*), and is a mere prolongation of the placenta.

In the beginning, the ovulum is a pulpy excrescence (Plate 5. fig. 16.), appearing to be perfectly homogeneous, with no trace of perforation or of envelopes. But, as it advances in growth, it is gradually (Plate 5. fig. 17 to 21.) enclosed in two sacs or integuments, which are open only at their apex; where, in both these sacs, a passage exists, called

CHAP. II. COMPOUND ORGANS IN FLOWERING PLANTS. 155

the *foramen* (Plate 5. fig. 21, *a.*); or, in the language of M. Mirbel, *Exostome* (fig. 25, *a.*), in the outer integument, and *Endostome* (fig. 25, *b.*), in the inner integument. The central part is a fleshy, pointed, pulpy mass, called the *nucleus*, or *nucelle* (Plate V. fig. 19, 20. *a*, 22. *b*, 23. *c*, 24. *d*, 25. *e*, 27. *e*).

The outermost of the sacs (Plate V. fig. 22. *c*, 23. *a*, 25. *c*) is called the *primine*. It is either merely a cellular coating, or it is traversed by numerous veins or bundles of tubes; these are sometimes very apparent, as in the Orange tribes; and M. Mirbel seems disposed to think that they often exist in a rudimentary state when they are not visible. Usually it is nearly as long as the secondine, but sometimes is remarkably shorter, as in the Euphorbia Lathyris when very young (Plate V. fig. 22.).

The outermost but one of the sacs (Plate V. fig. 23. *b*, 20. *b*, 25. *d*.) is called the *secondine*; it immediately reposes upon the primine, and often contracts an adhesion with it, so that the two integuments become confounded. In order to ascertain its existence, it is, therefore, often necessary to examine the ovulum at a very early period of its growth. It is probable that it always exists; but Myrica, Alnus, Corylus, Quercus, and Juglans have been named by M. Mirbel as plants in which the secondine is not perceptible (Plate V. fig. 24.). Its point is usually exerted beyond the foramen of the primine.

The *nucleus* (Plate V. fig. 22. *b*, 18, 19, 20. *a*, 24. *d*, 25. *e*) is a pulpy conical mass, enclosed by the primine and secondine, and often covered up by them; but frequently protruded beyond the latter, and afterwards, at a subsequent period of its growth, again covered by them. Sometimes its cuticle separates in the form of a third coating of the ovulum, called the tercine.

These three parts, the primine, the secondine, and the nucleus, have all an organic connection at some one point of their surface. That point is, in ovula whose parts do not undergo any alteration of direction in the course of their growth, at the base next the placenta; so that the nucleus is, like a cone, growing from the base of a cup, the base of which is connected with the hilum through another cup like itself (Plate V. fig. 23.). The axis of such an ovulum, which M. Mirbel calls *Orthotropous*, is rectilinear, as in Myrica, Cistus, Urtica, &c.; and the foramen is at the end of the ovulum most remote from the hilum.

But sometimes, while the base of the nucleus and that of the outer sacs continue contiguous to the hilum, the axis of the ovulum instead of remaining rectilinear is curved down upon itself (Plate V. fig. 26, 27.); so that the foramen, instead of being at the extremity of the ovulum most remote from the hilum, is brought almost into contact with it. Examples of this are found in Papilionaceous plants, Caryophylleous plants, Mignonette, &c. M. Mirbel, who first distinguished these, calls them *Campulitropous*. In both these modifications the base of the ovulum and the base of the nucleus are the same.

In a third class the axis of the ovulum remains rectilinear; but one of the sides grows rapidly, while the opposite side does not grow at all, so that the point of the ovulum is gradually pushed round to the base; while the base of the nucleus is removed from the hilum to the opposite extremity (Plate V. fig. 16—21.); and when this process is completed the whole of the inside of the ovulum is reversed; so that the apex of the nucleus, and consequently the foramen, corresponds with the base of the ovulum. Such ovula as these M. Mirbel terms *Anatropous*; they are very common: examples may be found in the Almond, the Apple, the Ranunculus, the Cucumber, &c. When the base of the nucleus is thus removed from the base of the ovulum, a communication between the two is always maintained by means of a vascular cord, called the *raphe* (Plate V. fig. 24. *e*, 25. *f*). This raphe, which originates in the placenta, runs up one side of the ovulum, until it reaches the base of the nucleus; and there it expands into a sort of vascular disk, which is called the *chalaza* (Plate V. fig. 24. *f*, 25. *g*.). As the chalaza is uniformly at the base of the nucleus, it will follow that, in Orthotropous and Campulitropous ovula it is confounded with the hilum; while it is only distinguished in Anatropous ones, in which alone it is distinctly to be recognised.

It has been remarked that the raphe, or vascular extension

of the placenta, always occupies the side next the ventral suture of the ovarium; and that when, as in Euonymus, it is turned towards the dorsal suture, that circumstance arises from an alteration in the position of the ovulum subsequent to its being fertilised.

It has already been stated that the passage through the primine and secondine is called the foramen; or the exostome, when speaking of that of the primine; and the endostome, in speaking of the secondine. Upon these M. Mirbel remarks, — " These two orifices are at first very minute; but they gradually enlarge; and when they have arrived at the maximum of dilatation they can attain, they contract and close up. This maximum of dilatation is so considerable in a great number of species, in proportion to the size of the ovulum, that, to give an exact idea of it, I would compare it not to a hole, as those express themselves who have hitherto spoken of the exostome and endostome, but to the mouth of a goblet or of a cup. It may therefore be easily understood that to perceive either the secondine or the nucleus, it is not necessary to have recourse to anatomy. I have often seen, most distinctly, the primine and secondine forming two large cups, one of which encompassed the other without entirely covering it, and the nucleus extending itself in the form of an elongated cone beyond the secondine, to the bottom of which its base was fixed."

In practical botany the detection of the foramen is often a matter of great importance; for it enables an observer to judge from the ovulum of the direction of the radicle of the future embryo: it having been ascertained by many observations that the radicle of the embryo is almost always pointed to the foramen. A partial exception to this law exists, however, in Euphorbiaceæ, in many of which M. Mirbel has noticed that, after fertilisation, the axis of the nucleus and the endostome is inclined five or six degrees, without the exostome changing its position; by this circumstance the foramen of the secondine and that of the primine cease to correspond, and the radicle, instead of pointing when formed to the exostome, is directed to a point a short distance on one side of it.

Besides the two external integuments, M. Mirbel has remarked the occasional presence of three others peculiar to the nucleus, which he calls the *tercine*, *quartine*, and *quintine*.

The former is the external coat of the nucleus, and is very generally, if not universally, present. As I am almost unacquainted either with it or the two latter, I can add nothing to the following remarks of M. Mirbel upon the subject : — " The quartine and quintine are productions slower to show themselves than the preceding. The quartine is not very rare, although no one has previously indicated it; as to the quintine, which is the *vesicula amnios* of Malpighi, the *additional membrane* of Mr. Brown, and the *sac of the embryo* of M. Adolphe Brongniart, I am far from thinking that it only exists in a small number of species, as Mr. Brown seems to suppose. If no one has noticed the quartine, it is, no doubt, because it has been confounded with the tercine; nevertheless these two envelopes differ essentially in their origin and mode of growth. I have only discovered the quartine in ovula of which the tercine is incorporated at an early period with the secondine; and I think that it is only in such cases that it exists. At its first appearance it forms a cellular plate, which lines all the internal surface of the wall of the cavity of the ovulum; at a later period it separates from the wall, and only adheres to the summit of the cavity : at this period it is a sac, or rather a perfectly close vesicle. Sometimes it rests finally in this state, as in Statice; in other cases it fills with cellular tissue, and becomes a pulpy mass; under this aspect it is seen in Tulipa gesneriana. All this is the reverse of what takes place in the tercine; for this third envelope always begins by being a mass of cellular tissue, (and at that time it has the name, as we have seen, of nucleus,) and generally finishes by becoming a vesicle.

" I have remarked the fifth envelope, or quintine, in many species; its general characters are such as to prevent its being mistaken. Its complete developement takes place only in a nucleus which remains full of cellular tissue, or in a quartine that has filled with the same. At the centre of the tissue is organised, as in a womb, the first rudiment of the quintine; it is a sort of delicate intestine, which holds by one end to the summit of the nucleus, and by the other end to the chalaza.

The quintine swells from top to bottom; it forces back on all sides the tissue that surrounds it, and it often even invades the place occupied by the quartine or the nucleus. A very delicate thread, the suspensor, descends from the summit of the ovulum into the quintine, and bears at its extremity a globule which is the nascent embryo."

It is apparently this quintine that Mr. Brown describes, in the ovulum of the Orchis tribe, as a thread consisting of a simple series of short cells, the lowermost joint or cell of which is probably the original state of what afterwards, from enlargement and deposit of granular matter, becomes the opaque speck, or rudiment, of the future embryo. (*Observ. on the Organs, &c. of Orch. and Asclepiad.* pp. 18, 19.)

" The existence," continues M. Mirbel, " of a cavity in the quartine, or, indeed, the destruction of the internal tissue of the nucleus, at the period when the quintine developes, becomes the cause of some modifications in the manner of existence of this latter integument. The quintine is never seen, in certain Cucurbitaceæ, adhering to the chalaza: it is nevertheless evident that the adhesion has existed. The quintine, distended at its upper part, and suspended like a lustre from the top of the cavity, still presents at its lower end a portion of a rudimentary intestine become distinct; the separation occurred very early, in consequence of the tearing of the tissue of the nucleus.

The quintine of Statice is reduced to a sort of cellular placenta, to the lower surface of which the embryo is attached. This abortion of the quintine arises from the quartine having a large internal cavity, which prevents the young quintine from placing itself in communication with the chalaza, and taking that developement which it acquires in a multitude of other species."

The fluid matter contained within the nucleus is called the liquor amnios, and is supposed to be what nourishes the embryo during its growth.

When an ovulum grows erect from the base of the ovarium, it is called *erect;* when from a little above the base, *ascending;* when it hangs from the summit of the cavity, it is *pendulous;* and when from a little below the summit, it is *suspended.*

160 ORGANOGRAPHY. BOOK I.

13. *Of the* Fruit.

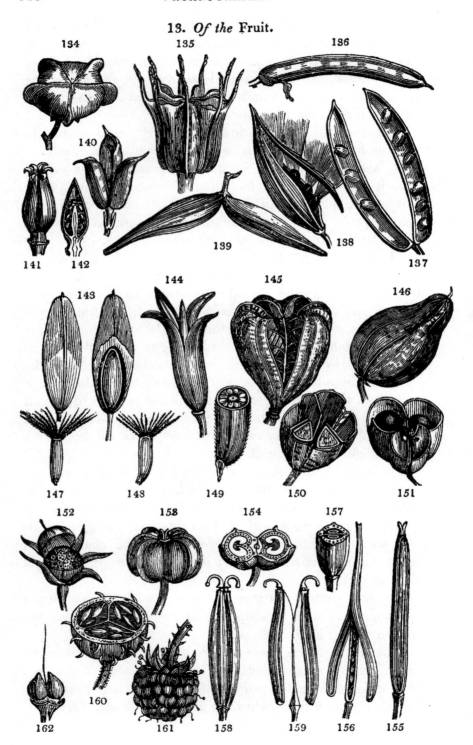

CHAP. II.   COMPOUND ORGANS IN FLOWERING PLANTS.   161

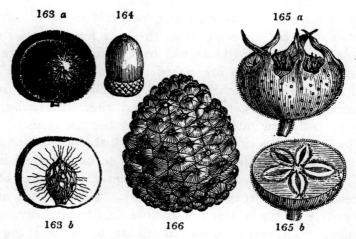

The *fruit* is the ovarium or pistillum arrived at maturity; but, although this is the sense in which the term is strictly applied, yet in practice it is extended to whatever is combined with the ovarium when ripe. Thus the pine-apple fruit consists of a mass of bracteæ, calyces, corollæ, and ovaria; that of the nut, the acorn, and many others, of the superior dry calyx and ovarium; that of the apple of succulent superior calyx, and corolla, and ovarium; and that of the strawberry-blite of a succulent inferior calyx and dry ovarium.

The fruit being the matured ovarium, it should exhibit upon some part of its surface the traces of a style or stigma; and this mark will, in many cases, enable the student to distinguish minute fruits from seeds. Many fruits were formerly called *naked seeds*, such as those of Umbelliferæ, Labiatæ, and Boragineæ, and the grain of corn; but now, that attention has been paid to the gradual developement of organs, such errors have been corrected. In cases where a trace of the style cannot be discovered, anatomy will generally show whether a minute body is a seed or fruit, by the presence, in the latter case, of two separable and obviously organically distinct coatings to the nucleus of the seed; but in other cases, when the pericarpium and the integuments of the seeds are combined in a single covering, and when no trace of style remains, as sometimes happens, nothing can be determined as to the exact nature of a given body without following it back in its growth to its young state. This, however, may be stated, that naked seeds, properly so called, are not known to exist in more than three or four orders in the whole vegetable kingdom; viz. in Coniferæ and Cycadeæ, where the ovula also are naked, and in Peliosanthes Teta and Leontice, in which the ovula, originally enclosed in an ovarium, rupture it at an early period after fertilisation, and subsequently continue naked until they become seeds.

Such being the case, it follows that all the laws of structure which exist in the ovarium are equally to be expected in the fruit; and this fact renders a repetition in this place of the general laws of formation unnecessary. Nevertheless, as, in the course of the advance of the ovarium to maturity, many changes often occur which contribute to conceal the real structure of the fruit, it is in all cases advisable, and in many absolutely necessary, to examine the ovarium, in order to be certain of the exact construction of the fruit itself. These changes are caused by the abortion, non-developement, obliteration, addition, or union of parts. Thus the three-celled six-ovuled ovarium of the oak and the hazel becomes, by the non-developement of two cells and five ovules, a fruit with one seed; the three-celled ovarium of the cocoa-nut is converted into a one-celled fruit by the obliteration of two cells and their ovula; and the two-celled ovarium of some Pedalineæ becomes many-celled by a division and elongation of the placentæ.

In a very early state the ovarium of the Lychnis and of the primrose consists of five cells, each with a placenta having a number of ovula; by degrees the dissepiments are ruptured and obliterated by the rapid growth of the shell of the ovarium; and it finally becomes a fruit with only one cell, and a large fungous placenta in the middle. In Cathartocarpus fistula a one-celled ovarium changes into a fruit, having each of its many seeds lodged in a separate cell, in consequence of the formation of numerous horizontal membranes which intercept the seeds. A still more extraordinary confusion of parts takes place in the fruit of the pomegranate after the ovarium is fertilised; and many other cases might be mentioned.

Every fruit consists of two principal parts, the *pericarpium* and the *seed*, the latter being contained within the former.

When the ovarium is inferior, or coheres with the calyx, the latter and the pericarpium are usually so completely united as to be inseparable and undistinguishable: in such cases it is usual to speak of the pericarpium without reference to the calyx, as if no such union had taken place. Botanists call a fruit, the pericarpium of which adheres to the calyx, an inferior fruit (*fructus inferus*); and that which does not adhere to the calyx, a superior fruit (*fructus superus*). But M. Desvaux has coined other words to express these ideas: a superior fruit he calls *autocarpien;* an inferior fruit, *heterocarpien;* terms wholly unnecessary and unworthy of adoption.

Every thing which in a ripe fruit is on the outside of the real integuments of the seed belongs to the pericarpium. It consists of three different parts, the *epicarpium*, the *sarcocarpium*, and the *endocarpium;* terms contrived by Richard, and very useful in practice.

The *epicarpium* is the external integument or skin; the *endocarpium*, called *putamen* by Gærtner, the inner coat or shell; and the *sarcocarpium*, the intermediate flesh. Thus, in the peach, the separable skin is the epicarpium, the pulpy flesh the sarcocarpium, and the stone the endocarpium or putamen. In the apple and pear the epicarp is formed by the cuticle of the calyx, and the sarcocarpium is confluent with the remainder of the calyx in one fleshy body.

The pericarpium is extremely variable in size and texture, varying from the dimension of a single line in length to the magnitude of two feet in diameter; and from the texture of a delicate membrane to the coarse fabric of wood itself, through various cartilaginous, coriaceous, bony, spongy, succulent, or fibrous gradations.

The *base* of the pericarpium is the part where it unites with the peduncle; its *apex* is where the style was: hence the organic and apparent apices of the fruit are often very different, especially in such as have the style growing from their sides, as in Rosaceæ and Chrysobalaneæ, Labiatæ and Boragineæ.

When a fruit has arrived at maturity its pericarpium either continues perfectly closed, when it is *indehiscent*, as in the hazel-nut, or separates regularly round its axis, either wholly or partially, into several pieces: the separation is called *dehiscence*, and such pieces *valves;* and the axis from which the valves separate in those cases where there is a distinct axis, is called the *columella*.

When the dehiscence takes place through the dissepiments it is said to be *septicidal;* when through the back of the cells it is called *loculicidal;* if along the inner edge of a simple fruit it is called *sutural;* if the dissepiments are separated from the valves the dehiscence is named *septifragal*.

In *septicidal* dehiscence the dissepiments divide into two plates and form the sides of each valve, as in Rhododendron, Menziesia, &c. Formerly botanists said that in this sort of dehiscence the valves were alternate with the dissepiments, or that the valves had their margins turned inwards. This may be understood from *fig*. 167., which represents the relative position of parts in a transverse section of a fruit with septicidal dehiscence; *v* being the valves, *d* the dissepiments, and *a* the axis.

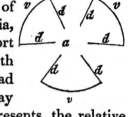

In *loculicidal* dehiscence the dissepiments form the middle of each valve, as in the lilac, or in the diagram 168., where the letters have the same value as above. In this it was formerly said that the dissepiments were opposite the valves.

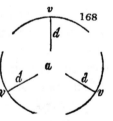

In *septifragal* dehiscence the dissepiments adhere to the axis and separate from the valves, as in Convolvulus; or in the diagram 169., lettered as before.

In *sutural* dehiscence there are no dissepiments, the fruit being composed of only one carpellum, as the *pea*.

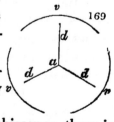

Besides these regular forms of valvular dehiscence, there is a mode which obtains in a very few plants, called *circumscissile*. This occurs by a transverse circular separation, as in Anagallis; in Jeffersonia it only takes place half round the fruit.

Valvular dehiscence, which is by far the most common mode by which pericarpia open, must not be confounded with either *rupturing* or *solubility*, — irregular and unusual contrivances of nature for facilitating the dispersion of seeds. In valvular dehiscence the openings have a certain reference to the cells, as has been already shown; but neither rupturing nor solubility bear any distinct relation to the cells. *Rupturing* consists in a spontaneous contraction of a portion of the pericarpium, by which its texture is broken through and holes formed, as in Antirrhinum and Campanula. *Solubility* arises from the presence of certain transverse contractions of a one-celled pericarpium, through which it finally separates into several closed portions, as in Ornithopus.

For the nature of the placenta and funiculus umbilicalis, see the observations under ovarium. These parts, which are mere modifications of each other, essentially appertain to the pericarpium, in which the former often acquires a spongy dilated substance, occasionally dividing the cells by spurious dissepiments, and often giving to the fruit an appearance much at variance with its true nature. In some seeds, as Euonymus europæus, it becomes exceedingly dilated around each seed, forming an additional envelope, called *arillus*. The true character of this organ was unknown till it was settled by Richard: before his time the term was applied, not only in its true sense to an enlargement of the placenta, but also to the endocarpium of certain Rubiaceæ and Rutaceæ, to the testa of Jasminum, Orchideæ, and others, and even to the perianthium of Carex. A very remarkable instance of the arillus is to be found in the nutmeg, in which it forms the part called the *mace*, surrounding the seed. It is never developed until after the impregnation of the ovulum.

Having thus explained the structure of the pericarpium, it is in the next place necessary to enquire into the nature of its modifications, which in systematic botany are of considerable importance. It is on the one hand very much to be regretted that the terms employed in this department of the science, which is that of Carpology, have been often used so vaguely as to have no exact meaning; while, on the other hand, they have been so exceedingly multiplied by various writers, that the phraseology of Carpology is a mere chaos. In practice but a small number of terms is actually employed; but it cannot be doubted that, if it were not for the inconvenience of overburdening the science with terms, it would conduce very much to clearness of description if botanists would agree to make use of some very precise and uniform nomenclature.

What, for instance, can be more embarrassing than to find the term *nut* applied to the superior plurilocular pericarpium of Verbena, the gland of Corylus, and the achenia of Rosa and Borago; and that of *berry* to the fleshy envelope of Taxus, the polyspermous inferior fruit of Ribes, the succulent calyx of Blitum, and several other things?

So much discordance, indeed, exists in the application of terms expressive of the modifications of fruit, that it is quite indispensable to give the definitions of some of the most eminent writers upon the subject in their own words, in order that the meaning attached by those authors to carpological terms, when employed by themselves, may be clearly understood.

In the phraseology of *writers antecedent to Linnæus*, the following are the only terms of this description employed; viz. —

1. *Bacca,* a berry, any fleshy fruit.
2. *Acinus,* a bunch of fleshy fruit, especially a bunch of grapes.
3. *Cachrys,* a cone, as of the pine tree.
4. *Pilula,* a cone like the Galbulus of modern botanists.
5. *Folliculus* (Fuchs), any kind of capsule.
6. *Grossus,* the fruit of the fig unripe.
7. *Siliqua,* the coating of any fruit.

In his " Philosophia Botanica " Linnæus gives the following definitions of the terms he employs: —

1. *Capsula,* hollow, and dehiscing in a determinate manner.
2. *Siliqua,* two-valved, with the seeds attached to both sutures.
3. *Legumen,* two-valved, with the seeds attached to one suture only.
4. *Conceptaculum,* one-valved, opening longitudinally on one side, and distinct from the seeds.
5. *Drupa,* fleshy without valves, containing a nut.

CHAP. II. COMPOUND ORGANS IN FLOWERING PLANTS. 167

6. *Pomum*, fleshy without valves, containing a capsule.
7. *Bacca*, fleshy without valves, containing naked seeds.
8. *Strobilus*, an amentum converted into a pericarpium.

Gærtner has the following, with definitions annexed to them : —

1. *Capsula*, a dry, membranous, coriaceous, or woody pericarpium, sometimes valveless, but more commonly dehiscing with valves. Its varieties are, —

a. *Utriculus*, an unilocular one-seeded capsule, very thin and transparent, and constantly valvular; as in Chenopodium, Atriplex, Adonis.

b. *Samara*, an indehiscent, winged, one or two-celled capsule; as Ulmus, Acer, Liriodendron.

c. *Folliculus*, a double one-celled, one-valved, membranous, coriaceous capsule, dehiscing on the inside, and either bearing the seed on each margin of its suture, or on a receptacle common to both margins; as Asclepias, Cinchona, and Vinca.

2. *Nux*, a hard pericarpium, either indehiscent or never dividing into more than two valves; as in Nelumbium, Boragineæ, and Anacardium.

3. *Coccum* is a pericarpium of dry elastic pieces or *coccula*, as in Diosma, Dictamnus, Euphorbia.

4. *Drupa* is an indehiscent pericarpium with a variable rind, very different in substance from the *putamen*, which is bony; as in Lantana, Cocos, Sparganium, Gaura, &c.

5. *Bacca*, any soft pericarpium, whether succulent or otherwise; provided it does not dehisce into regular valves, nor contain a single stone adhering to it. Of this the following are kinds : —

a. *Acinus*, a soft, succulent, semi-transparent unilocular berry, with one or two hard seeds, as the grape, Rivina, Rhipsalis, Rubus, Grossularia, &c.

b. *Pomum*, a succulent or fleshy, two or many celled berry, the dissepiments of which are fleshy or bony, and coherent at the axis; as Pyrus, Cratægus, Cydonia, Sapota, and others.

c. *Pepo*, a fleshy berry with the seeds attached at a distance from the axis upon the parietes of the pericarp; as Cucumis, Stratiotes, Passiflora, Vareca, and others.

To the term bacca all other succulent fruits are referred which belong to neither Acinus, Pomum, nor Pepo; as Garcinia, Caryophyllus, Cucubalus, Hedera.

6. *Legumen*, the fruit of Leguminosæ.

7. *Siliqua* and *Silicula*, the fruit of Cruciferæ.

Willdenow defines those employed by him in the following manner:—

1. *Utriculus*, a thin skin enclosing a single seed. Adonis, Galium, Amaranthus.

2. *Samara*, a pericarpium containing one, or at most two seeds, and surrounded by a thin membrane, either along its whole circumference or at the point, or even at the side. Ulmus, Acer, Betula.

3. *Folliculus*, an oblong pericarpium bursting longitudinally on one side, and filled with seeds. Vinca.

4. *Capsula*, a pericarpium consisting of a thin coat containing many seeds, often divided into cells, and assuming various forms. Silene, Primula, Scrophularia, Euphorbia, Magnolia.

5. *Nux*, a seed covered with a hard shell which does not burst. Corylus, Quercus, Cannabis.

6. *Drupa*, a nut covered with a thick succulent or cartilaginous coat. Prunus, Cocos, Tetragonia, Juglans, Myristica, Sparganium.

7. *Bacca*, a succulent fruit containing several seeds, and not dehiscing. It encloses the seeds without any determinate order, or it is divided by a thin membrane into cells. Ribes, Garcinia, Hedera, Tilia. Rubus has a compound bacca.

8. *Pomum*, a fleshy fruit that internally contains a capsule for the seed. It differs from the celled berry in having a perfect capsule in the heart. Pyrus.

9. *Pepo*, a succulent fruit which has its seeds attached to the inner surface of the rind. Cucumis, Passiflora, Stratiotes.

10. *Siliqua*, a dry elongated pericarp, consisting of two halves held together by a common permanent suture. Cruciferæ. *Silicula* is a small form of the same.

11. *Legumen*, a dry elongated pericarpium, consisting of

two halves or valves externally forming two sutures. Leguminosæ.

12. *Lomentum,* a legumen divided internally by spurious dissepiments, not dehiscing longitudinally, but either remaining always closed, as Cassia Fistula, or separating into pieces at transverse contractions along its length, as Ornithopus.

The following are enumerated as *spurious fruits.*

13. *Strobilus,* an Amentum the scales of which have become woody. Pinus.

14. *Spurious capsule.* Fagus, Rumex, Carex.

15. *Spurious nut.* Trapa, Coix, Mirabilis.

16. *Spurious drupe.* Taxus, Anacardium, Semecarpus.

17. *Spurious bacca.* Juniperus, Fragaria, Basella.

By this author the names of fruits are, perhaps, more loosely and inaccurately applied than by any other.

Professor Link objects to applying particular names to variations in anatomical structure; observing, "that botanists have strayed far from the right road in distinguishing these terms by characters which are precise and difficult to seize. Terms are only applied to distinct parts, as the leaf, peduncle, calyx, and stamens, and not to modifications of them. Who has ever thought of giving a distinct name to a labiate or papilionaceous corolla, or who to a pinnated leaf?" But this sort of reasoning is of little value if it is considered that the fruit is subject to infinitely greater diversity of structure than any other organ, and that names for these modifications have become necessary, for the sake of avoiding a minute explanation of the complex differences upon which they depend. Besides, to admit, as Professor Link actually does, such names as capsula, &c. is abandoning the argument; and when the following definitions, which this learned botanist has proposed, are considered, I think that little doubt need exist as to whether terms should be employed in the manner recommended by himself, or with the minute accuracy of the French. According to Professor Link, the following are the limits of Carpological nomenclature: —

1. *Capsula,* any dry membranous or coriaceous pericarp.
2. *Capsella,* the same, if small and one-seeded.

3. *Nux*, externally hard.
4. *Nucula*, externally hard, small, and one-seeded.
5. *Drupa*, externally soft, internally hard.
6. *Pomum*, fleshy or succulent, and large.
7. *Bacca*, fleshy or succulent, and small.
8. *Bacca sicca*, fleshy when unripe, dry when ripe, and then distinguishable from the capsule by not being brown.
9. *Legumen*,
10. *Siliqua*, } the pericarps of certain natural orders.
11. *Amphispermium*, a pericarpium which is of the same figure as the seed it contains.

In more recent times there have been three principal attempts at classing and naming the different modifications of fruit; namely, those of Richard, Mirbel, and Desvaux. These writers have all distinguished a considerable number of variations, of which it is important to be aware for some purposes, although their nomenclature is not much employed in practice. But, in proportion as the utility of a classification of fruit consists in its theoretical explanation of structure rather than in a strict applicability to practice, it becomes important that it should be founded upon characters which are connected with internal and physiological distinctions rather than with external and arbitrary forms. Viewing the subject thus, it is not to be concealed that, notwithstanding the undoubted experience and talent of the writers just mentioned, their carpological systems are essentially defective. Besides this, each of the three writers has felt himself justified in contriving a nomenclature at variance with that of his predecessors, for reasons which it is difficult to comprehend.

I have attempted to adjust the synonyms of carpological writers, and have also ventured to propose a new arrangement, in which those names which seem to be most legitimate are retained in every case, their definitions only being altered; previously to which I shall briefly explain the methods of Messrs. Richard, Mirbel, and Desvaux.

THE ARRANGEMENT OF RICHARD.

Class 1. Simple fruits.
§ 1. Dry.

    * Indehiscent.
    ** Dehiscent.
  § 2. Fleshy.
Class 2. Multiplied fruits.
Class 3. Aggregate or compound fruits.

The Arrangement of Mirbel.

Class 1. Gymnocarpiens. Fruit not disguised by the adherence of any other organ than the calyx.
  Ord. 1. *Carcerulaires.* Pericarpium indehiscent, but sometimes with apparent sutures, generally dry, superior or inferior, mostly unilocular and monospermous, sometimes plurilocular and polyspermous.
  Ord. 2. *Capsulaires.* Pericarpium dry, superior or inferior, opening by valves, but never separating into distinct pieces or cocci.
  Ord. 3. *Dieresiliens.* Pericarpium superior or inferior, dry, regular, and monocephalous (that is, having one common style), composed of several distinct pieces arranged systematically round a central real or imaginary axis, and separating at maturity.
  Ord. 4. *Etairionaires.* Pericarps several, irregular, superior, one or many-seeded, with a suture at the back.
  Ord. 5. *Cenobionaires.* A regular fruit divided to the base into several acephalous pericarpia; that is to say, not marked on the summit by the stigmatic scar, the style having been inserted at their base.
  Ord. 6. *Drupacées.* Pericarpium indehiscent, fleshy externally, bony internally.
  Ord. 7. *Bacciens.* Succulent, many-seeded.
Class 2. Angiocarpiens. Fruit seated in envelopes not forming part of the calyx.

The Arrangement of Desvaux.

Class 1. Pericarpium dry.
  Ord. 1. Simple fruits.

§ Indehiscent.
§ § Dehiscent.
Ord. 2. Dry compound fruits.
Class 2. Pericarpium fleshy.
Ord. 1. Simple fruits.
Ord. 2. Compound fruits.

In explanation of the principles upon which the classification of fruit which I now venture to propose is founded, it will of course be expected that I should offer some observations. In the first place, I have made it depend primarily upon the structure of the ovarium, by which the fruit is of necessity influenced in a greater degree than by any thing else, the fruit itself being only the ovarium at maturity. In using the terms simple and compound, I have employed them precisely in the sense that has been attributed to them in my remarks upon the ovarium; being of opinion that, in an arrangement like the following and those which have preceded it, in which theoretical rather than practical purposes are to be served, the principles on which it depends should be conformable to the strictest theoretical rules of structure. A consideration of the fruit without reference to the ovarium necessarily induces a degree of uncertainty as to the real nature of the fruit; the abortion and obliteration, to which almost every part of it is more or less subject, often disguising it to such a degree that the most acute carpologist would be unable to determine its true structure from an examination of it in a ripe state only. In simple fruits are stationed those forms in which the ovaria are multiplied so as to resemble a compound fruit in every respect except their cohesion, they remaining simple. But, as the passage which is thus formed from simple to compound fruits is deviated from materially when the ovaria are placed in more than a single series, I have found it advisable to constitute a particular class of such under the name of aggregate fruit. Care must be taken not to confound these with the fourth class containing collective fruits, as has been done by more carpologists than one. While the true aggregate fruit is produced by the ovaria of a single flower, a collective fruit, if aggregate, is produced by the ovaria of many flowers; a most important difference. As

CHAP. II. COMPOUND ORGANS IN FLOWERING PLANTS. 173

the pericarpium is necessarily much affected by the calyx when they adhere so as to form a single body, it is indispensable, if a clear idea is to be attached to the genera of carpology, that inferior or superior fruits should not be confounded under the same name; for this reason I have in all cases founded a distinction upon that character.

In order to facilitate the knowledge of the limits of the genera of carpology, the following analytical table will be found convenient for reference. It is succeeded by the characters of the genera in as much detail as is necessary for the perfect understanding of their application.

Class I. Fruit simple. APOCARPI.
    One or two-seeded.
        Membranous, - - - - UTRICULUS.
        Dry and bony, - - - ACHENIUM.
        Fleshy externally, bony internally, - - DRUPA.
    Many-seeded.
        Dehiscent.
            One-valved, - - - FOLLICULUS.
            Two-valved, - - LEGUMEN.
        Indehiscent, - - - - LOMENTUM.

Class II. Fruit aggregate. AGGREGATI.
    Ovaria elevated above the calyx.
        Pericarpia distinct, - - - ETÆRIO.
        Pericarpia cohering into a solid mass, - SYNCARPIUM.
    Ovaria enclosed within the fleshy tube of the calyx, - CYNARRHODUM.

Class III. Fruit compound. SYNCARPI.
    Sect. 1. Superior.
        A. Pericarpium dry externally.
            Indehiscent.
                One-celled, - - - CARYOPSIS.
                Many-celled.
                      Dry internally.
                          Apterous - - - CARCERULUS.
                          Winged, - - SAMARA.
                      Pulpy internally, - AMPHISARCA.
            Dehiscent.
                By a transverse suture, - - PYXIDIUM.
                By elastic cocci, - - REGMA.
                By a longitudinal suture, - CONCEPTACULUM.
                By valves.
                    Placentæ opposite the lobes of the stigma.
                      Linear, - - - SILIQUA.
                      Roundish, - - SILICULA.

                Placentæ alternate with the lobes of
                    the stigma.
                        Valves separating from the replum, CERATIUM.
                        Replum none, - - CAPSULA.
        B. Pericarpium fleshy.
            Indehiscent.
                Sarcocarpium separable, - - HESPERIDIUM.
                Sarcocarpium inseparable, - - NUCULANIUM.
            Dehiscent, - - - - - TRYMA.
Sect. 2. Inferior.
    A. Pericarpium dry.
        Indehiscent.
            Cells two or more, - - - CREMOCARPIUM.
            Cell one.
                Surrounded by a cupulate involucrum, GLANS.
                    Destitute of a cupula, - - CYPSELA.
            Dehiscent or rupturing, - - - DIPLOTEGIA.
    B. Pericarpium fleshy.
        Epicarpium hard.
            Seeds parietal, - - - PEPO.
            Seeds not parietal, - - - BALAUSTA.
        Epicarpium soft.
            Cells obliterated; or unilocular, - BACCA.
            Cells distinct, - - - - POMUM.

Class IV. Collective fruits.   ANTHOCARPI.
    Single.
        Perianthium indurated, dry, - - - DICLESIUM.
        Perianthium fleshy, - - - - SPHALEROCAR-
    Aggregate.   PIUM.
        Hollow, - - - - - SYCONUS.
        Convex.
            An indurated amentum, - - - STROBILUS.
            A succulent spike, - - - SOROSIS.

Class I. Fruit simple.   APOCARPI.
Ovaria strictly simple; a single series only produced by a single flower.

I. UTRICULUS, *Gærtner.* — (Cystidium, *Link.*)
    One-celled, one or few-seeded, superior, membranous, frequently dehiscent by a transverse incision. This differs from the *pyxis* in texture, being strictly simple, *i. e.* not proceeding from an ovarium with obliterated dissepiments.
        Example. — Amaranthus, Chenopodium.

II. ACHENIUM; (Akenium, of *many*; Spermidium; Xylodium, *Desv.*; Thecidium, *Mirb.*; Nux, *Linn.*)
    One-seeded, one-celled, superior, indehiscent, hard, dry, with the integuments of the seed distinct from it.
    Linnæus includes this among his seeds, defining it " semen tectum epidermide osseâ." I have somewhere seen it named Spermidium; a good term if it were wanted. M. Desvaux calls the nut of Anacardium a *Xylodium*.
        Examples. — Lithospermum, Borago.

III. Drupa. — *Drupe.* —*fig.* 163.

One-celled, one or two-seeded, superior, indehiscent, the outer coat (*naucum*) soft and fleshy, and separable from the inner or *endocarpium* (the stone), which is hard and bony; proceeding from an ovarium which is perfectly simple. This is the strict definition of the term drupa, which cannot strictly be applied to any compound fruit, as that of Cocos, certain Verbenaceæ, and others, as it often is. Fruits of the last description are generally carcerules with a drupaceous coat. The *stone* of this fruit is the *Nux* of Richard, but not of others.

Examples. — Peach, Plum, Apricot.

IV. Folliculus. — *Follicle* (Hemigyrus, *Desvaux*; Plopocarpium, *Desv.*), *fig.* 140.

One-celled, one or many-seeded, one-valved, superior, dehiscent by a suture along its face, and bearing its seeds at the base, or on each margin of the suture. This differs from the legumen in nothing but its having one valve instead of two. The Hemigyrus of Desvaux is the fruit of Proteaceæ, and differs from the follicle in nothing of importance. When several follicles are in a single flower, as in Nigella and Delphinium, they constitute a form of fruit called Plopocarpium by Desvaux, and admitted into his Etærio by Mirbel.

Examples. — Pæonia, Banksia, Nigella.

V. Legumen. — *Pod* (Legumen, *Linn.*; Gousse, *Fr.*), *fig.* 136, 137.

One-celled, one or many-seeded, two-valved, superior, dehiscent by a suture along both its face and its back, and bearing its seeds on each margin of the ventral suture. This differs from the follicle in nothing except its dehiscing by two valves. In Astragalus two spurious cells are formed by the projection inwards of either the dorsal or ventral suture, which forms a sort of dissepiment; and in Cassia a great number of transverse diaphragms (phragmata) are formed by projections of the placenta. Sometimes the legumen is indehiscent, as in Cathartocarpus, Cassia fistula, and others; but the line of dehiscence is in such species indicated by the presence of sutures. When the two sutures of the legumen separate from the valves, they form a kind of frame called *replum*, as in Carmichaelia.

Examples. — Bean, Pea, Clover.

VI. Lomentum. — (Legumen lomentaceum, *Rich.*)

Differs from the legumen in being contracted in the spaces between each seed, and there separating into distinct pieces, or indehiscent, but divided by internal spurious dissepiments, whence it appears at maturity to consist of many articulations and divisions.

Example. — Ornithopus.

Class II. Fruit aggregate. AGGREGATI.

Ovaria strictly simple; more than a single series produced by each flower.

VII. Etærio, *Mirb.*— (" Polychorion, *Mirb.*;" Polysecus, *Desvaux*; Amalthea, *Desv.*; Erythrostomum, *Desvaux*), *fig.* 161.

Ovaries distinct; pericarpia indehiscent, either dry upon a dry receptacle, as Ranunculus, dry upon a fleshy receptacle, as strawberry, or fleshy upon a dry receptacle, as Rubus. The last is very near the syncarpium, from which it differs in the ovaria not coalescing into a single mass. It is Desvaux's Erythrostomum. This term is applied less strictly by M. Mirbel, who admits into it dehiscent pericarpia, not placed upon an elevated receptacle, as Delphinium and Pæonia; but the fruit of these plants is better understood to be a union of several follicules within a single flower. If there is no elevated receptacle, we have Desvaux's Amalthea. The parts of an Etærio are Achenia.

Examples. Ranunculus, Fragaria, Rubus.

VIII. SYNCARPIUM. — (Syncarpium, *Rich.*; Asimina, *Desv.*)
Ovaries cohering into a solid mass, with a slender receptacle.
Examples. Annona, Magnolia.

IX. CYNARRHODUM. — (Cynarrhodum, *Officin. Desvaux.*)
Ovaries distinct; pericarpia hard, indehiscent, enclosed within the fleshy tube of a calyx.
Examples. Rosa, Calycanthus.

Class III. Fruit compound. SYNCARPI.
Ovaria compound.
Sect. 1. Fruit superior.
A. Pericarpium dry.

X. CARYOPSIS. — (Cariopsis, *Rich.*; Cerio, *Mirb.*)
One-celled, one-seeded, superior, indehiscent, dry, with the integuments of the seed cohering inseparably with the endocarpium, so that the two are undistinguishable; in the ovarium state evincing its compound nature by the presence of two or more stigmata; but nevertheless unilocular, and having but one ovulum.
Examples. Wheat, Barley, Maize.

XI. REGMA, *Mirb.*; — (Elaterium, *Rich.*; Capsula tricocca, *L.*)
Three or more celled, few-seeded, superior, dry, the cells bursting from the axis with elasticity into two valves. The outer coat is frequently softer than the endocarpium or inner coat, and separates from it when ripe; such regmata are drupaceous. The cells of this kind of fruit are called *cocci*.
Example. Euphorbia.

XII. CARCERULUS, *Mirb.*; — (Dieresilis, *Mirb.*; Cænobio, *Mirb.*; Synochorion, *Mirb.*; Sterigmum, *Desvaux*; Microbasis, *Desvaux*; Polexostylus, *Mirb.*; Sarcobasis, *Dec., Desv.*; Baccaularius, *Desv.*)
Many-celled, superior; cells dry, indehiscent, few-seeded, cohering by a common style round a common axis. From this the Dieresilis of Mirbel does not differ in any essential degree. The same writer calls the fruit of Labiatæ (*fig.* 162.), which Linnæus and his followers mistake for naked seeds, Cænobio: it differs from the Carcerulus in nothing but the low insertion of the style into the ovaria, and the distinctness of the latter.
Examples. Tilia, Tropæolum, Malva.

XIII. SAMARA, *Gærtn.*; — *Key.* (Pteridium, *Mirb.*; Pterodium, *Desv.*), *fig.* 143.

Two or more celled, superior; cells few-seeded, indehiscent, dry; elongated into wing-like expansions. This is nothing but a modification of the Carcerule.
Examples. Fraxinus, Acer, Ulmus.

XIV. Pyxidium (Pyxidium, *Ehr., Rich., Mirb.*; Capsula circumscissa, *L.*), *fig.* 152.

One-celled, many-seeded; superior, or nearly so; dry, often of a thin texture; dehiscent by a transverse incision, so that when ripe the seed and their placenta appear as if seated in a cup, covered with a lid. This fruit is one-celled by the obliteration of the dissepiments of several carpella, as is apparent from the bundles of vessels which pass from the style through the pericarpium down into the receptacle.
Example. Anagallis.

XV. Conceptaculum (Conceptaculum, *Linn.;* Double Follicule, *Mirb.*), *fig.* 138, 139.

Two-celled, many-seeded, superior, separating into two portions, the seeds of which do not adhere to marginal placentæ, as in the folliculus, to which this closely approaches, but separate from their placentæ, and lie loose in the cavity of each cell.
Examples. Asclepias, Echites.

XVI. Siliqua, *Linn. fig.* 155, 156, 157.

One or two-celled, many-seeded, superior, linear, dehiscent by two valves separating from the replum; seeds attached to two placentas adhering to the replum, and *opposite* to the lobes of the stigma. The dissepiment of this fruit is considered a spurious one formed by the projecting placentas, which sometimes do not meet in the middle; in which case the dissepiment or phragma has a slit in its centre, and is said to be *fenestrate*.

XVII. — Silicula. *Linn.*

This differs from the latter in nothing but its figure, and in containing fewer seeds. It is never more than four times as long as broad, and often much shorter.
Examples. Thlaspi, Lepidium, Lunaria.

XVIII. Ceratium. — (Capsula siliquiformis, *Dec.;* Conceptaculum, *Desv.*)

One-celled, many-seeded, superior, linear, dehiscent by two valves separating from the replum; seeds attached to two spongy placentæ adhering to the replum, and *alternate* with the lobes of the stigma. Differs from the siliqua in the lobes of the stigma being alternate with the placentæ, not opposite. This, therefore, is regular, while that is irregular in structure.
Examples. Glaucium, Corydalis, Hypecoum.

XIX. Capsula, *Capsule, fig.* 145, 146. 150, 151. 134, 135.

One or many-celled, many-seeded, superior, dry, dehiscent by valves, always proceeding from a compound ovarium. The valves are variable in their nature: usually they are at the top of the fruit, and equal in number to the cells; sometimes they are twice the number; occasionally they resemble little pores or holes below the summit, as in the Antirrhinum.
Examples. Digitalis, Primula, Rhododendron.

XX. Amphisarca. — (Amphisarca, *Desv.*)

Many-celled, many-seeded, superior, indehiscent; indurated or woody externally, pulpy internally.

Examples. Omphalocarpus, Adansonia, Crescentia.

B. Pericarpium fleshy.

XXI. TRYMA. — (Tryma, *Watson.*)

Superior, by abortion one-celled, one-seeded, with a two-valved indehiscent endocarpium, and a coriaceous or fleshy, valveless sarcocarpium.

Example. Juglans.

XXII. NUCULANIUM. — (Nuculanium, *Rich.;* Bacca, *Desvaux.*)

Two or more celled, few or many-seeded, superior, indehiscent, fleshy, of the same texture throughout, containing several seeds, improperly called *nucules* by the younger Richard. This differs scarcely at all from the berry, except in being superior.

Examples. Grape, Achras.

XXIII. HESPERIDIUM. — (Hesperidium, *Desv. Rich.*)

Many-celled, few-seeded, superior, indehiscent, covered by a spongy separable rind; the cells easily separable from each other, and containing a mass of pulp, in which the seeds are imbedded. The pulp is formed by the cellular tissue, which forms the lining of the cavity of the cells: this cellular tissue is excessively enlarged and succulent, is filled with fluid, and easily coheres into a single mass. The external rind is by M. De Candolle supposed to be an elevated discus of a peculiar kind, analogous to that within which the fruit of Nelumbium is seated; and perhaps its separate texture and slight connexion with the cells of the fruit seem to favour this supposition. But it is difficult to reconcile with such an hypothesis the continuity of the rind with the style and stigma, which is a sure indication of the identity of their origin; and it is certain that the shell of the ovarium and the pericarpium are the same. The most correct explanation of this structure is to consider the rind a union of the epicarp and sarcocarp, analogous to that of the drupa.

Example. Orange.

Sect. 2. Fruit inferior.

A. Pericarpium dry.

XXIV. GLANS (Glans, *Linn. Desv.;* Calybio, *Mirb.;* Nucula, *Desvaux*), *fig.* 164.

One-celled, one or few-seeded, inferior, indehiscent, hard, dry; proceeding from an ovarium containing several cells and several seeds, all of which are abortive but one or two; seated in that kind of persistent involucre called a cupule. The pericarpium is always crowned with the remains of the teeth of the calyx; but they are exceedingly minute, and are easily overlooked. Sometimes the gland is solitary, and quite naked above, as in the common oak; sometimes there is more than one completely enclosed in the cupule, as the beech and sweet chestnut.

Examples. Quercus, Corylus, Castanea.

XXV. CYPSELA (Akena, *Necker;* Akenium, *Rich.;* Cypsela, *Mirb.;* Stephanoum, *Desv.*), *fig.* 147, 148.

One-seeded, one-celled, indehiscent, with the integuments of the seed not cohering with the endocarpium; in the ovarium state evincing its compound nature by the presence of two or more stigmata; but nevertheless unilocular, and having but one ovulum. Such is the true structure of the Achenium; but, as that term is often applied to the simple superior fruits, called Nux by

Linnæus, I have thought it better, in order to avoid confusion, to adopt the name Cypsela.

Examples. All Compositæ.

XXVI. CREMOCARPIUM (Cremocarpium, *Mirb.*; Polakenium, or Pentakenium, *Rich.*; Carpadelium, *Desv.*), *fig.* 153, 154. 158, 159.

Two to five-celled, inferior; cells one-seeded, indehiscent, dry, perfectly close at all times; when ripe separating from a common axis. M. Mirbel confines the application of Cremocarpium to Umbelliferæ; but it is better to let it apply to all fruits which will come within the above definition. It will then be the same as Richard's Polakenium, excluding those forms in which the fruit is superior. The latter botanist qualifies his term Polakenium according to the number of cells of the fruit: thus when there are two cells it is *diakenium*, three *triakenium*, and so on. M. De Candolle calls the half of the fruit of Umbelliferæ *mericarp.*

Examples. Umbelliferæ, Aralia, Galium.

XXVII. DIPLOTEGIA (Diplotegia, *Desv.*), *fig.* 144.

One or many-celled, many-seeded, inferior, dry, usually bursting either by pores or valves. This differs from the Capsule only in being adherent to the calyx.

Examples. Campanula, Leptospermum.

B. Pericarpium fleshy.

XXVIII. POMUM, *Apple* or *Pome.* — (Melonidium, *Rich.*; Pyridium, *Mirb.*; Pyrenarium, *Desvaux*; Antrum, *Mœnch.*) *fig.* 165.

Two or more celled, few-seeded, inferior, indehiscent, fleshy; the seeds distinctly enclosed in dry cells, with a bony or cartilaginous lining, formed by the cohesion of several ovaria with the sides of the fleshy tube of a calyx, and sometimes with each other. These ovaria are called parietal by M. Richard. Some forms of Nuculanium and this differ only in the former being distinct from the calyx.

Examples. Apple, Cotoneaster, Cratægus.

XXIX. PEPO. — (Peponida, *Rich.*)

One-celled, many-seeded, inferior, indehiscent, fleshy; the seeds attached to parietal pulpy placentæ. This fruit has its cavity frequently filled at maturity with pulp, in which the seeds are imbedded; their point of attachment is, however, never lost. The cavity is also occasionally divided by projections of the placenta into spurious cells, which has given rise to the belief that in Pepo Macrocarpus there is a central cell, which is not only untrue but impossible.

Examples. Cucumber, Melon, Gourd.

XXX. BACCA, *Berry* (Bacca, *L.*; Acrosarcum, *Desvaux*), *fig.* 160.

Many-celled, many-seeded, inferior, indehiscent, pulpy; the attachment of the seeds lost at maturity, when they become scattered in the substance of the pulp. This is the true meaning of the term berry; which is, however, often otherwise applied, either from mistaking nucules for seeds, or from a misapprehension of the strict limits of the term.

Example. Ribes.

XXXI. BALAUSTA. — (Balausta, *Officin. Rich.*)

Many-celled, many-seeded, inferior, indehiscent; the seeds with a pulpy coat, and attached distinctly to their placentæ. The rind was called Malicorium by Ruellius.

Example. Pomegranate.

Class IV. Collective Fruits. ANTHOCARPI.
Fruit of which the principal characters are derived from the incrassated floral envelopes.

XXXII. DICLESIUM. — (Dyclesium, *Desvaux;* Scleranthum, *Mœnch;* Cataclesium, *Desvaux;* Sacellus, *Mirb.*)

Pericarpium indehiscent, one-seeded, enclosed within an indurated perianthium.

Examples. Mirabilis, Spinacia, Salsola.

XXXIII. SPHALEROCARPUM. — (Sphalerocarpum, *Desv.;* Nux baccata *of authors.*)

Pericarpium indehiscent, one-seeded, enclosed within a fleshy perianthium.

Examples. Hippophäe, Taxus, Blitum, Basella.

XXXIV. SYCONUS. — (Syconus, *Mirb.*)

A fleshy rachis, having the form of a flattened disk, or of a hollow receptacle, with distinct flowers and dry pericarpia.

Examples. Ficus, Dorstenia, Ambora.

XXXV. STROBILUS, *Cone* (Conus, or Strobilus, *Rich., Mirb.;* Galbulus, *Gærtn.;* Arcesthide, *Desvaux;* Cachrys, *Fuchs;* Pilula, *Pliny*), *fig.* 166.

An amentum, the carpella of which are scale-like, spread open, and bear naked seeds; sometimes the scales are thin, with little cohesion; but they often are woody, and cohere into a single tuberculated mass.

The Galbulus differs from the Strobilus only in being round, and having the heads of the carpella much enlarged. The fruit of the Juniper is a Galbulus, with fleshy coalescent carpella. Desvaux calls it Arcesthide.

Example. Pinus.

XXXVI. SOROSIS. — (Sorosis, *Mirb.*)

A spike or raceme converted into a fleshy fruit by the cohesion in a single mass of the ovaria and floral envelopes.

Examples. Ananassa, Morus, Artocarpus.

14. *Of the* Seed.

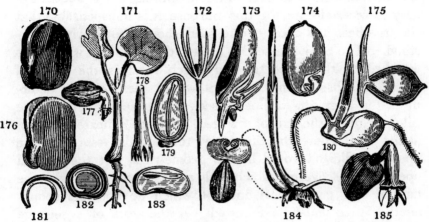

As the fruit is the ovarium arrived at maturity, and is therefore subject to the same laws of structure as the latter; so is the seed the ovulum in its most perfect and finally organised state, and constructed upon exactly the same plan as the ovulum. But as the fruit, nevertheless, often differs from the ovarium in the suppression, or addition, or modification of certain portions, so is the seed occasionally altered from the precise structure of the ovulum, in consequence of changes of like nature.

The seed is a body enclosed in a pericarpium, is clothed with its own integuments, and contains the rudiment of a future plant. It is the point of developement at which vegetation stops, and beyond which no increase, in the same direction with itself, can take place. In a young state it has already been spoken of under the name of ovulum; to which I also refer for all that relates to the insertion of seeds.

That side of a seed which is most nearly parallel with the axis of a compound fruit, or the ventral suture or sutural line of a simple fruit, is called the *face,* and the opposite side the *back.* In a compound fruit with parietal placentæ, the placenta is to be considered as the axis with respect to the seed; and that part of the seed which is most nearly parallel with the placenta, as the face. Where the raphe is visible, the face is indicated by that.

When a seed is flattened lengthwise it is said to be *com-*

pressed, when vertically it is *depressed;* a difference which it is of importance to bear in mind, although it is not always easy to ascertain it: for this purpose it is indispensable that the true base and apex of the seed should be clearly understood. The base of a seed is always that point by which it is attached to the placenta, and which receives the name of *hilum;* the base being found, it would seem easy to determine the apex, as a line raised perpendicularly upon the hilum, cutting the axis of the seed, ought to indicate the apex at the point where the line passes through the testa: but the apex so indicated would be the geometrical, not the natural apex; for discovering which with precision in all seeds, the natural and geometrical apex of which do not correspond, another plan must be followed. If the testa of a seed be carefully examined, it will usually be found that it is composed in great part of lines representing rows of cellular tissue, radiating from some one point towards the base, or, in other words, of lines running upwards from the hilum and meeting in some common point. This point of union or radiation is the true apex, which is not only often far removed from the geometrical apex, but is sometimes even in juxtaposition with the hilum, as in mignonette: in proportion, therefore, to the obliquity of the apex of the seed will be the curve of its axis, which is represented by a line passing through the whole mass of the seed from the base to the apex, accurately following its curve. If the lines above referred to are not easily distinguished, another indication of the apex resides in a little brown spot or areola, hereafter to be mentioned under the name of chalaza. Where there is no indication either externally or internally of the apex, it may then be determined geometrically.

The integuments of a seed are called the *testa;* the rudiment of a future plant, the *embryo* (Plate VI. fig. 1. *b,* &c.); and a substance interposed between the embryo and the testa, the *albumen* (fig. 1. *a,* 5. *a,* &c.).

The testa, called also *lorica* by Mirbel, *perisperme* and *episperme* by Richard, and *spermodermis* by De Candolle, according to some consists, like the pericarpium, of three portions; viz. 1. the external integument, *tunica externa* of

Willdenow, *testa* of De Candolle; 2. the internal integument, *tunica interna* of Willdenow, *endopleura* of De Candolle, *hilofère* and *tegmen* of Mirbel; and, 3. of an intervening substance answering to the sarcocarpium, and called *sarcodermis* by De Candolle: this last is chiefly present in seeds with a succulent testa, and by many is considered a portion of the outer integument, which is the most accurate mode of understanding it.

The *outer integument* is either membranous, coriaceous, crustaceous, bony, spongy, fleshy, or woody; its surface is either smooth, polished, rough, or winged, and sometimes is furnished with hairs, as in the cotton and other plants, which, when long and collected about either extremity, form what is called the *coma* (sometimes also, but improperly, the pappus). It consists of cellular tissue disposed in rows, with or without bundles of vessels intermixed: in colour it is usually of a brown or similar hue: it is readily separated from the inner integument.

In Maurandya Barclayana it is formed of reticulated cellular tissue; in Collomia linearis and others it is caused by elastic spirally twisted fibres enveloped in mucus, and springing outwards when the mucus is dissolved; in Casuarina it (or the inner integument) contains a great quantity of spirally fibrous cellules. In the genus Crinum it is of a very fleshy, succulent character, and has been mistaken for albumen, from which it is readily known by its vascularity. According to Mr. Brown, a peculiarly anomalous kind of partition, which is found lying loose within the fruit of Banksia and Dryandra, without any adhesion either to the pericarpium or the seed, is a state of the outer integument. It is said that in those genera the inner membrane (secondine) of the ovulum before fertilisation is entirely exposed, the primine being dimidiate and open its whole length; and that the outer membranes (primines) of the two collateral ovula, although originally distinct, finally contract an adhesion by their corresponding surfaces, and together constitute the anomalous dissepiment. But it may be reasonably doubted whether the integument here called secondine is not primine, and the supposed primine arillus.

The inner membrane (secondine) of the ovulum, however, in general appears to be of greater importance as connected with fecundation, than as affording protection to the nucleus at a more advanced period. For in many cases, before impregnation, its perforated apex projects beyond the aperture of the testa, and in some plants puts on the appearance of an obtuse, or even dilated stigma; while in the ripe seed it is often either entirely obliterated, or exists only as a thin film, which might readily be mistaken for the epidermis of a third membrane, then frequently observable.

"This third coat (tercine) is formed by the proper membrane or cuticle of the nucleus, from whose substance in the unimpregnated ovulum it is never, I believe, separable, and at that period is very rarely visible. In the ripe seed it is distinguishable from the inner membrane only by its apex, which is never perforated, is generally acute and more deeply coloured, or even sphacelated."

M. Mirbel has, however, justly remarked that the primine and the secondine are, in the seed, very frequently confounded; and that therefore the word testa is better employed, as one which expresses the outer integument of the seed without reference to its exact origin, which is practically of little importance. The tercine is also, no doubt, often absent. He observes, that these mixed integuments often give rise to new kinds of tissue; that in Phaseolus vulgaris the testa consists, indeed, of three distinct layers, but of those the *innermost* was the primine; and that the others, which represent nothing that pre-existed in the ovulum, have a horny consistence, and are formed of cylindrical cellules, which elongate in the direction from the centre to the circumference. And this is probably the structure of the testa of many Leguminosæ.

It sometimes happens that the endopleura (or tercine?) thickens so much as to have the appearance of albumen, as in Cathartocarpus fistula. In such a case as this it is only to be distinguished from albumen by gradual observation from the ovulum to the ripe seed.

With regard to the quartine and quintine, one of them is occasionally present in the form of a fleshy sac that is interposed between the albumen and the ovulum, and envelopes the latter. It is what was called the *vitellus* by Gærtner, and which M. Richard, by a singular prejudice, considered a dilatation of the radicle of the embryo, to his macropodal form of which he referred the embryo of such plants. Instances of this are found in Nymphæa and its allies, and also in Scitamineæ, peppers, and Saururus. Mr. Brown, who first ascertained the fact, considers this sac to be always of the same nature and origin, and as the *vesicula colliquamenti* or *amnios* of Malpighi.

The end by which the seed is attached to the placenta is called the *hilum* or *umbilicus* (Plate VI. fig. 5. *c*, 17. *e*, 11. *c*, &c.); it is frequently of a different colour from the rest of the seed, not uncommonly being black. In plants with small seeds it is exceedingly minute, and recognised with difficulty; but in some it is so large as to occupy fully a third part of the whole surface of the seed, as in the horse-chestnut, Sapoteæ, and others. Seeds of this kind have been called *nauca* by Gærtner. In grasses the hilum is indicated by a brownish spot situated on the face of the seed, and is called by M. Richard *spilus*. The centre of the hilum, through which the nourishing vessels pass, is called by Turpin the *omphalodium*. Sometimes the testa is enlarged in the form of irregular lumps or protuberances about the umbilicus; these are called *strophiolæ* or *carunculæ;* and the umbilicus, about which they are situated, is said to be strophiolate or carunculate. M. Mirbel has ascertained that in Euphorbia Lathyris the strophiola is the fungous foramen of the primine; and it is probable that such is often the origin of this tubercle: but at present we know little or nothing upon the subject.

The foramen in the ripe seed constitutes what is called the *micropyle:* it is always opposite the radicle of the embryo; the position of which is therefore to be determined without dissection of the seed, by an inspection of the micropyle, — often a practical convenience.

In some seeds, as the asparagus, Commelina, and others (*fig.* 185.), there is a small callosity at a short distance from the hilum: this callosity gives way like a little lid at the time of germination, emitting the radicle, and has been named by Gærtner the *embryotega*.

At the apex of the seed in the orange and many other

plants may be perceived upon the testa a small brown spot, formed by the union of certain vessels proceeding from the hilum: this spot is the *chalaza* (Plate VI. fig. 11. *b*). In the orange it is beautifully composed of dense bundles of spiral vessels and spiral ducts, without woody fibre. The vessels which connect the chalaza with the hilum constitute a particular line of communication, called the *raphe :* in most plants it consists of a single line passing up the face of the seed; but in many Aurantiaceæ and Guttiferæ it ramifies elegantly in every direction upon the surface of the testa.

The raphe is always a true indication of the face of the seed; and it is very remarkable that the apparent exceptions to this rule only serve to confirm it. Thus, in some species of Euonymus in which the raphe appears to pass along the back, an examination of other species shows, that the ovula of such species are in fact resupinate; so that with them the line of vascularity representing the raphe is turned away from its true direction by peculiar circumstances. In reality, the chalaza is the place where the secondine and the primine are connected; so that in *orthotropous* seeds, or such as have the apex of the nucleus at the apex of the seed, and in which, consequently, the union of the primine and secondine takes place at the hilum, there can be no apparent chalaza, and consequently no raphe: the two latter can only exist as distinct parts in anatropous seeds, when the base of the nucleus corresponds to the geometrical apex of the seed. Hence, also, there can never be a chalaza without a raphe, nor a raphe without a chalaza.

Something has already been said about the *arillus* (*fig.* 186. and 187.) when speaking of the ovulum; but it more properly comes under consideration along with the ripe seed. As a general rule it may be stated, that every thing proceeding from the placenta and not forming part of the seed is referable to the arillus. Even in plants like Hibbertia volubilis and Euonymus europæus, in which it is of unusual dimensions, it is scarcely visible in the unimpregnated ovarium; and it is stated by Mr. Brown, that he is not acquainted with any case in which it covers the foramen of the testa before impregnation.

The mass enclosed within the true testa or outer integument is called the *nucleus;* and consists either of *albumen* and *embryo*, or of the latter only.

The *albumen* (*perispermium*, Juss.; *endospermium*, Rich.; *medulla seminis*, Jungius; *secundinæ internæ*, Malpighi) (Plate VI. fig. 5. *a*, 1. *a*, 9. *a*, &c.), when present, is a body enclosing the embryo, and interposed between it and the integument of the seed: it is of various degrees of hardness, varying from fleshy to bony, or even stony, as in some palms. It is in all cases destitute of vascularity, and has been usually considered as the amnios in an indurated state; but Mr. Brown is of opinion that it is formed by a deposition or secretion of granular matter in the cellules of the amnios, or in those of the nucleus itself. The albumen is often absent, frequently much smaller than the embryo, but also occasionally of much greater size. This is particularly the case in monocotyledones, in some of which the embryo scarcely weighs a few grains, while the albumen weighs many ounces, as in the cocoa-nut. It is almost always solid, but in Annonaceæ and the nutmeg tribe it is perforated in every direction by dry cellular tissue, which appears to originate from the remains of the nucleus in which the albumen has been deposited: in this state it is said to be *ruminated*.

The *embryo* (or *corculum*) (Plate VI. fig. 1. *b*, &c.) is a fleshy body, occupying the interior of the seed, and constituting the rudiment of a future plant. It is usually solitary, but there are instances of the presence of several in one seed. It was originally developed within the innermost membrane of the ovulum. In most plants one embryo only is found in each seed. It nevertheless occurs, not unfrequently, that more than one is developed within a single testa, as occasionally in the orange and the hazel nut, and very commonly in Coniferæ, Cycas, the onion, and the misseltoe. Now and then a union takes place of these embryos.

It is divided into three parts; viz. the *radicle* (Plate VI. fig. 2. *b*, &c.) (*rhizoma* or *rostellum*); *cotyledons* (fig. 2. *a*, &c.),

and *plumula* (or *gemmula*) (fig. 2. *c*); from which is also by some distinguished the *cauliculus* or *neck* (*collet, scapus, scapellus*, or *tigelle*). Mirbel admits but two principal parts; viz. the cotyledons, and what he calls the *blastême*, which comprises radicle, plumula, and cauliculus.

Upon certain remarkable differences in the structure of the embryo, modern botanists have divided the whole vegetable kingdom into three great portions, which form the basis of what is called the natural system. These are, 1. Dicotyledons; 2. Monocotyledons; and, 3. Acotyledons. In order to understand exactly the true nature of the embryo in each of these, it will be requisite first to describe it fully as it exists in dicotyledons, and then to explain its organisation in the two others.

If a common Dicotyledonous embryo (Plate VI. fig. 2.), that of the apple for example, be examined, it will be found to be an obovate, white, fleshy body, tapering and solid at the lower end, and compressed and deeply divided into two equal opposite portions at the upper end; the lower tapering end is the *radicle*, and the upper divided end consists of two *cotyledons*. Within the base of the cotyledons is just visible a minute point, which is the *plumula*. The imaginary line of division between the radicle and the cotyledons is the *cauliculus*. If the embryo be now placed in circumstances favourable for *germination*, the following phenomena occur: the radicle will become elongated downwards, forming a little root; the cauliculus will extend upwards; the cotyledons will elevate themselves above the earth and unfold; and the plumula will lengthen upwards, and give birth to a stem and leaves. Such is the normal or proper appearance of a dicotyledonous embryo.

The *exceptions* to it chiefly consist, 1. in the *cohesion* of the cotyledons in a single mass, instead of their unfolding; 2. in an increase of their *number*; 3. in their occasional *absence*; and, 4. in their *inequality*. A *cohesion* of the cotyledons takes place in those embryos, which Gærtner called *pseudomonocotyledonous*, and Richard *macrocephalous*. In Hippocastanum, the horse-chestnut, the embryo consists of a homogeneous undivided mass, with a curved horn-like prolongation of one side directed towards the hilum. If a section be made in the direction of the axis of the horn-like prolongation through the whole mass of the embryo, a slit will be observable above the middle of the horn, at the base of which lies a little conical body. In this embryo the slit indicates the division between the two bases of a pair of opposite confluent cotyledons; the conical body is the plumula, and the horn-like prolongation is the radicle. In Castanea nearly the same structure exists, except that the radicle, instead of being curved and exserted, is straight, and enclosed within the projecting base of the two cotyledons; and in Tropæolum, which is very similar to Castanea in structure, the bases of the cotyledons are slit into four little teeth enclosing the radicle. The germination of these seeds indicates more clearly that the cotyledonary body consists of two and not of one cotyledon; at that time the bases of the cotyledons, which had been previously scarcely visible, separate and elongate, so as to extricate the radicle and plumula from the testa, within which they had been confined. In *number* the cotyledons vary from two to a much more considerable number. Ceratophyllum has constantly four, of which two are smaller than the others; in Coniferæ they vary from two to more than twelve.

Instances of the *absence* of cotyledons occur, 1. In Cuscuta (Plate VI. fig. 19.), to which they may be supposed to be denied in consequence of the absence of leaves in that genus; 2. in Lentibulariæ; 3. in Cyclamen, in which the radicle enlarges exceedingly: to these a fourth instance has by some been added in Lecythis, of which M. Richard gives the following account. The kernel is a fleshy almond-like body, so solid and homogeneous that it is extremely difficult to discover its two extremities until germination takes place: at that period one of the ends forms a little protuberance, which subsequently bursts through the integuments of the seed, and extends itself as a root; the other end produces a scaly plumula, which in time forms the stem. The great mass of the kernel is supposed by M. Richard to be an enlarged radicle. I, however, see no reason for calling the two-lobed part of the embryo (Plate VI. fig. 17. *c*) a plumula; it is merely cotyledons. An *inequality* of cotyledons is the most unusual circumstance with dicotyledons, and forms a distinct approach to the structure of monocotyledons: it occurs in Trapa and Sorocea, in which they are extremely disproportionate. In Cycas they are also rather unequal; but the structure of that plant is essentially dicotyledonous.

The embryo of MONOCOTYLEDONS (Plate VI. fig. 1. B. &c.) is usually a solid, cylindrical, undivided, homogeneous body, slightly conical at each extremity, with no obvious distinction of radicle, plumula, or cotyledons. In *germination* the upper end swells and remains within the testa (fig. 10. C. *b*, &c.); the lower lengthens, opens, and emits from within one or more radicles; and a thread-like green body is protruded from the upper part of the portion, which is lengthened beyond the testa. Here the portion remaining within the testa is a single cotyledon; that which lengthens, producing radicles from within its point, is the cauliculus and radicle; and the thread-like protruded green body is the plumula. If this is compared with the germination of dicotyledons, an obvious difference will be at once perceived in the manner in which the radicles are produced: in monocotyledons they are emitted from within the substance of the radicular extremity, and are actually sheathed at the base by the lips of the passage through which they protrude; while in dicotyledons they appear at once from the very surface of the radicular extremity, and consequently have no sheath at their base. Upon this difference in economy, Richard proposed to substitute the term Endorhizæ for monocotyledons, and Exorhizæ for dicotyledons. Some consider the former less perfect than the latter; endorhizæ being *involute*, or imperfectly developed, exorhizæ evolute, or fully developed. Dumortier adds to these names *endophyllous* and *exophyllous;* because the young leaves of monocotyledons are evolved from within a sheath (*coleophyllum* or *coleoptilum*), while those of dicotyledons are always naked. The sheath at the base of the radicle of monocotyledons is called the *coleorhiza* by Mirbel. Another form of monocotyledonous embryo is that of Aroideæ and their allies, in which the plumula is not so intimately combined with the embryo as to be undistinguishable, but is indicated externally by a little slit above the base (Plate VI.

fig. 6. B. *e*), within which it lies until called into developement by germination.

The *exceptions* to what has been now described ought, like those of dicotyledons, rather to be called remarkable modifications. Much stress has been laid upon them by several writers, who have thought it requisite to give particular names to their parts. To me, however, it appears far more advisable to explain their analogies without the unnecessary creation of new and bad names. In *Gramineæ* (Plate VI. fig. 4.) the embryo consists of a lenticular body lying on the outside of the base of the albumen on one side, and covered on its inner face by that body, and on its outer face by the testa: if viewed on the face next the testa, a slit will be observed of the same nature as that in the side of the embryo of Aroideæ; opening this cleft a small conical projection is discovered, pointing towards the apex of the seed. If the embryo be then divided vertically through the conical projection, it will be seen that the latter (*c*) is a sheath including other little scales resembling the rudiments of leaves; that that part of the embryo which lies next the albumen (*d*), and above the conical body, is solid; and that the lower extremity of the embryo (*e*) contains within it the indication of an internal radicle, as in other monocotyledons. In this embryo it is to be understood that the conical projection is the *plumula;* that part of the embryo lying between it and the albumen, a single scutelliform *cotyledon;* and the lower point of the embryo, the *radicle*. In wheat there is a second small cotyledon on the outside of the embryo, inserted a little lower down than the scutelliform cotyledon. This last is called *scutellum* by Gærtner, who considered it of the nature of vitellus. The late M. Richard considered the scutelliform cotyledon a particular modification of the radicle, which he called *hypoblastus;* the plumula a form of cotyledon, called *blastus;* the anterior occasional cotyledon a peculiar appendage, named *epiblastus;* and the radicle a protuberance of the cauliculus, called *radiculoda*. He further, in reference to this peculiar opinion, termed embryos of this description *macropodous*. In these ideas, however, Richard was manifestly wrong, as is now well known.

From what has been stated, it is apparent that dicotyledons are not absolutely characterised by having two cotyledons, nor monocotyledons by having only one. The real distinction between them consists in their endorhizal or exorhizal germination, and in the cotyledons of dicotyledons being opposite or verticillate, while they are in monocotyledons solitary or alternate. Some botanists have, therefore, recommended the substitution of other terms in lieu of those in common use. M. Cassini suggests *isodynamous* or *isobrious* for dicotyledons, because their force of developement is equal on both sides; and *anisodynamous* or *anisobrious* for monocotyledons, because their force of developement is greater on one side than on the other. Another writer, M. Lestiboudois, would call dicotyledons *exoptiles*, because their plumula is naked; and monocotyledons *endoptiles*, because their plumula is enclosed within the cotyledons; but there seems little use in these proposed changes, which are moreover as open to objections as the terms in common use.

The ACOTYLEDONOUS embryo is not exactly, as its name seems to indicate, an embryo without cotyledons; for, in that case, Cuscuta would be acotyledonous. On the contrary, it is an embryo, which does not germinate from two fixed invariable points, namely, the plumula and the radicle, but indifferently from any point of the surface; as in some of the Arum tribe, and in all flowerless plants.

For further illustrations of the embryo, consult Plate VI. and the explanation of its figures.

The *direction* of the embryo is either *absolute* or *relative*. Its absolute direction is that which it has independently of the parts that surround it. In this respect it varies much in different genera; it is either straight (Plate VI. fig. 5.), arcuate (fig. 9.), or falcate, uncinate, or coiled up (fig. 8.) (*cyclical*), folded up, spiral (fig. 19.), or bent at right angles (Plate V. fig. 28.) (*gromonical*, Link), serpentine, or in figure like the letter S (*sigmoid*).

Its relative position is determined by the relation it bears to the chalaza and micropyle of the seed; or, in other words, upon the relation that the integuments, the raphe, chalaza, hilum, micropyle, and radicle bear to each other. If the sacs of the ovulum are in no degree inverted, but have their common point of origin at the hilum, there being (necessarily) neither raphe nor chalaza visible, the radicle will in that case be at the extremity of the seed most remote from the hilum, and the embryo *inverted* with respect to the seed, as in Cistus, Urtica, and others, where it is said to be *antitropal*. But if the ovulum undergoes that remarkable extension of one side already described in speaking of that organ, in which the sacs are so inverted that their orifice is next the hilum, and their base at the apex of the ovulum, then there will be a raphe and chalaza distinctly present; and the radicle will, in the seed, be at the end next the hilum, and the embryo will be erect with respect to the seed, or *orthotropal*, as in the apple, plum, &c. On the other hand, supposing that the sacs of the embryo suffer only a partial degree of inversion, so that their foramen is neither at the one extremity nor the other, there will be a chalaza and a short raphe; and the radicle will point neither to the apex nor to the base of the seed, but the embryo will lie, as it were, across it, or be *heterotropal*, as is the case in the primrose. When an embryo is so curved as to have both apex and radicle presented to the hilum, as in Reseda, it is *amphitropal*.

In the words of Gærtner, an embryo is *ascending* when its apex is pointed to the apex of the fruit; descending, if to the base of the fruit; *centripetal*, if turned towards the axis of the fruit; and *centrifugal*, if towards the sides of the fruit: those embryos are called wandering, or *vagi*, which have no evident direction.

The *cotyledons* are generally straight, and placed face to face; but there are numberless exceptions to this. Some are separated by the intervention of albumen (Plate VI. fig. 11.); others are naturally distant from each other without any intervening substance. Some are straight, some waved, others arcuate or spiral. When they are folded with their back upon the radicle, they are called *incumbent;* if their edges are presented to the same part, they are *accumbent;* terms chiefly used in speaking of Cruciferæ.

15. *Of* Naked Seeds.

By naked seeds has been understood, by the school of Linnæus, small seed-like fruit, like that of Labiatæ, Boragineæ, grasses, and Cyperaceæ. But as these are distinctly covered by pericarpia, as has been shown above, the expression in the sense of Linnæus is obviously incorrect, and is now abandoned. Hence it has been inferred that there is no such thing in existence as a naked seed; that is to say, a seed which bears on its own integuments the organ of impregnation. To this proposition botanists had assented till the year 1825, when Mr. Brown demonstrated the existence of seeds strictly naked; that is to say, from their youngest state destitute of pericarpium, and receiving impregnation through their integuments without the intervention of style or stigma, or any stigmatic apparatus. That learned botanist has demonstrated that seeds of this description are uniform in Coniferæ and Cycadeæ, in which no pericarpial covering exists. But we have no knowledge at present of such an economy obtaining in other plants, as a constant character. It does however happen, as the same observer has pointed out, that in particular species the ovarium is ruptured at an early period by the ovula, which thus, when ripe, become truly naked seeds; remarkable instances of which occur in Ophiopogon spicatus, Leontice thalictroides, and Peliosanthes Teta.

CHAPTER III.

OF THE COMPOUND ORGANS IN FLOWERLESS PLANTS.

We have now passed in review all the different organs which exist in the most perfectly formed plants; that is to say, in those whose reproduction is provided for by the complicated apparatus of sexes and of fertilising organs. Let us next proceed to consider those lower tribes, some of which are scarcely distinguishable from animals, where there is no evident trace of sexes, in which nothing constructed like seeds is to be detected, and which seem to have no other provision made for the perpetuation of their races than a dissolution of their cellular system. In what I may have to say about them, I shall not, however, do any thing more than give a mere enumeration and description of their organs. All speculative considerations are in this case left out of view: those who wish to be informed upon such points may consult the " Introduction to the Natural System of Botany."

1. *Ferns.*

Filices, or ferns, are plants consisting of a number of leaves or *fronds,* as they are called, attached to a stem which is either subterraneous or elongated above the ground, sometimes rising like a trunk to a considerable height. They are the largest of known vegetables in which no organs of fructification analogous to those of phænogamous plants have been discovered. Their petioles, or *stipes (rachis,* W.; *peridroma,* Necker), consist of sinuous strata of indurated, very compact, woody fibre, connected by cellular tissue; and the wood of those which have arborescent trunks is formed by the cohesion of the bases of such petioles round a hollow or solid cellular axis. The organs of reproduction are produced from the back or under side of the fronds. In *Polypodiaceæ,* or what are more commonly called dorsiferous ferns, they originate, either upon the cuticle or from beneath it, in the form of spots at the anastomoses, margins, or extremities of the veins. As they increase in growth they assume the appearance of small heaps of granules, called *sori*; if examined beneath the microscope these granules, commonly called *capsules, thecæ,* or *conceptacles,* are found to be little brittle compressed bags formed of cellular membrane, partially surrounded by a thickened longitudinal ring (*gyrus, annulus, gyroma*), which at the vertex loses itself in the cellularity of the membrane, and at the base tapers into a little pedicel: the thecæ burst with elasticity by aid of their ring, and emit minute particles named *sporules,* from which new plants are produced; as from seeds, in vegetables of a higher order. Interspersed with these thecæ are often intermixed articulated hairs; and, in those genera in which the thecæ originate beneath the cuticle, the sori, when mature, continue covered with the superincumbent portion of the cuticle, which is then called the *indusium* or *involucrum* (*membranula,* Necker; *glandulæ squamosæ,* Guettard). In Trichomanes and Hymenophyllum the thecæ are seated within the dilated cup-like extremities of the lobes of the frond, and are attached to the vein which passes through their axis, which is then called their *receptacle.* In another tribe, called *Gleicheneæ,* the thecæ have a transverse complete, instead of a vertical incomplete ring, and they are nearly destitute of pedicels; in a third tribe the sori occupy the whole of the under surface of the frond, which becomes contracted, and wholly alters its appearance: the thecæ have no ring, and the cellular tissue of their membrane is not reticulated, but radiates regularly from the apex.

In these it has been in vain endeavoured to discover traces of organs of fecundation. Nevertheless, as it was difficult for sexualists to believe that plants of so large a size were destitute of such organs, it has been considered indispensable that they should be found; and, accordingly, while all seem to agree in considering the thecæ as female organs, a variety of other parts have been dignified by the title of male organs: thus, Micheli and Hedwig found them in certain stipitate glands of the frond; Stæhelin, Hill, and Schmidel, in the elastic ring; Kœlreuter, in the indusium; Gleichen, in the stomata; and Von Martius, in certain membranes enclosing the spiral vessels. None of these opinions are now adopted.

In Ophioglosseæ, a remarkable tribe of ferns, the fertile frond is rolled up in two lines parallel with its axis or midrib, and at maturity opens regularly by transverse valves along its whole length, emitting a fine powder, which, when magnified, is found to consist of particles of the same nature as the sporules found in the thecæ of other ferns: here there are no thecæ, the metamorphosed frond probably performing their functions. Such is my view of the structure of Ophiogosseæ; but by other botanists it is described as a dense spike of two-valved capsules, dehiscing transversely.

2. *Equisetaceæ.*

In these the organs of reproduction are arranged in an amentum, consisting of scales bearing on their lower surface an assemblage of cases, called *thecæ, folliculi,* or *involucra,* which dehisce longitudinally inwards. In these thecæ are contained two sorts of granules; the one very minute and lying irregularly among a larger kind, each of which is wrapped in two filaments, fixed by their middle, rolled spirally, having either extremity incrassated, and uncoiling with elasticity. By Hedwig the apex of the larger granules was supposed to be a stigma, and the thickened ends of the filaments anthers, the small granules being the pollen. At any rate it is certain that the larger granules, round which the elastic filaments are coiled, are the reproductive particles.

3. *Lycopodiaceæ.*

These are leafy plants with the habit of gigantic mosses. Their leaves and stem have the same structure as those plants, except that the former are sometimes provided with stomata, and the latter with vessels. Their organs of reproduction are of two kinds: the one kidney-shaped two-valved cases, called *thecæ, conceptacles,* or *capsules,* destitute of internal divisions, and filled with minute powder-like granules, which, in consequence of lateral compression, from being spherical, acquire the figure of irregular polygons; the other three or four-valved thecæ, of a similar appearance, containing three or four roundish fleshy bodies, each of which is at least fifty times larger than the granules contained in the first kind of theca, and is said by Brotero to burst with elasticity, — an observation which requires verification. The first kind of theca is found in all species of Lycopodiaceæ; the second is only found simultaneously in a few. The contents of both are believed to be sporules; but no satisfactory explanation has yet been offered of the cause of their difference in size, and probably also in structure. I would suggest that the powder-like grains are true sporules, and that the large ones are buds or viviparous organs, as has already been stated by Haller and Willdenow. A writer in the " Transactions of the Linnean Society" has figured and described the growth of the larger grains of Lycopodium denticulatum, and he considers that they exhibit the germination of a dicotyledonous plant; but, independently of any mistrust which may attach to the account, it is obvious enough that his own drawings and description represent a mode of germination analogous, not to that of dicotyledons, but rather to that of monocotyledons, but also reducible to the laws which govern the incipient vegetation of a bud.

The powder-like sporules are inflammable, and have been supposed by Haller, Linnæus, and others to be pollen, while the larger have been considered seeds; and to a part of the surface of the theca the office of stigma has been attributed. The thecæ themselves have been fancied to be male apparatus by Kœlreuter and Gærtner.

4. *Marsileaceæ.*

This very curious little order consists of plants differing from each other so much, that, although consisting of only four genera, it is necessary to subdivide it into two distinct tribes. As I have never had an opportunity of examining these plants in a fresh state, I beg to cite the observations of M. Adolphe Brongniart, who appears to have given them an especial attention.

In Marsileaceæ, properly so called, says this botanist,

which consist of the two genera, Marsilea and Pilularia, we remark at the base of the leaves certain involucra of a coriaceous, thick substance, and either indehiscent or opening into several valves, divided internally into cells by membranous dissepiments. Each of these cells contains two other cells, inserted on a part of its inner coating: of these one sort is ovaria, or rather grains, composed of an external transparent membrane which swells with humidity, and becomes a thick layer of gelatinous substance; the other is an internal, hard, and coriaceous membrane, of a yellow colour, and indicating on its surface a particular point, through which the embryo is protruded upon being developed. The other organs are more numerous, and consist of membranous bags, slightly swelling from humidity, opening at the summit, and enclosing in the middle of a gelatinous mucus many spherical globules, which are much smaller than the grains. Their leaves develope in a gyrate manner, like ferns.

In the second section of this order, to which the name Salvinieæ may be given, and which consists of the genera Salvinia and Azolla, we find at the base of the leaves membranaceous involucra of two sorts, and containing different organs. One kind includes a bunch of grains of an ovate figure, containing only one embryo in Salvinia, and from six to nine in Azolla. The integument of these grains is thin, reticulated, brownish, and does not swell in water like that of true Marsileaceæ: the pedicel which supports them appears, in Salvinia, to communicate laterally with the grain. The other involucra, which are supposed to be male organs, have a very complex structure, and have been well observed by Mr. Brown. In Salvinia they contain a great number of spherical granules, attached by long pedicels to a central column: these granules are much smaller than the grains; their surface is reticulated in like manner, and they do not burst by the action of water. All the species are floaters, and their leaves are not gyrate when developing, but are more like those of Lycopodiaceæ.

Thus far M. Brongniart. With respect to the nature of these two kinds of grains or granules, it has been thought, as is obvious from the foregoing remarks, that the smaller are males and the larger females; which has been supposed to be proved by the experiments of M. Savi of Pisa. This observer introduced into different vessels, 1. the granules; 2. the grains; and, 3. the two intermixed. In the two first nothing germinated; in the third the grains floated to the surface and developed themselves perfectly. These observations have, however, been repeated by M. Duvernoy without the same result. And it must be remarked that, if the functions of these grains and granules be what has been attributed to them, the male power of action and the female powers of reception cannot exist till both are discharged from the membranes or involucra, in which they are contained and placed in contact in water. Is it impossible that the granules or supposed male organs should be only grains in an imperfectly developed condition?

5. *Mosses.*

In the structure of these plants neither vessels nor woody fibre are employed; and from henceforward those organs disappear from the organisation of all the tribes to be noticed. Their stem consists of elongated cellular tissue, from which arise leaves composed, in like manner, entirely of cellular tissue without woody fibre; the nerves, as they are called, or, more properly speaking, costæ, which are found in many species, being formed by the approximation of cellules more elongated than those that constitute the principal part of the leaf. The leaves are usually a simple lamina; but in Polytrichum and a few others they are furnished with little plates, called lamellæ, running parallel with the leaf, and originating from the upper surface. At the summit of some of the branches of many species are seated certain organs, which are called male flowers, but the true nature of which is not understood. They are possibly organs of reproduction of a particular kind, as both Mees and Haller are recorded to have seen them produce young plants.

Agardh says they have only the form of male organs; and that they really appear to be gemmulæ. By Hedwig they were called *spermatocystidia*.

But, whatever may be the nature of these organs, there is

no doubt of the reproductive functions of the contents of what is named the *theca* or *capsule*, which is a hollow urn-like body, containing *sporules:* it is usually elevated on a stalk, named the *seta*, with a bulbous base, surrounded by leaves of a different form from the rest, and distinguished by the name of *perichætial* leaves. If this theca be examined in its youngest state, it will be seen to form one of several small sessile ovate bodies (*pistillidia*, Agardh; *prosphyses*, Ehrhart; *adductores*, Hedwig), enveloped in a membrane tapering upwards into a point; when abortive they are called *paraphyses*. In process of time the most central of these bodies swells, and bursts its membranous covering, of which the greatest part is carried upwards on its point, while the seta on which the theca is supported lengthens. This part, so carried upwards, is named the *calyptra:* if it is torn away equally from its base, so as to hang regularly over the theca, it is said to be *mitriform;* but if it is ruptured on one side by the expansion of the theca, which is more frequently the case, it is denominated *dimidiate.* When the calyptra has fallen off or is removed, the theca is seen to be closed by a lid terminating in a beak or rostrum: this lid is the *operculum*, and is either deciduous or persistent. If the interior of the theca be now investigated, it will be found that the centre is occupied by an axis, called the *columella;* and that the space between the columella and the sides of the theca is filled with sporules. The brim of the theca is furnished with an elastic external ring, or *annulus*, and an interior apparatus, called the *peristomium:* this is formed of two distinct membranes, one of which originates in the outer coating of the theca, the other in the inner coat; hence they are named the outer and inner peristomia. The nature of the peristomium is practically determined at the period of the maturity of the theca. At this time both membranes are occasionally obliterated; but this is an unfrequent occurrence: sometimes one membrane only remains, either divided into divisions, called teeth, which are always some multiple of four, varying from that number as high as eighty, or stretching across the orifice of the theca, which is closed up by it; this is sometimes named the *tympanum*. Most frequently both membranes are present, divided into teeth, from differences in the number or cohesion of which the generic characters of mosses are in a great measure formed. For further information upon the peristomium I must refer to Mr. Brown's remarks upon Lyellia, in the 12th volume of the Linnean Transactions.

The interior of the theca is commonly unilocular; but in some species, especially of Polytrichum, it is separated into several cells by dissepiments originating with the columella.

If at the base of the theca there is a dilatation or swelling on one side, this is called a *struma;* if it is regularly lengthened downwards, as in most of the Splachnums, such an elongation is called an *apophysis*.

The only material exception to this description of Musci exists in Andræa, in which the theca is not an urn-like case, but splits into four valves, cohering by the operculum and base. From the foregoing description, it will be apparent that the organs of reproduction of mosses cannot be said to be analogous to the parts of fertilisation of perfect plants. I must not, however, omit the opinion of other botanists upon this subject. The office of males has been supposed by Micheli to be performed by the paraphyses; by Linnæus and Dillenius, by the thecæ; by Palisot de Beauvois, by the sporules; by Hill, by the peristomium; by Kœlreuter, by the calyptra; by Gærtner, by the operculum; and, finally, Hedwig has supposed the males to be the staminidia. The female organs were thought by Dillenius and Linnæus to be assemblages of staminidia; by Micheli and Hedwig, the young thecæ; and, by Palisot de Beauvois, the columella.

For some suggestions as to the analogy that is borne between the organs of mosses and of other plants, see MORPHOLOGY hereafter.

6. *Hepaticæ.*

These differ remarkably from each other in the modifications of their organs of reproduction, while they have a striking resemblance in their vegetation. This latter, which bears the name of *frond* or *thallus*, is either a leafy branched tuft, as in mosses, with the cellular tissue particularly large, and the leaves frequently furnished with lobes, and appendages at the base, called *stipulæ* or *amphigastria*; or a sinuous flat mass of green vegetable matter lying upon the ground.

In Jungermannia, that part which is most obviously connected with the reproduction of the plant, and which bears an indisputable analogy to the theca of mosses, is a valvular brown case, called the capsule or conceptacle, elevated upon a white cellular tender seta, and originating from a hollow sheath or perichætium arising among the leaves. This conceptacle contains a number of spiral loose fibres (*elateres*), enclosed in membranous cases, among which sporules lie intermixed: when fully ripe, the membranous case usually disappears, the spiral fibres, which are powerfully hygrometric, uncurl, and the sporules are dispersed. When young, the conceptacle is enclosed in a membranous bag, which it ruptures when it elongates, but which it does not carry upwards upon its point, as mosses carry their calyptra. This part, nevertheless, bears the latter name.

Besides the conceptacles of Jungermannia, there are two other parts which are thought to be also intended for the purpose of reproduction: of these one consists of spherical bodies, scattered over the surface of some parts of the frond, and containing a granular substance; the other is a hollow pouch, formed out of the two coats of a flat frond, and producing from its inside, which is the centre of the frond, numerous granulated round bodies, which are discharged through the funnel-shaped apex of the pouch.

There are also other bodies situated in the axillæ of the perichætial leaves, called anthers, (or *spermatocystidia*, by Hedwig, and *staminidia*, by Agardh,) which " are externally composed of an extremely thin, pellucid, diaphanous membrane," — " within they are filled with a fluid, and mixed with a very minute granulated substance, generally of an olivaceous or greyish colour: this, when the anther has arrived at a state of maturity, escapes through an irregularly shaped opening, which bursts at the extremity."

In *Monoclea* and *Targionia* organs very analogous to those of Jungermannia are formed for reproduction.

In *Marchantia* the frond is a lobed flat green substance, not dividing into leaves and stems, but lying horizontally upon the ground, and emitting roots from its under surface. The organs of reproduction consist, *firstly*, of a stalked fungilliform receptacle, carrying on its apex a calyptra, and bearing *thecæ* on its under side; *secondly*, of a stalked receptacle, plane on the upper surface, with oblong bodies imbedded vertically in the disk, and called anthers; *thirdly*, " of little open cups (*cystulæ*), sessile on the upper surface of the fronds, and containing minute green bodies (*gemmæ*), which have the power of producing new plants." The first kind is usually considered a female flower, its sporules intermixed with elateres: the second male, and the third viviparous apparatus. In the opinion of many modern botanists, the granules of both the two first are sporules: about the function of the last there is no difference of opinion.

In *Anthoceros*, while the vegetation is the same as in Marchantia, the organs of reproduction are very different. They consist of a subulate column, issuing from a perichætium perpendicular to the frond, and opening halfway into two valves, which discover, upon opening, a subulate columella, to which sporules are attached without any elateres. There are also cystulæ upon the frond, in which are enclosed pedicellate, reticulated bodies, called anthers.

Sphærocarpus consists of a delicate roundish frond, on the surface of which are clustered several cystulæ, each of which contains a transparent spherule filled with sporules.

In *Riccia* the spherules are not surrounded by cystulæ, but immersed in the substance of the frond.

7. *Lichenes.*

These have a lobed frond or thallus, the inner substance of which consists wholly of reproductive matter, which breaks through the upper surface in certain forms, which have been called fructification. These forms are twofold; firstly, *shields*, or *scutella*, which are little coloured cups or lines with a hard disk, surrounded by a rim, and containing *asci*, or tubes filled with sporules; and, secondly, *soredia*, which are heaps of pulverulent bodies scattered over the surface of the thallus. The nomenclature of the parts of lichens has been excessively extended beyond all necessity: it is, however, absolutely indispensable that it should be fully understood by those who wish to read the systematic writers upon the subject: —

1. *Apothecia*, are shields of any kind.
2. *Scutellum*, is a shield with an elevated rim, formed by the thallus. *Orbilla* is the scutellum of Usnea.
3. *Pelta*, is a flat shield without any elevated rim, as in the genus Peltidea.
4. *Tuberculum*, or *cephalodium*, is a convex shield without an elevated rim.
5. *Trica*, or *Gyroma*, is a shield, the surface of which is covered with sinuous concentric furrows.
6. *Lirella*, is a linear shield, such as is found in Opegrapha, with a channel along its middle.
7. *Patellula*, an orbicular sessile shield, surrounded by a rim which is part of itself, and not a production of the thallus, as in Lecidea. *D. C.*
8. *Globulus*, a round deciduous shield, formed of the thallus, and leaving a hollow when it falls off, as in Isidium. *D. C.*
9. *Pilidium*, an orbicular hemispherical shield, the outside of which changes to powder, as in Calycium. *D. C.*
10. *Podetia*, the stalk-like elongations of the thallus, which support the fructification in Cenomyce.
11. *Scypha* (*oplarium*, Neck.), is a cup-like dilatation of the Podetium, bearing shields on its margin.
12. *Soredia* (*globuli*, *glomeruli*), are heaps of powdery bodies lying upon any part of the surface of the thallus; the bodies of which the soredia are composed are called *conidia* by Link, and *propagula* by others.
13. *Cystula*, or *Cistella*, a round closed apothecium, filled with sporules, adhering to filaments which are arranged like rays around a common centre, as in Sphærophoron.
14. *Pulvinuli*, are spongy, excrescence-like bodies, sometimes rising from the thallus, and often resembling minute trees, as in Parmelia glomulifera. *Greville.*
15. *Cyphellæ*, are pale tubercle-like spots on the *under* surface of the thallus, as in Sticta. *Grev.*
16. *Lacunæ*, are small hollows or pits on the *upper* surface of the thallus. *Grev.*

17. *Nucleus proligerus*, is a distinct cartilaginous body, coming out entire from the Apothecia, and containing the sporules. *Grev.*
18. *Lamina proligera*, is a distinct body containing the sporules, separating from the apothecia, often very convex and variable in form, and mostly dissolving into a gelatinous mass. *Grev.*
19. *Fibrillæ*, are the roots.
20. *Excipulus*, is that part of the thallus which forms a rim and base to the shields.
21. *Nucleus*, is the disk of the shield which contains the sporules and their cases.
22. *Asci*, are tubes, in which the sporules are contained while in the nucleus.
23. *Thallodes*, is an adjective used to express an origin from the thallus: thus, *margo thallodes* signifies a rim formed by the thallus, *excipulus thallodes* a cup formed by the thallus.
24. *Lorulum*, is used by Acharius to express a filamentous, branched thallus.
25. *Crusta* is a brittle crustaceous thallus.
26. *Gongyli*, are the granules contained in the shields, and have been thought to be the sporules by which lichens are propagated: but this is doubted by Agardh.

8. *Algæ* and *Characeæ*.

These, with fungi, constitute the lowest order of vegetable developement: they vary in size from mere microscopic objects to a large size, and are composed of cellular tissue in various degrees of combination; some are even apparently animated, and thus form a link between the two great kingdoms of organised matter. Their sporules are either scattered through the general mass of each plant, or collected in certain places which are more swollen than the rest of the stem, and sometimes resemble the pericarpia of perfect plants. The terms used in speaking of the parts of Algæ are the following:—

1. *Gongylus;* a round hard body, which falls off the mother plant, and produces a new individual: this is found in Fuci. *W.*
2. *Thallus;* the plant itself.
3. *Apothecia;* the cases in which the organs of reproduction are contained.
4. *Peridiola,* Fr.; the membrane by which the sporules are immediately covered.
5. *Granula;* large sporules, contained in the centre of many Algæ; as in Gloionema of Greville. *Crypt. fl.* 6. 30.
6. *Pseudoperithecium;* ⎫ terms used by Fries to express such
7. *Pseudohymenium;* ⎬ coverings of Sporidia as resemble
8. *Pseudoperidium;* ⎭ in figure the parts named perithecium, hymenium, and peridium in other plants: see those terms.
9. *Sporidia;* granules which resemble sporules, but which are of a doubtful nature. It is in this sense that Fries declares that he uses the word: vide *Plant. homonom.* p. 294.
10. *Phycomater,* Fries; the gelatine in which the sporules of Byssaceæ first vegetate.
11. *Vesiculæ;* inflations of the thallus, filled with air, by means of which the plants are enabled to float.
12. *Hypha,* Willd.; the filamentous, fleshy, watery thallus of Byssaceæ.
13. *Nucula;* one of the apothecia of Characeæ; described by Greville to be a sessile, oval, solitary, spirally striated body, with a membranous covering, and the summit indistinctly cleft into five segments, containing sporules.
14. *Globules;* the second organ of Characeæ; the excellent observer last quoted describes it as a minute round body of a reddish colour, composed externally of a number of triangular (always?) scales, which separate, and produce its dehiscence. The interior is filled with a mass of elastic, transversely undulated filaments. The scales are composed of radiating hollow tubes, partly filled with minute coloured granules, which freely escape when the tubes are injured: their nature is wholly unknown, and, I believe, hitherto unnoticed.

15. *Coniocysta;* tubercle-like closed apothecia, containing a mass of sporules.

9. *Fungi.*

The structure of these plants is yet more simple than that of Algæ, consisting of little besides cellular tissue, among which sporules lie scattered. Some, of the lowest degree of developement, are composed only of a few cellules, of which one is larger than the rest, and contains the sporules; others are more highly compounded, consisting of myriads of cellules, with the sporules lying in cases, or *asci*. Notwithstanding the extreme simplicity of these plants, writers upon fungi have contrived to multiply the terms relating to them in a remarkable manner. The following are all with which I am acquainted :—

1. The *pileus*, or *cap*, is the uppermost part of the plant of an Agaricus, and resembles an umbrella in form.
2. The *stipes*, is the stalk that supports the pileus.
3. The *volva*, or wrapper, is the involucrum-like base of the stipes of Agaricus. It originally was a bag enveloping the whole plant, and was left at the foot of the stipes when the plant elongated and burst through it.
4. The *velum*, or veil, is a horizontal membrane, connecting the margin of the pileus with the stipes: when it is adnate with the surface of the pileus, it is a *velum universale;* when it extends only from the margin of the pileus to the stipes, it is a *velum partiale*.
5. The *annulus*, is that part of the veil which remains next the stipes, which it surrounds like a loose collar.
6. *Cortina*, is a name given to a portion of the velum which adheres to the margin of the pileus in fragments.
7. The *hymenium*, is the part in which the sporules immediately lie; in Agaricus, it consists of parallel plates, called *lamellæ*, or *gills;* these are *adnate* with the stipes, when the end next it cohere with it: when they are adnate, and at the same time do not terminate abruptly at the stipes, but are carried down it more or less, they are *decurrent;* if they do not adhere to the stipes, they are said to be *free*.

CHAP. III. COMPOUND ORGANS IN FLOWERLESS PLANTS. 209

8. *Stroma*, is a fleshy body to which flocci are attached; as in Isaria and Cephalotrichum.
9. *Flocci*, are woolly filaments found mixed with sporules in the inside of many Gastromyci. The same name is also applied to the external filaments of Byssaceæ.
10. *Orbiculus*, is a round flat hymenium contained within the peridium of some fungi; as Nidularia. *W.*
11. *Nucleus*, is the central part of a perithecium.
12. *Sporangium*, is the external case of Lycoperdon and its allies.
13. *Sporangiola*, are cases containing sporidia.
14. *Perithecium*, is a term used to express the part which contains the reproductive organs of Sphæria and its co-ordinates.
15. *Peridium*, is also a kind of covering of sporidia; *peridiolum* is its diminutive.
16. *Ostiolum*, is the orifice of the perithecium of Sphæria.
17. *Sphærula*, is a globose peridium, with a central opening, through which sporidia are emitted, mixed with a gelatinous pulp.
18. *Capillitium*, is a kind of purse or net, in which the sporules of some fungi are retained; as in Trichia. *W.*
19. *Trichidium*, or *pecten*, is a tender, simple, or sometimes branched hair, which supports the sporules of some fungi; as Geastrum. *W.*
20. *Asci*, are the tubes in which the sporidia are placed; *ascelli* or *thecæ* are the same thing.
21. *Sporidia*, are the immediate covering of sporules; *sporidiola* are *sporules*.
22. *Thallus*, or *thalamus*, is the bed of fibres from which many fungi arise.
23. *Mycelia*, are the rudiments of fungi, or the matter from which fungi are produced.

BOOK II.

PHYSIOLOGY; OR, PLANTS CONSIDERED IN A STATE OF ACTION.

GENERAL CONSIDERATIONS.

WE have thus far considered plants as inert bodies, having certain modifications of structure, and formed upon a plan, the simplicity and uniformity of which is among the most beautiful proofs of the boundless power and skill of the Deity.

Our next business is to enquire into the nature of their vital actions, and to consider those phenomena in which the analogy that undoubtedly exists between plants and animals is most striking; in a word, to make ourselves acquainted with the exact nature of the laws of vegetable life.

In explaining these things, it is not my purpose to notice all the different speculations that ingenious men have from time to time brought forward: for this would be incompatible with the plan of my work, and would be far more curious than useful. On the contrary, I propose, in the first place, to give a summary exposition of the principal phenomena of vegetation, and then to support the statement by a detailed account of the more important proofs of all doubtful points.

I am the more anxious that this should be understood, because I know how prone the world is to misconstruction: I therefore beg it to be remembered, that when particular opinions are here passed over in silence, it is because I do not think them sufficiently proved to be recorded consistently with the plan I have prescribed to myself.

If we place a seed, — that of an apple, for instance, — in earth at the temperature of 32° Fahr., it will remain inactive till it finally decays. But if it is placed in moist earth above the temperature of 32°, and screened from the action of light, its integument gradually imbibes moisture and swells, oxygen is absorbed, carbonic acid expelled, and the vital action of the embryo commences. It elongates downwards by the radicle, and upwards by the cotyledons; the former penetrating the soil, the latter elevating themselves above it, acquiring a green colour by the deposition of carbon absorbed from the atmosphere in the light, and unfolding in the form of two opposite roundish leaves. This is the first stage of vegetation: the young plant consists of little more than cellular tissue; only an imperfect developement of vascular and fibrous tissue being discoverable, in the form of a sort of cylinder of bundles, lying just in the centre. The part within the cylinder, at its upper end, is now the medulla, without it the bark; while the cylinder itself is the preparation for the medullary sheath, and consists of vertical fibres passing through cellular tissue, which separates them horizontally in every direction.

The young root is now absorbing from the earth its nutriment, which passes up to the summit of the plant by the cellular substance of the medulla, and is thence impelled into the cotyledons, where it is aërated and evaporated: such of it as is not fixed in the cotyledons passes down through the bark into the root.

Forced onwards by the current of sap, which is continually impelled upwards from the root, the plumula next ascends in the form of a little twig, at the same time sending roots in the form of fibres downwards in the centre of the radicle, which become the earliest portion of wood that is deposited: these fibres, by their action, now compel the root to emit little ramifications. Previously to the elongation of the plumula its apex has acquired the rudimentary state of a leaf: this continues to develope as the plumula elongates, until, when the first internodium of the latter ceases to lengthen, the leaf has actually arrived at its complete formation. When fully grown it repeats in a much more perfect manner the functions previously performed by the cotyledons: it aërates the sap that it receives, and returns the superfluous portion of it downwards through the bark to the root; it also sends fibres down between the medullary sheath and the bark, thus forming the first stratum of wood in the new stem. During these operations, while the plumula is ascending, its leaf forming and acting, and the woody matter created by it descending, the cellular tissue of the stem is forming, and expanding horizontally to make room for the new matter forced into it; so that developement is going on simultaneously both in a horizontal and perpendicular direction. This process may not inaptly be compared to that of weaving, the warp being the perpendicular, and the weft the horizontal, formation. In order to enable the leaf to perform its functions of aeration completely it is traversed by veins originating in the medulla, and has delicate evaporating pores (*stomata*), which communicate with a highly complex pneumatic system that extends to almost every part of the plant.

After the production of its first leaf by the plumula, others are successively produced around the axis at its elongating point, all constructed alike, connected with the stem or axis in the same manner, and performing precisely the same functions as have been just described. At last the axis ceases to elongate; the old leaves gradually fall off; the new leaves, instead of expanding after their formation, retain their rudimentary condition, harden, and fold over one another, so as to be a protection to the delicate point of elongation; or, in other words, become the scales of a bud. We have now a shoot with a woody axis, and a distinct pith and bark; and of a more or less conical figure. At the axilla of every leaf a bud had been generated during the growth of the axis; so that the shoot, when deprived of its leaves, is covered from end to end with little, symmetrically arranged, projecting points, which are the buds. The cause of the figure of the perfect shoot being conical is, that, as the wood originates from the base of the leaves, the lower end of the shoot, which has the greatest number of strata, because it has the greatest number of leaves above it, will be the thickest; and the upper end, which has had the fewest leaves to distend it by their deposit, will have the least diameter. Thus that part of the stem which has two leaves above it will have wood formed by two successive deposits; that which has nine leaves above it will have wood formed by nine successive deposits; and so on: while the extreme vital point, as it can have no deposit of matter from above, will have no wood, the extremity being merely covered by the rudiments of leaves hereafter to be developed.

If at this time a cross section be examined, it will be found that the interior is no longer imperfectly divided into two portions, namely, medulla and skin, as it was when first examined in the same way, but that it has distinctly two, internal, perfect, concentric lines, the outer indicating a separation of the bark from wood; and the inner, a separation of the wood from the medulla: the latter too, which in the first observation was fleshy, and saturated with humidity, is become distinctly cellular, and altogether or nearly dry.

With the spring of the second year and the return of warm weather vegetation recommences.

The uppermost, and perhaps some other, buds which were formed the previous year gradually unfold, and pump up sap from the stock remaining in store about them; the place of the sap so removed is instantly supplied by that which is next it; an impulse is thus given to the fluids from the summit to the roots; new sap is absorbed from the earth, and sent upwards through the wood of last year; and the phenomenon called the flow of the sap is fully completed, to continue with greater or less velocity till the return of winter. The axis of the buds elongates upwards, forming leaves and buds in the same way as the parent shoot: in like manner also each bud sends down its roots, in the form of fibres within the bark and above the wood of the shoot from which it sprang; thus forming on the one hand a new layer of wood, and on the other a fresh deposit of bark. In order to facilitate this last operation, the old bark and wood are separated in the spring by the exudation from both of them of the glutinous, slimy substance called cambium; which appears to be expressly intended, in the first instance, to facilitate the descent of the subcortical fibres of the growing buds; and, in the second place, to generate the cellular tissue by which the horizontal dilatation of the axis is caused, and which maintains a communication between the bark and the centre of the axis. These lines of communication have, by the second year, become sufficiently developed to be readily discovered, and are in fact the medullary rays spoken of in the last book. It will be remembered that there was a time when that which is now bark constituted a homogeneous body with the medulla; and that it was after the leaves began to come into action that the separation which now exists between the bark and medulla took place. At the time when they were indissolubly united they both consisted of cellular tissue, with a few spiral vessels upon the line indicative of future separation. When a deposit of wood was formed from above between them they were not wholly divided the one from the other, but the deposit was effected in such a way as to leave a communication by means of cellular tissue between the bark and the medulla; and, as this formation is at all times coætaneous with that of the wood, the communication so effected between the medulla and bark is quite as perfect at the end of the third year as it is at the beginning of the first; and so it will continue to be to the end of the growth of the plant. The sap which has been sucked into circulation by the unfolding leaves is exposed, as in the previous year, to the effect of air and light; is then returned through the petiole to the stem, and sent downwards through the bark, to be from it either conveyed to the root, or distributed horizontally by the medullary rays to the centre of the stem. At the end of the year the same phenomena occur as took place the first season: wood is gradually deposited by slower degrees, whence the last portion is denser than the first, and gives rise to the appearance called the annual zones: the new shoot or shoots are prepared for winter, and are again elongated cones, as was the first; and this latter has acquired an increase in diameter proportioned to the quantity of new shoots which it produced, new shoots being to it now what young leaves were to it before.

The third year all that took place the year before is repeated: sap is absorbed by the unfolding leaves; and its loss is made good by new fluids introduced by the roots and transmitted through the alburnum or wood of the year before; new wood and liber are deposited by matter sent downwards by the buds; cambium is exuded; the horizontal developement of cellular tissue is repeated, but more extensively; wood towards the end of the year is formed more slowly, and has a more compact character; and another ring appears indicative of this year's increase.

In precisely the same manner as in the second and third years of its existence will the plant continue to vegetate, till the period of its decay, each successive year being a repetition of the phenomena of that which preceded it.

After a certain number of years the tree arrives at the age of puberty: the period at which this occurs is very uncertain, depending in some measure upon adventitious circumstances, but more upon the idiosyncrasy, or peculiar constitution of the individual. About the time when this alteration of habit is induced, by the influence of which the sap or blood of the plant is to be partially directed from its former courses into channels in which its force is to be applied to the production of new individuals rather than to the extension of itself;—about this time it will be remarked that certain of the young branches do not elongate, as had been heretofore the wont of others, but assume a short stunted appearance, probably not growing two inches in the time which had been previously sufficient to produce twenty inches of increase. Of these little stunted branches, called *spurs*, the terminal bud acquires a swollen appearance, and at length, instead of giving birth to new leaves, produces from its bosom a cluster of flower-buds, or *alabastra*, which had been enwrapped and protected from injury during the previous winter by several layers of imperfect leaves, now brought forth as bracteæ. Sap is impelled into the calyx through the pedicel by gentle degrees, is taken up by it, and exposed by the surface of its tube and segments to air and light; but having very imperfect means of returning, all that cannot be consumed by the calyx is forced onwards into the circulation of the petals, stamens, and pistillum. The petals unfold themselves of a dazzling white tinged with pink, and expose the stamens; at the same time the disk changes into a saccharine substance, which nourishes the stamens and pistillum, and gives them energy to perform their functions.

At a fitting time, the stigmatic surface of the pistillum being ready to receive the pollen, this is injected upon it from the anthers, which have remained in approximation to it for that particular purpose. When the pollen touches the stigma, the grains adhere firmly to it by means of its viscid surface, then emit a delicate membranous tube, which pierces into the stigmatic tissue, lengthens there, and conveys the vivifying matter contained in the pollen towards the ovula, which it finally enters by means of their foramen. This has no sooner occurred than the petals and stamens fade and fall away, their ephemeral but important functions being accomplished. All the sap which is afterwards impelled through the peduncle can only be disposed of to the calyx and ovarium, where it lodges: both these swell and form a young fruit, which continues to grow as long as any new matter of growth is supplied from the parent plant. After a certain period the juices of the fruit cease to be increased by the addition of new matter, its surface performs the functions of leaves in exposing the juice to light and air; finally the surface loses its green colour, assumes the rich, ruddy glow of maturity; the juices cease to be influenced by light; the peduncle is no longer a passage for fluids, but dries up and becomes unequal to supporting the fruit, which at last falls to the earth. Here, if not destroyed by animals, it lies and decays: in the succeeding spring its seeds are stimulated into life, strike root in the mass of decayed matter that surrounds them, and spring forth as new plants to undergo all the vicissitudes of their parent.

Such are the progressive phenomena in the vegetation, not only of the apple, but of all trees that are natives of northern climates, and of a large part of the herbage of the same countries,—modified, of course, by peculiarities of constitution, as in annual and herbaceous plants, and in those the leaves of which are opposite and not alternate; but all the more essential circumstances of their growth are the same as those of the apple tree.

If we reflect upon these phenomena, our minds can scarcely fail to be deeply impressed with admiration at the perfect simplicity and, at the same time, faultless skill with which all the machinery is contrived upon which vegetable life depends. A few forms of tissue, interwoven horizontally and perpendicularly, constitute a stem; the developement, by the first shoot that the seed produces, of buds which grow upon the same plan as the first shoot itself, and a constant succession of the same phenomenon, causes an increase in the length and breadth of the plant; an expansion of the bark into a leaf, within which ramify veins proceeding from the seat of nutritive matter in the new shoot, the provision of air-passages in its substance, and of evaporating pores on its surface, enables the crude fluid sent from the roots to be elaborated and digested until it becomes the peculiar secretion of the species; the contraction of a branch and its leaves forms a flower; the disintegration of the internal tissue of a petal forms an anther; the folding inwards of a leaf is sufficient to constitute a pistillum; and, finally, the gorging of the pistillum with fluid which it cannot part with causes the production of a fruit.

In hot latitudes there exists another race of trees, of which palms are the representatives, and in the north there are many herbs, in which growth, by addition to the outside, is wholly departed from, the reverse taking place; that is to say, their diameter increasing by addition to the inside. As the seeds of such plants are formed with only one cotyledon, they are called monocotyledonous; and their growth being from the inside, they are also named endogenous. In these plants the functions of the leaves, flowers, and fruit are in nowise different from those of dicotyledons; their peculiarity consisting only in the mode of forming their stems. When a monocotyledonous seed has vegetated it usually does not disentangle its cotyledon from the testa, but simply protrudes the radicle; the cotyledon swelling, and remaining firmly encased in the seminal integuments. The radicle shoots downwards to become root; and afterwards a leaf is emitted from the side of the collum, which elongated at the same time as the radicle. This first leaf is succeeded by another facing it, and arising from its axilla; the second produces a third facing it, and arising also from its axilla; and, in this manner, the production of leaves continues, until the plant, if caulescent, is ready to produce its stem. Up to this period no stem having been formed, it has necessarily happened that the bases of the leaves hitherto produced have been all upon the same plane; and as each has been produced from the bosom of the other without any such intervening space, as occurs in dicotyledonous plants, it would have been impossible for the matter of wood, if any had been formed, to be sent downwards around the circumference of the plant: it would, on the contrary, have been necessarily deposited in the centre. In point of fact, however, no deposit of wood like that of dicotyledons takes place, either now or hereafter. The union of the bases of the leaves has formed a fleshy stock, cormus, or *plate*, which, if examined, will be found to consist of a mass of cellular tissue, traversed by perpendicular bundles of vascular tissue and woody fibre, taking their origin in the veins of the leaves, of which they are manifest prolongations downwards; and there is no trace of bark, medullary rays, or central pith: the whole body being a mass of pith, woody fibre, and vascular tissue mixed together. To understand this formation yet more clearly, consider for a moment the internal structure of the petiole of a dicotyledon: it is composed of a bundle or bundles of vascular tissue encased in woody fibre, surrounded on all sides with pith, or, which is the same thing, parenchyma. Now suppose a number of these petioles to be separated from their laminæ, and to be tied in a bunch parallel with each other, and, by lateral pressure, to be squeezed so closely together that their surfaces touch each other accurately, except at the circumference of the bunch. If a transverse section of these be made, it will exhibit the same mixture of bundles of woody fibre and parenchyma, and the same absence of distinction between medulla, wood, and bark, which has been noticed in the cormus, or plate, of monocotyledons.

As soon as the plate has arrived at the necessary diameter it begins to elongate upwards, leaving at its base those leaves that were before at its circumference, and carrying upwards with it such as occupied its centre; at the same time, new leaves continue to be generated at the centre, or, as it must now be called, at the apex of the shoot.

As fresh leaves are developed, they thrust aside to the circumference those which preceded them, and a stem is by degrees produced. Since it has not been formed by additions made to its circumference by each successive leaf, it is not conical, as in dicotyledons; but, on the contrary, as its increase has been at the centre, which has no power to extend its limits, being strictly confined by the circumference which, when once formed, does not afterwards materially alter in dimensions, it is, of necessity, cylindrical: and this is one of the marks by which a monocotyledon is often to be known in the absence of other evidence. The centre being but little acted upon by lateral pressure, it remains loose in texture, and, until it becomes very old, does not vary much from the density acquired by it shortly after its formation; but the tissue of the circumference being continually jammed together by the pressure outwards of the new matter formed in the centre, in course of time becomes a solid mass of woody matter, the cellular tissue once intermingled with it being almost obliterated, and appearing among the bundles it formerly surrounded, like the interstices around the minute pebbles of a mosaic gem.

Such is the mode of growth of palms, and of a great proportion of arborescent monocotyledons. But there are others in which this is in some measure departed from. In the common asparagus the shoots produce a number of lateral buds, which all develope and influence its form, as the buds of dicotyledons; so that the cylindrical figure of monocotyledons is exchanged for the conical; its internal structure is strictly endogenous. In grasses a similar conical figure prevails, and for the same reason; but they have this additional peculiarity, that their stem, in consequence of the great rapidity of its growth, is fistular, with transverse phragmata at its nodi. It is not certain whether the subsequent internal growth of the stem is ever sufficient to fill up the central cavity; but, from a specimen of a bamboo in my possession, I incline to think that the lower part of grass stems does sometimes become filled up with solid matter.

Upon one or other of the two plans now explained are all flowering plants developed; but in flowerless plants it is different. In arborescent ferns the stem consists of a cylinder of hard sinuous plates connected by parenchyma, and surrounding a hollow axis, which sometimes becomes filled up with solid matter. It would seem, in these plants, as if the stem consisted of a mere adhesion of the petioles of the leaves in a single row; but we are, at present, too little acquainted with them in a living state to form any fixed opinion upon the subject.

In mosses and some Hepaticæ the stem seems to consist of nothing but an axis formed of the united bases of the leaves; and their growth may be considered analogous to that of an annual shoot of a dicotyledon without its wood. The remainder of flowerless plants are principally mere horizontal expansions of cellular tissue, analogous to nothing that is known in the other parts of the vegetable kingdom.

CHAPTER I.

ELEMENTARY ORGANS.

That of these the CELLULAR TISSUE is the most important is apparent by its being the only one of the elementary organs that is uniformly present in plants; and by its being the chief constituent of all those compound organs that are most essential to the preservation of species.

It transmits fluids in all directions. In most cellular plants no other tissue exists, and yet there a circulation of sap takes place; it constitutes the whole of the medullary rays, conveying the elaborated juices from the bark towards the centre of the stem; all the parenchyma in which the sap is diffused upon entering the leaf, and by which it is exposed to evaporation, light, and atmospheric action, consists of cellular tissue; nearly all the bark in which the descending current of the sap takes place is also composed of it; and in endogenous plants, where no bark exists, there appears to be no other route that the descending sap can take than through the cellular substance in which the vascular system is imbedded. It is, therefore, readily permeable to fluid, although it has no visible pores.

In all cases of wounds, or even of the developement of new parts, cellular tissue *is first generated:* for example, the granulations that form at the extremity of a cutting when imbedded in earth, or on the lips of incisions in the wood or bark; the extremities of young roots; scales, which are generally the commencement of leaves; pith, which is the first part created when the stem shoots up; nascent stamens and pistilla; ovula; and, finally, many rudimentary parts;—all these are at first, or constantly, formed of cellular tissue alone.

It may be considered *the flesh of vegetable bodies:* the matter which surrounds and keeps in their place all the ramifications or divisions of the vascular system is cellular tissue. In this the plates of wood of exogenous plants, the fibres of endogenous plants, the veins of leaves, and, indeed, the whole of the central system of all of them, are either imbedded or enclosed.

The action of impregnation appears to take place exclusively through its agency. Pollen is only cellular tissue in a particular state; when it bursts, the vivifying particles it contains are a still more minute state of the same tissue: the coats of the anther are composed entirely of it; and the tissue of the stigma, through which impregnation is conveyed to the ovula, is merely a modification of the cellular. The ovula themselves, with their sacs, at the time they receive the vivifying influence, are a semitransparent congeries of cellules.

It is, finally, *the tissue in which alone amylaceous or saccharine secretions are deposited.* These occur chiefly in tubers, as in the potato and arrow-root; in rhizomata, as in the ginger; in soft stems, such as those of the sago-palm and sugar-cane; in albumen, as that of corn; in pith, as in the Cassava; in the disk of the flower, as in Amygdalus; and, finally, in the bark, as in all exogenous plants; and cellular tissue is the principal, or exclusive, constituent of these.

WOODY FIBRE is apparently destined merely for the conveyance of fluid upwards or downwards, from one end of a body to another, and for giving firmness and elasticity to every part.

That it is intended *for the conveyance of fluid* in particular channels seems to be proved, 1. from its constituting the principal part of all wood, particularly of that which is formed in stems the last in each year, and in which fluid first ascends in the ensuing season; 2. from its presence in the veins of leaves where a rapid circulation is known to take place, forming in those plants both the adducent and reducent channels of the sap; and, 3. from its passing downwards from the leaves into the bark, thus forming a passage through which the peculiar secretions may, when elaborated, arrive at the stations where they are finally to be deposited. Mr. Knight is clearly of opinion that they convey fluid either upwards or downwards; in which I fully concur with him: the power of cuttings to grow when inverted seems, indeed, a conclusive proof of this. Dr. Dutrochet, however, endeavours to prove that they merely serve for a downward conveyance.

With regard *to its giving firmness and elasticity to every part*, we need only consider its surprising tenacity, as evinced in hemp, flax, and the like; and its constantly surrounding and protecting the ramifications of the vascular system, which has no firmness or tenacity itself. To this evidence might be added, the admirable manner in which it is combined to answer such an end. It consists, as has been seen, of extremely slender tubes, each of which is indeed possessed of but a slight degree of strength; but being of different lengths, tapering to each extremity, and overlapping each other in various degrees, these are consolidated into a mass that considerable force is insufficient to break. Any one, who will examine a single thread of the finest flax with a microscope that magnifies 180 times, will find, that that which to the eye appears a single thread, is in reality composed of many distinct fibres.

The real nature of the functions of the VASCULAR SYSTEM has been the subject of great difference of opinion; and may, indeed, be said to be so still. *Spiral vessels* have been most commonly supposed to be destined for the conveyance of air; and it seems difficult to conceive how any one accustomed to anatomical observations, and who has remarked their dark appearance when lying in water, can doubt that fact. Nevertheless, many others, and among them Dr. Dutrochet, assert that they serve for the transmission of fluids upwards from the roots. This observer states, that if the end of a branch be immersed in coloured fluid, it will ascend in both the spiral vessels and ducts; but that in the former it will only rise up to the level of the fluid in which the branch is immersed, while, through the latter, it will travel into the extremities of the branches. It has, however, been asked with much justice, how the opinion that spiral vessels are the sap-vessels is to be reconciled with the fact of their non-existence in multitudes of plants in which the sap circulates freely. To which might have been, or perhaps has been, added the questions, why they do not exist in the wood, where a movement of sap chiefly takes place in exogenous trees? and also, how it happens that their existence is almost constantly connected with the presence of sexes, if they are only sap-vessels? And further, it has always been remarked, that if a transverse section of a vine, for instance, or any other plant, be put under water, bubbles of air rise through the water from the mouths of the spiral vessels. But then, it has been urged, that coloured fluids manifestly rise in the spiral vessels; a statement that has been admitted, when the spiral vessels are wounded at the part plunged in the colouring fluid, but denied in other circumstances. Indeed, to any observer acquainted with the difficulty of microscopic investigations, the obscurity that practically surrounds a question of this sort must be apparent enough.

The subject has, however, been investigated with much care by Dr. L. W. Theodore Bischoff, who instituted some very delicate and ingenious experiments, for the purpose of determining the real contents and office of the spiral vessels. It is impossible to find room here for a detailed account of his experiments, for which the reader is referred to his thesis, *De vera Vasorum Plantarum Spiralium Structura et Functione Commentatio:* Bonnæ, 1829. It must be sufficient to state, that, by accurate chemical tests, by the most careful purification of the water employed from all presence of air, and by separating bundles of the spiral vessels of the gourd (*Cucurbita Pepo*), and of some other plants, from the accompanying cellular substance, he came to the following conclusions, which, if not exactly, are probably substantially, correct: " That plants, like all other living bodies, require, for the support of their vital functions, a free communication with air; and that it is more especially oxygen, which, when absorbed by the roots from the soil, renders the crude fluid fit for the nourishment and support of a plant, just as blood is rendered fit for that of animals. But, for this purpose, it is not sufficient that the external surface should be surrounded by the atmosphere; other aeriferous organs are provided, in the form of spiral vessels, which are placed internally, and convey air containing an unusual proportion of oxygen, which is obtained through the root, by their own vital force, from the earth and water.

In a hundred parts of this air twenty-seven to thirty parts are of oxygen, which is in part lost during the day by the surface of plants under the direct influence of the solar rays."

With such evidence of the aeriferous functions of the spiral vessels it is difficult to contend; and, indeed, it seems probable that this question is settled as far as spiral vessels, properly so called, are concerned. But there are many vessels abounding in the wood, to which they give a porous appearance when cut across, and which I have called *ducts*, that, although perhaps mere modifications of the spiral vessel, are, nevertheless, so far distinct as to convey air at one period of their existence, and fluid at another. In the vine, for instance, the true spiral vessels of the medullary sheath and of the herbaceous parts, always filled with air, must be carefully distinguished from the ducts of the wood; which are, undoubtedly, filled during the principal flow of the sap with that fluid, although they finally become dry and empty. And it may be further remarked, that the dotted ducts of such plants as Phytocrene gigantea, or water-vine, so well represented by Mr. Griffith in Dr. Wallich's *Plantæ Asiaticæ Rariores*, are apparently the principal conduits in that curious plant, as they are in Gramineæ, and other monocotyledons, of the fluid absorbed from the earth.

So that, while true spiral vessels may be admitted as undoubted vehicles of air, ducts of all kinds, and especially dotted ducts, cannot be doubted to be the passages through which fluid is conveyed when great rapidity is required. I have already stated that, although all these vessels are, in the present state of our anatomical knowledge, considered as equally belonging to the vascular system, yet that the dotted will rather be referred eventually to the cellular, and then their lymphatic office will be unquestioned.

In regard to the functions of air-cells and lacunæ, it may be sufficient to remark, that in all cases in which they form a part of the vital system, as in water plants, they are cavities regularly built up of cellular tissue, and uniform in figure in the same species; while, on the other hand, where they are not essential to vitality, as in the pith of the walnut, the rice-paper plant, the stems of Umbelliferæ, and the like, they are ragged, irregular distensions of the tissue.

In the former case they are intended to enable plants to float in water; in the latter, they are caused by the growth of one part more rapidly than another.

CHAPTER II.

OF THE ROOT.

It is the business of the root to absorb nutriment from the soil, and to transmit it upwards into the stem and leaves; and also to fix the plant firmly in the earth. Although moisture is, no doubt, absorbed by the leaves of all and the stems of many plants, yet it is certain that the greater part of the food of plants is taken up by the roots; which, hence, are not incorrectly considered vegetable mouths.

But it is not by the whole surface of the root that the absorption of nutriment takes place; it is the spongioles almost exclusively to which that office is confided: and hence their immense importance in vegetable economy, the absolute necessity of preserving them in transplantation, and the certain death that often follows their destruction. This has been proved, in the following manner, by the celebrated Senebier:—he took a radish, and placed it in such a position that the extremity only of the root was plunged in water: it remained fresh several days. He then bent back the root, so that its extremity was curved up to the leaves: he plunged the bent part in water, and the plant withered soon; but it recovered its former freshness upon relaxing the curvature, and again plunging the extremity of the root into the water.

This explains why forest trees, with very dense umbrageous heads, do not perish of drought in hot summers or dry situations, when the earth often becomes mere dust for a considerable distance from their trunk, in consequence of their foliage turning off the rain: the fact is obviously that the roots near the stem are inactive, and have little or nothing to do as preservatives of life except by acting as conduits, while the functions of absorption go on through the spongioles, which, being at the extremities of the roots, are placed beyond the influence of the shadow, and extend wherever moisture is to be found. This property prevents a plant from exhausting the earth in which it grows; for, as the roots are always spreading further and further from the main stem, they are continually entering new soil, the nutritious properties of which are unexhausted.

It is generally believed that roots increase *only* by their extremities, and that, once formed, they never undergo any subsequent elongation. This was first noticed by Du Hamel, who passed fine silver threads through young roots at different distances, marking on a glass vessel corresponding points with some varnish: all the threads, except those that were within two or three lines of the extremity, always continued to answer to the dots of varnish on the glass vessel, although the root itself increased considerably in length. Variations in this experiment, which has also been repeated in another way by Mr. Knight, produced the same result. It is possible that this peculiarity may be universal in exogenous plants; but it certainly is not constant in endogenous plants; and I doubt very much whether it is not confined to roots with a woody structure. From the following experiments it will be seen that in Orchideæ the root elongates independently of its extremity. On the 5th of August I tied threads tightly round the root of a Vanilla, so that it was divided into three spaces, of which one was 7 inches long, another 4 inches, and the third, which was the free growing extremity, 1 inch and $\frac{3}{8}$. On the 19th of September the first space measured $7\frac{1}{8}$ inches, the second $4\frac{3}{8}$ inches, and the third or growing extremity $2\frac{1}{8}$ inches. A root of Aerides cornutum was, on the 5th of August, divided by ligatures into spaces, of which the first measured 1 foot 3 inches, the second $2\frac{1}{4}$ inches, the third $3\frac{1}{8}$ inches, and the fourth, or growing end, 1 inch and $\frac{1}{8}$. On the 19th September the first space measured 1 foot $3\frac{1}{2}$ inches, the second $2\frac{3}{4}$ inches, the third $3\frac{1}{4}$ inches, and the fourth $4\frac{3}{4}$ inches.

Occasionally roots appear destined to act as reservoirs of nutriment, on which those of the succeeding year may feed when first developed, as is the case in the Orchis, the Dahlia, and others. But it must be remarked, that the popular notion extends this circumstance far beyond its real limits, by including among roots bulbs, tubers, and other forms of stem in a state of anamorphosis.

By some botanists, and among them by M. De Candolle, it has been thought that roots are developed from special organs, which are to them what leaf-buds are to branches; and this function has been assigned to those little glandular swellings so common on the willow, called *lenticular glands* by Guettard, and *lenticelles* by De Candolle.

According to Mr. Knight, the energies of a variety artificially produced exist longer in the system of the root than in that of the stem; so that it is more advisable to propagate old varieties of fruit trees from cuttings of the root than from those of the stem.

The roots not only absorb fluid from the soil, but they return a portion of their peculiar secretions back again into it; as has been found by Brugmans, who ascertained that some plants exude an acid fluid from their spongioles; and also by Mr. Macare, who has proved that to excrete superabundant matter from the roots is a general property of the vegetable kingdom. This, taken together with the fact that plants cannot digest their own secretions, explains why soil is so deteriorated by one species having long grown in it, that it will not support other individuals of the same species, until the fecal matter deposited in it shall have been decomposed. This is the solution of the necessity of the rotation of crops.

CHAPTER III.

OF THE SAP.

For the sustenance of plants a fluid is necessary which is absorbed by the roots from the earth, then sent upwards into the stem, afterwards impelled into the leaves, whence it descends through the liber, transferring itself to the inmost parts of the wood. This fluid, which constitutes the blood of plants, is called the sap. When first introduced into the system, and even when altered in some degree, by having dissolved the various substances it encounters in its passage, it is *true sap;* afterwards, when its nature has been more changed by elaboration in the leaves, it becomes what is called the *proper juice.*

If the sap be examined in its most simple state, it will be found to consist of water, mucilage, and sugar. As the two last can scarcely have been absorbed directly from the earth, it is inferred that as soon as the fluids, taken up by the roots, enter the system they suffer some chemical decomposition, the result of which is the production of mucilage and sugar. In addition to the supply of sap which is obtained by the roots, a certain quantity is, no doubt, also absorbed from the atmosphere by the leaves; as is evident from succulent plants, which will continue to grow and acquire weight long after their roots are severed from the earth. This absorption on the part of the leaves chiefly takes place at night, or in cloudy weather; while perspiration, on the other hand, goes on in the daytime in bright weather.

With regard to the chemical nature and changes of the sap, I cannot do better than give the statement of Link, with some necessary alterations. *The food of plants must be composed of oxygen, hydrogen, carbon, and azote.* Water, consisting of oxygen and hydrogen alone, is not sufficient. Many experiments, indeed, have been instituted to prove that pure water is a sufficient food, especially by Van Helmont, Eller,

Bonnet, Du Hamel, and others; but it is probable, as Wallerius has inferred, that the water out of which plants are formed already contains the necessary chemical principles. To this it is objected, that plants grown in water alone never arrive at perfection or mature their seeds. But this is not strictly true: they do perfect their seeds; but it is not surprising that crude water should be insufficient for purposes which are fully answered by water properly mixed and tempered.

That the extractive matter contained in earth was the real food of plants, was long ago stated by Woodward and Kylbel; and most physiologists have adopted this opinion. But it has been estimated by Theodore de Saussure that a plant when dried does not derive more than a twentieth part of its weight from extractive matter and carbonic acid dissolved in water: now, supposing this calculation not to be very accurate, yet it is probable that it is not far from the truth; and it at least serves to show that extractive matter and carbonic acid are not alone sufficient for the nutriment of plants.

Nevertheless, if neither extractive matter nor carbonic acid can be considered to constitute exclusively the food of plants, it is at least quite certain that they not only cannot exist without the latter, but that it forms by far the greater part of their food. It is well known that roots cannot perform their functions unless within the reach of the atmosphere. This arises from the necessity for their feeding upon carbonic acid, which, after having been formed by the oxygen of the atmosphere combining with the carbon in the soil, is then received into the system of the plant, to be impelled upwards, dissolved in the sap till it reaches the leaves, where it is decomposed by light, the oxygen liberated, and the carbon fixed. It has also been ascertained that, feed plants as you will, they will neither grow nor live whether you offer them oxygen, hydrogen, azote, or any other gaseous or fluid principle, unless carbonic acid is present.

Those principles are called foreign to plants which cannot be referred to either hydrogen, oxygen, carbon, or azote: such, for example, are carbonate of soda, sulphate of soda, nitrate of soda, the carbonates of potash, lime, and magnesia, phosphate of lime, chlorides of soda and potash, and the oxides of aluminum, silicium, iron, and manganese; and even, occasionally, phosphorus. De Saussure has demonstrated that the chemical principles which are present in soil are also to be found in the ashes of a plant that has grown in it; but he admits they undergo certain chemical changes, in consequence of the organic powers of vegetation. Dr. John, also, has ascertained that several salts, when taken up by the roots, undergo peculiar changes; as, for instance, potash into soda. To these, many well known observations may be added; plants growing in saline places contain so much salt, that it is perceptible to the taste. Potash also is yielded by the ashes of plants growing in salt soil; so that there is no doubt that this substance is produced by the detraction by organic power of the chlorine of chloride of potash.

There are, however, some experiments which, if they could be depended on, would materially weaken these hypotheses. Schrader grew barley and rye in well washed flower of sulphur, moistened with distilled water: they were afterwards analysed, and found to contain silex, lime, and magnesia, as well as oxides of iron and manganese. The same plants produced in earth did not yield a greater weight of ashes than those grown in sulphur; and these experiments are confirmed by those of Braconnot, as recorded in the Annales de Chimie, vol. lxi. p. 187. White mustard was grown in well washed litharge and flower of sulphur, moistened with distilled water; and its ashes yielded oxide of silicium and aluminum, carbonate of lime, and oxide of iron. Both these observers conclude that these foreign principles were produced by the organic power of vegetation; but Gehlen suggests that, as many principles are dissolved in oxygen and dispersed through the atmosphere, they may be communicated to plants by that medium. As it seems, however, certain that chemical principles do undergo changes from organic powers, it is not improbable that all the foreign principles of plants were originally suspended in water; but that plants have no capability of doing more than decompose such compound principles as they may absorb.

To reconcile the experiments of Saussure and John with

those of Schrader and Braconnot, it is suggested by Professor Link that the power and necessity of taking up matter from the soil varies in different plants: that some depend wholly upon it for their formation, others less, and some not at all. Thus, Salsolas grow but in saline soil, and in soil destitute of salt become languid and weak; other plants will only vegetate in calcareous earth: Trifolium pratense prefers gypsum, and succulent plants scarcely require soil at all.

That some plants have the power of secreting one kind of accessory principle, and others another kind from the same food, is clear from the fact, that, if wheat and peas be grown in the same water, earth, or medium, the former will uniformly deposit silex in their cuticle, and the latter never.

The course which is taken by the sap after entering a plant is the next subject of consideration. The opinion of the old botanists was, that it ascended from the roots between the bark and the wood: but this has been long disproved by modern investigators, and especially by the experiments of Mr. Knight. If a trunk is cut through in the spring, at the time the sap is rising, this fluid will be found to exude more or less from all parts of the surface of the section, except the hardest heart-wood, but most copiously from the alburnum. If a branch is cut half through at the same season, it will be found that, while the lower face of the wound bleeds copiously, scarcely any fluid exudes from the upper face; from which, and other facts, it has been fully ascertained that the sap rises through the wood, and chiefly through the alburnum. Observations of the same nature have also proved that it descends through the liber. But the sap is also diffused laterally through the cellular tissue, and this with great rapidity; as will be apparent upon placing a branch in a coloured infusion, which will ascend and descend in the manner just stated, and will also disperse itself laterally in all directions round the principal channels of its upward and downward route.

Corti, in 1774, Fontana, L. C. Treviranus, and especially Professor Amici, have made some most curious observations upon the movement of the sap in Chara. If a portion of Chara flexilis, or of any of the transparent species, or of any crustaceous kind, the opaque cuticle of which is first scraped away, be examined, a current of sap will be distinctly seen in each cellule setting from joint to joint, flowing down one side and returning up the other, without any membrane intervening to separate the opposing currents. Each cellule has a movement of its own, independent of that of the cellules above and below it. Sometimes the movement stops, and then goes on again after a brief interval If a cellule is divided into two by a ligature passed round it, a separate movement is seen in each of the divisions. This motion is rendered distinctly obvious by the numerous minute green granules which float in the transparent fluid, and which follow the course of the currents.

Another sort of motion, which it is probable is common to all descriptions of plants, has been seen by Link and others, particularly by Schultz, in the Chelidonium, Ficus Carica, and other plants. It is described as an exceedingly rapid motion of the fluid, which rushes out of one set of vessels, apparently tubes of woody fibre, into another, in a constant, uninterrupted stream. This has been denied, it is true, by Dutrochet; but he does not appear to have succeeded in seeing that which is nevertheless visible enough, if proper precautions are taken. Mirbel and Cassini both confirm the statements of Schultz; and to their testimony I may be permitted to add my own. In Alisma plantago and the transparent stipules of Ficus elastica I have distinctly seen powerful currents, such as are described, rushing along the tubes, like a stream of water down an inclined channel.

That there must be an exceedingly rapid flow of sap in many plants, is evident from the great loss they often experience by perspiration, — all which must be made good by fluid absorbed by the roots. A young vine-leaf, in a hot day, perspires so copiously, that, if a glass be placed next its under surface, it is presently covered with dew, which, in half an hour, runs down in streams. Hales computed the perspiration of plants to be seventeen times more than that of the human body. He found a sun-flower lose one pound four ounces, and a cabbage one pound three ounces a day by perspiration. Guettard asserts that the young shoots of

Cornus mascula lose twice their own weight a day. This perspiration is regulated by the number of the stomata: hence evergreens, in which they are small, and less numerous than in deciduous or herbaceous plants, perspire much less.

With this function are connected all the phenomena that attend transplantation. If a growing plant is removed from one situation to another in the summer, it will die; because its spongioles will be so much destroyed as to be incapable of absorbing fluid from the soil as fast as it is given off by the leaves; and hence the system will be emptied of fluid. But if a plant is growing in a pot, it may be transplanted at any season of the year; because its spongioles, being uninjured, will be able to counterbalance the loss caused by perspiration, as well after transplantation as before, if not better.

With regard to the vessels through which this universal diffusion of the sap takes place, it has already been stated that its upward course is always through the woody fibre, and probably also through the ducts; and that it passes downwards through the woody fibre. But there can be no reasonable doubt that it is also dispersed through the whole system by means of some permeable quality of the membranes of the cellular tissue, which is invisible to our eyes, even aided by the most powerful glasses. It has also been suggested that the sap finds its way upwards, downwards, and laterally through the intercellular passages which exist at the points of union of every individual elementary organ. That such a channel of communicating the sap is employed by Nature to a certain extent I do not doubt, especially in those plants in which the intercellular passages are very large; but whether this is an universal law, or has only a partial operation, is quite unknown, and is not perhaps susceptible of absolute proof. Link seems disposed to deny any conveyance of fluids through the intercellular passages.

The accumulation of sap in plants appears to be attended with very beneficial consequences, and to be deserving of the especial attention of gardeners. It is well known how weak and imperfect is the inflorescence of the turnip tribe, forced to flower before their fleshy root is formed; and how vigorous it is after that reservoir of accumulated sap is completed. Mr.

Knight, in a valuable paper upon this subject, remarks that the fruit of melons, which sets upon the plant when very young, uniformly falls off; while, on the contrary, if not allowed to set until the stem is well formed, and much sap accumulated for its support, it swells rapidly, and ripens without experiencing any deficiency of food in the course of its growth. In like manner, if a fruit tree is by any circumstance prevented bearing its crop one year, the sap that would have been expended accumulates, and powerfully contributes to the abundance and perfection of the fruit of the succeeding year. And again, the plan recommended by Mr. Knight, of always planting large tubers of the potatoe, is another proof of the importance of plenty of accumulated sap to the vigorous growth of all plants.

The cause of the motion of the sap is a subject which has long excited great curiosity, and has given rise to numberless conjectures. It was for a long time believed that there was a sort of circulation of the sap of plants, to and from a common point, analogous to that of the blood of animals; but this was disproved by Hales, and is not now believed. This excellent observer, whose " Statics " are an eternal monument of his industry and skill, thought that the motion of the sap, the rapidity of which he had found to be greatly influenced by weather, depended upon the contraction and expansion of the air, which exists in great quantities in the interior of plants. Others have ascribed the motion to capillary attraction. Mr. Knight was once of opinion that it depended upon a hygrometrical property of the plates of silver grain (medullary rays), which traverse the stem in all directions. A number of other theorists have called to their aid a supposed irritability of the vessels; but no contraction of the vessels has ever yet been noticed, and certainly does not take place in Chara, where the motion has been most distinctly observed. Du Petit Thouars suggests that it arises thus: — in the spring, as soon as vegetation commences, the extremities of the branches and the buds begin to swell: the instant this happens a certain quantity of sap is attracted out of the circumjacent tissue for the supply of those buds; the tissue, which is thus emptied of its sap, is filled instantly by that beneath or about it: this is in its turn replenished by the next; and thus the whole mass of fluid is set in motion, from the extremities of the branches down to the roots. Du Petit Thouars is therefore of opinion that the expansion of leaves is not the effect of the motion of the sap, but, on the contrary, is the cause of it; and that the sap begins to move at the extremities of the branches before it stirs at the roots. That this is really the fact, is well known to foresters and all persons accustomed to the felling or examination of timber in the spring. Some good observations upon this were communicated to Mr. Loudon's *Gardener's Magazine*, by Mr. Thomson, gardener at Welbeck; who, however, drew a wrong inference from them.

Amici is of opinion that the motion in Chara depends upon galvanic action; and Dutrochet has since formed a theory of all the motions of fluids in plants depending upon the same agency. He found that small bladders of animal and vegetable membrane, being filled with a fluid of greater density than water, securely fastened, and then thrown into water, acquired weight; he also remarked, that if the experiment was reversed, by filling with water and immersing them in a denser fluid, the contrary took place, and that the bladders lost weight: he took a small bladder, and filled it with milk, or gum arabic dissolved in water; to the mouth of this bladder he adapted a tube, and then plunged the bladder in water: in a short time the milk rose in the tube, whence he inferred that water had been attracted through the sides of the bladder. This experiment was also reversed, by filling the bladder with water, and plunging it in milk: the fluid then fell in the tube, whence he inferred that water had been attracted through the coat of the bladder into the milk. From these, and other experiments, M. Dutrochet arrived at the inference that, if two fluids of unequal density are separated by an animal or vegetable membrane, the denser will attract the less dense through the membrane that divides them: and this property he calls *endosmose*, when the attraction is from the outside to the inside; and *exosmose*, when it operates from the inside to the outside. In pursuing this investigation he remarked, that if an empty bladder is immersed in water, and the negative pole of a galvanic battery introduced into it, while the positive pole is applied to the water on the outside, a passage of fluid takes place through the membrane, as had previously happened when the bladder contained a fluid denser than water; by reversing the experiment, the reverse was found to take place: from all which Dutrochet deduces the following theory, that when two fluids of unequal density are separated by an intervening membrane, the more dense is negatively electrified, and the less dense positively electrified; in consequence of which two electric currents of unequal power set through the membrane, carrying fluid with them; that which sets from the positive pole, or less dense fluid, to the negative pole, or more dense fluid, being much the most powerful: and that the fluids of plants being more dense than those which surround them, a similar action takes place between them and the water in the soil, by means of which the latter is continually impelled into their system. Philosophers do not seem disposed to admit the legitimacy of M. Dutrochet's conclusion, that this transmission takes place by means of *galvanic agency*; but that the phenomenon is correctly described by the ingenious author, and that it is constantly operating in plants, is beyond all dispute. It is by endosmose that vapour is absorbed from the atmosphere, and water from the earth; that sap is attracted into fruits by virtue of their greater density; and probably that buds are enabled to empty the tissue that surrounds them when they begin to grow: it will, perhaps, be found the most ready explanation of most of the phenomena connected with the movement of fluids.

CHAPTER IV.

OF THE PITH, WOOD, AND BARK.

Various are the notions that from time to time have been entertained about the PITH. The functions of brain, lungs, stomach, nerves, spinal marrow, have by turns been ascribed to it. Some have thought it the seat of fecundity, and have believed that fruit trees deprived of pith became sterile; others supposed that it was the origin of all growth; and another class of writers, we cannot say observers, have declared that it was the channel of the ascent of sap. It is, however, no part of the plan of this work to refute this and similar exploded speculations.

It is probable that its real and only use is to serve in the infancy of a plant for the reception of the sap, upon which the young and tender vessels that surround it are to feed when they are first formed; a time when they have no other means of support. M. Dutrochet considers it to act not only as a reservoir of nutriment for the young leaves, but also to be the place in which the globules, which he calls nervous corpuscles, are formed out of the elaborated sap. (*L'Agent Immédiat*, &c., p. 44, &c.)

The MEDULLARY SHEATH seems to perform a far more important part in the economy of plants; it diverges from the medulla whenever a leaf is produced, and, passing through the petiole, ramifies among the cellular tissue of the lamina, where it appears as veins: hence veins are always composed of bundles of woody fibre and spiral vessels. So situated, the veins are in the most favourable position that can be imagined for absorbing the fluid that, in the first instance, is conducted to the young pith, and that is subsequently impelled upwards through the woody fibre. So essential is the medullary sheath to vegetation in the early age of a branch, that, as is well known, although the pith and the bark, and even the young wood, may be destroyed, without the life of a young shoot being much affected; yet, if the medullary sheath be cut through, the pith, bark, or wood being left, the part above the wound will perish.

The BARK acts as a protection to the young and tender wood, guarding it from cold and external accidents. It is also the medium in which the proper juices of the plant in their descent from the leaves are finally elaborated, and brought to the state which is peculiar to the species. It is from the bark that they are horizontally communicated to the medullary rays, by them to be deposited in the tissue of the wood. Hence, the character of timber is almost wholly dependent upon the influence of the bark, as is apparent from a vertical section of a grafted tree, through the line of union of the stock and scion. This line will be found so exactly drawn, that the limits of the two are determined in the oldest specimens as accurately as if they were fixed by rule and line: the woody tissue will be found uninterruptedly continuous through the one into the other, and the bark of the two indissolubly united; but the medullary rays emanating from the bark of each will be seen to remain as different as they were, while the stock and scion were distinct individuals.

As the bark, when young, is green like the leaves, and as the latter are manifestly a mere dilatation of the former, it is highly probable, as Mr. Knight believes, that the bark exercises an influence upon the fluids deposited in it wholly analogous to that exercised by the leaves, which will be hereafter explained. Hence it has been named, with much truth, the universal leaf of a vegetable.

The business of the MEDULLARY RAYS is, no doubt, exclusively to maintain a communication between the bark, in which the secretions receive their final elaboration, and the centre of the trunk, in which they are at last deposited. This is apparent from tangental sections of dicotyledonous wood manifesting an evident exudation of liquid matter from the wounded medullary rays, although no such exudation is elsewhere visible. In endogenous plants, in which there appears no necessity for maintaining a communication between the centre and circumference, there are no medullary rays. These rays also serve to bind firmly together the whole of the internal and external parts of a stem, and they give the peculiar character by which the wood of neighbouring species may be distinguished. If plants had no medullary rays, their wood would probably be, in nearly allied species, undistinguishable; for we are scarcely aware of any appreciable difference in the appearance of fibrous or vascular tissue. But the medullary rays differing in abundance, in size, and in other respects, impress characters upon the wood which are extremely marked. Thus, in the cultivated cherry, the plates of the medullary rays are very thin, the adhesions of them to the bark are very slight, and hence a section of the wood of that plant has a pale, smooth, homogeneous appearance; but in the wild cherry the medullary plates are much thicker, they adhere to the bark by deep broad spaces, and are arranged with great irregularity, so that a section of the wood of that variety has a deeper colour, and a twisted, knotty, very uneven appearance. As the medullary rays develope only horizontally, when two trees in which they are different are grafted or budded together, the wood of the stock will continue to preserve its own peculiarity of grain, notwithstanding its being formed by the woody matter sent down by the scion; for it is the horizontal developement that gives its character to the grain, and not the perpendicular fibres which are incased in it.

The wood is at once the support of all the deciduous organs of respiration, digestion, and impregnation, the deposit of the secretions peculiar to individual species, and also the reservoir from which newly forming parts derive their sustenance until they can establish a communication with the soil. Regarding the precise manner in which it is created, there has been great diversity of opinion. Linnæus thought it was produced by the pith; Grew, that the liber and wood were deposited at the same time in a single mass which afterwards divided in two, the one half adhering to the centre, the other to the circumference; Malpighi conceived that the wood of one year was produced by an alteration of the liber of the previous season. Duhamel believed that it was deposited by the secretion already spoken of as existing between the bark and wood, and called cambium: he was of opinion that this cambium was formed in the bark, and became converted into both cellular tissue and woody fibre; and he demonstrated the fallacy of those theories according to which new wood is produced by the wood of a preceding year. He removed a portion of bark from a plum tree; he replaced this with a similar portion of a peach tree, having a bud upon it. In a short time a union took place between the two. After waiting a sufficient time to allow for the formation of new wood, he examined the point of junction, and found that a thin layer of wood had been formed by the peach bud, but none by the wood of the plum, to which it had been tightly applied. Hence he concluded that alburnum derives its origin from the bark, and not from the wood. A variety of similar experiments was instituted with the same object in view, and they were followed by similar results. Among others, a plate of silver was inserted between the bark and the wood of a tree at the beginning of the growing season. It was said, that if new wood was formed by old wood it would be subsequently found pushed outwards, and continuing to occupy the same situation; but that if new wood was deposited by the bark, the silver plate would in time be found buried beneath new layers of wood. In course of time the plate was examined, and was found enclosed in wood.

Hence the question as to the origin of the wood seemed settled; and there is no doubt that the experiments of Duhamel are perfectly accurate and satisfactory as far as they go. It soon, however, appeared that, although they certainly proved that new wood is not produced by old wood, it was not equally clear that it originated from the bark. Accordingly a new set of experiments was instituted by Mr. Knight, for the purpose of throwing a still clearer light upon the production of the wood. Having removed a ring of bark from above and below a portion of the bark furnished with a leaf, Mr. Knight remarked that no increase took place in the wood above the leaf, while a sensible augmentation was observable in the wood below the leaf. It was also found that if the upper part of a branch is deprived of leaves, the branch will die down to the point where leaves have been left, and below that will flourish. Hence an inference is drawn that the wood is not formed out of the bark as a mere deposit from it, but that it is produced from matter elaborated in the leaves and sent downwards,—either through the vessels of the inner bark, along with the matter for forming the liber by which it is subsequently parted with; or that it and the liber are transmitted distinct from one another, the one adhering to the alburnum, the other to the bark. I know of no proof of the former supposition: of the latter there is every reason to believe the truth. Mr. Knight is of opinion that two distinct sets of vessels are sent down, one belonging to the liber, the other to the alburnum; and if a branch of any young tree, the wood of which is formed quickly, be examined when it is first bursting into leaf, these two sets may be distinctly seen and traced. Take, for instance, a branch of lilac in the beginning of April and strip off its bark: the new wood will be distinctly seen to have passed downwards from the base of each leaf, diverging from its perpendicular course, so as to avoid the bundle of vessels passing into the leaf beneath it; and if the junction of a new branch with that of the previous year be examined, it will be found that all the fibres of wood already seen proceeding from the base of the leaves, having arrived at this point, have not stopped there, but have passed rapidly downwards, adding to the branch an even layer of fibrous matter or young wood; and turning off at every projection which impedes them, just as the water of a steady but rapid current would be diverted from its course by obstacles in its stream. Now, if the new wood were a mere deposit of the bark, the latter, as it is applied to every part of the old wood, would deposit the new wood equally over the whole surface of the latter, and the deviation of the fibres from obstacles in their downward course could not occur. This, therefore, in my mind, places the question as to the origin of the wood beyond all further doubt. Mirbel, who formerly advocated the doctrine of wood being deposited by bark, has, with the candour of a man of real science, fairly admitted the opinion to be no longer tenable; and he has suggested in its room that wood and bark are independent formations,—which is no doubt true,—but, he adds, created out of cambium, in which it is impossible to concur; for this reason. All the writers hitherto mentioned or adverted to have considered the formation of wood only with reference to exogenous trees, and to such only of them as are the common forest plants of Europe. Had they taken into account exotic trees or any endogenous plants, they would have seen that none of their theories could possibly apply to the formation of wood in that tribe. In many exogenous plants of tropical countries wood is not deposited in regular circles all round the axis, but only on one side of the stem, or along certain lines upon it: were it a deposit from the bark, or a metamorphosis of cambium, it would necessarily be deposited with some kind of uniformity. In endogenous trees there is no cambium, and yet wood is formed in abundance; and the new wood is created in the centre, and not in the circumference: so that bark can have, in such cases, nothing whatever to do with the creation of wood.

No doubt aware of most of the difficulties in the way of the common theories of the formation of wood, M. Du Petit Thouars, an ingenious French physiologist, who had possessed opportunities of examining the growth of vegetation in tropical countries, constructed a theory, which, although in many points similar to the one proposed, but not proved, by his countryman, De la Hire, is nevertheless, from the facts and illustrations skilfully brought by the French philosopher to his aid, to be considered legitimately as his own. The attention of Du Petit Thouars appears to have been first especially called to the real origin of wood by having remarked, in the Isle of France, that the branches which are emitted by the truncheons of Dracæna (with which hedges are formed in that colony) root between the rind and old wood, forming rays of which the axis of the new shoot is the centre. These rays surround the old stem; the lower ones at once elongate greatly towards the earth, and the upper ones gradually acquire the same direction; so that at last, as they become disentangled from each other, the whole of them pass downwards to the soil. Reflecting upon this curious fact, and upon a multitude of others which I have no space to detail, he arrived at the conclusion, that it is not merely in the property of increasing the species that buds agree with seeds, but that they emit roots in like manner; and that the wood and liber are both formed by the downward descent of bud-roots, at first nourished by the moisture of the cambium, and finally imbedded in the cellular tissue which is the result of the organisation of that secretion. That first tendency of the embryo, when it has disengaged itself from the seed, to send roots downwards and a stem and leaves upwards, and to form buds in the axillæ of the latter, is in like manner possessed by the buds themselves; so that plants increase in size by an endless repetition of the same phenomenon.

Hence a plant is formed of multitudes of buds or fixed embryos, each of which has an independent life and action: by its elongation upwards forming new branches and continuing itself, and by its elongation downwards forming wood and bark; which is therefore, in Du Petit Thouars's opinion, a mass of roots.

A great deal of opposition has been offered to this view, especially among this writer's own countrymen; but it is remarkable that many of his antagonists have been from a class of naturalists of whom it may be said, that they are better known in consequence of the celebrity of the object of their attack than for any reputation of their own. To this, however, there are some exceptions, as, for instance, MM. Mirbel and Desfontaines, two of the most learned botanists of France. This theory, nevertheless, seems the only one that is adapted at once to the explanation of the real cause of the many anomalous forms of exogenous stems which must be familiar to the recollection of all botanists, and that, at the same time is equally applicable to the exogenous and endogenous modes of growth; a condition which, it will be readily admitted, is indispensable to any theory of the formation of wood that may be proposed. It also offers the simplest explanation of the phenomena that are constantly occurring in the operations of gardening.

It has recently been a subject of discussion in the Academy of Sciences at Paris; when M. Poiteau supported the theory, and MM. Mirbel, Cassini, and Desfontaines opposed it. The arguments used by the latter were two, both of which are undoubted fallacies. The first was, that if a large ring of bark be taken from the stem of a sycamore, and be replaced by a similar ring entirely destitute of buds from a red maple, the new bark will graft itself with the sycamore, and in time red maple wood will be formed beneath it. They said this ligneous production could not be derived from the buds of the red maple, because the ring of bark was devoid of any; nor could it proceed from the buds of the sycamore, because they would produce sycamore wood. But it is obvious that, in this experiment, the character of the red maple wood was derived from its medullary rays, which first formed an adhesion with those of the sycamore, and afterwards an independent horizontal formation, through which the fibres of the sycamore descended without altering its character. The other case was, that if a large ring of bark be taken from the trunk of a vigorous elm or other tree without being replaced with any thing, new beds of wood will be found in the lower as well as upper part of the trunk; while no ligneous production will appear on the ring of wood left exposed by the removal of the bark. Now this is so directly at variance with the observations of others, that it is impossible to receive it as an objection until its truth shall have been demonstrated. It is well known that if the least continuous portion of liber be left upon the surface of a wound of this kind, that portion is alone sufficient to establish the communication between the upper and lower lips of the wound; but, without some such slight channel of union, it is directly contrary to experience that the part of a trunk below an annular incision should increase by the addition of new layers of wood until the lips of the wound are united, unless buds exist upon the trunk below the ring.

The secretion called *cambium*, in the opinion of those who believe wood and bark to be independent simultaneous formations from the surface of the old wood and bark, is the matter which finally becomes organised as such: in Du Petit Thouars's theory it is a matter of organisation only as far as regards the origin of the cellular tissue of the medullary rays, and of the bark; while the superfluity of its moisture is a provision made by nature for the nutriment of the young fibres that descend through it.

CHAPTER V.

OF THE LEAVES.

Leaves are at once organs of respiration, digestion, and nutrition. They elaborate the crude sap impelled into them from the stem, parting with its water, adding to it carbon, and exposing the whole to the action of air; and while they supply the necessary food to the young fibres that pass downwards from them, and from the buds in the form of alburnum and liber, they also furnish nutriment to all the parts immediately above and beneath them. There are many experiments to show that such is the purpose of the leaves. If a number of rings of bark are separated by spaces without bark, those which have leaves upon them will live much longer than those which are destitute of leaves. If leaves are stripped off a plant before the fruit has commenced ripening, the fruit will fall off and not ripen. If a branch is deprived of leaves for a whole summer, it will either die or not increase in size perceptibly. The presence of cotyledons, or seminal leaves, at a time when no other leaves have been formed for nourishing the young plant, is considered a further proof of the nutritive purposes of leaves: if the cotyledons are cut off, the seed will either not vegetate at all, or slowly and with great difficulty; and if they are injured by old age, or any other circumstance, they produce a languor of habit which only ceases with the life of the plant, if it be an annual. This is the reason why gardeners prefer old melon and cucumber seeds to new ones: in the former the nutritive power of the cotyledons is impaired, the young plant grows slowly, a languid circulation is induced from the beginning; by which excessive luxuriance is checked, and fruit formed rather than leaves or branches.

Various are the secretions of plants that take place through the leaves: in those of monocotyledonous tropical plants in our hot-houses, nothing is more common than to see drops of water forming upon them from the effect of perspiration; in

Limnocharis Plumieri there is a large pore terminating the veins of the apex of the leaf, from which water is constantly distilled. The pitchers of Nepenthes, which are only a particular kind of leaves, secrete water enough to fill half their cavity. But, besides this more subtle fluid, secretions of a grosser quality take place in plants. The honey dew, which is so often attributed to insects, is one instance of the perspiration of a viscid, saccharine substance; the *manna* of the ash is another; and the gum ladanum that exudes from the Cistus ladaniferus is a third instance of this kind of perspiration.

It is believed that absorption takes place indifferently by either the upper or under surface of the leaf, but that some plants absorb more powerfully by one surface than by the other. Bonnet found that while the leaves of Arum, the kidney-bean, the lilac, the cabbage, and others, retained their verdure equally long whichever side was deprived of the power of absorption, the Plantago, some Verbascums, the marvel of Peru, and others, lost their life soonest where the *upper* surface was prevented from absorbing; and that in a number of trees and shrubs the leaves were killed very quickly by preventing absorption by the *lower* surface. From this there is only one safe conclusion to be drawn; that the absorbing surface of leaves varies in different species, and depends upon their peculiar organisation.

Leaves usually are so placed upon the stem that their upper urface is turned towards the heavens, their lower towards the earth; but this position varies occasionally. In some plants they are imbricated, so as to be almost parallel with the stem; in others they are deflexed till the lower surface becomes almost parallel with the stem, and the upper surface is far removed from opposition to the heavens. A few plants, moreover, invert the usual position of the leaves by twisting the petiole half round, so that either the two margins become opposed to earth and sky, or the lower surface becomes uppermost: this is especially the case with plants bearing phyllodia, or spurious leaves.

At night a phenomenon occurs in plants which is called their sleep: it consists in the leaves folding up and drooping, as those of the sensitive plant when touched. This scarcely happens perceptibly except in compound leaves, in which the leaflets are articulated with the petiole, and the petiole with the stem: it is supposed to be caused by the absence of light, and will be farther spoken of under the head of irritability.

After the leaves have performed their functions, they fall off: this happens at extremely unequal periods in different species. In some they all wither and fall off by the end of a single season; in others, as the beech and hornbeam, they wither in the autumn, but do not fall off till the succeeding spring; and, in a third class, they neither wither nor fall off the first season, but retain their verdure during the winter, and till long after the commencement of another year's growth: these are our evergreens. Mirbel distinguishes leaves into three kinds, as characterised by their periods of falling: —

1. *Fugacious* or *caducous*, which fall shortly after their appearance; as in Cactus opuntia.

2. *Deciduous* or *annual*, which fall off in the autumn; as Pyrus malus.

3. *Persistent, evergreen,* or *perennial,* which remain perfect upon the plant beyond a single season; as Ilex aquifolium, Prunus pseudo-cerasus, &c.

With regard to the cause of the fall of the leaf a number of explanations have been given, which may be found in *Willdenow's Principles of Botany,* p. 336. Sir James Smith was of opinion with Vrolik, that it is evidently a sloughing or casting off diseased or worn out parts, and in this I agree with them; but neither of these authors afford any explanation of the *cause* of this sloughing; nor do I think that it has been satisfactorily accounted for by any one, except Du Petit Thouars. If you will watch the progress of a tree, — of the elder for example, — says this writer, you will perceive that the lowest leaves upon the branches fall long before those at the extremities. The cause of this may be, perhaps, explained upon the following principle. In the first instance, the base of every leaf reposes upon the medulla of the branch to the sheath of which it is attached. But, as the branch increases in diameter by the acquisition of new wood, the space between the base of the leaf and the medulla becomes sensibly augmented. It has, therefore, been necessary that the fibres by which the leaf is connected with the medulla should lengthen, in order to admit the deposition of wood between the bark and the medulla. Now how does this elongation take place? As the bundles of fibres which run from the medulla into the leaf-stalk are at first composed only of spiral vessels, it is easy to conceive that they may be susceptible of elongation by unrolling. And in this seems to lie the mystery of the fall of the leaf; for the moment will come when the spiral vessels are entirely unrolled, and incapable of any further elongation: they will, therefore, by the force of vegetation, be stretched until they snap, when the necessary communication between the branch and the leaf is destroyed, and the latter falls off.

It is highly probable that this interruption of the necessary communication between the leaves and their support is the principal cause of the fall of the leaf in all cases; and it seems to be strongly supported by the following phenomena: 1. the early fall of the lower leaves of plants during the period of vegetation; 2. the fall of the leaves of evergreen trees, and of those whose foliage withers and persists during winter, at the period when new wood is formed by the operation of new shoots in the spring; 3. the ultimate fall of those leaves which are subject to no fixed periods of defoliation; 4. the persistence of the leaves of stunted trees, which have not formed wood enough during a single season to cause a rupture of the conducting vessels of the petioles, as the beech and hornbeam. The well known indication by which gardeners judge of their probable success in transplanting a tree or other plant in leaf, may also be considered a further proof of the justness of Du Petit Thouars's theory. On such occasions, if the leaves wither and hang upon the branches, the omen is unfavourable; but if they are cast off, it is a certain indication of success. Here the action of the spiral vessels may be understood to be impaired by the interruption of the regular transmission of oxygenated air through them, caused by the act of transplantation: as soon as the energies of the plant are renewed, a sudden increase of diameter supervenes, the communication is cut off between the leaves and the stem by the too rapid extension of the spiral vessels, and the leaves fall off.

Respiration takes place by the power the leaves possess of inspiring and expiring oxygen and decomposing carbonic acid. They have been found to vitiate the atmosphere at night by inhaling oxygen abundantly, and exhaling a small quantity of carbonic acid; and to restore the air to its purity in the sun's rays, by decomposing their carbonic acid and parting with their oxygen.

It was long since remarked by Priestley, that if leaves are immersed in water and placed in the sun, they part with oxygen. This fact has been subsequently demonstrated by a great number of curious experiments, to be found in the works of Ingenhouz, Saussure, Senebier, and others. Saussure found that plants in cloudy weather, or at night, inhaled the oxygen of the surrounding atmosphere, but exhaled carbonic acid if they continued to remain in obscurity. But, as soon as they were exposed to the rays of the sun, they respired the oxygen they had previously inhaled, in about the same quantity as they received it, and with great rapidity. Dr. Gilly found that grass leaves exposed to the sun in a jar for four hours produced the following effect: —

At the beginning of the Experiment there were in the Jar: —		At the close of the Experiment there were: —	
Of nitrogen	10.507	Of nitrogen	10.507
Of carbonic acid	5.7	Of carbonic acid	.37
Of oxygen	2.793	Of oxygen	7.79
	19.000		18.667

Heyne tells us that the leaves of Bryophyllum calycinum in India, are acid in the morning, tasteless at noon, and bitter in the evening; Link himself found that they readily stained litmus paper red in the morning, but scarcely produced any such effect at noon. The same phenomenon is said also to occur in other plants, as Cacalia ficoides, Sempervivum arboreum, &c. This stain in the litmus paper could not have arisen from the presence of carbonic acid, as that gas will not alter blue paper, but it must have been caused by the oxygen inhaled at night. It has also been found that this last power is retained even at noon, if the plant is not exposed to the sun. A similar explanation may be given of a phenomenon remarked by Pajot de Charmes, who found that the flowers of Cichorium intybus were daily changed from blue to white, according to the action of light. It is also well known that fruit is more acid in the morning than in the evening.

As the decomposition of carbonic acid gas is thus evidently an important part of the act of respiration, it might be supposed that to supply a plant with a greater abundance of carbonic acid than the atmosphere will usually yield would be attended with beneficial consequences. To ascertain this point several experiments have been instituted; the most important of which are those of Saussure, who found that, *in the sun,* an atmosphere of pure carbonic acid gas, or even air, containing as much as sixty per cent., was destructive of vegetable life; that fifty per cent. was highly prejudicial; and that the doses became gradually less prejudicial as they were diminished. From eight to nine per cent. of carbonic acid gas was found more favourable to growth than common air. This, however, was only in the sun: any addition, however small, to the quantity of carbonic acid naturally found in the air was prejudicial to plants placed in the shade.

Nitrogen, *per se,* is incapable of affording any support to the developement of plants, as was proved by Saussure, who found that, five days after immersion in pure nitrogen, the buds of poplars and willows were in a state of decay. This is remarkable, considering how large a proportion of the air we breathe consists of nitrogen.

While oxygen and carbon are thus essential to vegetation when not administered in excess, almost all other gases are more or less deleterious. Drs. Turner and Christison found that so small a quantity as $\frac{1}{10,000}$ of sulphurous acid gas, — a proportion so minute as to be imperceptible to the smell, — was sufficient to destroy the life of leaves in forty-eight hours. The same observers state, in an excellent paper in Brewster's Journal for January, 1828, the effects of other gases upon plants. I much regret that want of space prevents my giving their experiments in detail: the results, which are as follows, are very important. — Hydrochloric or muriatic acid gas was found to produce effects not inferior,—nay, even superior,—to those of the sulphurous acid. It was found that so small a quantity as a fifth of an inch, although diluted with 10,000 parts of air, destroyed the whole vegetation of a plant of considerable size in less than two days. "Nay, we afterwards found that a tenth part of a cubic inch in 20,000 volumes of air had nearly the same effects. In twenty-four hours the leaves of a laburnum were all curled in on the edges, dry and discoloured; and, though it was then removed into the air, they gradually shrivelled and died. Like the sulphurous acid, the hydrochloric acid gas acts thus injuriously in a proportion which is not perceptible to the smell. Even a thousandth part of hydrochloric acid gas is not distinctly perceptible; a ten thousandth made no impression on the nostrils whatever, although great care was taken to dry thoroughly the vessels used in making the mixtures."

"Chlorine may be expected to have the effects of hydrochloric acid gas; and so indeed it has, but they appear to be developed more slowly. Two cubic inches in two hundred parts of air did not begin to affect a mignonette plant for three hours; half a cubic inch in a thousand parts of air did not injure another in twenty-four hours: but when the plants did become affected, the same drooping, bleaching, and desiccation were observed."

"Nitrous acid gas is probably as deleterious as the sulphurous and hydrochloric acid gases. In the proportion of a hundred and eightieth, it attacked the leaves of a mignonette plant in ten minutes; and half a cubic inch in 700 volumes of air caused a yellowish green discolouration in an hour, and drooping and withering in the course of twenty-four hours. The leaves were not acid on the surface."

"The effects of sulphureted hydrogen are quite different from those of the acid gases. The latter attack the leaves at the tips first, and gradually extend their operation towards the leaf-stalks; when, in considerable proportion, their effects began in a few minutes, and if the quantity was not great, the parts not attacked generally survived, if the plants were removed into the air. The sulphureted hydrogen acts differently: two cubic inches in 230 times their volume of air had no effect in twenty-four hours. Four inches and a half in eighty volumes of air caused no injury in twelve hours; but, in twenty-four hours, several of the leaves, without being injured in colour, were hanging down perpendicularly from the leaf-stalks, and quite flaccid; and though the plant was then removed into the open air, the stem itself soon began also to droop and bend, and the whole plant speedily fell over and died. When the effects of a large quantity, such as six inches in sixty times their volume, were carefully watched, it was remarked that the drooping began in ten hours, at once from the leaf-stalks; and the leaves themselves, except that they were flaccid, did not look unhealthy. Not one plant recovered, any of whose leaves had drooped before it was removed into the air."

"The effects of ammonia were precisely similar to those of sulphureted hydrogen just related, except that after the leaves drooped they became also somewhat shrivelled. The progressive flaccidity of the leaves; the bending of them at their point of junction with the footstalk, and the subsequent bending of the stem; the creeping, as it were, of the languor and exhaustion from leaf to leaf, and then down the stem, were very striking. Two inches of gas in 230 volumes of air began to operate in ten hours. A larger quantity and proportion seemed to operate more slowly."

"Cyanogen appears allied to the two last gases in property, but is more energetic. Two cubic inches diluted with 230 times their volume of air affected a mignonette plant in five hours; half a cubic inch in 700 volumes of air affected another in twelve hours; and a third of a cubic inch in 1700 volumes of air affected another in twenty-four hours. The leaves drooped from the stem without losing colour; and removal into the air, after the drooping began, did not save the plants.

"Carbonic oxide is also probably of the same class, but its power is much inferior. Four cubic inches and a half, diluted with 100 times their volume of air, had no effect in twenty-four hours on a mignonette plant. Twenty-three cubic inches, with five times their volume of air, appeared to have as little effect in the same time; but the plant began to droop when it was removed from the jar, and could not be revived.

" Olefiant gas, in the quantity of four cubic inches and a half, and in the proportion of a hundredth part of the air, had no effect whatever in twenty-four hours.

" The protoxide of nitrogen, or intoxicating gas, the last we shall mention, is the least injurious of all those we have tried; indeed, it appears hardly to injure vegetation at all. Seventy-two cubic inches were placed with a mignonette plant in a jar of the capacity of 509 cubic inches for forty-eight hours; but no perceptible change had taken place at the end of that time."

Hydrogen seems to act unequally upon vegetation. Saussure found that a plant of Lythrum salicaria, after five weeks, had caused no alteration in a known volume of hydrogen by which it was surrounded, and had not itself experienced any apparent effect. Sir Humphrey Davy, however, states that some plants will grow in an atmosphere of hydrogen, while others quickly perish under such treatment.

In order to enable plants to perform the functions of respiration, and to expose their juices to the action of the atmosphere, leaves are furnished with special organs which are admirably adapted to the end in view. To prevent too rapid an evaporation beneath the solar rays, the leaf is overspread with a cuticle, which, as has been stated, is a hollow plate of membrane enclosing air; and at the same time to furnish them with the means of parting with superfluous moisture, at periods when the cuticle offers too much resistance, the stomata act like valves, and open to permit its passage: or when, in dry weather, the stem does not supply fluid in sufficient quantity from the soil for the nourishment of the leaves, these same stomata open themselves at night, and allow the entrance of atmospheric moisture, closing when the cavities of the leaf are full. In submersed leaves, in which no variation can take place in the condition of the medium in which they float, both cuticle and stomata would be useless, and accordingly neither exists. For the purpose of exposing the fluids contained in the leaves to the influence of the air, the cuticle would frequently offer an insufficient degree of surface. In order, therefore, to increase the quantity of surface that is exposed, the tissue of the leaf is cavernous, each stoma opening into a cavity beneath it, which is connected with multitudes of intercellular passages. Nor are the stomata and the cavernous parenchyma of the leaf the only means provided for the regulation of its functions. Hairs, no doubt, perform no mean office in their economy. In some cases these processes seem destined only for protection against cold, as in those plants in which they only clothe the buds and youngest leaves, falling away as soon as the tender parts have become hardened; but it can hardly be doubted that in many others they are absorbent organs, intended to collect humidity from the atmosphere. In succulent plants, or in such as grow naturally in shady places, where moisture already exists in abundance, they are usually wanting; but in hot, dry, exposed places, where it is necessary that the leaf should avail itself of every means of collecting its food, there they abound, lifting up their points and separating at the approach of the evening dews, but again falling down, and forming a layer of minute cavities above the cuticle, as soon as the heat of the sun begins to be perceived.

CHAPTER VI.

OF THE BRACTEÆ, CALYX, COROLLA, AND DISK.

The bracteæ, when but slightly removed from the colour and form of leaves, no doubt perform functions similar to those of the latter organs; and when coloured and petaloid, it may be presumed that they perform the same office as the corolla. Nothing, therefore, need be said of them separately.

With regard to the calyx, corolla, and disk, I shall chiefly follow M. Dunal's statements in his ingenious pamphlet, *Sur les Fonctions des Organes floraux colorés et glanduleux.* 4to. Paris, 1829. The calyx seems, when green, to perform the functions of leaves, and to serve as a protection to the petals and sexual organs; when coloured its office is undoubtedly the same as that of the corolla.

The common notion of the use of the corolla is, that, independently of its ornamental appearance, it is a protection to the organs of fecundation: but, if it is considered that the stamens and pistils have often acquired consistence enough to be able to dispense with protection *before the petals are enough developed* to defend them, it will become more probable that the protecting property of the petals, if any, is of secondary importance only.

Among the many speculations to which those interesting ornaments have given birth is one, that the petals and nectaria are the agents of a secretion which is destined to the nutrition of the anthers and young ovula. These parts are formed in the flower-bud long before they are finally called into action: in the almond, for example, they are visible some time before the spring, beneath whose influence they are destined to expand. In that plant, just before the opening of the flower, the petals are folded up; the glandular disk that lines the tube of the calyx is dry and scentless; and its colour is at that time dull, like the petals at the same period. But as soon as the atmospheric air comes in direct contact with these parts, the petals expand and turn out of the calyx, the disk enlarges, and the aspect of both organs is altered. Their compact tissue gradually acquires their full colour and velvety surface; the surface of the disk, which before was dry, becomes lubricated by a thick liquid, exhaling that smell of honey which is so well known. At this time the stamens perform their office. No sooner is that effected than they wither, the petals dry up and fall away, the secretion from the disk gradually dries up, and, in the end, the disk perishes along with the other organs to which it appertained. If the disk of an almond flower be broken before expansion, it will be seen that the fractured surface has the same appearance as that of those parts which in certain plants contain a large quantity of fæcula, as the tubers of the potato, Cyperus esculentus, &c. This led M. Dunal to suspect that the young disks also contained fæcula. This he afterwards ascertained, by experiment, to be the fact in the spadix of Arum italicum before the dehiscence of the anthers; but, subsequently to their bursting, no trace of fæcula could be discovered. Hence he inferred that the action of the air upon the humid fæcula of the disk had the effect of converting it into a saccharine matter fit for the nutrition of the pollen and young ovula; just as the fæcula of the albumen is converted in germination into nutritive matter for the support of the embryo.

In support of this hypothesis M. Dunal remarks, that the conditions requisite for germination are analogous to those which cause the expansion of a flower. The latter open only in a temperature above 32° Fahr., that of 10° to 30° centig. (50° to 86° Fahr.) being the most favourable; they require a considerable supply of ascending sap, without the watery parts of which they cannot open; and, thirdly, flowers, even in aquatic plants, will not develope in media deprived of oxygen.

Thus the conditions required for germination and for flowering are the same: the phenomena are in both cases also very similar.

When a germinating seed has acquired the necessary degree of heat and moisture, it abstracts from the air a portion of its oxygen, and gives out an equal quantity of carbonic acid gas; but, as one volume of the latter gas equals one volume of oxygen, it is evident that the seed is, in this way, deprived of a part of its carbon. Some changes take place in the albumen and cotyledons; and, finally, the fæcula that they contained is replaced by saccharine matter. In like manner a flower, while expanding, robs the air of oxygen, and gives out an equal volume of carbonic acid; and a sugary matter is also formed, apparently at the expense of the fæcula of the disk or petals.

The quantity of oxygen converted into carbonic acid gas in germination is, *cæteris paribus*, in proportion to the weight of the seed; but some seeds absorb more than others. Theodore de Saussure has shown that exactly the same phenomenon occurs in flowers.

Heat is a consequence of germination; the temperature is also augmented during flowering, as has been proved by Theodore de Saussure in the Arum, the gourd, the Bignonia radicans, Polyanthes tuberosa, and others.

The greater part of the saccharine matter produced during germination is absorbed by the radicle, and transmitted to the first bud of the young plant. M. Dunal is of opinion that the sugar of the nectary and petals is in like manner conveyed to the anthers and young ovula, and that the free liquid honey which exists in such abundance in many flowers, is a secretion of superabundant fluid, since it can be taken away, as is well known, without injury to the flower.

This opinion will probably be considered the better founded, if it can be shown that the disengagement of caloric and destruction of oxygen are in direct relation to the developement of the glandular disk, and also are most considerable at the time when the functions of the anthers are most actively performed.

In no plants, perhaps, is the glandular disk more developed than in Arums; and it is here that the most remarkable degree of developement of caloric has been observed. Senebier found that the bulb of a thermometer, applied to the surface of the spadix of Arum maculatum, indicated a temperature 7° higher than that of the external air. M. Hubert remarked this in a still more striking degree upon Arum cordifolium at the Isle of France. A thermometer placed in the centre of five spadixes stood at 111°, and in the centre of twelve, at 121°, although the temperature of the external air was only 66°. The greatest degree of heat in these experiments was at sunrise. The same observer found that the male parts of six spadixes, deprived of their glandular part, raised the temperature only to 105°; and that the same number of female spadixes only to 86°; and, finally, that the heat was wholly destroyed by preventing the spadix from coming in contact with the air.

From experiments of Saussure it seems certain that the disengagement of heat, and, consequently, destruction of oxygen, is chiefly caused by the action of the anthers, or at least of the organs of fecundation, as appears from the following table:—

Names.	Duration of the Experiment.	Oxygen destroyed.		
		By the bud.	By the flower during its expansion.	By the flower in withering.
Passiflora serratifolia	12 hours.	6 times its vol.	12	7
Hibiscus speciosus	24	6	8,7	7
Cucurbita maxima, male flower	24	7,4	12	10
Arum italicum, spadix cold	24	5 to 6		
——— spadix hot			30	
——— 24 hours after				5

It was also found that flowers in which the stamens, disk, pistil, and receptacle only were left, consumed more oxygen than those that had floral envelopes, as is shown by the following table:—

Species.	Duration of the Experiment.	Oxygen destroyed.	
		By the flowers entire.	By the usual organs only.
Cheiranthus incanus	24 hours.	11·5 times their vol.	18 times their vol.
Tropæolum majus	24	8·5	16·3
Cucurbita maxima, male	10	7·6	16·0
Hypericum calycinum	24	7·5	8·5
Hibiscus speciosus	12	5·4	6·3
Cobæa scandens	24	6·5	7·5

And it is here to be noticed, that those whose sexual apparatus destroyed the most oxygen have the greatest quantity of disk, and *vice versâ*; with the exception of Cobæa scandens, in which the disk is very firm and persistent, and, probably, therefore acts very slowly.

When the cup-shaped disk of the male flowers of the gourd was separated from the anthers, the latter only consumed 11·7 times their volume of oxygen in the same space of time which was sufficient for the destruction of sixteen times their volume when the disk remained. The spatha of Arum maculatum consumed, in twenty-four hours, five times its volume of oxygen; the termination of the spadix thirty times; the sexual apparatus 132 times, in the same space of time.

An entire Arum dracunculus, in twenty-four hours, destroyed thirteen times its volume of oxygen; without its spatha fifty-seven times; cut into four pieces, its spatha destroyed half its volume of oxygen; the terminal appendix twenty-six times; the male organs 135 times; the female organs ten times.

The same ingenious observer also ascertained that double flowers, that is to say those whose petals replace sexual organs, vitiate the air much less than single flowers, in which the sexual organs are perfect.

Is it not then, concludes M. Dunal, probable that the consequence of all these phenomena is the elaboration of a matter destined to the nutriment of the sexual organs? since the production of heat and the destruction of oxygen are in direct relation to the abundance of glandular surface, and since these phenomena arrive at their maximum of intensity at the exact period when the anthers are most developed, and the sexual organs in the greatest state of activity.

M. De Candolle has remarked, that the colouring matter of the corolla is probably very different from that of the leaves, since it will not etiolate by the exclusion of light. It may, however, be doubted whether this is strictly correct; for it is well known that forced flowers have little colour without full exposure to light, and that the intensity of colour in blossoms produced in stoves in winter is far less than that of the same species flowering in the open air in summer.

CHAPTER VII.

OF THE STAMENS AND PISTILLUM.

Having already, in the last chapter, explained the chemical action of the stamens and pistilla, I shall now confine myself to the consideration of their physical effect upon each other.

The duty of the stamens is to produce the matter called pollen, which has the power of fertilising the pistillum through its stigma. The stamens are therefore the representatives in plants of the male sex, the pistillum of the female sex.

The old philosophers, in tracing analogies between plants and animals, were led to attribute to the former distinction of sexes, chiefly in consequence of the practice among their countrymen of artificially fertilising the female flowers of the date with those which they considered male, and also from the existence of a similar custom with regard to figs. This opinion, however, was not accompanied by any distinct idea of the respective functions of particular organs; nor was it generally applied, although Pliny, when he said that " all trees and herbs are furnished with both sexes," may seem to contradict this statement; the fact is, that his was rather an empirical notion than an opinion depending upon philosophical deduction. Nor does it appear that any more distinct evidence existed of the universal sexuality of vegetables till about the year 1676, when it was for the first time distinctly pointed out by Grew. Claims are, indeed, laid to a priority of discovery over this great observer by Cæsalpinus, Malpighi, and others; but there is nothing so precise in their works as we find in the declaration of Grew, " that the attire (meaning stamens) do serve as the male for the generation of the seed." It would not be useful, if I had the space, to enter into any detailed account of the gradual advances which these opinions made in the world, nor to trace the progress of discovery of the precise nature of the several parts of the stamens and pistillum. Suffice it to say that, in the hands of Linnæus, the doctrine of the sexuality of plants was finally established, never again to be seriously controverted; for the denial of this fact, which has been since occasionally made by a few men, such as Alston, Smellie, and Schelver, has merely exposed the weakness of such hypercritics. We know that the powder which is contained in the case of the anthers, and which is called pollen, must generally come in contact with the viscid surface of the stigma, or no fecundation can take place. It is possible, indeed, without this happening, that the fruit may increase in size, and that the seminal integuments may even be greatly developed; the elements of all these parts existing before the action of the pollen can take effect: but, under such circumstances, whatever may be the developement of either the pericarpium or the seeds, no embryo can be formed. I have said that it is generally indispensable that the pollen and the stigma should come in contact. To this, however, there is a notable exception in Orchideous plants, in which nature seems to have specially guarded against the pollen coming in contact with the stigma by locking it up in cells from which it is not readily disengaged; and to have provided, in the form of glands and other apparatus of the stigma, peculiar means of conveying the impregnating matter to the stigma without actual contact between the latter and the pollen. Mr. Brown has, indeed, attempted to show that even in these plants actual contact between the two parts is necessary, as, indeed, Monsieur Adolphe Brongniart had done before him; but the evidence to the contrary is so strong that we pause for proof before we admit the conjectures and statements that have been brought forward upon this subject. Another order, that of Asclepiadeæ, has also been included in the number of those in which fertilisation takes place through peculiar glands, without actual contact between stigma and pollen; but Mr. Brown states, that in this tribe the grains of pollen are enclosed in a kind of sac, the most prominent part of the convex edge of which is applied to the stigma when fecundation is about to occur; and then a number of extremely slender threads, each of which is the pollen tube of a single grain, are emitted from this edge into the tissue of the stigma: almost every grain in the sac is said to produce its tube, and the tubes to be directed from all parts of it towards the point of dehiscence.

This universality of sexes in vegetables must not, however, be supposed to extend further than what are usually called, chiefly from that circumstance, perfect plants. In cryptogamic plants, beginning with ferns, and proceeding downwards to fungi, there are either no sexual organs whatever, or the males are so imperfectly developed as to be invisible or of no effect.

The exact mode in which the pollen took effect was for a long time an inscrutable mystery, and is even now not fully explained. It was generally supposed, that, by some subtle process, a material vivifying substance was conducted into the ovula through the style; but nothing certain was known upon the subject until the observations of Amici and of Adolphe Brongniart had been published. It is now known, that a short time after the application of the pollen to the stigma each grain of the former emits a tube of extreme tenuity, not exceeding the 1500dth or 2000dth of an inch in diameter, which pierces the conducting tissue of the stigma, and finds its way down to the region of the placenta, including within it the active molecules found in the grain; no one has actually seen the tubes pass further than the placenta; but there appears to be good reason for supposing that the vivifying matter communicated by the pollen tubes to the placenta is by some unknown means transmitted by the latter to the foramen of the ovulum, through which it finally passes into the nucleus, there to become the new embryo. In order to facilitate the contact between the placenta and the foramen of the ovulum, some very curious contrivances have been remarked. In Euphorbia Lathyris the apex of the nucleus is protruded far beyond the foramen, so as to lie within a kind of hood-like expansion of the placenta: in all campulitropous ovula the foramen is bent downwards, by the unequal growth of the two sides, so as to come in contact with the conducting tissue; and in Statice Armeria, Daphne Laureola, and some other plants, the surface of the conducting tissue actually elongates and stops up the mouth of the ovulum, while impregnation is taking effect. Again, in Helianthemum and Cistus, which, like the rest of their tribe, have the foramen of the ovulum very remote from the placenta, Mons. Adolphe Brongniart has observed, that, at the period of impregnation, the ovula are bent down in such a way as to present their foramina to the conducting tissue; and that when, owing to the shortness of the umbilical cords this is impossible, other not less ingenious means of establishing the contact have been provided by nature.

The function of the stigma has been already shown to be to catch the grains of pollen. This is a mere mechanical operation, and is effected by a viscid secretion, which is generally exuding from the surface of the stigma; which is also generally covered with minute papillæ, or sometimes with fringes, all of which, undoubtedly, aid the action of the organ. In some plants, as in Compositæ, Campanulaceæ, Lobeliaceæ, &c. the style is furnished with a sort of brush, with which it sweeps out the cells of the anthers, and collects the pollen either upon its own stigma, or scatters it upon that which is next it.

CHAPTER VIII.

OF THE FRUIT.

The fruit, which is mechanically destined as a mere protection to the seed, by which its race is to be maintained, is also, next to the wood, the most important part in the productions of vegetation. It constitutes the principal part of the food, especially in winter, of birds and small animals; it is often more ornamental than the flowers themselves, and it contributes most materially to the necessities and luxuries of mankind. When ripe, it falls from the plant, and, borne down by its weight, lies on the ground at the foot of the individual that produced it: here its seeds vegetate, when it decays, and a crop of new individuals arises from the base of the old one; but, as plants produced in such a manner would soon choke and destroy each other, nature has provided a multitude of ways for their greater dispersion. Many are carried to distant spots by the animals which eat them; others, provided with a sort of wings, such as the Samara, and the pappus of Compositæ, fly away upon the wind to seek a distant station; others scatter their seeds abroad by an explosion of the pericarpium, caused by a sudden contraction of the tissue; others falling upon the surface of streams, are carried along by the current; while others are dispersed by a variety of methods which it would be tedious to enumerate. The fruit, during its growth, is supported at the expense of the sap generally; but most especially of that which had been previously accumulated for its maintenance. This is less apparent in perennial or ligneous plants than in annual ones, but is capable of demonstration in both. Mr. Knight has well observed, that in annual fruit-bearing plants, such as the melon, if a fruit is allowed to form at a very early period of the life of the plant, as, for instance, in the axilla of the third leaf, it rarely sets or arrives at maturity, but falls off soon after beginning to swell, from want of an accumulation of food for its support; while, if the same plant is not allowed to bear fruit until it has provided a considerable supply of food, as will be the case after the leaves are fully formed, and have been some little time in action, the fruit which may then set swells rapidly, and speedily arrives at the highest degree of perfection of which it may be susceptible. And in woody trees, also, a similar phenomenon occurs: it is well known to gardeners, that, if a season occurs in which trees in a state of maturity are prevented bearing their usual crops, the succeeding year their fruit is unusually fine and abundant; owing to their having a whole year's extra stock of accumulated sap to feed upon.

The cause of the fruit attracting food from surrounding parts is probably to be sought in the phenomenon called endosmose. All the sap that may be at first impelled into the fruit by the action of vegetation, not being able to find an exit, collects within the fruit, and, in consequence of evaporation, becomes gradually more dense than that in the surrounding tissue: it will then begin to attract to itself all the more aqueous fluid that is in communication with it; and the impulse once given in this way to the concentration of the sap in particular points will continue until the growth of the fruit is completed, and its tissue so much gorged as to be incapable of receiving any more food, when it usually falls off.

M. Berard, of Montpelier, has made some curious observations upon the chemical actions of fruit, the substance of which is as follows (See *Annales de Chimie*, vol. xvi. pp. 152. 225.: —

Fruit does not act like leaves on the air. The result of its action, as well in light as in darkness, is, at every instant of its formation, a loss of carbon by the fruit, which combines with the oxygen of the air, and forms carbonic acid. This loss of carbon is essential to the ripening of the fruit; for, when the fruit is placed in an atmosphere deprived of oxygen, this function becomes suspended, the ripening is stopped, and, if the fruit remains attached to the tree, it dries up and dies.

A fruit which happens naturally to be enclosed in a shell may, nevertheless, ripen, because the membrane which forms the husk is permeable to the air. The communication between the external and internal air is so free, that the two portions are always of uniform composition; so that when the air thus contained is analysed, it is always found to be of the same composition as atmospheric air.

When fruits, separated from the tree, but capable of completing their own ripening, are placed in media free from oxygen, they do not ripen: the power, however, is only suspended, and may be re-established by placing the fruit in an atmosphere capable of taking carbon from it. But, if the fruit remain too long in the first situation, although it preserves the same external appearance nearly, it has entirely lost the power of ripening.

Hence it results, that most fruits, and especially those that do not require to remain on the tree, may be preserved for some time, and the pleasure they afford us thus prolonged. The most simple process consists in placing at the bottom of a bottle, a paste formed of lime, sulphate of iron, and water, and afterwards to introduce the fruit, it having been pulled a few days before it would have been ripe. Such fruits are to be kept from the bottom of the bottle, and, as much as possible, from each other; and the bottle to be closed by a cork and cement. The fruits are thus placed in an atmosphere free from oxygen, and may be preserved for a longer or shorter time, according to their nature; peaches, prunes, and apricots, from twenty days to a month: pears and apples for three months. If they are withdrawn after this time, and exposed to the air, they ripen extremely well; but, if the times mentioned are much exceeded, they undergo a particular alteration, and will not ripen at all.

Ripe fruit exposed to the air rots and decays: in this case, it first changes the oxygen of the surrounding air into carbonic acid, and then liberates from itself a large quantity of the same acid gas. It appears that the presence of oxygen gas is necessary to the rotting or decay of fruits: when it is absent, a different change takes place.

When the fruit cannot ripen, except on the tree, its ripening is not produced by a chemical change of the substances it contained whilst still green, but by the change of new substances furnished to it by the tree; and when it appears to lose the acid taste it had in its unripe state, it is because that taste is hidden by the large quantity of sugar it receives in ripening.

In the fruits which ripen off the tree the quantity of sugar is also found considerably to increase; and, in this case, it must be formed at the expense of the substances previously in the fruit. Gum and lignin are the only principles the proportion of which diminish at the same time; it is, therefore, natural to conclude, that it is the portions of these substances which have disappeared that have been converted into sugar; and, as the lignin contains most carbon, it is natural to suppose it is from it the oxygen takes the carbon to form carbonic acid, that change so indispensable to ripening.

Finally, the alteration the lignin suffers in the ripening continues during the decay of the fruit. It becomes brown, and its decomposition occasions the formation of much carbonic acid: sugar is also decomposed at this time, and it is to its disappearance that the peculiar taste of decayed fruits is to be attributed. The sugar, in its decomposition, also gives rise, no doubt, to the formation of carbonic acid.

CHAPTER IX.

OF THE SEED.

The action of the seed is confined to that phenomenon which occurs when the embryo that it contains is first called into life, and which is named germination.

Excepting at this time the seed is an inert mass, often containing nutritious matter, or some of the secretions of the species, and covered with a skin, which is in many cases so efficient a protection to the young embryo, that the seeds are enabled to resist, not only, as is well known, the powerful action of the gastric juice, after having been taken into the stomach of animals, but even that of exceedingly high temperatures. Spallanzani has stated, that he caused seeds to germinate after immersion in boiling water; and a case is mentioned by Duhamel, in which seeds retained their vitality after an exposure to 235° of Fahrenheit. These are, however, rare cases.

The earliest indication of germination consists in the parts of the seeds swelling, in consequence of the absorption of water by their cuticle, and a chemical change taking place in the nature of its juices. This is said to depend upon a loss of carbon, and an addition of oxygen, without which latter in abundance, it is believed that seeds cannot germinate at all. It is this which causes the starch of barley to be converted into sugar in malt. For this reason it has been recommended, that muriatic acid, chlorine, and various metallic oxides, should be used in inducing the rapid germination of seeds. Humboldt, Willdenow, and others, have declared their use to be advantageous. Link says the same; others deny the utility; and I believe that their action is very doubtful.

It is well known, that seeds will not germinate in the light. This is caused by light decomposing the carbonic acid gas, expelling the oxygen, and fixing the carbon, whence all the parts become hardened; a condition under which vegetation cannot proceed.

If seeds are sown as soon as they are gathered, they generally vegetate at the latest in the ensuing spring; but if they are dried first, it often happens that they will lie a whole year or more in the ground without altering. This character varies extremely in different species: the power of preserving their vitality is also extremely variable; some will retain their germinating powers many years, in any latitude, and under almost any circumstances. Clover will come up from soil newly brought to the surface of the earth, in places in which no clover had been previously known to grow in the memory of man. Many of the rarest plants in our gardens have been raised from old seeds, taken off specimens in herbaria: others perish so soon, that a few days' exposure is sufficient to destroy them. This is particularly the case with such as contain much oil.

As those conditions which are necessary to the germination of seeds are, heat, moisture, and darkness, it follows, that, in order to prevent this occurring, these three conditions are, if possible, to be obviated. Thus, in packing seeds for travelling to a long distance, it is found, by experience, that no mode is so suitable for their conveyance as being packed loosely in coarse canvass bags, hung to the ceiling of the cabin of a ship; where they are exposed to light and air, and where they are protected from damp: this is much better than enveloping them in wax, or mixing them with sugar, as has been sometimes done; both pernicious practices. It has been thought, that if seeds were mixed with charcoal, as their carbon would, under such circumstances, always be in excess to their oxygen, their vegetation might be safely suspended. But it has not been found, from experience, that any practical advantage arises from this method; seeds perishing in charcoal as quickly as in most other media. The best material is the English coarse brown paper, made from old tarred rope, in which a large quantity of tar is incorporated. No material will preserve seeds so long a time as this; while cartridge paper offers them no protection whatever.

The germination of seeds may be retarded by excluding them from light, and surrounding them with moist earth, rammed very hard: mango-seed, packed in hard clay, will thus travel safely from the West Indies; it was thus that the Araucaria Dombeyi was first brought to England from Chili; and many other seeds, which cannot otherwise be transported will live a long time under such circumstances.

As soon as the chemical changes now spoken of have taken effect, the embryo swells and bursts its envelopes, protruding its radicle, which pierces the earth, deriving its support at first from the cotyledons or albumen, but subsequently absorbing nutriment from the earth, and communicating it upwards to the young plant. The manner in which the embryo clears itself from its integuments differs in various species; sometimes it dilates equally in all directions, and bursts through its coat, which thus becomes ruptured in every direction; more frequently the radicle passes out at the hilum, or near it, or at a point apparently provided by nature for that purpose, as in Canna, Commelina, &c. If the radicle has a coleorhiza or rootsheath, this is soon perforated by the radicle contained within it, which passes through the extremity; as in grasses, and most monocotyledonous plants. The cotyledons either remain under ground, sending up their plumula from their centre, as the oak; or from the side of their elongated cauliculus, as monocotyledons; or they rise above the ground, acquire a green colour, and perform the ordinary functions of leaves, as in the radish and most plants. In the mangrove germination takes place in the pericarpium before the seed falls from the tree; a long thread-like radicle is emitted, which elongates till it reaches the soft mud in which such trees usually grow, where it speedily strikes root, and separates from its parent. Trapa natans has two very unequal cotyledons; of these, the larger sends out a very long petiole, to the extremity of which are attached the radicle, the plumula, and the smaller cotyledon (Mirbel). Cyclamen germinates like a monocotyledon: its single cotyledon does not quit the seed till the end of germination; and its radicle thickens into a fleshy knob, which roots from its base (Mirbel). The Cuscuta, which has no cotyledons, strikes root downwards, and lengthens upwards, clinging to any thing near it, and performing all the functions of a plant without either leaves or green colour.

In monocotyledons the cotyledon always remains within the seminal integuments; while its base lengthens and emits a plumula. In Cycas, which has two cotyledons, the seminal integuments open, and the radicle escapes. The cotyledons remain within the integuments, in the oak; but their petioles lengthen and liberate the plumula.

CHAPTER X.

OF COLOUR, SMELL, AND TASTE.

Upon this obscure subject Link has some good observations, the substance of which is as follows: —

The ordinary membranes and juices of plants are destitute of colour: they acquire colour by losing oxygen and acquiring carbon. Mucilage, sugar, starch, gluten, and other principles of that kind, are entirely devoid of colour or white. So also are vegetable membranes freed from foreign matter. The addition of oxygen makes no difference; for all the pure acids of plants are white: but azote seems to possess a great power of discolouration.

Young roots are white; but their epidermis becomes burned, as it were, by the surrounding atmosphere parting with its hydrogen in forming water, and thus acquires a brown or blackish cast. Some remarks upon this mode of staining bodies are to be found in the *Annales de Chimie*, vol. v. p. 80., by Fourcroy; and in volume vi. p. 238. of the same work, by Berthollet. In woody roots the colouring principle generally becomes red.

Young stems are green like the leaves: as they grow older they are scorched like the roots. Within the stem a colouring principle is formed, which seems to own its colour to extractive matter in itself colourless; leaves are green from loss of oxygen. If light is absent, and oxygen accumulates in the green parts, those lose their colour; but become green again upon exposure to light. Humboldt has some excellent remarks upon this in the *Journal de Physique*, vol. xl. p. 151. He has there demonstrated, that not only light, but every other agent which attracts oxygen from plants, will equally produce green colour.

When leaves become sickly, they admit an unusual quantity of oxygen, in consequence of which the green principle changes to yellow or red, as we see in sickly or dying leaves.

Where plants have lost their colour, from being placed in darkness, they contain a smaller quantity of principles soluble in alcohol than when in a green state, as has been shown by Senebier in his *Physiologie Végétale*, vol. iv. p. 278. Hence, principles deprived of oxygen which are soluble in alcohol are produced only under exposure to light.

The colour of flowers depends upon their degree of disoxydation, and may be arranged thus: 1. red changing into blue; 2. blue; 3. yellow; and, 4. scarlet, which is a modification of yellow: green is very uncommon.

Fruit is at first green: it is afterwards either scorched by dryness, or, by the retention of oxygen, becomes sweet or acid.

Mineral colours are known to depend upon particular proportions of water: as, for instance, sulphate of copper. Can they depend in plants upon a similar cause; that is to say, upon particular proportions of oxygen and hydrogen?

The odours of plants depend upon a certain volatile oil, which varies extremely in different species. It is more or less volatile, more or less soluble in alcohol, and more or less miscible in water; on which accounts there is a great difference in the readiness with which it yields to distillation. This is especially apparent in the oil of roses, which, in hot countries, or in hot weather, passes over very readily; but, in cold seasons, inseparably combines with the water used in distillation.

There is no doubt that particles of the odoriferous oil are continually flying off; and it has even been asserted, that those of Dictamnus albus will inflame upon the application of fire.

In proportion to the more or less volatile nature of their oil are the odours of plants more or less diffused. Hence in the same plant several scents may be distinguished, some at a distance, some on a near approach, as in the flowers of Vicia Faba. Some odours are not perceptible until the parts of a plant are rubbed, or they are more copiously given out when this is done. Some are most apparent in the recent state, others in the dried, as is the case with the melilots: in these the odoriferous juices are combined with too much water, which it is necessary to separate by drying before the fragrance is apparent.

Some flowers scatter their fragrance only at night: this is particularly the case with Cruciferæ, having dingy yellowish brown petals. In these the volatile oil is either so rapidly dispersed in the day-time as to be imperceptible; while it is more slowly and densely evolved in the night: or it may be decomposed in the day-time by the extrication of oxygen.

Plants in hot countries are more fragrant than those of cold countries, as is apparent in the rose; but their fragrance is sometimes so much dispersed by heat as to be imperceptible. Nocca found the Calendula lose its smell in a hot-house; and the common horehound is destitute of smell in Portugal.

The *tastes* of plants are sweet, acid, bitter, astringent, or austere and acrid. Vegetable membrane and resins communicate no taste to the palate, as they do not dissolve in the mouth. Mucilage has no flavour; sweetness is derived from the presence of sugar, and acidity of acids; bitterness usually denotes the presence of extractive matter. The extractive principle may be compared with neutral salts, which are usually bitter, as sulphate of potash, lime, and magnesia combined with acids and others; for, as a neutral salt consists of a metal, inflammable matter, and oxygen, so extractive matter consists of carbon, hydrogen, and oxygen. The astringency of minerals is only perceptible in those neutral salts that contain acid in excess, as alumine, sulphate of iron, sulphate of copper, and others. With these we may compare the astringent principle of vegetables, which also contains acid in excess; for it not only combines with the earths, but also stains litmus paper red: the carbon and hydrogen it contains may be compared with a metal and combustible matter. That kind of taste which is called herbaceous is caused by a mixture of mucilage with a little astringency. An acrid taste is almost peculiar to volatile oils.

CHAPTER XI.

OF THE DIRECTIONS TAKEN BY THE ORGANS OF PLANTS.

The substance of all that is known upon this subject has been combined with some excellent observations of his own by Dr. Dutrochet in a memoir, of which I shall avail myself in the following remarks: —

"The general phenomena of nature," says this writer, "which are daily before our eyes, are often those which mankind considers the least attentively. Those who are unaccustomed to reflect upon such subjects can scarcely believe that there is any very extraordinary mystery in the ascent of the stems of vegetables, or in the descent of their roots; and yet this is one of the most curious circumstances connected with vegetable life. The downward direction of the roots may appear easy of explanation: it may be said that, like all other bodies, they have a tendency towards the centre of the earth, in consequence of the known laws of gravity; but on what principle, then, is to be explained the upward tendency of the stem, which is in direct opposition to those laws? and here lies the difficulty. Dodart is the first who appears to have paid attention to this circumstance: he pretends to explain the turning backwards of seeds sown in an inverted position by the following hypothesis: he assumed that the root is composed of parts that contract by humidity; and that the stem, on the contrary, contracts by dryness. For this reason, according to him, it ought to happen that, when a seed is sown in an inverted position, the radicle will turn back towards the earth, which is the seat of humidity; and that the plumula, on the contrary, turns to the sky, or rather atmosphere,—a drier medium than the earth. The experiments of Du Hamel are well known, in which he attempted to force a radicle upwards and a plumula downwards, by enclosing them in tubes, which prevented the turning back of these parts. It was found that, as the radicle and plumula could not take their natural direction, they became twisted spirally. These experiments, while they prove that the opposite tendencies of the radicle and plumula cannot be altered, still leave us in ignorance of the cause of such tendencies. We are equally ignorant of the cause of the directions of the leaves. Bonnet believed that he could explain that phenomenon upon the hypothesis of Dodart just referred to, with respect to the radicle and plumula. According to him the lower surface of the leaves is, like the radicle, composed of fibres which contract by humidity; and the upper, like the plumula, of fibres that contract by dryness. As a proof of these assertions, Bonnet manufactured some artificial leaves; the upper surface of which was parchment, which contracts by dryness, and the lower of linen, which relaxes by moisture. These leaves were submitted to the action of dryness and humidity; and Bonnet found they were affected much in the same way as true leaves,—so easy is it to find proofs to support a favourite hypothesis."

In consequence of the unsatisfactory nature of these and other theories, more modern physiologists have been satisfied with inscribing the particular directions taken by plants among the *vital phenomena* of vegetation. And this is, perhaps, as much as we are likely to ascertain relating to it, and all similar manifestations of the overruling power of nature. Dutrochet, however, being of opinion that some more direct explanation of this phenomenon is to be found, instituted a variety of experiments of a novel kind. " Seeing," he remarks, " that the stem is always directed towards heaven, and the roots towards the earth, we cannot but believe that there is some relation between the cause of gravitation and that of the life of vegetables: the constant direction of the stem towards the light leads us also to suppose that this agent performs some important part in determining the directions of the parts of plants. The stem must be placed in the midst of the atmosphere in order to develope itself; the roots, on the contrary, require to lie within the earth. Hence, it may be inferred, that several causes concur to produce the phenomena in question."

Dutrochet filled with earth a box, the bottom of which was perforated with many holes: he placed seed of the kidney-bean in these holes, and suspended the box in the air, at about eighteen feet from the earth. Here the seeds, being placed in holes pierced through the bottom of the box, received the influence of the atmosphere and light from below; while the humid earth was placed above them. If the cause of the different directions of the radicle and plumula consisted in an affinity of the former for humidity, and of the latter for the atmosphere, the radicle ought to shoot upwards, and the plumula downwards; but this did not take place. The radicles, on the contrary, found their way downwards out of the box into the atmosphere, where they quickly dried up and perished; and the plumulas forced their way backwards into the earth. This experiment was afterwards modified, by increasing the quantity of earth above the seeds, and by some other contrivances; but the result was always the same: it was uniformly found, that there was no affinity between the radicle and the seat of moisture sufficient to counteract the natural downward tendency of the roots. It was also inferred, that there existed no more positive affinity between the stems and the atmosphere than between the roots and water.

There are certain parasitic plants which strike their roots into the stems of other plants, and which always grow at right angles with the stem to which they are fixed. The seed of the miseltoe will germinate in any direction, either upwards, downwards, or laterally. The first movement made by this plant consists in an extension of its cauliculus, which derives its support from the cotyledons, and which terminates at the radicular end, in a small green tubercle of a paler colour than the radicle itself. When the seed is fixed upon a branch by its natural glue, this incipient movement is effected at right angles with the branch; the young shoot is then curved backwards, and the radicular extremity descends to the surface of the branch, to which it adheres by expanding into a kind of disk. From this expansion the roots are emitted, and penetrate the interior of the branch whereon the seed of the miseltoe is fixed: its stem takes the directions above mentioned with reference to the centre of the branch on which it is fixed, and not with reference to the earth; so that, with regard to the latter, it is sometimes ascending, sometimes descending, sometimes horizontal. The same phenomena occur if the germination takes place upon dead wood or inorganic substances: a number of seeds were glued to the surface of a cannon ball; all the radicles were directed towards the centre of the ball. Hence it is obvious that the tendency of the miseltoe is not towards the surface of its nutrition, but it obeys the attraction of the body upon which it grows. The miseltoe, which does not grow on the earth, obeys the attraction of any other body; while those plants which naturally grow in the earth obey no other attraction than that of the earth. Parasitical fungi, those which constitute mouldiness; aquatics, which originate on stones, all grow perpendicular to the body that produces them, and will therefore be placed in all kinds of positions with respect to the earth.

The tendency downwards of the roots, and upwards of the stem, is chiefly observable in the ascending and descending caudex; that is to say, in the axes of the vegetable considered as a whole. The lateral emissions of this axis always deviate from its direction in a greater or less degree: we know that the roots produced by the tap root, and the branches which proceed from the side of the principal stem, scarcely ever take a direction absolutely vertical. This is probably due to several causes, one of which is undoubtedly the general tendency of all the parts of plants to take a direction perpendicular to the plane of the body on which they grow. The branches of trees are to those which produce them what the miseltoe is to the branch on which it vegetates: but, as there is a double attraction operating upon all branches,—that is to say, an attraction towards the stem and an attraction upwards, in consequence of the general law to which they all submit,— it results that a middle direction is taken, and, instead of one branch continuing to grow at right angles with another, it soon abandons that direction, and points its extremity towards the sky.

It has been hitherto seen, that the roots of vegetables are positively attracted by the body on which they grow: it appears, however, from the following experiment, that this attraction is influenced essentially by the mass of the body. Thus, if a seed of miseltoe is made to vegetate on a thread, the radicle turns itself in all sorts of ways, and exhibits no signs of attraction to the thread. Dutrochet made a seed of miseltoe germinate on a thread: he then glued it upon one of the points of a fine needle, fixed like that of a compass, balancing it by a bit of wax at the other end of the needle: he next placed a piece of wood at about half a line distance from the radicle, and then covered the whole apparatus with a glass, placed under such conditions that it was impossible that any cause could move the needle: in five days the embryo began to bend, and direct its radicle towards the bit of wood without the needle's changing its position, although it was extremely moveable upon its centre: in two days more the radicle was directed perpendicularly to the bit of wood with which it had come in contact, and still the needle had not stirred. This proves, says Dr. Dutrochet, that the direction of the radicle of the miseltoe towards a neighbouring body is not the immediate result of any attraction on the part of such a body; but that it is the result of a spontaneous movement of the embryo, in consequence of the attracting influence exerted upon its radicle, which is thus the mediate or occasional cause of the phenomenon. It is obvious, indeed, that the inflexion of the stem of the embryo of the miseltoe could not be due to the immediate attraction on the part of the bit of wood; for an exterior power sufficient to produce this inflexion would much more readily have produced a change in the direction of the needle, to one of whose points the seed was fixed: there can, therefore, be no doubt that the movement was *spontaneous;* that is to say, that it was caused by an internal vital cause, put in action by the influence of an exterior agent. This spontaneous direction of the radicle of the miseltoe under the influence of attraction proves incontestably that attraction only influenced its nervous powers, and not its ponderable matter: and the same is undoubtedly the case with terrestrial plants. The unknown power of attraction is only the accidental cause of the ascent of the stem, and of the descent of the roots, and not the immediate cause: in this case, attraction only operates as an agent for exciting nervous action. Other evidence exists to confirm most incontestably this important conclusion, that the visible movements of vegetation are all *spontaneous;* being brought into action by the influence of an external agent, but not movements originating with that agent.

Light is another cause of no less power than that just described: it is well known that a plant, placed in a room from which the light is excluded, except at a single aperture, directs its stem and leaves towards that aperture, and no longer takes a perpendicular position. The same tendency of the stems towards the light takes place in the open air. As light is diffused nearly equally around all bodies exposed to it, they will naturally assume a direction towards the heavens; so that light thus becomes an aid to gravitation. It might even be believed that light alone was a sufficient cause of the perpendicular position of the stems of vegetables, if experience did not prove the contrary. Dutrochet laid horizontally on the ground, in a dry and dark place, the stems of Allium Cepa and Allium Porrum, taken up with their bulbs. These plants, although taken out of the ground, continue to live for a long time; their stems became curved, and their upper end took a direction towards the heavens. This happened in about ten days; but, being repeated in the open air, three days were sufficient to produce the direction. In the first experiment, light being wholly excluded, gravitation only could have operated in giving the stem a perpendicular direction, — that power being the only one which is known to act in a direction perpendicular to the horizon. Modifications of this experiment were instituted, to be certain that humidity had no effect, and the same result was obtained. In the prosecution of these investigations, it also appeared that it was not merely the summit of the stem which had a tendency to a perpendicular direction, but that all the moveable parts of the plant possessed a similar disposition, provided they were coloured.

Stems are sometimes directed towards the earth, in which they attempt to bury themselves like roots; a phenomenon worthy of the greatest attention, not only on its own account, but for the sake of the circumstances connected with it. Many vegetables, besides their above-ground stems, have also subterranean stems: these creep horizontally in the interior of the earth, without manifesting any tendency towards the sky: they are white, like roots, of which they assume the course and the station. Sometimes, however, they are pink, as in Sparganium erectum; in such cases it is the cuticle that is coloured, and not the subjacent parenchyma: but, whenever the point of their stems approaches the surface of the soil, it becomes green, and, from that moment, they acquire an upward tendency. Is it hence to be inferred that there is some recent connection between the colours of the parts of vegetables and the directions they assume?

"Generally," M. Dutrochet proceeds to remark, " stems are directed towards the light, which is in accordance with their colour, which is usually green; while the roots have usually a tendency to avoid the light, which coincides with their want of colour. The colour of the roots is, in fact, nothing but that of the vegetable tissue; and can by no means be compared to that of the petals of some plants, which arises from the presence of a white colouring matter. Light, which is the principal, but not sole, cause of the colour of stems and their organs, has no power of infusing colour into the roots, as may be easily seen by roots growing in glasses of water; in spite of the influence of the light they constantly remain colourless; and this does not depend upon immersion in water, because leaves developed in that medium are nevertheless green. Although roots have, in general, no tendency towards the light, yet such a disposition does become manifest, provided the terminal shoot of a root becomes slightly green, as occasionally happens. Having induced some seeds of Mirabilis Jalapa to germinate in damp moss, I remarked that the young roots, when about as long as the finger, were terminated by a shoot of a slightly green colour. Wishing to know whether these roots would turn towards the light, I placed them in a glass vessel filled with water, having a wooden cover pierced with holes to receive the roots and fix the seeds. I enveloped the vessel in black cloth, leaving only a narrow vertical slit, through which light could enter the interior. This slit was exposed to the rays of the sun; and, a few hours after, I found that all my roots had hooked back their points towards the slit through which light was introduced. The same experiment was tried with colourless roots; but no alteration in their direction was produced. From this it appears evident that colour is one of the conditions that determine the directions of vegetables and their parts towards the light, and consequently towards the sky. This is so true, that colourless stems are known to assume the directions of roots. In the Sagittaria sagittifolia this is particularly obvious. Shoots are produced from the axillæ of all the radical leaves which grow at the bottom of the water. These shoots have their points directed towards the sky, like those of all vegetables. The young stems, which are produced by these shoots, are entirely colourless, like roots; and, instead of taking a direction towards the sky, as coloured stems would do, they lead downwards, pointing towards the centre of the earth. In order to take this position, the young shoot forces its way through the substance of the petiole which covers it; thus overcoming a mechanical obstacle in its tendency towards the earth. This subterranean stem next takes a horizontal course, and does not assume any tendency towards the sky until the points become green. M. Dutrochet has also remarked a similar phenomenon in roots. It is well known that exposed stems of many plants produce roots: when green, they turn upwards, as in Pothos and Cactus phyllanthus; when colourless, they point downwards. Hence it is to be inferred that stems do not descend merely because they are stems, but because their parenchyma is coloured; and that roots descend not in their quality of roots, but because their parenchyma is colourless. It seems, however, that although this law is uniform in its operation in all terrestrial plants, yet that a deviation, or apparent deviation, from it exists in the parasitic miseltoe. The radicle of this plant, which is of a paler green than the other parts, instead of turning towards the light, avoids it with so much pertinacity, that it is impossible to induce it to take such a direction; so that it seems to be repelled by light. Dutrochet does not seem to be able to satisfy himself of the reason of this exception: but it appears to be by no means difficult to account for. We have seen that, in the direction of its radicle, nature has enabled it to fulfil its functions as a parasitic plant by the attraction of the body on which it is placed, rather than by the much more powerful attraction of the earth. In order to ensure this particular tendency, without possessing which the existence of the miseltoe would be put in hazard, its root has received from the same all-powerful hand a disposition so much greater than other plants to avoid light, and to bury itself in the obscurity of the interior of a tree, as to be sufficient to overcome the influence of its green colouring matter.

The next direction of the parts of plants, which may be called *special*, is that of the upper surface of the leaves towards the sky, and of the lower towards the earth. This disposition is so powerful, that, if the usual direction of a leaf is inverted, the petiole will twist so as to enable it to recover itself. This phenomenon has been noticed by Bonnet, whose explanation has been already given (p. 278.), but which is obviously inadmissible. There is always a natural difference between the two faces of the leaf: the upper is always the most deeply coloured; a difference which will be found constant in all cases. The face with the deepest colour turns towards the sky or light, and, with the weakest colour, towards the earth or obscurity; and this is so constant a law, that it will be found that, if that surface of the leaf which is naturally inferior is more deeply coloured than the superior, the petiole will be twisted round by the greater affinity of the lower surface for the light, which will thus become uppermost, the leaf presenting the appearance of an inverted leaf. This may be seen in many grasses, but not in Zea Mays, Triticum repens, and Agrostis rubra. Hence it is to be concluded, that the upper surface of the leaf is not turned towards the heavens merely in consequence of its quality of being the upper surface, but because it is generally the most deeply coloured.

The same law influences the directions of the petals, in which the upper surface, — that which is turned towards the heavens, — is always the most highly coloured: this, indeed, is sometimes not very apparent, but is nevertheless constant. Even in white petals, — such, for example, as those of Lilium album, — the upper face will be found of a dense but brilliant white, while the lower is of a much paler hue. The white colour of the petals, Dutrochet proceeds to remark, like all the other colours of plants, is due to a particular kind of colouring matter deposited in the parenchyma lying below the cuticle. Thus the whiteness of the flowers of plants is not dependent upon the absence of colour, like the roots and etiolated stems: in the former a white colouring matter exists; in the latter the whiteness is caused by absence of colour. Some apparent exceptions to this law,— such as the outside of many monopetalous flowers being paler than the inside, as in Digitalis purpurea, Fritillaria latifolia, and others,— Dutrochet thinks may be explained thus: — They, no doubt, are due to the tendency of the less coloured part to avoid the light, which is manifested by bearing down the flower so as to approach the seat of obscurity as nearly as possible: all such flowers being always nodding. This tendency is aided by the weakness of the peduncle, which seems to have been specially provided for enabling such flowers to retire from the light. In papilionaceous plants, the inside of the vexillum, which is most deeply coloured, always turns itself towards the light; and the alæ twist themselves half round, to effect the same object. The ovaria often take a different direction after the fall of the corolla than they had before. Thus, during flowering, the ovarium of Digitalis purpurea was nodding like the flower, the direction of which it was compelled to follow. Immediately after the fall of the corolla, it turns upwards towards the light, to which it is attracted by its green colour. A contrary phenomenon is presented by the ovarium of Convolvulus arvensis. The flower is turned towards the sky: as soon as it has fallen, the ovarium takes a direction towards the earth, bending down the peduncle. This cannot be due to the weight of the ovarium, which is much lighter than its peduncle, but must depend upon its disposition to avoid the light, on account of its pallid hue, which is nearly the same as that of the root. In Convolvulus sepium, on the contrary, in which the ovarium is equally pale, its erect position is maintained, and the influence of *decoloration* counteracted by the greater affinity to the light of two large green bracteæ in which it is enveloped.

From the following and some other experiments, Dutrochet infers that the direction of leaves to the light is not mechanically caused by the operation of an external agent, but is due to a spontaneous motion, put in action by the influence of external agency. He took a leaf and cut off its petiole, the place of which was supplied by a hair, hooked by one end upon the leaf, and having a piece of lead attached to its opposite extremity. They were plunged in a vessel of water: the weight of the lead carried the leaf to the bottom of the water, where it stood erect in consequence of its lightness inducing it to attempt to ascend. Being exposed in a window, so that the under surface was turned to the light, no alteration took place in its position. Now, as from Bonnet's experiments, it is certain that leaves immersed in water act exactly as if surrounded with air, it is to be inferred that the external influence of the light is of no effect, unless aided by a spontaneous power within the vegetable which was destroyed by the removal of the petiole. Leaves immersed in water under similar circumstances, with their petioles and stem uninjured, turned towards the light as they would have done in the open air. The power thus supposed to exist in all probability depends upon the same nervous matter which has elsewhere been shown to exist, in a greater or less degree, in all plants, and especially in those called sensitives.

Those who desire more information upon this very curious subject should consult Dutrochet's work, *Sur la Motilité des Végétaux*.

CHAPTER XII.

OF IRRITABILITY.

The vitality of plants seems to depend upon the existence of an irritability, which, although far inferior to that of animals, is, nevertheless, of an analogous character.

This has been proved by a series of interesting experiments by M. Marcet, of Geneva, upon the exact nature of the action of mineral and vegetable poisons. The subject of his observations was the common kidney-bean; and in each experiment a contrast was formed between the plant operated upon and another watered with spring water. A vessel containing two or three bean plants, each with five or six leaves, was watered with two ounces of water, containing twelve grains of oxide of arsenic in solution. At the end of from twenty-four to thirty-six hours the plants had faded, the leaves drooped, and had even begun to turn yellow. Attempts were afterwards made to recover the plants, but without success. A branch of a rose tree was placed in a solution of arsenic; and in twenty-four hours ten grains of water and 0.12 of a grain of arsenic had been absorbed. The branch exhibited all the symptoms of unnatural decay. In six weeks a lilac tree was killed, in consequence of fifteen or twenty grains of moistened oxide of arsenic having been introduced into a slit in one of the branches. Mercury, under the form of corrosive sublimate, was found to produce effects similar to those of arsenic; but no effect was produced upon a cherry tree, by boring a hole in its stem, and introducing a few globules of liquid mercury. Tin, copper, lead, muriate of barytes, a solution of sulphuric acid, and a solution of potash, were found to be all equally destructive of vegetable life; but it was ascertained, by means of sulphate of magnesia, that those mineral substances which are innocuous to animals are harmless to vegetables also. In the experiments with vegetable poisons, the bean plants were carefully taken from the earth, and their roots immersed in the solutions used. It had been previously ascertained that plants so transplanted and placed in water under ordinary circumstances would remain in excellent health for six or eight days, and continue to vegetate as if in the earth. A plant was put into a solution of nux vomica at nine in the morning: at ten o'clock the plant seemed unhealthy; at one the petioles were all bent in the middle; and in the evening the plant was dead. Ten grains of an extract of Cocculus suberosus, dissolved in two ounces of water, destroyed a bean plant in twenty-four hours; Prussic acid produced death in twelve hours, laurel water in six or seven hours, a solution of belladonna in four days, alcohol in twelve hours.

From the whole of his experiments, M. Marcet concludes, — 1st, That metallic poisons act upon vegetables nearly as they do upon animals: they appear to be absorbed and carried into different parts of a plant, altering and destroying the vessels by corrosive powers. 2dly, That vegetable poisons, especially those which have been proved to destroy animals by their action upon the nervous system, also cause the death of plants: whence he infers that there exists in the latter a system of organs which is affected by poisons, nearly as the nervous system of animals.

These facts have been confirmed by other experiments of M. Macaire, which will be mentioned presently under the head of Poisons.

Irritability, in the common acceptation of the term in botany, means those extreme cases of excitability in which an organ exhibits movements altogether different from those we commonly meet with in plants. Of this kind of irritability there are three distinct classes; namely, those which depend upon atmospheric phenomena, spontaneous motions, and such as are caused by the touch of other bodies.

Among the cases of irritability excited by particular states of the atmosphere, the singular phenomenon called, by Linnæus, the sleep of plants is the most remarkable. In plants with compound leaves, the leaflets fold together while the petiole is recurved at the approach of night; and the leaflets again expand and raise themselves at the return of day.

In others the leaves converge over the flowers, as if to

shelter those more delicate organs from the chill air of night. The flowers of the crocus and similar plants expand beneath the bright beams of the sun, but close as soon as they are withdrawn. The Œnotheras unfold their blossoms to the dews of evening, and wither away at the approach of day. Some Silenes roll up their petals in the day, and expand them at night. The florets of numerous Compositæ, and the petals of the genus Mesembryanthemum, are erect in the absence of sun, but become reflexed when acted upon by the sun's beams; and many other such phenomena are familiar to every observer of nature. It is probable, indeed, that a different effect is produced upon all plants by day and night, although it is less visible in some than in others: thus plants of corn, in which there is little indication of sleep when grown singly, exhibit that phenomenon very distinctly when observed in masses; their leaves become flaccid, and their ears droop at night. These effects have been generally attributed to the action of light; and it is probable that that agent contributes very powerfully to produce them; for a flower removed from the shade will often expand beneath a lamp, just as it will beneath the sun itself. De Candolle found that he could induce plants to acknowledge an artificial day and night, by alternate exposure to the light of candles. There must, however, be some cause beyond light, of the nature of which no opinion has yet been formed: many flowers will close in the afternoon while the light of the sun is still playing upon them, and the petals of others will fold up under a bright illumination.

Spontaneous movements are far more uncommon than those which have just been described. In Megaclinium falcatum, the labellum, which is connected very slightly with the columna, is almost continually in motion; in a species of Pterostylis, shown me by Mr. Brown, I observed a kind of convulsive action of the labellum; the filaments of Oscillatorias are continually writhing like worms in pain; several other Confervas exhibit spontaneous movements: but the most singular case of the kind is that of Hedysarum gyrans. "This plant has ternate leaves: the terminal leaflet, which is larger than those at the side, does not move, except to sleep; but the lateral ones, especially in warm weather, are in continual motion, both day and night, even when the terminal leaflet is asleep. External stimuli produce no effect; the motions are very irregular; the leaflets rise or fall more or less quickly, and retain their position for uncertain periods. Cold water poured upon it stops the motion, but it is immediately renewed by warm vapour."

To this class of irritability ought, perhaps, to be referred the curious phenomenon well known to exist in the fruit of Momordica elaterium, the spirting cucumber. In this plant the peduncle, at a certain period, when the fruit has attained its perfect maturity, is expelled, along with the seeds and the mucus that surrounds them, with very considerable violence. Here, however, endosmosis appears to offer a satisfactory explanation. According to Dutrochet, the fluid of the placentary matter in this fruit gradually acquires a greater density than that which surrounds it, and begins to empty the tissue of the pericarpium: as the fruit increases in size the same operation continues to take place; the pulpy matter in the centre is constantly augmenting in volume at the expense of the pericarpium; but, so long as growth goes on, the additio n of new tissue, or the distention of old, corresponds with the increase of volume of the centre. At last growth ceases, but endosmosis proceeds; and then the tissue that lines the walls of the central cell is pressed upon forcibly by the pulp that it encloses, until this pressure becomes so violent that rupture must take place somewhere. The peduncle, being articulated with the fruit, at length gives way, and is expelled with violence; at the same time the cellules of tissue lining the cavity all simultaneously recover their form, the pressure upon them being removed, and instantly contract the space occupied by the mucous pulp; the consequence of which is that it also is forced outwards at the same time as the peduncle. It has been found by measurement, that the diameter of the central cavity is less after the bursting of the fruit than before.

Movements produced by touch, or by external violence, are very frequent. The sensitive plant (Mimosa pudica), which will rapidly fold up its leaves as if in a state of sleep, is, perhaps, the most familiar instance: but many others also exist. If the centre of the leaf of the Dionæa muscipula is irritated, the sides collapse, so as to cross the ciliæ of their margin, like the teeth of a steel-trap for catching animals. Roth is recorded to have seen something of the same kind in Drosera rotundifolia. If the bottom of the stamens of the common berberry is touched on the inside with the point of a needle, they spring up against the pistillum. The valves of Impatiens noli-tangere, when the fruit is ripe, separate and spring back with great elasticity when touched. In this case the phenomenon is apparently capable of explanation upon a similar principle to the Momordica elaterium. In the fruit of Impatiens the tissue of the valves consists of cellules, that gradually diminish in size from the outside to the inside; and the fluids of the external cellules are the densest. The latter gradually empty the inner cellules and distend themselves, so that the external tissue is disposed to expand, and the internal to contract, whenever any thing occurs to destroy the force that keeps them straight. This at last happens by the disarticulation of the valves, the peduncle, and the axis; and then each valve rapidly rolls inwards with a sudden spontaneous movement. M. Dutrochet proved that it was possible to invert this phenomenon by producing exosmose: for that purpose he threw fresh valves of Impatiens into sugar and water, which gradually emptied the external tissue, and, after rendering the valves straight, at length curved them backwards.

The column of the genus Stylidium, which in its quiescent position is bent over one side of the corolla, if slightly irritated, instantly springs with a jerk over to the opposite side of the flower. In Kalmia the anthers are retained in little niches of the corolla; and, as soon as they are by any cause extricated, the filaments, which had been curved back, recover themselves with a spring. In certain orchideous plants, of the tribe called Vandeæ, the caudicula to which the pollen masses are attached will often, upon the removal of the anther, disengage themselves with a sudden jerk.

An elaborate exposition of the phenomena accompanying the movements of the sensitive plant has been given by Dutrochet, in his *Recherches, &c. sur la Structure intime des Animaux et des Végétaux, et sur leur Motilité,* which should be consulted by all readers desirous of studying the irritability of that very remarkable plant.

CHAPTER XIII.

OF THE EFFECT OF POISONS UPON VEGETATION.

When treating of irritability, some experiments of M. Marcet were quoted which had a reference to this subject. It is one, however, of sufficient importance to demand a chapter by itself.

At present we know scarcely any thing of the causes of the diseases of the vegetable kingdom; but it can scarcely be doubted, that to ascertain what are the specific effects of deleterious matters upon the vital powers of plants, is to lay the foundation of an acquaintance with their pathology.

M. Marcet's experiments proved that narcotic and irritating poisons produced an effect upon vegetables altogether analogous to that which they produce upon animals. The very valuable experiments with gases by Drs. Turner and Christison, mentioned formerly (p. 252.), lead to the same conclusion. These gentlemen remark, that " the phenomena, when compared with what was observed in the instances of sulphurous and hydrochloric acid, would appear to establish, in relation to vegetable life, a distinction among the poisonous gases nearly equivalent to the difference existing between the effects of the irritant and the narcotic poisons on animals. The gases which rank as irritants in relation to animals seem to act locally on vegetables, destroying first the parts least plentifully supplied with moisture. The narcotic gases, — including under that term those that act on the nervous system of animals, — destroy vegetable life by attacking it throughout the whole plant at once. The former, probably, act by abstracting the moisture of the leaves; the latter, by some unknown influence on their vitality. The former seem to have upon vegetables none of that sympathetic influence upon general life, which in animals follows so remarkably injuries inflicted by local irritants.

A similar result was arrived at by M. Macaire, whose very curious and instructive experiments are recorded in the *Bibliothèque Universelle*, xxxi. 244., and which I think of sufficient importance to be detailed at length.

The first plant used was the Berberis vulgaris. The six stamina of the flowers of this plant have the property of rapidly approaching the pistil when touched by the point of an instrument. The motion occurs at the base of the stamens. When cold, the motion is sometimes retarded. When put into water or solution of gum, the flowers may be preserved many days, possessing their irritability. The petals and stamens close at night to open again in the morning. Putting the stem of this plant into dilute prussic acid for four hours, occasioned the loss of the contractile property by irritation; the articulation became flexible, and might be inclined in any direction by the instrument. The leaves had scarcely begun to fade. On placing the expanded flowers on the prussic acid, the same effect took place, but much more rapidly.

The experiment being repeated, with an aqueous solution of opium, a similar effect was produced in nine hours.

Dilute solutions of oxide of arsenic and arseniate of potash were used: the stamens lost the power of approaching the pistil; but they were stiff, hard, withdrawn backwards, and could not have their direction altered without fracture. It seemed like an irritation, or a vegetable inflammation.

Solution of corrosive sublimate more slowly produced the same effects.

Sensitive plant (Mimosa pudica). — Experiments were now made with this vegetable. When a leaf of this plant is cut, and allowed to fall on pure water, the leaflets generally contract rapidly; but after a few moments expand, and are then susceptible of contraction by the touch of any other body. They may thus be preserved in a sensible state two or three days. If the section be made with a very sharp instrument, and without concussion, the leaves may be separated without any contraction. The branches of this plant may be preserved for several days in fresh water. Gum-water also effects the same purpose.

When a cut leaf of this plant falls upon a solution of corrosive sublimate, the leaf rapidly contracts, and the leaflets curl up in an unusual manner, and do not again expand. When put into pure water, the sensibility does not return, but the whole remains stiff and immovable. A little solution of corrosive sublimate being put into a portion of pure water, containing an expanded branch of the plant, gradually caused curling up of the leaves, which then closed and fell. If the solution be very weak, the leaves open on the morrow, and are still sensible, but ultimately contract, twist, and remain stiff till they die. Solutions of arsenic and arseniate of potash produce the same effects.

A leaf of the sensitive plant was in a cold diluted solution of opium: in a few moments it opened out as in water, and, after half an hour, gave the usual signs of contractibility. In six hours it was expanded, and had a natural appearance, but could not be excited to move. The leaflets were flexible at the articulations, and offered a singular contrast to the state of irritation produced by corrosive sublimate. Pure water did not recover the plant. A large branch, similarly situated, expanded its leaves; but in half an hour had lost much of its sensibility: the leaflets, though alive, seemed asleep, and required much stimulating to cause contraction. In one hour the contractions ceased: in two hours the branch was dead.

A leaf placed in prussic acid (Scheele's strength) contracted, then slightly dilated, but was quite insensible, and the articulations were flexible: water did not recover it. If the acid be very weak, the leaflets dilate and appear to live, but are insensible. A drop of the acid placed on two leaflets of a healthy plant gradually causes contraction of the other leaflets, pair by pair. Solutions of opium and corrosive poisons have no effect when applied this way. After some time they dilate, but are insensible to external irritation: the sensibility returns in about half an hour; but the leaflets appear as if benumbed.

The plant exposed to the vapour of prussic acid is affected in the same way: ammonia appears to favour the recovery of the plant.

A cup containing dilute prussic acid was so placed that one or two leaves, or sometimes a branch, of a healthy plant could be plunged into the liquid, or left to repose on its surface. The leaflets remained fresh and extended, but were almost immediately insensible. Being left in this state for two hours, they were expanded; and no irritation could cause their contraction, though otherwise there was no appearance of an unnatural state. At five o'clock in the evening the leaves were left to themselves. At nine o'clock they were open and insensible. At midnight they were still open, whilst all the rest of the plant, and the neighbouring plants, were depressed, contracted, and in the state of sleep. On the morrow they resumed a little sensibility, but seemed benumbed.

In the same manner M. Macaire has interfered with other plants as to the state of sleep, and observes that prussic acid thoroughly deranges the botanical indications of time of Linnæus.

CHAPTER XIV.

OF THE DISEASES TO WHICH PLANTS ARE SUBJECT.

The diseases to which plants are subject are many and important: a few arise from mechanical causes, such as bruises or wounds, but the origin of the greater part is almost wholly unknown. It is probable, however, that some of them arise from a derangement of the circulation of the fluids, and an undue absorption of water, by which a brown colour and decay are produced.

Tabes, or *gangrene*, consists in a general languor of the system: the leaves and stem become flaccid, and the plant withers away; or, if it be succulent, becomes rotten. It is said by Link to arise from exposure to excessive cold, and a too rapid subsequent change to heat.

Anasarca, or *dropsy*, is a similar disease, peculiar to succulent plants, arising from an excessive introduction of water into the system. It produces rapid rottenness, and can only be stopped by destroying all the parts affected by it, and exposing the individual to a very dry atmosphere.

Scorching, or *insolation*, is a local disease attributable to exposure to too high a temperature. It is vulgarly supposed to be caused by drops of water collected on the surface of leaves, and destroying them by acting as burning lenses; but, as there is no focus formed when the water-drops lie on the leaves, it is obvious that no burning can be produced by them. It is often, I believe, caused by excessively rapid evaporation. *Marcor*, or *welting*, is a variety of this.

Chlorosis, or *etiolation*, is a kind of constitutional debility. The individual affected is pale, and destitute of a healthy green: the stems are weak, long, and slender; no flowers are produced; and the plant is readily killed. It is supposed to depend upon the accumulation of oxygen, and to be caused by various circumstances. The attacks of insects upon the roots, by which the motion of the sap is deranged, are often the real cause. A form of this, in which healthy, well-formed leaves, which apparently perform their natural functions, become perfectly white, is common in Camellia reticulata: neither the cause nor the cure of this has been ascertained. Cold is probably sometimes the cause of similar appearances.

Canker, or *caries*, exhibits itself continually in a brown discolouration of the medulla and parts adjacent, and externally in small brown dead spots, which gradually extend on all sides, until they surround the branch and kill it. These spots are always dry and hard, never containing any fluid. It is this which is so fatal to many of the apple and pear trees of this country. Its cause and mode of cure are equally unknown. Apparently healthy shoots will, if grafted on another stock, carry the disease with them, and, like the gout and scrofula in human constitutions, will sooner or later be sure to break out. The cure of the disease is, therefore, as far as we know, impracticable.

Carcinoma is a disease in which an unusual deposit of cambium takes place between the wood and bark: no wood is formed; but, instead, the cambium becomes putrid, and oozes out through the bark, which thus separates from the alburnum. The cause of this is probably to be sought in the soil. It is a very dangerous disease, and the elm is particularly liable to its attacks. Some fine trees of this kind perished, a few years since, in the avenue at Camberwell called "The Grove." As soon as the bark is separated from the wood, the intervening space is peopled by swarms of Scolytus destructor, and similar insects.

Extravasation, or *gumming*, consists in a discharge of thick sap from particular parts of the tree through the bark: the circumjacent surface withers but does not rot, as in Carcinoma.

Alburnitas is when a layer of soft wood is interposed between others of a harder texture: it is supposed to arise from a wet season.

Galls, or *tumid excrescences*, are local affections caused by the puncture of insects. They are produced by an excessive deposition of cellular tissue; and are of no consequence to the general health of the individual subject to them.

Albigo, ferrugo, and *uredo*, commonly called *mildew, smut, rust, brand*, and other names, are diseases caused by the presence of myriads of minute fungi of the genera Erycibe, Cæoma, Aspergillus, Puccinia, Uredo, and others. They are to plants what intestinal worms are to animals. Whether their presence is due to a languid state of the plant, which is thus rendered unusually susceptible of their attacks; whether the minute particles from which they are generated rise up from the earth through the vessels of the stem along with the sap, or whether they are originally contained in the seed and carried onwards with its growth, manifesting themselves whenever the plant arrives at a suitable state; finally, whether they are produced spontaneously, in consequence of the particular state of the atmosphere and tissue of the plant, are all points hereafter to be determined: nothing certain is known upon the subject at present. A very good account of the smut of barley is given by Mons. Adolphe Brongniart in the *Annales des Sciences*, vol. xx. p. 171.

Ergot, or *clavus*, is an excrescence from the seeds of grasses, of a brown or blackish colour; its nature or origin is undetermined. It does not depend upon the presence of parasitical fungi, and it possesses properties wholly foreign to the plant that bears it. That of the rye is frequently used successfully in medicine for the purpose of accelerating parturition.

Spotting, or *necrosis*, is chiefly found upon the leaves and soft parenchymatous parts of vegetables. It consists of small black spots, below which the substance of the plant decays: in many cases it no doubt arises from wet and cold, as in the cucumber and melon, which are wholly free from it in the warm weather of summer, but which are attacked immediately upon the arrival of the cold dewy nights of August and September.

Melligo and *salsugo*, by some reckoned diseases, are rather natural exudations of the juices of certain plants. Melligo produces the manna of the ash, the gum ladanum of the Cistus, &c.; *salsugo*, saline secretions of the same kind.

To these may be added, —

Suffocatio, or *choking up*, when every part diminishes in size.

Jeterus, or *jaundice*, a general yellowness.

Pernio, or *chilblains*, wounds caused by frost.

Rachitis, or premature falling of grain.

Verminatio, or being preyed upon by the larvæ of insects.

Phthiriasis, or *lousiness*, when the leaves and stems are infested by Aphides, or the like.

Squamatio, or *scaliness*, when scales are formed instead of leaves.

Exostosis, or *clubbing* of the roots.

Crispatura, or *curling*, in consequence of the leaves being punctured by insects.

Decoloration, or *loss of colour*.

Coloration, or *staining*.

Anthozusia, or change of the leaves into petals.

CHAPTER XV.

OF HYBRID PLANTS.

It is well known that, in the animal kingdom, if the male and female of two distinct species of the same genus breed together, the result is an offspring intermediate in character between its parents, but uniformly incapable of procreation unless with one of its parents; while the progeny of varieties of the same species, however dissimilar in habit, feature, or general characters, is in all cases as fertile as the parents themselves. A law very similar to this exists in the vegetable kingdom.

Two distinct species of the same genus will often together produce an offspring intermediate in character between themselves, and capable of performing all its vital functions as perfectly as either parent, with the exception of its being unequal to perpetuating itself by seed; or should it not be absolutely sterile, it will become so in the second, third, or, very rarely, fourth generation. It may, however, be rendered fertile by the application of the pollen of either of its parents; in which case its offspring assumes the character of the parent by which the pollen was supplied. This power of hybridising appears to be far more common in plants than in animals; for while only a few animal mules are known, there is scarcely a genus of domesticated plants in which this effect cannot be produced by the assistance of man, in placing the pollen of one species upon the stigma of another. It is, however, in general only between nearly allied species that this intercourse can take place; those which are widely different in structure and constitution not being capable of any artificial union. Thus the different species of strawberry, of certain tribes of Pelargonium, and of Cucurbitaceæ, intermix with the greatest facility, there being a great accordance between them in general structure and constitution; but no one has ever succeeded in compelling the pear to fertilise the apple, nor the gooseberry the currant. And as species that are very dissimilar appear to have some natural impediment which prevents their reciprocal fertilisation, so does this obstacle, of whatever nature it may be, present an insuperable bar to the intercourse of different genera. All the stories that are current as to the intermixture of oranges and pomegranates, of roses and black currants, and the like, may therefore be set down to pure invention.

By far the best series of observations that has been instituted with a view to determine the laws of hybridism was that of Kölreuter, who, about the year 1775, commenced a set of experiments, which he continued to prosecute for twenty years, upon species of the genera Digitalis, Verbascum, Solanum, Malva, Linum, Dianthus, and Mirabilis. It is upon those experiments, combined with the subsequent experience of others and my own observations, that the foregoing statement has been made.

It has, nevertheless, been asserted by divers experienced cultivators of the present day, that the conclusions drawn from the experiments of Kölreuter have been too hasty; and that if they apply to the genera that were the special subject of the attention of that observer, they are by no means applicable to plants in general. It has been urged, in proof of this statement, that many different species of African Gladioli, of Pelargonium, of South American Amaryllis, of Crinum, of Triticum, &c., breed freely together, and that their seedlings are as fertile as themselves.

I must confess that these instances are by no means such as to shake my confidence in the accuracy of the laws deduced from Kölreuter's experiments. In the first place, there is a degree of vagueness and looseness in all the cases that are specified, which is particularly striking if compared with the precision with which Kölreuter's experiments were conducted; secondly, in all the cases above mentioned, which I believe are the most remarkable, there is much room for doubt whether the supposed species upon which the argument is founded are any thing more than wild varieties of each other. The African Gladioli are known to intermix freely; but

Mr. Herbert, in his account of them, in the *Horticultural Transactions*, vol. iv. p. 16., admits that he cannot speak to the power of their mules to perpetuate themselves by seed. No botanist can fix positive characters to a large part of the reputed species of Pelargonium, or to the South American Amaryllises, which Mr. Herbert calls Hippeastra; many of the supposed species of Crinum seem to have no better claim to be so considered than the varieties that might be picked from a bed of tulips; and, lastly, the Tritica cærulescens, polonicum, and tomentosum, upon which Bellardi's experiments were founded, are plants with the history of which no man is acquainted, and which, in all probability, derive their origin from the Triticum æstivum, or common wheat.

All I think that can be conceded upon this subject is, that more hybrid plants are fertile to the third or fourth generation than Kölreuter supposed: that they will all, in time, revert to one or other of their parents, or become absolutely barren, there can be no doubt whatever.

The cause of the sterility of mule plants is at present entirely unknown. Sometimes, indeed, a deficiency of pollen may be assigned; but in many cases there is no perceptible difference in the healthiness of structure of the fertilising organs of a male plant and of its parents. I know of no person who has attempted to prove this by comparative anatomical observations, except Professor Henslow, of Cambridge; who, in an excellent paper upon a hybrid Digitalis, investigated anatomically the condition of the stamens and pistillum, both of his hybrid and its two parents, with great care and skill. The result of his enquiry was, that no appreciable difference could be detected.

Although this power of creating mule plants that are fertile for two or three generations incontestably exists, yet in wild nature hybrid varieties are far from common; or, at least, there are few well attested instances of the fact. Among the most remarkable cases are the Cistus Ledon, constantly produced between C. monspessulanus and laurifolius, and Cistus longifolius, between C. monspessulanus and populifolius, in the wood of Fontfroide, near Narbonne, mentioned by Mr. Bentham. Again, the same acute botanist ascertained that Saxifraga luteopurpurea of Lapeyronse, and S. ambigua of De Candolle, are only wild accidental hybrids between S. aretioides and calyciflora: they are only found when the two parents grow together; but there they form a suite of intermediate states between the two. Gentians having a similar origin have also been remarked upon the mountains of Europe. It is difficult not to believe that a great number of the reputed species of Salix, Rosa, Rubus, and other intricate genera, have also had a hybrid origin; but I am not aware that there is at present any positive proof of this.

In a practical point of view, I am inclined to believe that the power of obtaining mule varieties by art is one of the most important means that man possesses of modifying the works of nature, and of rendering them better adapted to his purposes. In our gardens some of the most beautiful flowers have such an origin; as, for instance, the roses obtained between R. indica and moschata, the different mule Potentillæ and Cacti, the splendid Azaleas raised between A. pontica and A. nudiflora coccinea, and the magnificent American-Indian Rhododendrons. By crossing varieties of the same species, the races of fruits and of culinary vegetables have been brought to a state as nearly approaching perfection as we can suppose possible. And if similar improvements have not taken place in a more important department,— namely, the trees that afford us timber,— our experience fully warrants our entertaining the belief that, if proper means were adopted, improved varieties of as much consequence might be introduced into our forests, as have already been created for our gardens.

In conducting experiments of this kind, it is well to know that, in general, the characters of the female parent predominate in the flowers and parts of fructification; while the foliage and general constitution are chiefly those of the male parent. Thus, in the celebrated mule Rhododendron, gained by Lord Carnarvon by fertilising R. arboreum with R. Catawbiense, the mule variety had the flowers and colour of R. arboreum, but more the leaves and hardiness of constitution of R. Catawbiense.

CHAPTER XVI.

OF FLOWERLESS PLANTS.

VERY little can be said to be *positively* known of the manner in which the organs of flowerless plants perform what are supposed to be their functions. We are entirely ignorant of the manner in which the stems of those that are arborescent are developed, and of the course taken by their ascending and descending fluids, — if, indeed, in them there really exist currents similar to those of flowering plants; which may be doubted. We know not in what way the fertilising principle is communicated to the sporules, or reproductive grains; the use of the different kinds of reproductive matter found in most tribes is entirely concealed from us; it is even suspected that some of the simplest forms (of Algæ and Fungi, at least) are the creatures of spontaneous growth; and, in fine, we seem to have discovered little that is positive about the vital actions of these plants, except that they are reproduced by their sporules, which differ from seeds in germinating from any part of their surface, instead of from two invariable points. Under these circumstances, it would be useless to dwell upon the subject: those who wish to make themselves acquainted with the speculations of botanists, are referred to the valuable writings of Hedwig on Mosses, of Sprengel upon Cryptogamic plants, of Dr. Hooker on Jungermanniæ, of Fries on Fungi, of Agardh, Greville, and Bory de St. Vincent upon Algæ, of Meyer upon Lichens, of Bischoff upon Equisetum, &c.; and, finally, to the Introduction to the Natural System of Botany by the author of the present work.

BOOK III.

TAXONOMY; OR, OF THE PRINCIPLES OF CLASSIFICATION.

CHAPTER I.

OF THE GENERAL OBJECTS OF CLASSIFICATION.

The objects of classification are twofold: firstly, to place natural bodies in such an arrangement, that the station of a given object can be certainly discovered upon the application of some particular mode of investigation; and, secondly, to arrange them so, that the relation they mutually bear to each other in the scale of the creation may be distinctly manifested. These objects may be attained either singly, or in conjunction with each other; hence two distinct kinds of arrangement have arisen: one called *artificial*, in which the sole purpose is to discover the names of individuals; and the other called *natural*, in which the natural affinity of beings is preserved, and, at the same time, a power is maintained of discovering any given individual by the application of a particular mode of analysis.

In order to attain either or both of these ends, it is necessary to possess a power, firstly, of analysing the facts we possess, and of reducing them from their most complex to their most simple state; and, secondly, of recombining them in various degrees, according to the peculiarities which our previous analysis has shown us that they exhibit. To effect this in

CHAP. I. GENERAL OBJECTS OF CLASSIFICATION.

the vegetable kingdom, it has been found expedient to divide plants into groups of different degrees of importance, called
    Classes,
        Subclasses,
            Orders,
                Tribes,
                    Genera,
                        Subgenera,
                            Species.

Of these, the first are characterised by peculiarities common to all those that succeed them; the second, by peculiarities of a less universal application; the third, by others of minor importance; and so on until we arrive at species, which form the ultimate point of analysis to which our investigations can extend. Species are created by Nature herself, and remain always the same, in whatever manner they may be combined: they form the basis of all classification, and are the only part of it which can be considered absolute. For although, in a natural system, all other combinations, whether genera, tribes, orders, or by whatever name they may be known, comprehend species agreeing much more with each other than with any thing else, and having a positive general resemblance in the majority of their features, yet no fixed limits can be assigned to any of them: on the contrary, they pass, by means of various intermediate species, into the other genera, tribes, orders, &c. to which they are most nearly allied. For this reason, viz. that no fixed limits can be assigned to orders, genera, &c., we find the ideas about them fluctuating with the degree of our knowledge; which is the true cause of those changes in the limits of genera, &c. which persons unacquainted with the subject are apt to consider arbitrary; but which, in skilful hands, are dependent upon a progressive advance in our knowledge of science.

Classifications are founded upon the modifications of various organs of plants, the constancy or mutability of form of which determine what is called the *value of characters;* upon these the different groups depend. It was formerly supposed that the organs of fructification were more constant in their characters, and less subject to variation, than any other part; and hence they were exclusively adopted as the basis of classification. But modern investigations have shown that characters drawn from the mode in which plants grow, and from certain anatomical peculiarities, are of much higher value: so that the organs of fructification are now chiefly employed for the distinction of genera, or of orders and tribes. And, even in these minor groups, the organs of vegetation are frequently of high importance.

What better characters, for example, can be found in the fructification of Cinchonaceæ than in the arrangement of their leaves, which are universally opposite, entire, and connected by intervening stipulæ; or how can the fructification of Myrtaceæ characterise that order more accurately than their simple, opposite, entire, exstipulate, dotted leaves, with an intramarginal vein? No better characters are afforded by the fructification than by the foliage of Stellatæ, Gramineæ, and many others.

CHAPTER II.

OF ARTIFICIAL ARRANGEMENTS.

If a classification is to be entirely artificial, it is sufficient to take some two or three parts of the fructification, and form a scheme of arrangement out of their modifications. It is of no consequence what these parts are, or how the object is effected, provided the end is attained, of ascertaining readily and certainly the name of a given object. That method will be the best which depends upon the most obvious and permanent characters. It must not, however, be expected that such a kind of classification will be ever, under the most favourable circumstances, in any degree perfect. This is impossible from the very nature of things: there is no part of either the fructification or vegetation of plants sufficiently constant in single characters to render such perfection practicable; to the most ingeniously devised characters exceptions will continually occur. The sexual system of Linnæus, is, in botany, the one to which the world has consented to ascribe the greatest utility.

This celebrated system depends upon a consideration of certain modifications in the arrangement, or difference in the number, of the stamens and styles. To acquire the power of referring plants to their places in this system, nothing more is requisite than just so much knowledge of structure as will enable the student to distinguish the one set of organs from the other, to count their number, and to determine to which of the modifications of arrangement admitted by Linnæus they are to be referred. Hence the great popularity acquired by the Linnæan system, which is at first sight so simple and precise as to leave nothing to be wished for. And it is in reality extremely useful in many respects, and very generally applicable; but, unfortunately, practical difficulties surround the student as he proceeds, which often lead to confusion and uncertainty. Nothing is more common than for plants to have their stamens cohering at the base into a kind of cup: such ought, if the Linnæan arrangement is rigidly observed, to be referred to Monadelphia; yet they are constantly placed by Linnæus himself in other classes. A plant will bear flowers differing in the number of their stamens upon the same individual, or upon different individuals; and species of the same genus also differ from each other in that respect: out of this arise, of necessity, the most perplexing difficulties, which no ingenuity or exercise of the reasoning faculties can overcome; because the system of Linnæus being in every respect artificial, depending wholly and absolutely upon the technical characters he employs, no check exists by which errors can be discovered, if they arise out of variations or uncertainty in their characters.

Besides this, there is another difficulty of equal importance. Unless the plant which is to be examined is in flower, it cannot be referred to its station; or if there is any imperfection in the developement of its stamens or styles, which is a very common case, the same difficulty opposes itself. The facilities, therefore, which have been attributed to the system of Linnæus by those who have adopted it, although undoubtedly great in one point of view, are by no means such as have been represented and believed.

The following are the characters of the classes and orders of the Linnæan System:—

Characters of the Classes.

Class I.	Stamen	1	-	-	-	Monandria.
II.		2	-		-	Diandria.
III.		3	-	-	-	Triandria.
IV.		4		-	-	Tetrandria.
V.		5	-	-	-	Pentandria.
VI.		6		-	-	Hexandria.
VII.		7	-	-	-	Heptandria.
VIII.		8		-	-	Octandria.

CHAP. II.   OF ARTIFICIAL ARRANGEMENTS.   311

Class IX.	Stamens	9 - - -	Enneandria.
X.		10 - -	Decandria.
XI.		12—19 - -	Dodecandria.
XII.		{ 20 or more, inserted into the calyx - }	Icosandria.
XIII.		{ 20 or more, inserted into the receptacle }	Polyandria.
XIV.		2 long and 2 short	Didynamia.
XV.		4 long and 2 short	Tetradynamia.
XVI.		{ united by their filaments into a tube }	Monadelphia.
XVII.		{ united by their filaments into two parcels - - }	Diadelphia.
XVIII.		{ united by their filaments into several parcels - - }	Polyadelphia.
XIX.		{ united by their anthers into a tube }	Syngnesia.
XX.		{ united with the pistillum - - }	Gynandria.
XXI.		{ and pistils in separate flowers, but both growing on the same plant - }	Monœcia.
XXII.		{ and pistils not only in separate flowers, but those flowers situated upon two different plants - - }	Diœcia.
XXIII.		{ and pistils separate in some flowers, united in others, either on the same plant, or two or three different ones - - }	Polygamia.

x 4

Class XXIV. Stamens { and pistils either not ascertained, or not to be discovered with any certainty, insomuch that the plants cannot be referred to any of the foregoing classes - - } Cryptogamia.

Characters of the Orders.

These depend upon the number of the styles, or of the stigmas, *if there be no style,* in the first thirteen classes; such are accordingly named, —

Monogynia	- -	Style 1
Digynia	- -	- 2
Trigynia	- -	- 3
Tetragynia	- -	- 4
Pentagynia	- -	- 5
Hexagynia	- -	- 6
Heptagynia	- -	- 7
Octogynia	- -	- 8
Enneagynia	- -	- 9
Decagynia	- -	- 10
Dodecagynia	- -	- 12
Polygynia	- -	- more than 12.

In the 14th class, Didynamia, the orders depend upon the nature of the ovarium. —In *Gymnospermia,* the first order, the ovarium is divided into four lobes, from the base of which proceeds a single style, and within each of which is contained a single seed. In *Angiospermia,* the 2d order, the ovarium is not lobed, and is usually two-celled, and many seeded.

In the 15th class, *Tetradynamia,* the orders are characterised by the form of the fruit. *Siliquosæ* have a long pod. *Siliculosæ* have a short one.

The orders of the 16th, 17th, and 18th classes, Monadelphia, Diadelphia, and Polyadelphia, depend upon the number of the stamens, and have the same nomenclature as the thirteen first classes.

The orders of Syngenesia are determined by the arrangement of their flowers, and by the sex of their florets: thus —

Polygamia, has florets crowded together in heads.

1. *Polygamia æqualis,* has each floret hermaphrodite, or furnished with perfect stamens and pistillum.

2. *Polygamia superflua,* has the florets of the disk hermaphrodite; those of the ray female only.

3. *Polygamia frustranea,* has the florets of the disk hermaphrodite; those of the ray sterile.

4. *Polygamia necessaria,* has the florets of the disk male, of the ray female.

5. *Polygamia segregata,* " has several florets, either simple or compound, but with a proper calyx, included within one common calyx."

Monogamia, has the flowers separate, not crowded in heads. This order is generally abolished by Linnæan botanists, but for no good reason.

The orders of the 20th, 21st, and 22d classes are distinguished by the number, &c. of the stamens;

The two orders of the 23d class depend upon whether the genera are monœcious or diœcious.

The last class, Cryptogamia, is divided into orders according to the principles of the natural system, and are, 1. Filices; 2. Musci; 3. Hepaticæ; 4. Algæ; 5. Fungi.

Various modifications and alterations of this method have, from time to time, been proposed by various writers; some reducing the whole of Monœcia, Diœcia, and Polygamia to other classes, others combining some of the smaller classes with the larger ones. Of these and similar alterations it is not necessary to say more in this place; but I must not pass by, with equal silence, the changes proposed by the late Professor Richard. These have not, indeed, been adopted by other writers: but this is, I conceive, to be attributed rather to the neglect into which artificial systems have now fallen, than to any other circumstance. As the changes are proposed upon philosophical principles, I shall notice them in the words of M. Achille Richard, for the information of those botanists who still interest themselves in the fate of the sexual system.

The 10 first classes are preserved without alteration.

The 11th, *Polyandria*, is characterised thus: stamens more than 10, inserted under the pistil, which is either simple or multiple; insertion hypogynous. This class, which replaces Dodecandria, answers in all respects to the Polyandria of Linnæus.

The 12th class is *Calycandria*, with the following character :— stamens more than 10, inserted on the calyx, the ovarium being superior or parietal; insertion perigynous. This answers in part to Dodecandria, and in part to Icosandria. It contains all the true Rosaceæ.

The 13th class is *Hysterandria*. Its character is, stamens more than 10, inserted upon an ovarium which is perfectly inferior; insertion consequently epigynous. It answers to a part of Icosandria, and contains Myrtus, Punica, &c.

These three classes are much more precise, and also maintain in a more perfect manner the natural relations of plants, than those of Linnæus; whose characters, depending upon the number of stamens, frequently lead the student into error.

In the 14th class, *Didynamia*, the orders designed by Linnæus give a false idea of structure; there being no such thing as naked seeds, as expressed by his name *Gymnospermia*. For this latter the name *Tomogynia* is proposed; and for *Angiospermia*, *Atomogynia*.

The 19th class, *Synantheria*, replaces *Syngenesia*. The stamens are united by the anthers only, so as to form a small tube; and the ovary is monospermous. It comprehends the plants called Compositæ only; and is divided into the following orders, in the room of those of Linnæus :—

1. *Carduaceæ;* head composed indifferently of male, female, or hermaphrodite florets; receptacle covered with numerous hairs; style slightly swollen beneath the stigma; connectivum sometimes lengthened beyond the anthers, so as to form a five-toothed tube. To this belong Carduus, Centaurea, &c.

2. *Corymbiferæ;* head flosculous or radiate; receptacle naked or covered with paleæ, one to each floret. In the preceding order there were several to each floret. Here belong Tussilago, Gnaphalium, Erigeron, &c.

3. *Chicoraceæ;* head composed of ligulate florets wholly, as in Lactuca, Cichorium, &c.

The 20th class, *Symphysandria,* is the *Syngenesia Monogamia* of Linnæus, distinguished from the last by its polyspermous ovarium: it contains Lobelia, Viola, &c.

Gynandria, Monœcia, and Diœcia are preserved without change.

24th class, *Anomalœcia,* is the same as *Polygamia* of Linnæus.

25th class, *Agamia,* is the Linnæan Cryptogamia.

From this explanation of Richard's proposed reformation of the sexual system, it is apparent that its fault is its unnecessary alteration of the established nomenclature of Linnæus; and that its merit consists in a better adjustment of the Dodecandrous, Icosandrous, Polyandrous, and Syngenesious classes; which are, undoubtedly, improvements of great importance.

By some writers the mode of analysing characters contrived by Lamarck, and continued by M. De Candolle in the Flore Française, has been referred to artificial systems; but scarcely with propriety. We can hardly call that a system which is merely a particular mode of analysis, without any arrangement consequent upon it: nevertheless, as it is useful to understand this mode of investigation, no better place can, perhaps, be found for explaining it than the present.

The mental operation by which one thing is distinguished from another consists in a continual contrasting of characters. For instance, in a mass of individuals we distinguish one set which is coloured, and another which is colourless; of those that are coloured we distinguish red, black, blue, and green; of the red, some are square, others are round; of the round, some are sculptured on their surface, others are even: — and so we proceed, analysing the subject by a constant series of contrasts, until we have arrived at the point beyond which no analysis can go.

Such is the character of Lamarck's method: take, for instance, the following instance from the Flore Française, 3d edit. vol. i. p. 257.

ANCHUSA.

1.	{ Flowers in spikes, heads, or dense panicles	-	-	-	2
	{ Flowers widely scattered, or in loose racemes	-	-	*laxiflora.*
2.	{ Leaves all alternate	-	-	-	3
	{ Each bunch of flowers with two opposite leaves	-	-	*sempervirens.*
3.	{ Flowers in racemes or panicles, leaves flat	-	-	4
	{ Flowers in heads, leaves undulated	-	-	*undulata.*
4.	{ Stem straight, from nine to twelve inches high	-	-	5
	{ Stems spreading, from six to nine inches high	-	-	*tinctoria.*
5.	{ Calyx divided as far as the middle	-	-	*angustifolia.*
	{ Calyx divided almost to the base	-	-	6
6.	{ Flowers violet; scales of the orifice bearded	-	-	*italica.*
	{ Flowers azure blue; scales slightly hairy	-	-	*Barrelieri.*

Thus it is obvious that one character is contrasted with another, until nothing more remains to contrast. The plan has been successfully adopted by Messrs. Hooker and Taylor, in their Muscologia Britannica; and I think it, or some similar plan, should be introduced into every work on systematic natural history, on account of the labour it saves the reader when casual investigation only is his object.

But I do not entirely approve of the method as it stands in the works of De Candolle. The mode of printing is particularly objectionable; the constant reference from number to number is fatiguing, and confuses; and the contrast of characters is not brought sufficiently before the eye. I have therefore used, in my own works, a slight modification, which at least is free from the objections now mentioned. Of this the following example is taken from the 43d page of my Synopsis of the British Flora.

ANALYSIS OF THE BRITISH GENERA OF CARYOPHYLLEÆ.

Sepals united in a cylindrical tube (*Sileneæ*).
    Stigmata 2.
        Calyx with bracteæ at the base - - - 1. DIANTHUS.
        Calyx naked at the base - - - 2. SAPONARIA.
    Stigmata 3 - - - - - 3. SILENE.
    Stigmata 5.
        Calyx-teeth simple - - - - 4. LYCHNIS.
        Calyx-teeth foliaceous - - - 5. AGROSTEMMA.

Sepals distinct, or cohering only at the base (*Alsineæ*).
    Capsule dehiscing with distinct valves.
        Valves 2 - - - - - 6. BUFFONIA.
        Valves 3 - - - - 7. CHERLERIA.
        Valves 6 - - - - - 8. SPERGULA.
        Valves 4 or 5.
            Capsule with four cells - - - 9. ELATINE.
            Capsule with one cell - - - 10. SAGINA.
    Capsule dehiscing at the apex with teeth.
        Petals entire.
            Sepals and petals 4 - - - 11. MŒNCHIA.
            Sepals and petals 5 - - - 12. ARENARIA.
        Petals toothed - - - - 13. HOLOSTEUM.

I must not dismiss this subject without remarking, that, however excellent this analytical method is, if well managed, it is of all the very worst if used injudiciously. One false step, either on the part of the author who frames it, or on that of the reader, instantly leads astray, and induces errors of the most serious kind. Every thing depends upon a judicious selection of contrasting characters on the one hand, and a scrupulous attention to them on the other. This method ought never to be used by itself, but merely as a kind of key to long generic or specific characters and descriptions.

CHAPTER III.

OF THE NATURAL SYSTEM.

A NATURAL method of arrangement differs essentially from an artificial one in this, — that it does not depend upon modifications of any one part more than of another. Its divisions are framed from a careful consideration of every, even the minutest, character that is appreciable; and consist of species, not arbitrarily collected by a few common signs, but agreeing with each other as far as possible in every material point of structure. Groups formed upon this principle will necessarily consist of species having a greater resemblance to each other than to any thing else; and, if skilfully constructed, will have so great a general resemblance, that a knowledge of the structure, habits, qualities, or other important peculiarities of a single species, gives an accurate general idea of all the others that the group contains. While an artificial mode of classification leads to nothing further than the determination of the name of a given species, and consists of assemblages of objects so incongruous and unlike, that, in the words of a modern philosopher, "our ingenuity is exercised to determine what can be the cause of their resemblance;" a natural system is essentially composed of collections of species so like, that our difficulty is rather in ascertaining their differences.

The classes, orders, &c. of a natural system depending thus upon characters impressed upon vegetation by the hand of Nature, and arising out of combinations of peculiarities that are uniformly the same, and of resemblances about which there can be no difference of opinion, it follows that there can be but one system properly called natural. If our knowledge of the subject were perfect, the mutual affinities, and the combinations of plants intŏ classes and orders, would be as fixed and unchangeable as the planetary system itself. But, in the actual state of botanical science, the only parts that can be considered approaching perfection are a few of the primary divisions, and the lowest groups or natural orders, as they are called. All that relates to the distribution of those orders, with relation to each other, is in the highest degree imperfect and fluctuating; a great defect in the subject viewed as a system, but of little practical importance. In practice, indeed, it may be doubted whether the real affinities of the minor natural assemblages of plants can ever be distinctly preserved in any arrangement for systematic purposes. In the words of Mr. Herschel, " the classifications by which science is advanced" (as distinguished from artificial systems of nomenclature) " cross and intersect one another, as it were, in every possible way, and have for their very aim to interweave all the objects of nature in a close and compact web of mutual relations and dependence." (*Discourse*, p. 140.) A natural class, or tribe, or order, or genus, or species, will have a great number of points of resemblance with others; some of which points will be of more importance, or more immediate, than others; and all of them affecting its position in a greater or less degree: but upon paper we are constrained to follow a lineal arrangement, in which only one point of agreement instead of many can be indicated. We are, therefore, able to point out by our classification, in a partial manner only the relations borne by bodies to each other: and in proportion as we fail in maintaining these affinities perfectly, do we recede from the course of nature and become artificial. For this there is clearly no help, if we wish that our classifications should be adapted to practical purposes. If the subject is to be treated as a mere operation of the mind, recourse can then be had to systems founded on the principles of Oken, Fries, and others, of which something will be said hereafter.

The earliest attempts at arrangement were natural, as far as the knowledge of their authors enabled them to make them so; and artificial systems only arose after a long series of years, out of the imperfect state of botanical science, which was not sufficient to render a really natural system practicable.

At last a French naturalist, M. Antoine Laurent de Jussieu, who with a profound knowledge of the subject combined great logical acuteness, conceived the idea of what is now universally recognised as the natural system of botany.

Properly speaking, this system is subject to no kind of artificial arrangement: it consists of certain groups called natural orders, all of which are, or should be, independent of each other; and the characters of which are derived indifferently from every part of the plant. But as it would be extremely embarrassing to the student to acquire a just notion of these groups, unless some mode were devised of analysing their characters, several plans have been invented by which the groups have been reduced to a sort of artificial arrangement, with greater or less violence to their mutual affinities. As all these plans must, as has been shown, necessarily be linear, the real affinities of plants must be very imperfectly indicated by them: they are, therefore, of no value whatever, except for the purpose of facilitating investigation. They must be understood to form no part of what must strictly be called the natural system; they may be varied at pleasure, according to the ingenuity of the botanist; and that will be the best which is most facile, and which, at the same time, offers the fewest interruptions to the series of mutual relations. At present I think there are few botanists who will deny that they are all extremely defective; and that one of the greatest services that could be rendered to systematic botany would be to devise some scheme by which the orders could be better and more naturally arranged under their primary classes. Whoever does this will have to divest himself of all the prejudices, and they are not a few, which have grown up with the system of Jussieu, and that have taken deep root in the minds of his followers: he must judge for himself upon every single point that may come before him, and he must forget that any such artificial arrangements have existed as those of Jussieu himself, De Candolle, and others. It is even to be expected that the organs of vegetation will be, for this purpose, employed even more than those of the fructification; and that anatomical characters analogous to those which characterise the really natural primary divisions of Vasculares and Cellulares, and of Exogenæ and Endogenæ, will be applied to the grouping, in subordinate masses, of the orders themselves.

CHAP. III.   OF THE NATURAL SYSTEM.   321

At present scarcely any attempts of this nature have been made, except by Agardh and Bartling; but the endeavours of those botanists, however meritorious, are far from coming up to what may be expected.

The first outline of the natural orders that are now adopted was given to the public by Jussieu in 1789, in the following form: —

INDEX OF THE CLASSES.

Acotyledones - - - - - Class I.
Monocotyledones
  { Stamens hypogynous - - - Class II.
  { Stamens perigynous - - - Class III.
  { Stamens epigynous - - - Class IV.
Dicotyledones.
  Apetalous
    { Stamens epigynous - - Class V.
    { Stamens perigynous - - Class VI.
    { Stamens hypogynous - - Class VII.
  Monopetalous
    { Corolla hypogynous - Class VIII.
    { Corolla perigynous - Class IX.
    { Corolla epigynous:
        With connate anthers - Class X.
        With distinct anthers - Class XI.
  Polypetalous
    { Stamens epigynous - - Class XII.
    { Stamens hypogynous - - Class XIII.
    { Stamens perigynous - - Class XIV.
  Diclinous, irregular - - Class XV.

List of the Orders.

Class I.
1. Fungi.
2. Algæ.
3. Hepaticæ.
4. Musci.
5. Filices.
6. Naiades.

Class II.
7. Aroideæ.
8. Typhæ.
9. Cyperoideæ.
10. Gramineæ.

Class III.
11. Palmæ.
12. Asparagi.
13. Junci.
14. Lilia.
15. Bromeliæ.
16. Asphodeli.
17. Narcissi.
18. Irides.

Class IV.
19. Musæ.
20. Cannæ.
21. Orchides.
22. Hydrocharides.

Class V.
23. Aristolochiæ.

Class VI.
24. Elæagni.
25. Thymelææ.
26. Proteæ.

27. Lauri.
28. Polygoneæ.
29. Atriplices.

Class VII.
30. Amaranthi.
31. Plantagines.
32. Nyctagines.
33. Plumbagines.

Class VIII.
34. Lysimachiæ.
35. Pediculares.
36. Acanthi.
37. Jasmineæ.
38. Vitices.
39. Labiatæ.
40. Scrophulariæ.
41. Solaneæ.
42. Boragineæ.
43. Convolvuli.
44. Polemonia.
45. Bignoniæ.
46. Gentianæ.
47. Apocyneæ.
48. Sapotæ.

Class IX.
49. Guaiacanæ.
50. Rhododendra.
51. Ericæ.
52. Campanulaceæ.

Class X.
53. Cichoraceæ.
54. Cynarocephalæ.
55. Corymbiferæ.

Class XI.
56. Dipsaceæ.
57. Rubiaceæ.
58. Caprifolia.

Class XII.
59. Araliæ.
60. Umbelliferæ.

Class XIII.
61. Ranunculaceæ.
62. Papaveraceæ.
63. Cruciferæ.
64. Capparides.
65. Sapindi.
66. Acera.
67. Malpighiæ.
68. Hyperica.
69. Guttiferæ.
70. Aurantia.
71. Meliæ.
72. Vites.
73. Gerania.
74. Malvaceæ.
75. Magnoliæ.
76. Anonæ.
77. Menisperma.
78. Berberides.
79. Tiliaceæ.
80. Cisti.
81. Rutaceæ.
82. Caryophylleæ.

Class XIV.
83. Sempervivæ.
84. Saxifragæ.
85. Cacti.
86. Portulaceæ.
87. Ficoideæ.
88. Onagræ.
89. Myrti.
90. Melastomæ.
91. Salicariæ.
92. Rosaceæ.
93. Leguminosæ.

94. Terebintaceæ.	97. Cucurbitaceæ.
95. Rhamni.	98. Urticæ.
Class XV.	99. Amentaceæ.
96. Euphorbiæ.	100. Coniferæ.

Appended to these is a list of genera, the characters of which were not at that time sufficiently well known to enable Jussieu to refer them to any of the preceding orders. They are arranged according to the following artificial plan: —

1. Monopetalous with a superior ovarium.
2. Monopetalous with an inferior ovarium.
3. Polypetalous with a superior ovarium.
4. Polypetalous with an inferior ovarium.
5. Apetalous, hermaphrodite, with a superior ovarium.
6. Apetalous, hermaphrodite, with an inferior ovarium.
7. Apetalous, diclinous, with a superior ovarium.
8. Apetalous, diclinous, with an inferior ovarium.

In this attempt, which may be truly pronounced the most important event which has occurred in botany, next to the universal reformation of natural history by Linnæus, advantage was taken of an arrangement used in the Trianon Garden in 1759, by Bernard de Jussieu, in which the natural affinities of plants had been seized in a manner that is perfectly surprising, if we consider how little was known at that time of those anatomical peculiarities upon which the natural affinities of the vegetable kingdom are now determined.

Since the date of 1789, alterations, additions, and improvements in the system of Jussieu have been constantly making. For these the world is chiefly indebted to Jussieu himself, our celebrated countryman Mr. Brown, De Candolle, the late Louis Claude Richard, Kunth, Auguste St. Hilaire, Von Martius, and a few others. These have given a new feature to the system, and have brought it to a far more advanced state of completeness than, perhaps, could have been expected in the short space of thirty or forty years. To show what that state is, and the existing notions relating to the limits of natural orders, would occupy much more space than could be spared in the present work. The student will find ample information upon the subject in my *Introduction to the Natural System of Botany*, published in 1830.

CHAPTER IV.

OF SPECULATIVE MODES OF ARRANGEMENT.

Besides the two principles of classification now explained, there is a third, which has arisen in the minds of certain German metaphysical naturalists, and which they conceive to be the true natural system; but which may more properly be termed a *speculative* attempt at forming a system by the mere force of reason, without attending to the facts upon which any system must depend. While other naturalists take facts as the basis of all arrangement, and accommodate their system to the data they possess, the authors of the speculative modes of classification first form a system from the idea that all matter is subject to the influence of certain universal causes, which prevail equally in every kingdom of nature; and then attempt to adjust facts to the arrangement thus formed by the mere force of imagination. Strange as this doctrine seems, and foreign as it may appear to every principle of sound philosophy, yet, as it is possible that some useful ideas may be elicited even from the wildest of such speculations, I shall avail myself of an excellent exposition of the subject by M. Choisy, of Geneva, to give some account of the doctrines in question.

The first principle, they say, to be inculcated is, that there is a fundamental difference between philosophical and empirical science: there is, says Nees Von Esenbeck, a speculative, and also an experimental, knowledge of things; the first is dependent upon a pure conception of nature, the last relies upon material observation. In consequence of thus distinguishing two methods, these philosophers consider themselves enabled to establish what they call *unity*, which is their leading dogma; and by which they mean the existence of some single power that overrules all other powers, and determines the structure and existence of every thing. By the aid of such facts as are applicable to their particular notions, each philosopher then constructs his edifice, well cemented together in all its parts, as he believes; but, in reality, possessing no solidity whatever. Their unity, — a vague and mysterious notion, which is only intelligible with respect to the Deity, — they think can be comprehended by man by other means than by a comparison of the admirable works of the Creator; assertions destitute of proof, and even contradicted by experience, are said, nevertheless, to be demonstrated by the tendency which they have to indicate a principle of union in the different parts of the material world. The paths which have been pursued to attain the unity which is sought for, and the systems that have been invented for that purpose, are many: the following are the principal. — First, They declare that certain general laws exist which control the universe. Secondly, All things are formed of the same elements, in different states of combination. Thirdly, The same principles and the same phenomena occur in the whole of each individual, and in all its parts, as occurs in the earth itself; and these are repeated down to the most minute phenomenon throughout the whole globe.

Among general laws, the most important is that of *polarity;* every unity arises in nature from a primitive, opposite, and double action, and this double action is found in every subject which is examined. The word *pole* is used to express the two elements out of which this double action arises; and polarity indicates their respective actions. Thus, we every where find the positive and negative, the real and the imaginary, the light and the dark, the cause and the effect; and, according to Oken, this polarity produces every simple power in nature, by the juxtaposition of the *plus* and *minus* of which it is composed: without this nothing would exist, and there would be no world. Polarity is manifested by *motion,* which depends entirely upon the primitive opposition of an internal cause, and an exterior point of impulse: in itself it exhibits an instance of unity; but it is the result of a double action, that of polarity. According to Nees, the system of the world would not go on if the law of polarity did not exist; we must have an axis and an equator, which unite in the motion of rotation; we must even distinguish two sorts of polarity, one of which belongs to the shorter and the other to the longer axis; each of these has its *major* or objective pole, and its *minor* or subjective pole: the point of meeting of the axis, and all which pertains thereto, represents death; and all which exists towards the poles is endued with life. The type of life, which is placed within the influence of fixed poles, is called *a year;* the type of life, which is within the influence of varying poles, is called *day* and *night:* the four poles, thus comprehending four elements of life, give rise to four kingdoms in animated nature; that is to say, to fungi, plants, animals, and man. Wilbrand is another author who has written a work specially to examine the law of polarity. He defines it to be "an opposition between two things which mutually support each other, and of which one is nothing without the other: the union of these two things is a third, which is the result, and which would not exist without their opposition." He has no difficulty in finding traces every where of this double polar action; he is neither embarrassed by magnetism nor by electricity; in them the positive and negative poles are evidences, which cannot be overlooked: he also concludes that all bodies ought to be susceptible of magnetism. Galvanism is also to be readily explained by the doctrine of polarity: but, in order to apply it equally to all parts of chemistry, Wilbrand supposes that in every chemical analysis there is a complex sort of synthesis; that, whenever repulsion occurs, it is accompanied by a contrary action of affinity. Thus, in oxidation there is synthesis, but, at the same time, a disengagement of heat, or analysis; in the formation of salts there is a polarity between the acid and its base: finally, if any other chemical action takes place, there is a tendency to neutralisation, and consequently a proof of the previous existence of opposite elements susceptible of combining with each other. Organic and inorganic matter are in opposition, as life and death are. There is a true polarity between animals and plants; they form two opposite ranks, marching parallel with each other, but with no communication, except at one extremity, where they seem to mix together. Vegetation offers contrasts not less remarkable, in the relation of the root to the bud, and of the stem to the leaf: the functions of the sexual organs, and the geographical distribution of plants, are still further proofs of the truth of this law of nature. The animal kingdom, in which we can distinguish males and females, respiration and circulation, arteries and veins, the irritable system and the sensible system, and, finally, the two symmetrical halves of the body, is rich in proofs of the oppositions required by polarity. Lastly, if we admit, with our author, such further proofs as are afforded by the contrast between the different colours of the solar spectrum; between light, an imponderable fluid, and the ponderable terrestrial influences which control it; between the different positions produced by diurnal, barometrical, and magnetic oscillations;—if we admit all these, they must be sufficient, it would seem, to convince the most incredulous, that the most general and universal law of all nature is that of polarity. Such are the reasonings on which the doctrines of polarity depend. Let us now view the theories of elemental influence.

These opinions, of which Schelling is the author, consist in taking some one of the imponderable fluids, such as light, heat, or magnetism, as the type of unity, and as the primitive source of the action of other elements, in thus explaining the formation of these latter; and, lastly, in referring the nature and action of all material bodies to the influence of polarity upon ponderable matter.

According to Schelling, unity consists in magnetism; which is, with him, the first principle of matter, the germ of vitality, and the common bond by which apparent differences are reconciled. Electricity and magnetism are not different, the former being a sort of diffuse magnetism; and all bodies which seem destitute of magnetism really possess it in the form of electricity. Light and heat are also modifications of the same matter, and may be presumed to be different only in the degree of their elasticity. Light is a combination of an ethereal substance and an oxygenic principle: the first, which may be considered the positive pole, produces the power of expansion; the second, which is the negative pole, determines its material existence. Heat proceeds from certain efforts at cohesion which take place between bodies, and from the tendency to fix the equilibrium of their attractive powers, by which tendency the calorific fluid is disengaged. The existence of this fluid cannot be understood without cohesion, which is itself a combination of a positive general principle, and identical with a special negative variable principle, which is absolutely distinguished by its longitudinal action, and relatively distinguished by its transverse action: thus, the act of cohesion between an azotic and a carbonic element is an absolute act of cohesion between a positive and a negative; while that which determines the union of a hydrogenic element and an oxygenic element is an act of relative cohesion.

Steffens's theory is almost the same as that of Schelling. In the view of this philosopher, magnetism and electricity are of the highest rank: he also admits four ponderable elements; only he thinks that azote and carbon are more particularly poles of magnetism, and hydrogen and oxygen those of electricity. He applies to geology and metallurgy what Schelling did not attempt to extend beyond physics and chemistry: finally, he views heat as a general indifference, a perfect saturation, an absolute cohesion.

An opposite theory is held by Oken, a disciple of Schelling, less submissive, and more bold in his opinions, than the last. A substance which he names ether exists every where; a substance which forms and fills the universe; absolute as God; identical with God; — but considered, with God, as the origin of the physical world. This ether is influenced by a tendency to centralisation, which constitutes *gravity;* every finite sphere possessing gravity is *matter;* but matter is eternal, — boundless, — filling all space, — is space itself, time, form, and motion: it is therefore a being endued with properties and relations almost wholly metaphysical, as far as has yet been shown. We must give it life and real physical action: this duty devolves upon light; and, according to the disciples of Oken, Schelling is egregiously mistaken in ascribing it to magnetism. Light is an active expansion of ether; it diffuses itself through all space from the centre to which its material properties give it a tendency; its action is, therefore, peripherical and negative on the one hand, and central and positive on the other. The sun of our system is the consequence of the central action, while the planets result from that of the circumference; and between these heavenly bodies there is a reciprocal action of polarity, moving in straight lines, and assuming a luminous appearance. Light, adds Oken, is not matter, for the sun and planets lose nothing; it is merely an expansion—a double action from the circumference to the centre, and from the centre to the circumference. Light cannot put the ether in motion without producing another result, that of heat: heat is, therefore, ether acted upon by light. Cold is ether unmoved and unexpanding: death, darkness, cold, are therefore synonymous terms. Every motion is an indication of fire: all things have been formed by fire, and all things will return to it; the only change, therefore, that bodies undergo in nature is that of their calorific state. Fire, which is the joint production of heat and light, is the first active element to be considered in material natural bodies. Air is the second element, and may be considered as ether surrounding the corporeal world: it derives also much power of motion from its properties of polarity; and it is gaseous. Water, the third element, is more fixed: it is liquid, and composed of hydrogen, an *azotic* principle destitute of vitality, and of oxygen, a *zootic* or vivifying principle. Finally, earth, the fourth element, is the type of solidity, ether centralised and reduced into a mass. To this element the carbonic principle is essential. The union of these four elements, and of the principles which belong to them, determine all appearances in the physical world. As to electricity, it only performs a very secondary part; it is the expansion of air and of other elements—the *centropherical* action of the sun and planets; which are therefore electrical with respect to each other. The sun has electricity in excess, the planets in deficiency. Magnetism is a struggle between ponderable matter and light; but it only appertains to the metals, and especially to iron: all metals possess it, virtually at least, although their magnetic properties may not be manifested unless they are approached by iron. The existence of a *magnetic* fluid is, therefore, not to be admitted, any more than the fluids called by the French *caloric* or *lumic:* all these pretended fluids are mere expansions or actions, and not bodies. The four ponderable elements, considered by themselves, each constitute a kingdom of nature, which may be called that of *single elements;* the combination of two, three, or four of them gives birth to the *mineral, vegetable,* and *animal* kingdoms. Such are the opinions of Oken; which one of his disciples, Blasche, pronounces unattackable.

There are also other means of attaining this unity, in which so much virtue is expected to be found. It may be attained by reducing every thing to a small number of principal agents; by the repetition of the same phenomena in a multitude of different ways, and especially by numerical analogies. It is declared essential to the perfect understanding of natural objects, that they should always be reducible to a determinate number of genera and families, beyond which number no division can take place. There is, however, considerable difficulty in determining what this number is. It is either wholly arbitrary, or it depends upon the number of elements, the existence of which is admitted; and upon the number of modifications of which they are susceptible. Thus, with Oken, 4 is the sacred number in the mineral kingdom; classes, orders, and genera are all affected by this number: only the families intermediate betweeen orders and genera are classed in tens. From this mode of subdivision it follows, that nature has produced minerals after certain constant and unchangeable laws; that their number is as fixed as their station; and, finally, that neither genera nor families are to be fabricated at will. The numbers 3, 4, and 10 appears to prevail in the vegetable kingdom, and 3 and its multiples among animals. In the zoological classification of Goldfuss, the number 4 prevails almost universally for both orders and families.

These doctrines of numbers have been carried even farther by Wagner and Goldbeck, who have attempted to restore the old Pythagorean doctrine, *that numbers are the principles of things.* They declare that, till their time, every observer or philosopher had been in error; that what the Europeans fancy to be science is no science at all; that even the notions of their German brethren were notoriously mere plays upon words: and that numbers, and numbers alone, can be the basis of systems. Man thinks, compares, acts only as a

CHAP. IV.   SPECULATIVE MODES OF ARRANGEMENT.   331

mathematician; every thing about him is numerical. It is chiefly from the mystical contemplation of 0 that the most important results are obtained. Dr. Goldbeck calls his book, " *The Meaning of* 0, *or the first Dawn of Light in the Horizon of Truth.*" It must however be confessed, that this sort of philosophy has found but little favour in the eyes of the world.

The influence of the planets is another notion which has gained some partisans, who see, in the action of beings, and especially of such as are organic, nothing but a repetition of the elliptical course of the planets, and who find in every thing its axis, its poles, and its focus. The ellipsis traced by the earth ought, according to Runge, to be repeated by every thing that it bears; all of which consequently move in an ellipsis. In the eyes of Kieser there is only one organic principle, which varies in various directions, and always terminates its elliptical cycle under the form of man, animal, or plant. The details of this assimilation of the phenomena of the planets with those which control living beings are highly curious.

The forces that determine the actions of all bodies, and in particular of the heavenly bodies, are not such as are usually admitted. Thus Oken objects altogether to the theories of Newton: with this philosopher, force of projection and mutual attraction are words devoid of meaning, as expressing no distinct ideas. But substitute for them polarity, original and reciprocal polar action, and nothing can be more clear and satisfactory than the ideas which are derived from those expressions. It is this polarity which determines and fixes the number, size, and distance of the different stars that compose our mundane system. Those which are not subject to these laws, such as comets, are not so because their degree of polarity is not fixed: so long as they retain any degree whatever they will circulate around the sun; they may be created from time to time by the polarisation of the luminous matter of the ether; and, consequently, when they reach any point in space in which this polarity entirely ceases, their course is stayed, and they are again resolved into ether: such is the end of comets which do not re-appear. Polarity can only take place from the sun to planets, satellites, or comets; it cannot take place between one sun and another, or from a smaller to a more considerable sun, of which the former is itself a planet. Nothing, therefore, can be more absurd than to suppose that the universe has a common centre, about which revolve other centres, themselves having bodies revolving round them. The universe is composed of a series of systems independent of each other.

The theory of the elements is the basis of both the physical and chemical opinions of these philosophers; who have therefore been particularly occupied in investigating those sciences which have the most direct relation with those elements. Thus the nature of light, and the theories of opaqueness and transparency, have given rise to much controversy between the two chief antagonists, Schelling and Oken. The first admits that there is much uncertainty as to the nature of the luminous fluid. To him the arguments of Newton and of Euler appear equal: he, nevertheless, has a leaning for those of the latter. He says that the luminous matter diffused between the sun and the earth is influenced by the former in such a way, that the oscillations caused by that influence are communicated to our atmosphere. He thinks that transparency is the result of an indifference, of a positive and negative saturation of the elements; and as combustion is nothing but the struggle of these elements, transparency ought to result from combustion: he considers water and glass as examples of the truth of this theory. Opaqueness results from bodies being removed from the influence of the sun by a different action of their elements. Oken considers that transparency is due to the free expansion of the luminous fluid, which can only take place under two conditions: first, the disorganisation of bodies which are transparent; and secondly, their being composed of two substances: for bodies absolutely identical, as metals, are necessarily opaque; the phenomena of reflection and refraction are explained by combining the power of the luminous fluid to expand with its tendency to centralise. A colour is a partial luminous expansion, a mixture of light and darkness. The object of the prism and the lens is not to create colours, but to render visible those clear-obscure rays which we say are coloured, and which escaped before on account of their minuteness.

With regard to the three kingdoms of nature, they are in immediate dependence upon the elements. Oken sets out from these, 1. *fire* (a mixture of heat, light, and weight); 2. *air* or condensed fire (a mixture of azote, oxygen, and carbon); 3. *water*, or condensed air (a mixture of hydrogen, oxygen, and carbon); 4. *earth*, or condensed water, with carbon in excess. Such are the elements which constitute the four bases of all natural beings. Earth is the only one of these four which is susceptible of change; fire, air, and water being always the same. The changes that occur in the earth are induced by the three other elements; so that the number of these changes can never exceed three. When the earth combines with one of the other elements, a binary alliance occurs, forming *minerals*, or terrestrial *bielementary* bodies; if air, earth, and water ally themselves, and are only influenced by fire, light, or heat, plants, or trielementary bodies, are generated; and where all four elements concur in the formation of a product, that product is an animal, or a quadrielementary body. Hence there are in nature four kingdoms, four great masses — the *elements, minerals, vegetables,* and *animals.*

This elemental theory of Oken, joined to the necessity of finding in every being subdivisions corresponding with the influence of the four elements, brings it to pass that the number 4 exercises a paramount influence over all his classifications. The number is, however, by no means admitted by his brother philosophers; Grohmann sees three non-intellectual masses or spheres, from which he distinguishes the intellectual or physical element. Each of these masses subdivides into three others, each increasing in dignity and originating from its predecessors. The inorganic mass is divided into tellurism, atmosphere, and light; the plant into root, stem, and flower; the animal, into abdomen, chest, and head. The forces of nature are expansion, contraction, and crystallisation. Organic forms are dependent upon sensibility, irritability, and the force of organisation.

Groh maintains the existence of a triple subdivision in the solar system; namely, magnetism, electricity, and halism, corresponding with the elements of fire, air, and water: this author, however, admits 7 as the divisor of vegetation, and 9 as that of animals.

The relations of the two kingdoms of organic matter have, of course, occupied the attention of the philosophers whose doctrines are under consideration. According to Oken, *organisation* is a planet upon our planet: the principle of life is galvanism; its base is the mass of organic matter. Beyond this, all organisation proceeds from a mucus, which is nothing but terrestrial matter modified by air and water. The original mucus, out of which every thing has proceeded, is the sea, living, in its very essence, in a perpetual organising motion. Love sprang from the foam of the sea; it is in it that the power of organisation resides, especially in hot countries. From those countries vegetables, animals, and man himself, derive their origin. All organic beings must necessarily die, because they only represent one of the poles of the organic world, and this world is eternal only as a whole. Spontaneous motion is the only means of distinguishing organic from inorganic matter; it is the evidence of the superiority of one over the other. Wherever, in the original universal mucus, the sea, earth, water, and air are united, there is an organic point. The first of these organic points, solid on the outside, and fluid within, are little vesicles, or infusoria. Plants and animals are nothing but metamorphoses of these infusoria: whence it follows that the theory of generation is synthetical, and not analytical. Besides, this organisation is twofold, the planetary and the solar, the obscure and the luminous, the plant and the animal. The plant, not excited to motion by the presence of nutritive juices, dies; animals move about to find the means of subsistence. Such, and such alone, is the difference of the two kingdoms.

Nees Von Esenbeck, being guided by the considerations of polarity which have already been alluded to, admits four kingdoms in organic matter; viz. *fungi*, or the *northern* and *terrestrial system; plants*, or the *southern* and *solar system; animals*, or the *circumferential* and *nocturnal system;* and *man*, or the *central* and *diurnal system*. Like Oken, this naturalist believes that the first organic elements are simple and elementary vesicles: but he goes still further, and determines that fungi are, in fact, the type and origin of the texture and structure of the vegetable kingdom. The notions of Kieser are also something of this kind.

In applying these opinions to the systematic arrangement of the vegetable kingdom, Nees Von Esenbeck separates it into two grand divisions, — fungi, and plants properly so called. They consist of organic vesicles of the most simple and primitive nature, which may be considered the basis of vegetable tissue, and which are capable of being extended in any and every direction, according to the predominance of the poles. Plants are organic beings, formed by the enlargement of those vesicles which had acquired the figure of a tube. Three is the original number of their parts and subdivisions; but, by the influence of metamorphosis of a more or less active nature, the numbers 2 and 5 are also produced: thus the divisor of the parts of acotyledones is 2 (a statement which some recent observations of Turpin made, as it would seem, without his being aware of the opinion of Nees, appear to confirm), of monocotyledones 3, of dicotyledones 5.

According to Oken a plant consists of four principal parts, — the parenchyma or pith, the stock, the flower, and the fruit: the first is divided into cells, veins or vessels, and tracheæ; the second into root, stem, and leaves; the third into seed, capsule, and corolla; the fourth is incapable of division. These ten fundamental organs give rise to the ten principal classes; viz.

A. *Parenchymatose*.

1. *Cellular.* Fungi consisting merely of cellules or pulverulent grains, or simple filaments.

2. *Venous.* Fungi with their cellules or filaments contained in a common vesicle.

3. *Tracheal.* Fungi with a triple envelope.

B. *Cauline*.

4. *Radical.* Plants with roots, but without a true stem; mosses, lichens, &c.

5. *Cauline.* Plants with roots and a perfect stem. Monocotyledones.

6. *Foliar.* Plants with roots, stems, and perfect leaves. Apetalous dicotyledones.

C. *Floral.*

7. *Seminal.* Plants with a naked seed and no capsule. Epigynous dicotyledons.

8. *Capsular.* Seed covered with a pericarpium; corolla monopetalous and hypogynous; monopetalous dicotyledones.

9. *Corolline.* Plants with a perigynous corolla; polypetalous perigynous dicotyledones.

D. *Fructual.*

10. *Fructual.* Dicotyledones with a polypetalous hypogynous corolla.

Each of these classes is subdivided into four orders, according to the same fundamental organs; and into ten tribes, according to the ten secondary organs; and each tribe into ten genera, according to the same principles: so that the number of genera in botany must of necessity be 750. But I need not pursue these fantasies any further.

Having thus shown what the general principles are upon which these philosophers would form their systems, I shall next proceed to detail the more particular principles in use among them; in doing which I cannot do better than reproduce a translation I some years since made, for the *Philosophical Magazine,* of the canons of Fries upon this subject. They are the following: —

§ 1. Nature is an universal complication of phenomena, existing and acting in all places and at all times; an infinite power made manifest by the successive evolutions of a finite power; the sum of the whole creation in a continuous state — all existent matter proceeding from perfection and pregnant with futurity.

In nature there is a perpetual struggle, an uninterrupted rotation. The powers of formation and destruction operate alternately, whence nature is always dead and regenerate. The human mind, viewing this last phenomenon in its most extensive and, at the same time, most satisfactory sense, calls eternity in a state of ceaseless variation by the name of NATURE.

§ 2. Nature must be considered as either perfect or approaching perfection (*vel ut naturans vel ut naturata*).

§ 3. The powers and the productions of nature are co-existent.

All power is as it were a law under which a given production holds its existence, but in such a manner that all power is the finite revelation of an infinite law. To act and to exist is the same thing.

Power therefore is nature without production; Production is matter without power. Neither exists in nature by itself.

§ 4. All the powers of nature are more or less perfect manifestations of one primitive power, which acts by its different productions, according to the same eternal, immutable, absolute laws. But the powers of nature act only by mutual reaction; so that each power of nature becomes in its products impeded, interrupted, or quiescent.

The most perfect primitive power appears nowhere in nature absolute; but more or less impeded. Hence the powers of nature are various, some one among them being more perfect or active than the rest (less impeded).

The existence of nature depends upon this kind of control, and successive evolution; every power which is absolute and independent of restraint becomes infinite, and ceases to be perceptible as a finite power—(Nature).

Powers of low degree act upon those of higher degree; but the lowest powers, when not struggling with higher, contain opposing principles in themselves; for example, attraction (repulsion), electricity, magnetism, &c.

This opposition, which pervades all nature, is called *Polarity*.

The more agreement there is between powers, the greater also the agreement between their productions.

The more perfect a power, the more complex its actions; the more perfect its productions.

The more complex are actions, with the more difficulty are their laws explained: thus, for example, the laws of affinity and of motion are almost ascertained; by no means those of vitality or of sensation.

§ 5. All things which exist in nature are a whole, and at the same time a part of a larger whole. They are capable of being themselves resolved into other wholes until the human mind sinks under ideas of sublimity and subtilty which are imperceptible to it,—of the universe and of atoms.

An atom is a whole (an individual), a plant is a whole, the

earth is a whole, the universe is a whole: hence all things which exist are parts of one highest whole.

The vital principle of every individual is one; the same vitality animates the universe; so that there is one and the same primitive power which is revealed, by divers phenomena, in divers degrees of perfection. § 4.

Let us imagine all nature to be an immense sphere; all the rays converging in the *centre*, where they finally become confluent in a point, which may be called the point of *identity*. This point comprehends the perfection of all the rays; for that the most perfect and most completely formed creations, as the sun, are always situated in the centre, is testified by all authority, by all experience.

The powers of nature diverging from each centre in polar opposition, are continually passing into opposite series. A new sphere is formed by each opposition, whence the highest (most perfect) sphere is again and again resolved into new spheres, which form wholes of themselves, and each of which, according as its power is a more or less perfect evolution, in itself reflects the whole in a more or less distinct degree.

The centres of these spheres may be exceedingly distant from each other, but their rays always impinge upon the rays of some other sphere: hence they are not the most perfect forms (*summa*) of each section which run into each other, but those which are least perfect (*infima*).

The different spheres, therefore, being dependent upon the same eternal laws, and only varying according to the idea peculiar to each sphere, answer the one to the other. Hence among all natural productions, a *more near* or *more remote* resemblance is perceptible: the one of them being such resemblance as exists between subjects contained in the same sphere:
— *Salts*, for example, which are formed of the same basis combined with different acids; the other being such resemblance as exists between subjects contained in different spheres of the same degree of evolution, as Isomorphous Salts, the bases of which are different, but the form the same, on account of the identical relation of their elements. The former is called *Affinity*, the latter *Analogy*.

§ 6. It is impossible for the human mind, itself a finite creation, to regard nature, whether her powers or her productions are considered, in the light of the whole manifestation of an infinite power, but only as parts or fragments of such manifestation. But to comprehend these as one whole, that is, as an eternal and immutable, yet ever varying body, or, as innumerable forms of one highest whole, is the end of all disquisition, the sum of which we call a *System*.

It is necessary not to confound with systems, properly so called, those indexes of nature which are incorrectly called *Artificial Systems*. Indexes have references only to names, systems to ideas. " Tum primùm homines res ipsas neglexerint, quum nimio studio nomina quærere inciperent."— Galen.

§ 7. A system contains within itself the seeds of some more complete evolution, but it does not admit of arbitrary alterations.

Not that any absolute system can ever be contrived ; for I am by no means of the opinion of those who expect that a system is to be as unchangeable as if it were petrified.

§ 8. If nature be closely pursued, a system is called *Natural ;* if this Ariadnean thread be not followed, it is called *Artificial* or factitious.

There is, however, no absolutely natural system ; such is only ideal: neither is there any merely artificial system ; because its principles must necessarily be borrowed from nature herself.

Besides, nature wholly disavows our sections, she being a whole ; all systems, therefore, as far as their arrangement is concerned, are necessarily artificial.

It is by the comparison of various systems with each other that our notions of such as are natural and such as are artificial are acquired, those having the former designation which press most closely upon the footsteps of nature. Hence it is that a system which is to-day called natural, becomes to-morrow, by the accession of new ideas, artificial ; as that of Tournefort, &c.

Is it not, then, a vain labour to search after a natural system, since such will never be found ; and are not all attempts at it rash, until every thing which is capable of observation shall have been observed ? If this were admitted, it would be useless to seek for perfection in any thing ; for we can never hope that our experience will be perfect ; and there will be no want of subjects for examination to a person who shall live a thousand ages hence. Such sublime truths as the present age shall strike out, are, therefore, not to be contemned because they will become more full and perfect hereafter.

§ 9. A system of nature proceeding from subjects of the most simple organisation to such as are more perfect, or from the circumference to the centre, is called a *Mathematical System.*

For mathematicians assume that nature herself proceeded from forms of the most simple kind to those which are more perfect; and that, therefore, is the most natural road, which nature herself has followed in forming her creations.

All natural bodies, indeed, originate in successive developement, yet in a continuous series within a determinate sphere. Every new sphere originates in a digression from a series which is otherwise continuous. Whenever a more perfect sphere is separated from one which preceded it, and has acquired a higher station than its parent, it may be itself pressed down by such new ones as emerge from itself; but the depressed sphere is also capable of continuation in its descent; and under this mode of developement the same principles and the same types are regenerated under more perfect forms in the higher spheres.

Those, therefore, are mistaken who assume that nature proceeded in a simple series to her most perfect productions. Thus, for example, all parasites, both animals and plants, must necessarily have been created later than their matrix (and should therefore be the most perfect parts of the creation). But Fungi, which are the latest in the series of vegetable developement, are the most simple of all in their structure.

In Minerals, of which the most simple are at the same time the most perfect, the Mathematical system may be employed, because it corresponds with the Philosophical. But in higher spheres, in which vitality must be considered, the laws of mathematics are of no avail.

§ 10. A system of nature which takes for the basis of its arrangement the order of developement of individuals is called Physiological.

But take care not to imagine that the first series of evolution is a simple one. As the evolution of the animal and vegetable kingdom may be said to have proceeded with nearly equal paces, so the different sections of vegetables cannot be said to have arisen out of a simple series, but out of parallel or radiant series. Many Algæ must have been created more recently than the most perfect plants, Entozoa than the most perfect animals. Whence it is to be inferred, 1st, That nature, properly speaking, can only be said to have proceeded from the most simple forms to those which are more compound, in theory (*de ideis*);

but, 2dly, to have often operated in an inverse order in her forms.

§ 11. *Philosophical systems* do not depend upon individual productions which are subject to continual variation, but upon eternal and unchangeable ideas. These always proceed from the centre to the circumference, or from the most perfect productions to those of a lower order.

This is the method of my Mycological system, and it agrees with the Mathematical system if the order be inverted.

A Philosophical system depends upon the laws of logic: for the laws of logic are by no means notions contrived by man, but eternal and immutable, and established by Nature herself. As the rotation of the heavenly bodies, discovered after the laws of mathematics, must necessarily follow those laws; so also no observation in nature can invalidate the laws of logic. For the laws of logic are the laws of nature.

It must be observed, however, that a system, although logically true, may be naturally false, because it may have been deduced from false principles; but every true system cannot deviate from the rules of logic.

§ 12. A Philosophical system is superior to all others.

It may at first appear, perhaps, of little moment, what way we follow in enumerating the productions of nature; but if one way is more certain and more facile than another, that is surely to be preferred.

To me it appears most advisable to commence with that which is most perfect, most completely developed, and therefore most easily understood; and thence to descend to forms of a more imperfect kind, and therefore of a more doubtful nature. The half-developed portions of the lower forms would never be understood, if they were not more completely developed in the higher forms. This is the path which is pointed out both by experience and common sense; the idea of a seed is not derived from an *Uredo*, nor that of a vegetable from an *Erineum;* but the reverse.

This is especially true of those lower spheres which bring up the rear: the last point of simplicity will never be attained, and will never be determined; although our microscopes are daily extending our views, the poles of vitality will never be reached. It is better, therefore, to set out from a *certain* point (the centre) than from an *uncertain* point (the circumference), which may be extended to infinity.

So it is more wise, in studying Man, to take our notions of humanity from those in whom it exists in the highest degree of perfection, rather than to search over-curiously for a man whose intellect is approximating to that of animals.

§ 13. In a systematic arrangement the higher forms are always to be taken before the lower.

The highest arrangement is always to be taken from the highest and most essential characters,—from each highest character originates a particular section,—and all the sections which are subordinate to this character are to be comprehended under its common title. The higher the distinction, the greater its dignity and importance.

Nature is always passing into series in polar opposition: hence a dichotomous mode of distribution is not only the most natural, but almost the only true one. Logic and nature, which are ever in accordance, prove this continually. Thus, for instance, natural bodies are more properly divided into organic and inorganic, than, overlooking this distinction, into minerals, plants, and animals; so also is the distribution of vegetables into cotyledonous and acotyledonous preferable to that of monocotyledonous and dicotyledonous.

But as the most sacred things are the most open to abuse, so also is the dichotomous disposition, which is of the highest value when nature is strictly followed, the most artificial of all when arbitrary distinctions take the place of those which are essential; as the analytical *index* of Lamarck. Many, for this reason, altogether object to such a form of arrangement; but the abuse of a thing does not destroy its use.

When the members of a bipartite section are again dichotomously divided upon analogous principles, four sections are created, of which the first and second, and the third and fourth, are in affinity; but the first and third, and the second and fourth, are in analogy.

But when this method of division becomes circuitous, a more direct path is undoubtedly to be discovered: hence other numbers are admitted, especially the quaternary (or double dichotomy), and also others in which dichotomy is understood.

There are other and most acute observers (Oken, MacLeay) who contend for other fixed fundamental numbers. Care must be taken, however, that no cabalistical or occult virtues are attributed to any particular number; in the higher spheres a higher number is, on account of the multiplicity of organs, admissible than can be used in the lower spheres; the only object of such a contrivance being to explain in what direction rays pass off from their centre, and at what points the rays of different spheres impinge upon each other. To do this a determinate number is required.

We must, moreover, avoid extending too precipitately any system whatever to specialities. We can proceed in no direction further than the power of arrangement acquired by what we positively know of nature admits. I certainly am not of the number of those who assume that infinity is to be circumscribed within strict limits; although I may be of opinion that infinity and universal harmony are better explained by them than according to any arbitrary rules of arrangement.

In the formation of sections and genera it is most especially necessary to beware that they do not depend upon characters alone; so that if the character should hereafter prove defective, the section or the genus may still remain unchanged. In this lies the difference between an artificial and natural arrangement; the former depending upon characters, the latter upon affinity. Hence Linnæus did not characterise his families of plants, nor Ehrenberg those of fungi, rightly perceiving that affinity is of the first importance, characters of secondary.

It is occasionally necessary to admit into a particular section a genus or species in which the most important character of such section does not exist, but then, its truly essential character cannot have been detected. Thus, when we say that Rosaceæ are dicotyledoneous, perigynous, polypetalous, &c., and refer to them *Alchemilla*, it will be easily seen that the really essential character of Rosaceæ remains to be discovered.

§ 14. *Every sphere (section) expresses a particular idea; hence its character is best expressed by a simple notion.*

But to effect this, it is necessary that the character which is really most essential shall have been detected. For if a section, of which the primary character is unknown, be circumscribed by a simple notion, the most arbitrary and artificial arrangement possible would be the result.

When the essential character is once detected, all others will be wholly dependent upon it (for when this character is changed the others are changed also), and those which do not depend upon it are accidental.

It must not, however, from this be understood that a system is to be applied to one part only of its subject: on the contrary, it embraces all parts, arranging them upon the same principles, — but when they diverge in opposite directions, one is to be chosen in preference to another.

§ 15. Physical or Physiological marks are capable of distinguishing spheres (sections) of the highest order only; but in those of the lowest they are always to be consulted.

Physiological characters, as being those which are most essential, are little subject to variation, and therefore will not suffice for distinguishing the lower spheres (orders, genera, species); they are, nevertheless, to be continually consulted as to origin, station, geographical distribution, &c. which illustrate the series of affinities in various ways.

§ 16. Essential characters are generally the most hidden, and demand acute investigation; the most superficial being those which are accidental.

Hence it is that accidental characters, or those of a lower order, are first seized, as being those which are most immediately under our eyes: thus the low distinctions of species and varieties are easily acquired by mere tyros, while the higher are within the comprehension of masters of the science alone.

The whole progress which has been made in natural history has been a succession of triumphs of the more essential characters over those accidental ones which had been previously received. Thus, in the following comparison, how much more important are those distinctions which are

ESSENTIAL than those which are	SUPERFICIAL.
1. Mammalia, Amphibia, Pisces, of Linnæus.	1. Quadrupeds, Serpents, Fishes, of old authors.
2. Monocotyledones, Dicotyledones, &c.	2. Trees, shrubs, herbs, &c.
3. Hymenomycetes, Gasteromycetes, &c.	3. Fungi stipitati, sessiles, claviformes.
4. Lichens from their fruit.	4. Lichens from their thallus.

The foregoing proposition must not, however, be inverted, by supposing that *the more hidden characters are, the more essential;* Natural History would then become not only micrological, but very difficult and erroneous. Where an object is easily distinguished by marks immediately under our eyes, microscopical differences are not to be sought after. Besides, characters indicated by highly magnifying microscopes are, in fact, as superficial as those seen by the naked eye.

§ 17. The primary powers of nature are arranged according to the following laws. They are these : —

A. TERRESTRIAL (*Tellustres*), acting together or in contact.
a. *Acting together and continuous in their productions.*
   1. *Sensibility*, or the power of motion, sensation, and consciousness. The object of *Psychology*.
   2. *Vitality*, or the power of absorbing heterogeneous matter, of assimilating it to an internal circulation, and of bringing forth progeny of the same nature as the parent. The object of *Physiology*.
b. *Acting in contact, and absolute in their productions.*
   3. *Affinity*. The object of *Chemistry*.
   4. *Electricity*. The object of *Physics*.

B. SIDEREAL (*Siderales*), acting from a great distance.
   a. Reproduction.   1. *Light.*
   b. Production.   2. *Attraction.*

§ 18. The productions of nature, which are co-existent with these, are also considered as, —

A. TERRESTRIAL; various in form, placed in juxtaposition or cohesion with each other, arranged both by terrestrial and sidereal influence, and composed of parts which, taken together, form a whole. *Natural bodies* properly so called; the objects of *Natural History*.
a. *Organic;* reproductive, composed of various definite organs, and formed by internal developement.
   1. ANIMALS, possessing sensation. The objects of *Zoology*.
   2. VEGETABLES, possessing vitality (not sensation). The objects of *Botany*.
b. *Inorganic;* productive, homogeneous, formed of particles in juxtaposition (and not possessing the qualities of organic bodies).
   3. MINERALS; ponderable. The objects of *Mineralogy*.
   4. ELEMENTS; imponderable. The objects of *Physics*.

B. SIDEREAL; a system of EARTHS, which are spheroidal, very distant from each other, subject to the influence of sidereal power alone, and composed of the heterogeneous, but individually entire, productions of nature. STARS, the objects of *Astronomy*.

a. Possesing light and attraction, reproductive, central.
   1. *Suns.*
b. Possessing attractive power, but no light of their own, productive, circumferential. 2. *Planets.*

VEGETABLES

are Living, insensible organic bodies.

§ 19. The end of life (and therefore of vegetation) is twofold: the preservation of the *individual* and of the *kind;* the former is called Nutrition, the latter Generation. Hence there is a twofold system of organs; namely, of nutrition and of multiplication. But the organs of nutrition are either prepared by the mother (*Germination*), or developed by the plant itself (*Vegetation*); so also the organs of multiplication are either confined to the plant (*Flowering*), or continued in a new individual (*Fructification*).

There are, therefore, four primary functions of vegetables: germination, vegetation, flowering, and fructification. This is the basis of my system.

§ 20. According to these two systems of organs, and four primary functions of life, the following arrangement is produced of—

VEGETABLES.

A. Organs of Nutrition.
a. *in Germination.*
  1. Cotyledoneous, *producing cotyledons.*
  2. Nemeous, *producing a thread.*

b. *in Vegetation.*
  1. Vascular, *having cellular tissue and spiral vessels.*
  2. Cellular, *having cellular tissue, and no spiral vessels.*

B. Organs of Multiplication.
c. *in Flowering.*
  1. Phænogamous, *having sexes or manifest flowers.*
  2. Cryptogamous, *having no sexes, and destitute of flowers.*

d. *in Fructification.*
  1. Spermideous, *bearing seeds.*
  2. Sporideous, *bearing sporules.*

§ 21. Upon the same principles COTYLEDONEOUS Vegetables (Vascular, Phanerogamous, and Spermideous) are divided according to —

A. Their Organs of Nutrition.
a. *in Germination.*
  1. Dicotyledoneous, *with a double expanded cotyledon.*
  2. Monocotyledoneous, *with a single enclosed cotyledon.*
b. *in Vegetation.*
  1. Exogeneous, *the trunk youngest at the circumference.*
  2. Endogeneous, *the trunk youngest, and softest in the centre.*

B. Organs of Multiplication.
c. *in Flowering.*
  1. (Androdynamous?)
  2. (Gynodynamous?)

d. *in Fructification.*
  1. Seminiferous. *Agardh.*
  2. Graniferous. *Agardh.*

§ 22. NEMEOUS Vegetables (Cellular, Cryptogamous, Sporideous, are also disposed according to —

A. Organs of Nutrition.
a. *in Germination.*
  1. Heteronemeous, *threads in germination copulating into a heterogeneous body.*
  2. Homonemeous, *threads in germination either separate or confluent into a homogeneous body.*
b. *in Vegetation.*
  1. Diplogeneous, *formed of regular connected cellules.*
  2. Haplogeneous, *formed of anomalous somewhat filamentose cellules.*

B. Organs of multiplication.
c. *in Flowering.*
  1. Cryptandrous, *something analogous to sexual distinction.*
  2. Anandrous (*Link*), *nothing analogous to sexual difference.*

d. *in Fructification.*
  1. ? Sporiferous. *Agardh.*
  2. ? Sporidiiferous. *Agardh.*

§ 23. The organs of Vegetation offer modes of subdivision in proportion to the lateness of their evolution.

Germination offers very few, Vegetation a greater number, Flowers many, Fruit very numerous modes.

Their dignity is the converse of this; the most essential modes depending upon germination and vegetation, the less essential upon flowering, and almost accidental modes upon the fruit (at least the pericarpium).

In this manner the vegetable kingdom, or rather world, is divided into two hemispheres by Germination, and into four quarters by Vegetation, into Classes by Flowers, and into Orders and Families by Fructification.

§ 24. Systems truly constructed upon these principles also comprehend all other essential differences, and at the same time explain them.

CHAPTER V.

OF THE VALUE OF CHARACTERS.

The basis of the natural system of botany consists in a just view of the mutual affinities of plants; and this is only to be obtained by determining the precise *value of the various characters* by which these affinities are indicated. Plants agree with or are distinguished from each other by various peculiarities of form or structure: all these peculiarities, of what kind soever they may be, are called characters. For example, a lichen is distinguished from a nettle by having a cellular instead of a vascular organisation; its cellularity is, therefore, one of the distinguishing characters of the lichen: it is also distinguished from mosses by its organs of reproduction being contained in scutella instead of thecæ; its scutella are also, therefore, a discriminative character. The sweetbriar-rose is distinguished from the dog-rose by its leaves being covered with sweet-scented glands; these, therefore, are a distinctive character of the sweetbriar-rose. The rose and the vetch are both constructed with pinnated leaves and stipulæ; pinnated leaves and stipulæ are, therefore, characters of agreement between the rose and vetch. The apple and pear have both an inferior fleshy fruit or pome; they agree, therefore, in the character of the fruit, and so on: in short, every permanent circumstance whatsoever, in which plants differ or agree, is a character. The value of these characters is, however, extremely different, as will presently be seen.

Those characters are of the highest value which are the most *constant;* and those are of the lowest value which are the most *inconstant.*

The value of a character is, in the first place, to be determined by the purpose to which it is to be applied. A distinction may be extremely important in characterising a large quantity of species; which will be useless when the object is to distinguish one species from another; because the universal existence of it in the whole mass of species which renders it, in a collective sense, important, destroys its value when the discrimination of species or subordinate groups is the object in view. Thus the peculiar state of the fibrous tissue of gymnospermous dicotyledons, by which they are anatomically distinguished from all other plants, is of no importance in discriminating one genus of that group from another, because it is common to them all. On this account, the more universal a character is, the better is it adapted for the most extensive collections of species; and the more circumscribed and unfrequent it is, the better it is adapted to the distinction of genera or species.

The permanence of a character, or, in other words, the absence of a tendency to variation in the modification of any organ, is another consideration, of the first importance in determining its value; those which are the most permanent being of the highest value, and those which are the most variable being of the lowest. For instance, the capsule and seed of a plant will be produced from generation to generation without sensible variation; this, therefore, is a constant and valuable character: but the colour of its flowers may be white or blue, and the surface of its leaves smooth or hairy; the colour, therefore, of the flower, and the surface of the leaves are, in such a case, inconstant and worthless characters. It is not easy to lay down any general rules upon this subject, the number of exceptions being so great as to invalidate almost any rule which can be proposed. It is, however, very important to know what the experience of botanists has taught them; and this I will now endeavour to explain.

Anatomical differences have been found of the highest value in characterising the primary divisions of the vegetable kingdom: the presence of spiral vessels is universal in flowering plants; while their absence is equally characteristic of flowerless plants. The presence of glandular swellings upon the walls of the fibrous tissue of gymnosperomous dicotyledons is a mark by which they may be always known. How far anatomical differences are important in fixing the limits of subordinate groups is not yet known; but much is to be expected from future enquiries.

The modes in which plants grow, which may be called their *physiological differences*, are of very great importance. Not only are exogenous and endogenous plants thus positively limited, but in the latter grasses are instantly known by their hollow stems and phragmata; and palms by their simple cylindrical stems, with a terminal bud only. This subject is like the last, still in its youngest infancy.

The *root*, whether true or spurious, is not often employed as a character, chiefly on account of the difficulty of observing it, and its being seldom preserved in herbaria. It will frequently distinguish varieties from each, as in the radish, and other garden productions: it sometimes affords good characters for species, as in the genus crocus, some grasses, and many herbaceous plants; in which latter it is especially important to observe whether it is creeping or fibrous. In the genus orchis sections are formed from differences in the roots, some consisting of an undivided lobe, others having their roots palmate, and others fasciculate; occasionally, but, I believe, only in monocotyledones, it is a good aid for discriminating certain natural orders; thus, Asphodeleæ are mostly characterised by their scaly bulbs, and Irideæ by their solid rhizomata, or cormi; but generally the root or root-like bodies are to be excluded from all characters higher than those of species.

The *stem* is of the greatest importance in many of the first characters of vegetation: its greater or less degree of *vigour* distinguishes the *varieties* of Phaseolus vulgaris, and those of numerous other plants; its *direction* is a common characteristic of *species*, as to whether it is erect or decumbent, or twining or trailing, and so on; its *form*, whether round or semiterete, or angular or triquetrous, &c. is always of value in distinguishing *species*: in Labiatæ and Stellatæ, it is an important character of the *order*, being universally square; its *substance*, whether fistular or solid, is occasionally, but not often, taken into account in distinguishing *species*; and, finally, its *internal anatomy* is of the greatest consequence in distinguishing the highest divisions of the vegetable kingdom. The more remarkable of these have been noticed in speaking of physiological characters; but, independently of those great modifications of growth, it is probable that many orders are capable of being distinguished by the manner in which the vascular and cellular systems are respectively arranged. We know that Calycantheæ have the peculiarity of possessing four additional woody axes lying in the bark; and there seems good reason for believing that other orders possess other distinctive characters.

The mode in which *the stem ramifies* will sometimes distinguish a species; but the presence or absence of the abortive branches called spines, is rarely indicative of more than wild and cultivated varieties of the same species.

The *hairs*, *prickles*, and other appendages of the external surface of plants, are chiefly used to mark species or varieties. Species are often excellently characterised by the presence or absence of hairs or prickles upon particular parts, and also by their direction and form: but these are sometimes fallacious characters, depending upon the situation in which plants grow, and other accidents. The form of hairs is of more consequence; thus, in Elæagneæ they are lepidote; in Euphorbiaceæ and Malvaceæ they are stellate or furcate; in Boragineæ they have a bulbous base; and the genus Indigofera is distinctly characterised by the hairs being attached by their middle.

Leaf-buds are scarcely ever noticed by botanists; and yet it is not improbable that the arrangement of their leaves before expansion may offer characters of moment, as in Salix; and it is certain that in some genera, Fraxinus, for instance, the colour and form of the buds afford the surest distinction of species.

Leaves afford a multitude of characters, some of which are used merely to distinguish *species* and *varieties*; others are of a *higher class*. To the former belong all those differences in margin and figure to which leaves are particularly subject; generally every decided difference in the mode of division or form of the leaf is accounted a distinction of a species; but this must not be understood as a uniformly constant character; for in many garden plants species vary with entire and cut, or even pinnatifid leaves, and with linear, lanceolate, or even broader leaves; differences which in such cases serve only to characterise varieties; this is the case with Tilia Europæa, Alnus glutinosa, Fagus sylvatica, Cytisus Laburnum, and many others: on which account differences in the degree of division, or in the outline of leaves, must not be trusted to as characters indicative of species, unless they are accompanied by other differences in other parts of the organs of either vegetation or fructification. The nature of the margin of a leaf is sometimes of very great importance in characterising natural orders; thus, in Cinchonaceæ and Myrtaceæ it is always perfectly entire; but in many orders it is of no value, as in Rosaceæ, in which leaves vary from being entire to serrate, and to almost every degree of division. The *degree of composition* sometimes affords characters of a high rank, applicable to the distinction of genera, or even of *natural orders;* thus, Berberis is known by the composition of its leaves, which is most apparent in the section called Mahonia, and indicated in those with simple leaves by an articulation at the apex of the petiole: in true Aurantiaceæ, the leaves are always compound, as is indicated by the leaflets being constantly articulated with the petiole; but in other orders the composition of the leaves is of scarcely any value, even for distinguishing genera; thus, in Rosaceæ, pinnated, pinnatifid, and simple leaves are found in the same genus, as Pyrus for instance. *Insertion* is a point of the *highest consequence*, and less liable to exceptions than most characters. It is a general rule, that there cannot be opposite and alternate leaves in the same natural order; and although to this there are exceptions, yet upon the whole it may be taken as a good character. Most monocotyledonous plants have alternate leaves, but in some they are approximated so much as to appear as if opposite. The *arrangement* of the veins often affords excellent characters: in most monocotyledonous plants, the veins are simple and parallel; in most dicotyledonous plants they are reticulated; in all true Myrtaceæ, the venæ externæ form a continuous line parallel with the margin, by which, I believe, they are to be positively distinguished; in most Amentaceæ, the venæ primariæ run directly to the teeth of the margin, in which they terminate, without bend or interruption; and numberless other cases may be instanced, in which characters of similar importance may be obtained from venation. *Dotting*, that is to say, the presence of receptacles of oil within the substance of a leaf, giving it a dotted appearance if viewed against the light, is generally a character of natural orders: thus Myrtaceæ are distinctly separated from Melastomaceæ, by their leaves being dotted; and Aurantiaceæ, and some others, are to be known by their dots, while Samydeæ have a mixture of dots and short transparent bars.

According to Mr. Brown, the *Stomata* are sometimes of importance in determining the limits and affinities of genera, or of their natural sections, according as they vary in figure, position, and size, with respect to the meshes of the cuticle. He remarks, that they occupy both surfaces of the leaf in all the Proteaceæ of Southern Africa, except Brabejum.

The presence or absence of *Stipulæ* is a character of the same kind of importance as the insertion of leaves: generally, stipulæ cannot be present and absent in the same natural order; and to this the exceptions are so few as not to invalidate the rule. All Rosaceæ, for instance, have stipulæ, except Lowea berberifolia and Spiræas, in which they are absent.

The *Bracteæ* are usually subordinate to other characters, and are generally employed as a means of distinguishing species, the distinctions of which are often conveniently expressed by the relative length of the bracteæ and pedicel, the form of the bracteæ, and similar characters; but in some plants they possess peculiarities of so marked a kind that they are used for distinguishing genera: in Gramineæ, for instance, bracteæ constitute the whole of the floral envelopes, and are exclusively used for generic characters. In Compositæ, the bracteæ, out of which the involucrum and appendages of the receptacle are formed, afford good distinctions of genera; and in Umbelliferæ, they afforded Linnæus generic characters which he thought good, although they are undervalued at present.

The *Inflorescence* is chiefly used to distinguish species; it seldom affords generic characters, but occasionally those of orders. In the latter case, however, it is only the great forms of inflorescence which can be taken into account; thus, a spike, raceme, or panicle, may all be expected to exist in the same natural order; but a cyme will scarcely be found with them; and the umbel and capitulum are usually confined exclusively to certain orders. Thus, the inflorescence of Betulineæ and Salicineæ is a catkin, of Compositæ a capitulum, and of Umbelliferæ an umbel; but in Asphodeleæ we have the spike, raceme, and panicle, along with the umbel.

With the *Calyx* the organs of fructification are said to begin. Slight modifications of its form and surface are used for characterising *species;* more decided peculiarities denote *genera;* thus, the species of Myosotis differ in having calyxes, open or closed, acute or obtuse, naked or covered with hairs; and the genera Polemonium and Phlox are distinguished by the former having an urceolate calyx, the latter a prismatic one. Characters in the calyx, of a still higher kind, distinguish some natural orders; especially that of its being adherent or not adherent to the ovarium: this is generally considered a point of the greatest importance; but there are very numerous exceptions to it, of which more will be said when we come to the ovarium. The position of the lobes of the calyx or sepals, with respect to the axis of inflorescence, is important in distinguishing some natural orders: Leguminosæ, for instance, in which the fifth lobe of the calyx is anterior with respect to the axis of inflorescence, are by that character alone separable from Rosaceæ, in which it is posterior with respect to the axis. The æstivation of the calyx should always be carefully noted; it is often different from that of the corolla, and is frequently the character of an order; thus, Thymelææ are distinguished from Proteaceæ by the æstivation of the calyx of the latter being valvate, of the former imbricate. All Malvaceæ, and the orders more immediately allied to them, have a valvate æstivation.

The *Corolla* offers many valuable characters: by the difference in its colour or odour, varieties are distinguished; species are known by variations in its form, surface, or proportions; genera and orders, by distinctions in its form, composition, insertion, and æstivation. For example, a monopetalous and polypetalous plant rarely co-exist in the same genus, or even order: hypogynous, perigynous, or epigynous corollas are in like manner incompatible with each other in the same order; but whether a corolla is rotate, infundibuliform, or hypocrateriform, is never of more than generic importance: its æstivation distinguishes orders, and, perhaps, occasionally genera; but is scarcely of so much consequence as the æstivation of the calyx. In Asclepiadeæ, the æstivation is imbricate; but Mr. Brown admits into the same order a valvate æstivation. The appendages of the corolla, such as are found in Caryophylleæ, Asclepiadeæ, Brodiæa, Narcissus, and others, seldom characterise any higher group than a genus; they are, however, admitted into the characters of some natural orders, as, for example, Olacineæ.

In considering the peculiarities of the *Stamens*, it will be found that their length, with respect to the other parts of fructification, and their surface, are scarcely ever of more than specific importance; nevertheless, in Boragineæ, their length with respect to the corolla has been taken as a generic character; and the nakedness or hairiness of the anthers distinguishes Chelone from Pentstemon. Their insertion, whether perigynous, hypogynous, or epigynous, usually distinguishes natural orders; but to this there are many exceptions; thus, in Papaveraceæ, which are characterised by having hypogynous stamens, Eschscholtzia has them perigynous; in Anonaceæ, which has hypogynous stamens, Eupomatia has them perigynous: in short, there is no doubt that the insertion of the stamens, although in many cases an important character, has been much overvalued. But the most valid circumstance to consider, with respect to the stamens, is their insertion with regard to the petals: commonly they are alternate with them; but in Primulaceæ, and some others, they are opposite to them; Rhamneæ and Celastrineæ are distinctly separated by that character, which will be seen to be of peculiar importance, when it is recollected that if stamens are opposite to the petals, they necessarily indicate not only the character, which is apparent, but also the *suppression* of the first series of stamens. The proportion that stamens bear to one another is also a character of value in dividing orders; as in Labiatæ, for example, and true Scrophularineæ and Orobancheæ, in which they are didynamous, with the fifth stamen abortive; the tetradynamous stamens of Cruciferæ are a principal feature of that order. They should always agree with the outer parts of fructification in number or proportion; that is to say, if the divisions of the calyx and corolla are five, the stamens should either be five or some multiple of it. If this is departed from, the flower becomes unsymmetrical, and an irregularity of structure is then indicated, which may characterise either a genus or an order: didynamous and tetradynamous stamens are of this kind. Their union into one or more bundles, as in Leguminosæ, Malvaceæ, Hypericineæ, &c. is a character, the importance of which is very variable.

The manner in which the *Anthers* are attached to their filament is sometimes a good generic character, as in the order of Amaryllideæ, for example; but the mode of their dehiscence and the number of their cells are characters appertaining to orders; Melanthaceæ and Irideæ, for instance, are readily known by their anthers bursting on the side next the corolla, instead of that next the pistillum; and one of the distinguishing characters of Malvaceæ is, to have unilocular anthers. The æstivation of anthers may occasionally furnish valuable characters; but this is a subject hitherto not much attended to. Their cohesion into a tube is the great character of Compositæ, Lobelia, and others.

The *Pollen* will sometimes furnish excellent characters; of course varying in importance in proportion to the degree in which the pollen recedes from its ordinary nature. Thus, Campanulaceæ have round pollen, Lobeliáceæ oval; here it is a character of an order. Some Onagrariæ have it cohering by filamentous processes, in others it is not coherent; in that order it is only of generic value. In Asclepiadeæ it coheres in fleshy masses, while in Apocineæ it is granular; here it has the power of distinguishing those two orders. In Orchideæ it coheres in various degrees, whence it has been found useful for characterising the subordinate groups of that natural order. It is probable that much remains to be done in applying the characters of pollen to purposes of discrimination.

The presence or absence of a *Disk* is usually a character of generic value only. It is scarcely sufficiently universal at any time to limit an order: its exact value is not sufficiently known.

In examining the *Pistillum*, the first thing to observe is whether it is superior or inferior. This is a point of primary importance, notwithstanding the numerous exceptions which exist with respect to it. Thus, in Anonaceæ with superior ovara, we have Eupomatia with an inferior one; in Gesnereæ it is variable, and in Saxifrageæ the genera differ in this character; but still it must always be looked upon as among the very best distinctions that can be used for limiting natural orders. Who, for example, has ever seen Cinchonaceæ, Umbelliferæ, Compositæ, or Orchideæ, with a superior ovarium? or where is the instance of Ranunculaceæ, Papaveraceæ, or Caryophylleæ, having an inferior ovarium? The next thing to remark is its internal structure, whether simple or compound, unilocular or plurilocular, with placentæ proceeding from the sides, the axis, or the centre. The value of these characters is uncertain: for the most part it is this. A simple and compound pistillum will not co-exist in the same genus, but may in the same order; as in Rosaceæ, in which all degrees of combination of pistilla are to be found. This law is, however, influenced by the prevalence of one state of pistillum over others in the same order. For example, in Leguminosæ, the pistillum is universally simple; for this reason, Moringa, in which it is compound, has been properly excluded: it would be difficult to reconcile a pistillum of more than two cells with Cruciferæ, or a simple pistillum with Scrophularineæ and Labiatæ. Yet in the genus Nicotiana, the type of which is a bilocular ovarium, we have a species (N. multivalvis) in which it is plurilocular; and the same happens with one species of Nolana. A unilocular and plurilocular pistillum may be found in the same order, but scarcely in the same genus: placentæ from the axis and the sides of an ovarium are so much at variance with each other, that when the latter is in excess, as in Papaveraceæ and Butomeæ, they cannot co-exist in the same natural order; but when the parietal placenta is only caused by a partial contraction of the dissepiments, as in Cyrtandraceæ, Pedalineæ, and Gesnerieæ, it differs so little from the placenta in the axis, that the two may be expected to be found in the same order, or perhaps in even the same genus. Separation of the styles, or their union in a single style, is occasionally used as a character for orders and genera; but it is not always a good one, and should at least be accompanied by others of more importance. The structure of the stigma is usually a character of genera, but not often of orders, unless in Goodenovieæ and the neighbouring tribes, in which the presence of a peculiar appendage, called the indusium, is so remarkable as to be used for a character of those orders.

The insertion of the *Ovula*, when they are definite in number, especially in simple pistilla, often affords excellent characters for orders. Thus, in Myristiceæ, the ovula are erect; in Laurineæ, pendulous; in Santalaceæ they are pendulous; in Elæagneæ, erect; Juncagineæ, with erect ovula, are so distinguished from Fluviales with pendulous ovula, and abundance of similar cases may be cited. The position of the foramen is of great importance, inasmuch as it indicates the direction of the radicle, but of this more will be said when I speak of the seed.

The character of the *Fruit* is of much the same importance as that of the pistillum. The points in which it chiefly differs is its texture and mode of dehiscence. Its form and surface are often used to distinguish species; its texture and dehiscence for characters of genera, and even of orders. With respect to its texture, that is to say, whether it is fleshy or dry, capsular or baccate, membranous or woody, it is seldom that these characters can be used for distinctions of orders; for instance, in Onagrariæ, Myrtaceæ, Melastomaceæ, and many others, there are genera, the fruit of some of which is baccate, of others capsular; and there can scarcely be a group of plants in which the fruit, if usually dry, may not be expected to be fleshy; or if fleshy, in which we ought not to look for it in a dry state. It is somewhat different with regard to dehiscence or indehiscence. In Cruciferæ, Ranunculaceæ, Rosaceæ, Leguminosæ, and many others, the fruit varies from perfect indehiscence with solitary seeds to dehiscence with many seeds; but there is no instance of dehiscence in Boragineæ,

Compositæ, or Labiatæ. The mode of dehiscence is never of higher value than for distinguishing genera; it never ought to be taken as the distinctive character of orders. Jussieu, it is true, divided Rhodoraceæ from Ericeæ, and Pediculares from Scrophularineæ, upon differences in dehiscence; but these principles are now abandoned.

The nature of the testa of the *Seed* is not often taken into account, except for genera, which sometimes derive their characters from the seed being winged or not winged; but Mr. Brown has employed it for characterising a few orders of monocotyledones.

Albumen is held by many botanists of great account; and there is no doubt that if it exists in excess, it becomes one of the most important of all characters by which natural orders can be distinguished: but as it pre-exists in all seeds, remains of it may also be expected to be found in all seeds: thus, in Rosaceæ, in which it is said to be absent, it unquestionably exists; when, therefore, albumen is taken as a character for a natural order, it must be present in great abundance; under such a restriction, no character can be more valuable. Its absence in Bignoniaceæ divides that order from Scrophularineæ, in which it is abundant. Its disproportion to the embryo is a good character for Orobancheæ, as distinguished from Scrophularineæ. The texture of the albumen is also taken into account in characterising orders; in some it is fleshy, in others oily, in others horny or bony, and in Anonaceæ and Myristiceæ it is remarkable for being ruminate.

The *Embryo*, as the miniature representation of the future plant, would naturally be expected to possess some distinctive characters of a higher nature than any others; and such is found to be the fact. The want of one cotyledon, and peculiar mode of germination in monocotyledones, distinguish that great class from dicotyledones: its structure divides Aroideæ from other monocotyledonous orders; its position separates Gramineæ from Cyperaceæ; its form characterises Amaranthaceæ and Chenopodeæ; the direction of its radicle with regard to the hilum is a great peculiarity of Cistineæ and many other orders.

Finally, the *Proportion* that one part bears to another is

sometimes of generic, and always of specific value; in the latter case it is of peculiar importance. Species are apt to vary in the size, colour, surface, or even form of many of the organs, but it is much more seldom that the proportion that one part bears to another changes.

Such is a general view of the relative value of the principal characters employed by botanists for distinguishing plants: many others are occasionally called into use; but the knowledge of them will be better acquired by experience than by precept. That the value assignable to each particular character is, in many cases, far from definite, is apparent. This arises out of the very nature of the subjects upon which the science of botany depends. The rules under which nature arranges the modifications of vegetable structure are, to our feeble apprehension, so uncertain, that it is beyond our power to reduce them within any fixed laws.

I cannot quit this part of my subject without saying a word or two upon the nature of *affinity* and *analogy*, to which much importance is attached by some naturalists. It is assumed that the first systematic division of which beings are susceptible, is into two series, agreeing with each other in general circumstances, and being in *analogy*; the component parts of each series according in particulars, and being in *affinity* with each other. The relations of *analogy*, we are told, are ideally represented by two lines running parallel with each other; and those of *affinity*, otherwise called *transition*, by circles connected with each other in inextricable entanglement. That this is in many cases a distinction that it is important to draw, is undoubted; but I confess it does not appear to me to possess the degree of value that is attached to it. Every student of the natural relations of plants must necessarily keep it constantly in view; for I presume that *analogy* is equivalent to *remote affinity*, and affinity to *immediate analogy*; differences perfectly well understood. But that no more absolute idea can attach to the two terms is surely evident from this, — that cases must be continually occurring in which they are convertible, and that their application depends altogether upon contingent circumstances. Thus, for example, the genera Berberis and Bocagea are in analogy if considered with reference to Berberideæ and Anonaceæ; but in affinity if viewed as a part of Thalamifloræ: so, again, Aroideæ and Aristolochiæ are in analogy with respect to Monocotyledones and Dicotyledones, but in affinity as to Vasculares.

Having thus explained the usual value of characters, let us next proceed to enquire what difference there is in their nature. Characters are either *positive* or *negative;* that is to say, either explanatory of something that is present, or of something that is absent. Thus, when we say that two plants are distinguished by the one having ovate leaves and the other cordate leaves, such a character would be positive, being founded upon differences present to our senses. Again, when one class of plants is said to have two cotyledons and another but one, that character is positive. But if we say that *A.* has ovate leaves, and *B.* leaves not ovate, the latter would be a negative character: it does not explain what sort of leaves *B.* possesses, but merely tells us that they are not like those of *A.* Hence it is obvious, that while positive characters offer distinct points of comparison between one thing and another, negative characters afford no room for comparison. When we know what the characters are that two objects possess, we have the means of comparing them with each other; but if we are only acquainted with the character of one, we, of course, have no means afforded us of comparison. For this reason, while positive characters can be used for combining objects, negative characters can be only used for separating them. When we say that a plant is monopetalous, or polypetalous, or apetalous, monandrous, polyandrous, or icosandrous, vascular or cellular, resemblances or differences are pointed out which have a real existence, and from which positive characters may be obtained; but when we speak of plants not being monopetalous, or polypetalous, not being monandrous or polyandrous, and so on, no information whatever is afforded beyond that simple fact, and no conclusion can be drawn from such statements. Negative characters should never, therefore, be employed, except in case of necessity; and they should, if used, be understood merely as standing in the place of better characters, to be subsequently supplied as the science advances towards perfection. Care, however, must be taken not to confound with negative characters, positive characters depending upon a negation, as has been done by a French botanist of no mean fame; such, for instance, as having *no* corolla, *no* calyx, *no* leaves; these are not negative characters at all, but positive, indicating, indeed, the absence of something; but that very absence is a positive character: thus, if we say *A.* is a plant furnished with a corolla, and *B.* is a plant destitute of a corolla, such a character would indicate distinctly both the *quality* and *difference* of both *A.* and *B.*, and consequently would be positive; but if we were to say *A.* has a rotate corolla and *B.* has not a rotate corolla, such a character would be negative, because, while we learn the quality of the corolla of *A.*, we know nothing about that of *B.* further than that it is not like *A.*'s; it may be hypocrateriform or infundibuliform, monopetalous or polypetalous, or may be even wholly wanting, for aught is shown by the character: this is what is really negative, and what ought to be avoided.

Characters may be also *prominent* or *obscure*, that is to say, readily seen or discoverable with difficulty; and botanists are divided in opinion both as to the respective value of such, and the propriety of introducing the latter into the distinctions of plants. But if we reflect a moment, and consider what the test of a character should be, there cannot be an instant's difficulty in determining which of these kinds of characters to adopt. The goodness of a character depends, first, upon its applicability, and, secondly, upon its permanency; those which are most applicable and most constant are in all cases to have the preference; therefore, in deciding between an obvious and an obscure character, we have nothing to do but to ascertain which is most applicable and most constant. If we find a character, however minute and difficult to seize, answering better to these qualities than others which are more obvious and easily seen, the former is undoubtedly to be preferred;

for it must be always borne in mind that *the great object is not to make science easy, but perfect.* Undoubtedly, if an obvious character and an obscure one are equally permanent and applicable, the former is to be preferred; because, if we can combine facility with certainty, it is very desirable that we should do so.

CHAPTER VI.

OF SPECIES, VARIETIES, GENERA, ORDERS, AND CLASSES.

The next point of consideration is the meaning that is attached to the words species, genus, order, and class, and the characters upon which those groups severally depend.

A *species* is a union of individuals agreeing with each other in all essential characters of vegetation and fructification, capable of reproduction by seed without change, breeding freely together, and producing perfect seed from which a fertile progeny can be reared. Such are the true limits of a species; and if it were possible to try all plants by such a test, there would be no difficulty in fixing them, and determining what is species and what is variety. But, unfortunately, such is not the case. The manner in which individuals agree in their external characters is the only guide which can be followed in the greater part of plants. We do not often possess the means of ascertaining what the effect of sowing their seed or mixing the pollen of individuals would be; and, consequently, this test, which is the only sure one, is, in practice, seldom capable of being applied. The determination of what is a species, and what a variety, becomes therefore wholly dependent upon external characters, the power of duly appreciating which, as indicative of specific difference, is only to be obtained by experience, and is, in all cases, to a certain degree, arbitrary. It is probable that, in the beginning, species only were formed; and that they have, since the creation, sported into varieties, by which the limits of the species themselves have now become greatly confounded. For example, it may be supposed that a Rose, or a few species of Rose, were originally created. In the course of time these have produced endless varieties, some of which, depending for a long series of ages upon permanent peculiarities of soil or climate, have been in a manner fixed, acquiring a constitution and physiognomy of their own. Such supposed varieties have again intermixed with each other, producing other forms, and so the operation has proceeded. But as it is impossible, at the present day, to determine which was the original or originals, from which all the Roses of our own time have proceeded, or even whether they were produced in the manner I have assumed; and as the forms into which they divide are so peculiar as to render a classification of them indispensable to accuracy of language; it has become necessary to give names to certain of those forms, which are called species. Thus it seems that there are two sorts of species: the one, called natural species, determined by the definition given above; and the other, called botanical species, depending only upon the external characters of the plant. The former have been ascertained to a very limited extent: of the latter nearly the whole of sytematic botany consists. In this sense a species may be defined to be " an assemblage of individuals agreeing in all the essential characters of vegetation and fructification." Here the whole question lies with the word essential. What is an essential character of a species? This will generally depend upon a proneness to vary, or to be constant in particular characters, so that one class of characters may be essential in one genus, another class in another genus; and these points can be only determined by experience. Thus, in the genus Dahlia, the form of the leaves is found to be subject to great variation; the same species producing from seed, individuals, the form of whose leaves vary in a very striking manner: the form of the leaves is, therefore, in Dahlia, not a specific character. In like manner, in Rosa, the number of prickles, the surface of the fruit, or the surface of their leaves, and their serratures, are found to be generally fluctuating characters, and cannot often be taken as essential to species. The determination of species is, therefore, in all respects, arbitrary, and must depend upon the discretion or experience of the botanist. It may, nevertheless, be remarked, that decided differences in the forms of leaves, in the figure of the stem, in the surface of the different parts, in the inflorescence, in the proportion of parts, or in the form of the sepals and petals, usually constitute good specific differences.

A *genus* is an assemblage of species, agreeing with each other in the essential characters of fructification. It is an artificial means of condensing our ideas of the forms of plants, by sinking characters of minor importance in such as are of greater. While species depend upon the endless modifications of the organs of vegetation, genera depend upon the less numerous varieties of the organs of fructification, to which those of vegetation are subordinate. But as the value of the characters derived from the organs of fructification is uncertain, and dependent upon no fixed rules, so the limits of genera are arbitrary, and depend upon the caprice of botanists. Hence we find that some are disposed to excessive combination, and others to excessive division, in their genera. The only rule that can be given is this, that as genera are destined to analyse and simplify our ideas by reducing variable characters to those which are less variable, and characters which are common to many plants to those which are common to a few, care must be taken that this end is effected by a perfect analysis of the characters of fructification. And this being the case, an excessive multiplication of genera, which ought to be only the result of a very careful analysis, is infinitely better than an excessive reduction of them. The former leads to precision of ideas, the latter to confusion of them. In general, a genus should not be formed upon a solitary species, unless that species is so distinctly characterised that it cannot be referred to any other known genus; but wherever two or more species can be found agreeing in particular modifications of the structure of their fructification, by which they also differ from others, such two or more species are to be taken as the representatives of a genus.

Strong peculiarities in the vegetation of plants may be sometimes used as generic characters, but those of the fructification should always be preferred if they can be found.

An *order* is an assemblage of genera, agreeing with each other in the higher characters of vegetation and fructification. Orders, like genera, are a contrivance for analysing and simplifying our ideas, by reducing their number; and, like genera, they are also to a certain degree arbitrary. But as orders depend upon modifications of structure of a less variable kind than either genera or species, so are their limits better understood. Orders are characterised by all the parts both of vegetation and fructification, but their essential distinctions depend upon a few only, such as the presence or absence of stipulæ; the insertion of the leaves, whether opposite or alternate; the degree of division of the calyx and corolla, station of the stamens, structure of the fruit, figure of the embryo, presence or absence of albumen, &c. &c.

Classes are merely orders of a higher kind, combined by a few characters common to and distinctive of many.

BOOK IV.

GLOSSOLOGY; OR, OF THE TERMS USED IN BOTANY.

In order to comprehend the language of botanists, it is necessary that the unusual terms or words which are employed in writing upon the subject, and which are either different from words in vulgar use, or which are in botany employed in a particular sense, should be fully explained.

It is a very common plan to mix up Glossology with Organography, or to confound the definition and explanation of those characteristic terms of the science which are universally applicable, with the description of particular organs: but this plan is attended with many inconveniences, and is far less simple than to treat of the two separately. It was an error into which Linnæus fell, in composing his admirable Philosophia Botanica; and is the more remarkable, if the logical precision with which that work is otherwise composed be considered. Instead of distinguishing those terms which have a general application to all plants or parts of plants, according to circumstances, from such as have a particular application, and relate only to special modifications, he placed under his definition of each organ those terms which he knew to be applicable to it; but, as it was not his practice to repeat terms after they had been once explained, it frequently happened that beginners in the science, finding a given term explained once only, and with reference to a particular organ, fell into the mistake of supposing that that term was applicable only to the organ under which it was explained. To avoid this difficulty, other botanists have collected under each organ all the terms which could by possibility be applied to it, and have repeated them over and over again without regard to previous definitions; as if they supposed it impossible to convey by words an idea of the meaning of any term whatever, without noticing at length every possible application of it. Thus, in Willdenow's *Principles of Botany*, the most common and simple terms are repeated five, six, and even seven times; and in a more modern work, of very high character (*Les Elémens de Physiologie Végétale et de Botanique*, by Mirbel), the same practice has been carried so far, that the application of the word *simple* is explained in twenty-three different instances.

The true principles of arranging the glossology of science have, however, been long before the public. In the year 1797 Professor Link, of Berlin, in his *Prodromus Philosophiæ Botanicæ*, distinguished the characteristic or common terms used in botany from those which applied only to particular organs; and his example was afterwards followed by Illiger, a learned German naturalist, who, in the year 1810, proposed a total reformation of the method of describing the terms employed in Natural History (see his *Versuch einer Systematischen vollständigen Terminologie für das Thierreich und Pflanzenreich*). Little attention, however, was paid to the principles of these writers till the year 1813; when the learned Professor De Candolle adopted them in his *Théorie Elémentaire de la Botanique*, with his accustomed skill and sagacity.

The characteristic terms of botany are those which have a general application to any or all the parts of plants, and must not be confounded with such as have a particular application only, which will be found under the organs to which they respectively belong: the former are either *individual* or *collective*; of which the first apply to plants, or parts of plants, considered abstractedly; the second to plants, or their parts, considered in masses. To these are to be added those syllables and marks which, either prefixed or affixed to a known term, occasion an alteration in its signification. These I call *terms of qualification*. In the following arrangement, those terms which are seldom used are marked with a †; and those are entirely omitted which are used in Botany in their common acceptation.

CHARACTERISTIC TERMS are either INDIVIDUAL or COLLECTIVE.
CHARACTERISTIC INDIVIDUAL TERMS are either ABSOLUTE or RELATIVE.

CHARACTERISTIC INDIVIDUAL ABSOLUTE TERMS relate to, —
1. *Figure.*
   A. with respect to general form.
   B.   outline.
   C.   the apex or point.
2. *Division.*
   A. With respect to the margin.
   B.   incision.
   C.   composition or ramification.
3. *Surface.*
   A. With respect to marking or evenness.
   B.   appendages.
   C.   polish.
4. *Texture.*
5. *Size.*
6. *Duration.*
7. *Colour.*
8. *Variegation.*
9. *Veining.*

CHARACTERISTIC INDIVIDUAL RELATIVE TERMS comprehend, —
1. *Estivation.*
2. *Direction.*
3. *Insertion.*
   A. with respect to the mode of attachment or of adhesion.
   B.   situation.

CHARACTERISTIC COLLECTIVE TERMS relate to, —
1. *Arrangement.*
2. *Number.*

CLASS I. OF INDIVIDUAL TERMS.

THE terms which are included in this class are applied to the parts of a plant considered by themselves, and not in masses: they are either *absolute,* being used with reference to their own individual quality; or *relative,* being employed to express the relation which is borne by plants, or their parts, to some other body. Thus, for example, when we say that a plant has a *lateral, ovate* spike of flowers, the term *lateral* is relative, being used to express the relation which the spike bears to the stem; and the term *ovate* is absolute, being expressive of the actual form of the spike: and, again, in speaking of a *rugose, terminal* capsule, *rugose* is absolute, *terminal* is relative.

I. *Of Individual Absolute Terms.*

These relate to figure, division, surface, texture, size, duration, colour, variegation, and veining.

1. Of Figure.

A. *With respect to general form.*

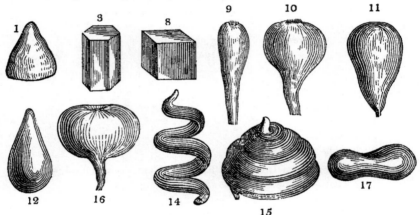

1. Conical (*conicus,* †*pyramidalis*); having the figure of a true cone; as the prickles of some roses, the root of carrot, &c.
2. Conoidal (*conoideus*); resembling a conical figure, but not truly one; as the calyx of Silene conoidea.
3. Prism-shaped (*prismaticus*); having several longitudinal angles and intermediate flat faces; as the calyx of Frankenia pulverulenta.
4. Globose (*globosus, sphæricus,* † *globulosus*); forming nearly a true sphere; as the fruit of Ligustrum vulgare, many seeds, &c.
5. Cylindrical (*cylindricus*); having nearly a true cylindrical figure; as the stems of grasses, and of most monocotyledonous plants.
6. Tubular (*tubulosus,* † *tubulatus*); approaching a cylindrical figure, and hollow; as the calyx of many Silenes, &c.
7. Fistulous (*fistulosus*): this is said of a cylindrical or terete

CLASS I.   INDIVIDUAL ABSOLUTE TERMS.   373

body, which is hollow, but closed at each end, as the leaves and stems of the onion.

8. Cubical (†*cubicus*); having or approaching the form of a cube: a very rare form, chiefly occurring in some seeds, as that of Vicia lathyroides.
9. Club-shaped (*clavatus*, †*claviformis*); gradually thickening upwards from a very taper base; as the appendages of the flower of Schwenkia, or the style of Campanula and Michauxia.
10. Turbinate, or top-shaped (*turbinatus*); inversely conical, with a contraction towards the point, as the fruit of some roses.
11. Pear-shaped (*pyriformis*); differing from turbinate in being more elongated, as in many kinds of pears.
12. † Tear-shaped (†*lachrymæformis*); the same as pear-shaped, except that the sides of the inverted cone are not contracted; as the seed of the apple.
13. † Strombus-shaped (†*strombuliformis*); twisted in a long spire, so as to resemble the convolutions of the shell called a Strombus; as the pod of Acacia strombulifera, or Medicago polymorpha.
14. Spiral (*spiralis*); twisted like a corkscrew.
15. Cochleate (*cochleatus*); twisted in a short spire, so as to resemble the convolutions of a snail-shell; as the pod of Medicago cochleata, the seed of Salicornia.
16. Turnip-shaped (*napiformis*); having the figure of a depressed sphere; as the root of the turnip, radish, &c.
17. † Placenta-shaped (†*placentiformis*); thick, round, and concave, both on the upper and lower surface; as the root of Cyclamen.

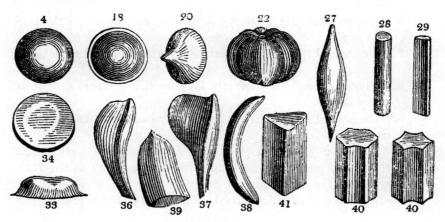

18. Lens-shaped (*lenticularis, lentiformis*); resembling a double convex lens, as the seeds of Amaranthus.
19. Buckler-shaped (*scutatus, scutiformis*), having the figure of a small round buckler, as the scales upon the leaves of Elæagnus; lens-shaped, with an elevated rim.
20. Bossed (*umbonatus*); round, with a projecting point in the centre, like the boss of an ancient shield, as the pileus of many species of Agaricus.
21. Gibbous (*gibbus, gibbosus*); very convex or tumid, as the leaves of many succulent plants: properly speaking, this term should be restricted to solid convexities.
22. †Melon-shaped, (†*meloniformis*); irregularly spherical, with projecting ribs; as the stem of Cactus melocactus: a bad term.
23. Spheroidal (*spheroideus*); a solid with a spherical figure, a little depressed at each end. *De Cand.*
24. Ellipsoidal (*ellipsoideus*); a solid with an elliptical figure. *De Cand.*
25. Ovoidal (*ovoideus*); a solid with an ovate figure, or resembling an egg. *De Cand.*
26. Shield-shaped (*clypeatus*); in the form of an ancient buckler; the same as scutate, No. 19.
27. Spindle-shaped (*fusiformis,* †*fusinus*); thick, tapering to each end; as the root of the long radish. Sometimes conical roots are called fusiform, but improperly.
28. Terete, or taper (*teres*); the opposite of angular; usually employed in contradistinction to that term, when speaking of long bodies. Many stems are terete.
29. Half terete (*semiteres*); flat on one side, terete on the other.
30. Compressed (*compressus*); flattened lengthwise, as the pod of a pea.
31. Depressed (*depressus*); flattened vertically, as the root of a turnip.
32. Plane (*planus*); a perfectly level or flat surface, as that of many leaves.
33. Cushioned (*pulvinatus*); convex, and rather flattened: seldom used.
34. Discoidal (*discoideus*); orbicular, with some perceptible thickness, parallel faces, and a rounded border; as the fruit of Strychnos Nux-vomica.
35. Curved (*arcuatus, curvatus*); bent, but so as to represent the

CLASS I.   INDIVIDUAL ABSOLUTE TERMS.   375

arc of a circle; as the fruit of Astragalus hamosus, Medicago falcata, &c.

36. Scimitar-shaped (*acinaciformis*); curved, fleshy, plane on the two sides, the concave border thick, the convex border thin; as the leaves of Mesembryanthemum acinaciforme.

37. Axe-shaped (*dolabriformis*); fleshy, nearly straight, somewhat terete at the base, compressed towards the upper end; one border thick and straight, the other enlarged, convex, and thin; as the leaves of Mesembryanthemum dolabriforme.

38. Falcate (*falcatus*); plane and curved, with parallel edges, like the blade of a reaper's sickle; as the pod of Medicago falcata: any degree of curvature, with parallel edges, receives this name.

39. Tongue-shaped (*linguiformis*); long, fleshy, plain, convex, obtuse; as the leaves of Sempervivum tectorum, and some aloes.

40. Angular (*angulosus*); having projecting longitudinal angles. We say *obtuse-angled* when the angles are rounded, as in the stem of Salvia pratensis; and *acute-angled* when they are sharp, as in many Carices. Some call these angles the *acies*.

41. Three-cornered (*trigonus*); having three longitudinal angles and three plain faces, as the stem of Carex acuta.

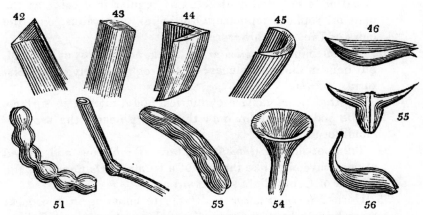

42. Three-edged (*triangularis*, *triqueter*); having three acute angles with concave faces, as the stems of many plants; generally used as a synonyme of trigonus.

43. Two-edged (*anceps*); compressed, with two sharp edges, as the stem of an Iris.

44. Keeled (*carinatus*); formed in the manner of the keel of a boat; that is to say, with a sharp projecting ridge, arising from a flat or concave central rib, as the glumes of grasses.

45. Channelled (*canaliculatus*); long and concave, so as to resemble a gutter or channel; as the leaves of Lygeum Spartum, Tradescantia virginica, &c.

46. Boat-shaped (*cymbiformis, navicularis*); having the figure of a boat in miniature; that is to say, concave, tapering to each end, with a keel externally, as the glumes of Phalaris canariensis: scarcely different from 44.

47. Whip-shaped (*flagelliformis*); long, taper, and supple, like the thong of a whip; as the stem of Vinca, and of many plants. This term is confined to stems and roots.

48. Rope-shaped (*funalis,* †*funiliformis*); formed of coarse fibres resembling cords, as the roots of Pandanus, and other arborescent monocotyledons. *Mirbel.*

49. Thread-shaped (*filiformis*); slender like a thread, as the filaments of most plants, and the styles of many.

50. Hair-shaped (*capillaris*); the same as filiform, but more delicate, so as to resemble a hair; it is also applied to the fine ramifications of the inflorescence of some plants, as grasses.

51. Necklace-shaped (*moniliformis,* †*nodosus,* Mirb.); cylindrical or terete, and contracted at regular intervals; as the pods of Sophora japonica, Ornithopus perpusillus, &c., the hairs of Dicksonia arborescens, &c.

52. Worm-shaped (*vermicularis*); thick, and almost cylindrical, but bent in different places; as the roots of Polygonum Bistorta. *Willd.*

53. Knotted (*torulosus*); a cylindrical body, uneven in surface, as the pod of Chelidonium: this is very nearly the same as moniliform.

54. Trumpet-shaped (*tubæformis, tubatus*); hollow, and dilated at one extremity, like the end of a trumpet, *De Cand.*; as the corolla of Caprifolium sempervirens.

55. Horned (*cornutus, corniculatus*); terminating in a process resembling a horn, as the fruit of Trapa bicornis. If there are two horns, the word *bicornis* is used; if three, *tricornis;* and so on.

56. Beaked (*proboscideus*); having a hard terminal horn, as the fruit of Martynia.

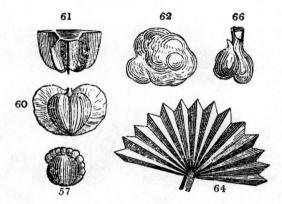

57. **Crested** (*cristatus*); having an elevated, irregular, or notched ridge, resembling the crest of a helmet. This term is chiefly applied to seeds, and to the appendages of the anthers of some Ericæ; such as E. triflora and comosa.
58. **Petal-like** (*petaloideus*); having the colour and texture of a petal; as one lobe of the calyx of Mussænda, the bracteæ of many plants, the stamen of Canna, the stigmata of Iris.
59. **Leaf-like** (*foliaceus*, †*foliiformis*, †*phylloideus*); having the texture or form of a leaf; as the lobes of the calyx of Rosa, the apex of the fruit of Fraxinus, the persistent petals of Melanorhæa.
60. **Winged** (*alatus*); having a thin broad margin; as the fruit of Paliurus australis, the seed of Malcomia, Bignonia, &c. In composition *pterus* is used; as *dipterus* for two-winged, *tripterus* for three-winged, *tetrapterus* for four-winged, &c.; *peripterus* when the wing surrounds any thing; *epipterus* when it terminates.
61. †**Mill-sail-shaped** (†*molendinaceus*); having many wings projecting from a convex surface; as the fruit of some umbelliferous plants, and of Moringa.
62. †**Knob-like** (†*gongylodes*); having an irregular, roundish figure.
63. **Halved** (*dimidiatus*); only half, or partially formed. A leaf is called dimidiate when one side only is perfect; an anther when one lobe only is perfect; and so on.
64. **Fan-shaped** (*flabelliformis*); plaited like the rays of a fan; as the leaf of Borassus flabelliformis.
65. **Grumous** (*grumosus*); in form of little clustered grains, as the root of Neottia Nidus-avis, *Mirb.*; rather as the fæcula in the stem of the Sago palm.
66. †**Testicular** (†*testiculatus*); having the figure of two oblong bodies, as the roots of Orchis mascula.

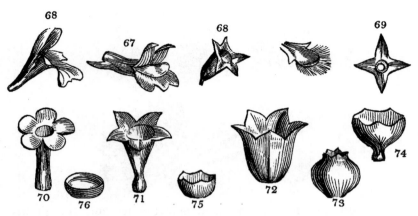

67. **Ringent, or personate** (*ringens, personatus*); a term applied to a monopetalous corolla, the limb of which is unequally divided; the upper division, or lip, being arched; the lower prominent, and pressed against it, so that when compressed, the whole resembles the mouth of a gaping animal; as the corolla of Antirrhinum.

68. **Labiate** (*labiatus*); a term applied to a monopetalous calyx or corolla, which is separated into two unequal divisions; the one anterior, and the other posterior with respect to the axis: hence bilabiate is more commonly used than labiate. Salvia, and many other plants, afford examples. It is often employed instead of ringent.

69. **Wheel-shaped** (*rotatus*); a calyx or corolla, or other organ, of which the tube is very short, and the segments spreading; as the corolla of Veronica and Galium.

70. **Salver-shaped** (*hypocrateriformis*); a calyx or corolla, or other organ, of which the tube is long and slender, and the limb flat; as in Phlox.

71. **Funnel-shaped** (*infundibularis, infundibuliformis*); a calyx or corolla, or other organ, in which the tube is obconical, gradually enlarging upwards into the limb, so that the whole resembles a funnel; as the corolla of Nicotiana.

72. **Bell-shaped** (*campanulatus,* †*campanaceus,* †*campaniformis*); a calyx, corolla, or other organ, in which the tube is inflated and gradually enlarged into a limb, the base not being conical; as the corolla of Campanula.

73. **Pitcher-shaped** (*urceolatus*); the same as campanulate, but more contracted at the orifice, with an erect limb; as the corolla of Vaccinium myrtillus.

74. **Cup-shaped** (*cyathiformis*); the same as pitcher-shaped, but

not contracted at the margin; the whole resembling a drinking cup; as the limb of the corolla of Symphytum.

75. †Cupola-shaped (†*cupuliformis*); slightly concave, with a nearly entire margin; as the calyx of Citrus, or the cup of an acorn.

76. Kneepan-shaped (*patelliformis*); broad, round, thick; convex on the lower surface, concave on the other: the same as *meniscoideus*, but thicker. The embryo of Flagellaria indica is patelliform.

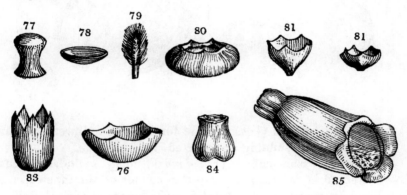

77. †Pulley-shaped (†*trochlearis*); circular, compressed, contracted in the middle of its circumference, so as to resemble a pulley; as the embryo of Commelina communis.

78. Scutelliform (*scutelliformis*); the same as patelliform, but oval: not round, as the embryo of grapes.

79. Brush-shaped (†*muscariformis*); formed like a brush or broom; that is to say, furnished with long hairs towards one end of a slender body, as the style and stigma of many Compositæ.

80. †Acetabuliform (†*acetabuliformis*, †*acetabuleus*); concave, depressed, round, with the border a little turned inwards; as the fruit of some lichens.

81. †Goblet-shaped (†*crateriformis*); concave, hemispherical, a little contracted at the base; as some Pezizas.

82. †Cotyliform (†*cotyliformis*); resembling *rotate*, but with an erect limb.

83. †Poculiform († *poculiformis*); cup-shaped, with a hemispherical base and an upright limb; nearly the same as campanulate.

84. †Pouch-shaped (†*scrotiformis*); hollow, and resembling a little double bag; as the spur of many Orchises.

85. †Foxglove-shaped (†*digitaliformis*); like campanulate, but longer, and irregular; as the corolla of Digitalis.

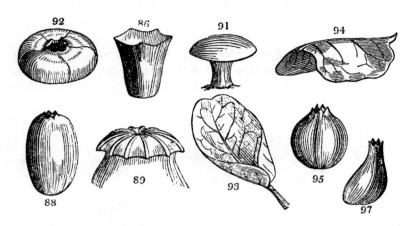

86. †Vase-shaped (†*vascularis*); formed like a flower-pot; that is to say, resembling an inverted truncate cone.
87. †Tapeworm-shaped (†*tænianus*); long, cylindrical, contracted in various places, in the manner of the tapeworm.
88. †Sausage-shaped (†*botuliformis*); long, cylindrical, hollow, curved inwards at each end; as the corolla of some Ericas.
89. †Umbrella-shaped (†*umbraculiformis*); resembling an expanded umbrella; that is to say, hemispherical and convex, with rays, or plaits, proceeding from a common centre; as the stigma of poppy.
90. †Meniscoid (†*meniscoideus*); thin, concavo-convex, and hemispherical, resembling a watch-glass.
91. Mushroom-headed (*fungiformis, fungilliformis*); cylindrical, having a rounded, convex, overhanging extremity; as the embryo of some monocotyledonous plants, as Musa.
92. †Nave-shaped (†*modioliformis*); hollow, round, depressed, with a very narrow orifice; as the ripe fruit of Gaultheria.
93. Hooded (*cucullatus*); a plane body, the apex or sides of which are curved inwards, so as to resemble the point of a slipper, or a hood; as the leaves of Pelargonium cucullatum, the spatha of Arum, the labellum of Pharus.
94. †Saddle-shaped (†*sellæformis*); oblong, with the sides hanging down, like the laps of a saddle; as the labellum of Cateleya Loddigesii.
95. Turgid (*turgidus*); slightly swelling.
96. Bladdery (*inflatus*); thin, membranous, slightly transparent,

CLASS I.   INDIVIDUAL ABSOLUTE TERMS.   381

swelling equally, as if inflated with air; as the calyx of Cacubalus.
97. Bellying (*ventricosus*); swelling unequally on one side, as the corolla of many labiate and personate plants.
98. Regular (*regularis*); in which all the parts are symmetrical: a rotate corolla is regular; the flower of a cherry is regular.
99. Irregular (*irregularis*); in which symmetry is destroyed by some inequality of parts: a labiate corolla, the flower of the horse-chesnut, and the violet, are irregular.
100. Abnormal (*abnormis*); in which some departure takes place from the ordinary structure of the family or genus to which a given plant belongs. Thus, Nicotiana multivalvis, in which the ovarium has many cells instead of two, is unusual or abnormal.
101. Normal (*normalis*); in which the ordinary structure peculiar to the family or genus of a given plant is in nowise departed from.

B. *With respect to outline.*

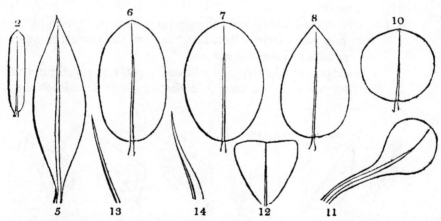

1. Outline (*ambitus, circumscriptio*); the figure represented by the margin of a body.
2. Linear (*linearis*); narrow, short, with the two opposite margins parallel; as the leaf of the Taxus.
3. †Band-shaped (†*fasciarius*); narrow, very long, with the two opposite margins parallel; as the leaves of Zostera marina.
4. Strap-shaped (*ligulatus, loratus*), narrow, moderately long, with the two opposite margins parallel; as the leaves of Amaryllis equestris.
5. Lanceolate (*lanceolatus*); narrowly elliptical, tapering to each

end; as the leaf of Plantago lanceolata, Daphne mezereum, &c.

6. Oblong (*oblongus*); elliptical, obtuse at each end; as the leaf of the hazel.
7. Oval (*ovalis, ellipticus*); elliptical, acute at each end; as the leaf of Cornus sanguinea.
8. Ovate, or †egg-shaped (*ovatus*); oblong or elliptical, broadest at the lower end, so as to resemble the longitudinal section of an egg; as the leaf of Stellaria media.
9. Orbicular (*orbicularis*); perfectly circular, as the leaf of Cotyledon orbiculare.
10. Roundish (*rotundus, subrotundus, rotundatus*); orbicular, a little inclining to be oblong; as the leaf of Lysimachia nummularia, Mentha rotundifolia.
11. Spatulate (*spatulatus*); oblong, with the lower end very much attenuated, so that the whole resembles a chemist's spatula; as the leaf of Bellis perennis.
12. Wedge-shaped (*cuneatus, cuneiformis,* †*cunearius*); inversely triangular, with rounded angles; as the leaf of Saxifraga tridentata.
13. Awl-shaped (*sabulatus*); linear, very narrow, tapering into a very fine point from a broadish base; as the leaves of Arenaria tenuifolia, Ulex europæus.
14. Needle-shaped (*acerosus*); linear, rigid, tapering into a fine point from a narrow base; as the leaves of Júniperus communis.

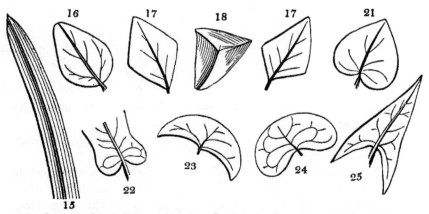

15. Sword-shaped (*ensiformis, gladiatus*); lorate, quite straight, with the point acute; as the leaf of an Iris.
16. †Parabolical (†*parabolicus*); between ovate and ellip-

CLASS I.   INDIVIDUAL ABSOLUTE TERMS.   383

tical, the apex being obtuse; as the leaf of Amaranthus Blitum.

17. Rhomboid (*rhombeus, rhomboideus*); oval, a little angular in the middle; as the leaf of Hibiscus rhombifolius.

18. Deltoid (*deltoides*); a solid, the transverse section of which has a triangular outline, like the Greek Δ; as the leaf of Mesembryanthemum deltoideum.

19. Triangular (*triangularis*); having the figure of a triangle of any kind; as the leaf of Betula alba.

20. Trapeziform (*trapeziformis, trapezoideus*); having four edges, those which are opposite not being parallel; as the leaf of Adiantum trapeziforme, Populus nigra.

21. Heart-shaped (*cordatus, cordiformis*); having two round lobes at the base, the whole resembling the heart in a pack of cards; as the leaf of Alnus cordifolia.

22. Eared (*auriculatus*); having two small rounded lobes at the base; as the leaf of Salvia officinalis.

23. Crescent-shaped (*lunatus, lunulatus,* †*semilunatus*); resembling the figure of the crescent; as the glandular apex of the involucral leaves of many Euphorbias.

24. Kidney-shaped (*reniformis,* †*renarius*); resembling the figure of a kidney; that is to say, crescent-shaped, with the ends rounded; as the leaf of Asarum europæum.

25. Arrow-headed (*sagittatus*); gradually enlarged at the base into two acute straight lobes, like the head of an arrow; as the leaf of Rumex acetosella.

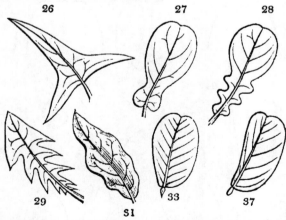

26. Halbert-headed (*hastatus*); abruptly enlarged at the base into two acute diverging lobes, like the head of an halbert; as the leaf of Arum maculatum.

384 GLOSSOLOGY. BOOK IV.

27. Fiddle-shaped (*panduratus, panduriformis*); obovate, with a deep recess or sinus on each side; as the leaves of Rumex pulcher.
28. Lyre-shaped (*lyratus*); the same as panduriform, but with several sinuses on each side, which gradually diminish in size to the base; as the leaf of Geum urbanum, Raphanus raphanistrum.
29. Runcinate, or hook-backed (*runcinatus*); curved in a direction from the apex to the base; as the leaf of Leontodon taraxacum.
30. Tapering (*attenuatus*); gradually diminishing in breadth.
31. Wavy (*undulatus*); having an uneven, alternately convex and concave margin, as the holly leaf.
32. Equal (*æqualis*); when both sides of a figure are symmetrical; as the leaf of an apple.
33. Unequal (*inæqualis*); when the two sides of a figure are not symmetrical; as the leaf of Begonia.
34. Equal-sided (*æquilaterus*); the same as equal.
35. Unequal-sided (*inæquilaterus*); the same as unequal.
36. Oblique (*obliquus*); when the degree of inequality in the two sides is slight.
37. Halved (*dimidiatus*); when the degree of inequality is so great that one half of the figure is either wholly or nearly wanting; as the leaf of many Bryonias.

C. *With respect to the apex, or point.*

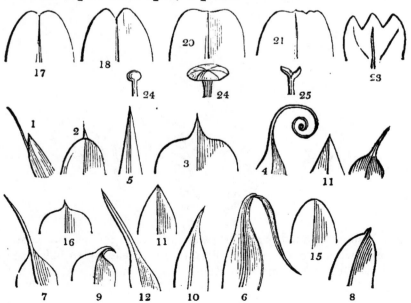

1. **Awned** (*aristatus*); abruptly terminated in a hard, straight, subulate point of various lengths; as the paleæ of grasses. The arista is always a continuation of the costa, and sometimes separates from the lamina below the apex.
2. **Mucronate** (*mucronatus*); abruptly terminated by a hard, short point; as the leaf of Statice mucronata.
3. **Cuspidate** (*cuspidatus*); tapering gradually into a rigid point. It is also used sometimes to express abruptly acuminate; as the leaf of many Rubi.
4. **Cirrhous** (*cirrhosus, apice circinatus*); terminated by a spiral or flexuose, filiform appendage; as the leaf of Gloriosa superba. This is due to an elongation of a costa.
5. **Pungent** (*pungens*); terminating gradually in a hard, sharp point; as the leaves of Ruscus aculeatus.
6. **Bristle-pointed** (*setosus*, †*setiger*); terminating gradually in a very fine, sharp point; as the leaves of many mosses.
7. **Hair-pointed** (*piliferus*); terminating in a very fine, weak point; as the leaves of many mosses.
8. **Pointletted** (*apiculatus*); terminating abruptly in a little point; differing from mucronate in the point being part of the limb, and not arising wholly from a costa.
9. **Hooked** (*uncinatus*, †*uncatus*); curved suddenly back at the point; as the leaves of Mesembryanthemum uncinatum.
10. **Beaked** (*rostratus, rostellatus*); terminating gradually in a hard, long, straight point; as the pod of radish.
11. **Acute, or sharp-pointed** (*acutus*); terminating at once in a point, not abruptly, but without tapering in any degree; as any lanceolate leaf.
12. **Taper-pointed** (*acuminatus*); terminating very gradually in a point; as the leaf of Salix alba.
13. †**Acuminose** (†*acuminosus*); terminating gradually in a flat, narrow end.
14. **Tail-pointed** (*caudatus*); excessively acuminated, so that the point is long and weak, like the tail of some animal; as the calyx of Aristolochia trilobata, the petals of Brassia caudata.
15. **Blunt** (*obtusus*); terminating gradually in a rounded end; as the leaf of Berberis vulgaris.
16. **Blunt with a point** (*obtusus cum acumine*); terminating abruptly in a round end, the middle of which is suddenly lengthened into a point; as the leaf of many Rubi.
17. **Retuse** (*retusus*); terminating in a round end, the centre of which is depressed; as the leaf of Vaccinium Vitis Idæa.

18. Emarginate (*emarginatus*); having a notch at the end, as if a piece had been taken out; as the leaf of Buxus sempervirens.
19. †*Accisus*; when the end has an acute sinus between two rounded angles. *Link*.
20. Truncate (*truncatus*); terminating very abruptly, as if a piece had been cut off; as the leaf of Liriodendron tulipifera.
21. Bitten (*præmorsus*, †*succisus*); the same as truncate, except that the termination is ragged and irregular, as if bitten off: the term is generally applied to roots; the leaf of Caryota urens is another instance.
22. †Dedaleous (†*dædaleus*); when the point has a large circuit, but is truncated and rugged. *W.*
23. Trident-pointed (*tridentatus*); when the point is truncated, and has three indentations (*W.*); as Saxifraga tridentata, Potentilla tridentata.
24. Headed (*capitatus*); suddenly much thicker at the point than in any other part: a term confined to cylindrical or terete bodies; as Mucor, glandular hairs, &c.
25. Lamellar (*lamellatus, lamellosus*); having two little plates at the point, as the style of many plants.
26. †Blunt (†*hebetatus*, De Cand.); having a soft, obtuse termination.
27. Pointless (*muticus*). This term is employed only in contradistinction to some other that indicates being pointed: thus, if, in contrasting two things, one was said to be mucronate, the other, if it had not a mucro, would be called pointless; and the same term would be equally employed in contrast with cuspidate or aristate, or any such. It is also used absolutely.

2*. Of Division.

A. *With respect to the margin.*

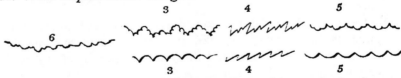

1. Entire (*integer*); properly speaking, this means having no kind of marginal division; but sometimes it has been used to indicate not pinnatifid, and also nearly destitute of marginal division.
2. Quite entire (*integerrimus*); perfectly free from division of the margin.
3. Crenated (*crenatus*); having convex teeth. When these teeth are themselves crenated, we say *bicrenate*.

CLASS I. INDIVIDUAL ABSOLUTE TERMS.

4. Sawed (*serratus*); having sharp, straight-edged teeth pointing to the apex. When these teeth are themselves serrate, we say *biserrate*, or *duplicato-serrate*.
5. Toothed (*dentatus*); having sharp teeth with concave edges. When these teeth are themselves toothed, we say *duplicato-dentate*, or doubly toothed, but not *bidentate*, which means two-toothed.
6. Gnawed (*erosus*); having the margin irregularly toothed, as if bitten by some animal.
7. Curled (*crispus*); having the margin excessively irregularly divided and twisted; as in many varieties of the garden endive, Mentha crispa, Ulmus cucullata.

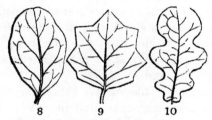

8. Repand (*repandus*, †*sinuolatus*); having an uneven, slightly sinuous margin; as the leaf of Solanum nigrum.
9. Angular (*angulatus, angulosus*); having several salient angles on the margin; as the leaf of Datura Stramonium.
10. Sinuate (*sinuatus*); having the margin uneven, alternately with deep concavities and convexities; as the leaf of Quercus robur.

B. *With respect to incision.*

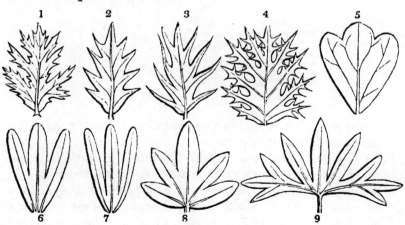

1. Torn (*lacerus*); irregularly divided by deep incisions
2. Cut (*incisus*); regularly divided by deep incisions.
3. Slashed (*laciniatus*); divided by deep, taper-pointed, cut incisions.
4. Squarrose-slashed (*squarroso-laciniatus*); slashed with minor divisions at right angles with the others.
5. Lobed (*lobatus*); partly divided into a determinate number of segments. We say *bilobus*, two-lobed, as in the leaf of Bauhinia porrecta; *trilobus*, three-lobed, as in the leaf of Anemone hepatica; and so on.
6. Split (*fissus*); divided nearly to the base into a determinate number of segments. We say *bifidus*, split in two; *trifidus*, in three; as in the leaf of Teucrium chamæpitys; and so on. When the segments are very numerous, *multifidus* is used.
7. Parted (*partitus*); divided into a determinate number of segments, which extend nearly to the base of the part to which they belong. We say *bipartitus*, parted in two; *tripartitus*, in three; and so on.
8. Palmate (*palmatus*); having five lobes, the midribs of which meet in a common point, so that the whole bears some resemblance to a human hand; as the leaf of Passiflora cærulea.
9. Pedate (*pedatus*); the same as palmate, except that the two lateral lobes are themselves divided into smaller segments, the midribs of which do not directly run into the same point as the rest; as the leaf of Arum dracunculus, Helleborus niger, &c.

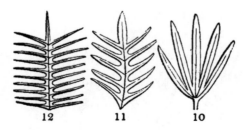

10. Fingered (*digitatus*); the same as palmate, but the segments less spreading, and narrower.
11. Pinnatifid (*pinnatifidus, pennatipartitus, pinnaticisus*); divided almost to the axis into lateral segments, something in the way of the side divisions of a feather; as Polypodium vulgare. M. De Candolle distinguishes several modifications of pinnatifidus: — 1. *Pinnatifidus*, when the lobes are divided

down to half the breadth of the leaf; 2. *pinnatipartitus*, when the lobes pass beyond the middle, and the parenchyma is not interrupted; 3. *pinnatisectus*, when the lobes are divided down to the midrib, and the parenchyma is interrupted; 4. *pinnatilobatus*, when the lobes are divided to an uncertain depth: *lyrate* and the like belong to this modification. He has similar variations of palmatus and pedatus; viz. *palmatifidus, palmatipartitus, palmatisectus, palmatilobatus;* and *pedatifidus, pedatipartitus, pedatisectus,* and *pedatilobatus.*

12. Comb-shaped (*pectinatus*); the same as pinnatifid; but the segments very numerous, close, and narrow, like the teeth of a comb; as the leaf of Lavandula dentata, all Mertensias.

C. *With respect to composition or ramification.*
  1. Simple (*simplex*); scarcely divided or branched at all.
  2. Quite simple (*simplicissimus*); not divided or branched at all.
  3. Compound (*compositus*); having various divisions or ramifications. As compared with the two following, it applies to cases of leaves in which the petiole is not divided; as in the orange.
  4. Decompound (*decompositus*); having various compound divisions or ramifications. In leaves it is applied to those the petiole of which bears secondary petioles; as in the leaf of Mimosa purpurea.

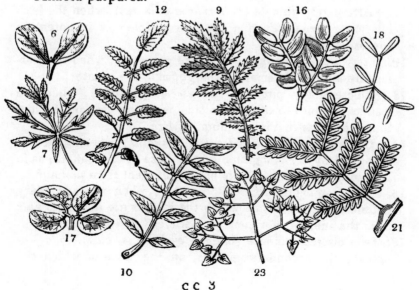

5. Supradecompound (*supradecompositus*); having various decompound divisions or ramifications. In leaves it is applied to such as have the primary petiole divided into secondary ones, and the secondary into a third set; as in the leaf of Daucus Carota.

6. †Bifoliolate (†*bifoliolatus, binatus*); when in leaves the common petiole is terminated by two leaflets growing from the same point; as in Zygophyllum Fabago. This term has the same application as *unijugus* and *conjugatus*. We say *trifoliolate*, or *ternate*, when the petiole bears three leaflets from the same point, as in Menyanthes trifoliata; †*quadrifoliolate*, if there are four from the same point, as in Marsilea quadrifolia; and *quinquefoliolate*, or *quinate*, if there are five from the same point, as in Potentilla reptans; and so on.

7. †Vertebrate (†*vertebratus*); when the leaf is contracted at intervals, there being an articulation at each contraction; as in Cussonia spicata. *Mirb.*

8. Pinnate (*pinnatus*); when simple leaflets are arranged on each side a common petiole; as in Polypodium vulgare.

9. Pinnate with an odd one (*impari-pinnatus*); when the petiole is terminated by a single leaflet or tendril; as in Pyrus aucuparia. If there is a tendril, as in the pea, it is called *cirrhose*.

10. Equally pinnate (*paripinnatus, abruptè pinnatus*); when the petiole is terminated by neither leaflet nor tendril, as Orobus tuberosus.

11. †Alternately pinnate (†*alternatim pinnatus*); when the leaflets are alternate upon a common petiole; as in Potentilla rupestris. *Mirb.*

12. Interruptedly pinnate (*interruptè pinnatus*); when the leaflets are alternately small and large, as in the potatoe.

13. †Decreasingly pinnate (†*decrescentè pinnatus*); when the leaflets diminish insensibly in size from the base of the leaf to its apex; as in Vicia sepium. *Mirb.*

14. †Decursively pinnate (†*decursivè pinnatus*); when the petiole is winged by the elongation of the base of the leaflets; as in Melianthus. *Mirb.* This is hardly different from pinnatifid.

15. Digitato-pinnate (*digitato-pinnatus*); when the secondary petioles, on the sides of which the leaflets are attached, part from the summit of a common petiole. *Mirb.*

16. Twin digitato-pinnate (*bidigitato-pinnatus, biconjugato-pinnatus*); the secondary petioles, on the sides of which the

leaflets are arranged, proceed in twos from the summit of a common petiole; as in **Mimosa** purpurea. *Mirb.*

17. Bigeminate (*bigeminatus, biconjugatus*); when each of two secondary petioles bears a pair of leaflets; as in Mimosa unguis Cati. *Mirb.*

18. Tergeminate (*tergeminus, †tergeminatus*); when each of two secondary petioles bears towards its summit one pair of leaflets, and the common petiole bears a third pair at the origin of the two secondary petioles; as in Mimosa tergemina. *Mirb.*

19. Thrice digitato-pinnate (*†tridigitato-pinnatus, ternato-pinnatus*); when the secondary petioles, on the sides of which the leaflets are attached, proceed in threes from the summit of a common petiole; as in Hoffmannseggia. *Mirb.*

20. † *Quadridigitato-pinnatus*, as in Mimosa pudica, and † *Multidigitato-pinnatus*, are rarely used, but are obvious modifications of the last.

21. Bipinnate (*bipinnatus, †duplicato-pinnatus*); when the leaflets of a pinnate leaf become themselves pinnate; as in Mimosa Julibrissin, Fumaria officinalis, &c.

22. Biternate (*biternatus, †duplicato-ternatus*); when three secondary petioles proceed from the common petiole, and each bears three leaflets; as in Fumaria bulbosa, Imperatoria Ostruthium, &c. *Mirb.*

23. Triternate (*triternatus*); when the common petiole divides into three secondary petioles, which are each subdivided into three tertiary petioles, each of which bears three leaflets; as the leaf of Epimedium alpinum.

24. Tripinnate (*tripinnatus*); when the leaflets of a bipinnate leaf become themselves pinnate; as in Thalictrum minus, or Œnanthe Phellandrium.

25. Paired (*conjugatus, unijugus, †unijugatus*); when the petiole of a pinnated leaf bears one pair of leaflets; as Zygophyllum Fabago. *Bijugus* is when it bears two pairs; as in Mimosa fagifolia: *trijugus, quadrijugus, quinquejugus*, &c. are also employed when required. *Multijugus* is used when the number of pairs becomes very considerable; as in Orobus sylvaticus, Astragalus glycyphyllus.

26. Branched (*ramosus*); divided into many branches: if the divisions are small, we say *ramulosus*.

27. Somewhat branched (*subramosus*); having a slight tendency to branch.

28. Excurrent (*excurrens*); in which the axis remains always in the centre, all the other parts being regularly disposed round it; as the stem of Pinus Abies.

29. Much-branched (*ramosissimus*); branched in a great degree.

30. †Disappearing (†*deliquescens*); branched, but so divided that the principal axis is lost trace of in the ramifications; as the head of an oak tree.

31. Dichotomous (*dichotomus*); having the divisions always in pairs; as the branches and inflorescence of Stellaria holostea: if they are in threes, we say *trichotomus*; as the stem of Mirabilis Jalapa.

32. Twin (*didymus*); growing in pairs, or divided into two equal parts; as the fruit of Galium.

33. Forked (*furcatus*); having long terminal lobes, like the prongs of a fork; as Ophioglossum pendulum.

34. Stellate (*stellatus*); divided into segments, radiating from a common centre; as the hairs of most malvaceous plants.

35. Jointed (*articulatus*); falling in pieces at the joints, or separating readily at the joints; as the pods of Ornithopus, the leaflets of Guilandina Bonduc: it is also applied to bodies having the appearance of being jointed; as the stem and leaves of Juncus articulatus.

36. Granular (*granulatus*); divided into little knobs or knots; as the roots of Saxifraga granulata.

37. †Byssaceous (†*byssaceus*); divided into very fine pieces, like wool; as the roots of some Agarics.

38. †Tree-like (†*dendroides*); divided at the top into a number of fine ramifications, so as to resemble the head of a tree; as Lycopodium dendroideum.

39. †Brush-shaped (†*aspergilliformis*); divided into several fin ramifications, so as to resemble the brush (*aspergillus*) used for sprinkling holy water in the ceremonies of the Catholic Church; as the stigmas of grasses.

40. Partitioned (*loculosus*, †*septatus*, †*phragmiger*); divided by internal partitions into cells; as the pith of the plant that produces the Chinese rice-paper. This is never applied to fruits.

41. Anastomosing (*anastomozans*); the ramifications of any thing which are united at the points where they come in contact are said to anastomose. This term is confined to veins.

42. Ruminate (*ruminatus*); when a hard body is pierced in various directions by narrow cavities filled with dry cellular matter; as the albumen of the nutmeg, and the Anona.

43. †Cancellate (†*cancellatus*); when the parenchyma is wholly absent, and the veins alone remain, anastomosing and forming a kind of net-work; as the leaves of Hydrogeton fenestralis.

44. Perforated (*pertusus*); when irregular spaces are left open in the surface of any thing, so that it is pierced with holes; as the leaves of Dracontium pertusum.

3*. Of Surface.

A. *With respect to marking, or evenness.*

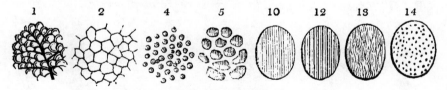

1. Rugose (*rugosus*); covered with reticulated lines, the spaces between which are convex; as the leaves of sage.

2. Netted (*reticulatus*); covered with reticulated lines which project a little; as the under surface of the leaves of most Melastomas, the seeds of Geranium rotundifolium.

3. †Half-netted (†*semireticulatus*); when, of several layers of any thing, the outer one only is reticulated; as in the roots of Gladiolus communis.

4. Pitted (*scrobiculatus*); having numerous small shallow depressions, or excavations; as the seed of Datisca cannabina, Passiflora, &c.

5. Lacunose (*lacunosus*); having numerous large, deep depressions, or excavations.

6. Honey-combed (*favosus, alveolatus*); excavated in the manner of a section of honeycomb; as the receptacle of many Compositæ, the seeds of Papaver.

7. †Areolate (†*areolatus*); divided into a number of irregular squares or angular spaces.
8. Scarred (*cicatrisatus*); marked by the scars left by bodies that have fallen off: a stem, for instance, is scarred by the leaves that have fallen.
9. Ringed (*annulatus*); surrounded by elevated or depressed bands; as the roots of some plants, the cupulæ of several oaks, &c.
10. Striated (*striatus*); marked by longitudinal lines; as the petals of Geranium striatum.
11. Lined (*lineatus*); the same as striatus.
12. Furrowed (*sulcatus*); marked by longitudinal channels; as the stem of Conium, of the parsnep, of Spiræa Ulmaria, &c.
13. †Aciculated (†*aciculatus*); marked with very fine irregular streaks, as if produced by the point of a needle.
14. Dotted (*punctatus*); covered by minute impressions, as if made by the point of a pin; as the seed of Anagallis arvensis, Geranium pratense.
15. Even (*æquatus*); the reverse of any thing expressive of inequality of surface.

B. *With respect to appendages or superficial processes.*

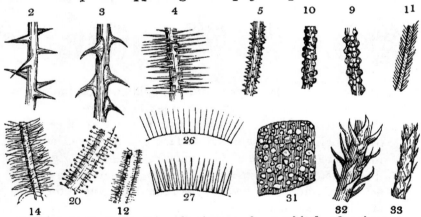

1. Unarmed (*inermis*); destitute of any kind of spines or prickles.
2. Spiny (*spinosus*); furnished with spines, as the branches of Cratægus Oxyacantha.
3. Prickly (*aculeatus*); furnished with prickles, as the stem of a rose.
4. Bristly (*echinatus*); furnished with numerous rigid hairs, or straight prickles; as the fruit of Castanea vesca.

5. Muricated (*muricatus*); furnished with numerous short, hard excrescences; as the fruit of the Arbutus Unedo.
6. Spiculate (†*spiculatus*); covered with fine fleshy, erect points.
7. Rough (*scaber, asper, exasperatus*); covered with hard, short, rigid points; as the leaves of Borago officinalis.
8. Roughish (*scabridus*); slightly covered with short, hardish points; as the leaf of Thymus Acinos.
9. Tubercled (*tuberculatus, verrucosus*); covered with little excrescences or warts; as the stem of Cotyledon tuberculata, the leaf of Aloe margaritifera.
10. Pimpled (*papillosus*, †*papulosus*); covered with minute tubercles or excrescences, of uneven size, and rather soft; as the leaves of Mesembryanthemum crystallinum.
11. Hairy (*pilosus*); covered with short, weak, thin hairs; as the leaf of Prunella vulgaris, Daucus Carota.
12. Downy (*pubens, pubescens*); covered with very short, weak, dense hairs. Pubescens is most commonly employed in Latin, but pubens is more classical; as the leaves of Cynoglossum officinale, Lonicera Xylosteum, &c.
13. Hoary (*incanus*); covered with very short, dense hairs, placed so closely as to give an appearance of whiteness to the surface from which they grow; as the leaf of Mathiola incana.
14. Shaggy (*hirtus, villosus*); covered with long, weak hairs; as Epilobium hirsutum.
15. Tomentose (*tomentosus*); covered with dense, rather rigid, short hairs, so as to be sensibly perceptible to the touch; as Onopordum Acanthium, Lavatera arborea, &c.
16. Velvety (*velutinus*); the same as the last, but more dense, so that the surface resembles that of velvet; as Cotyledon coccineus.
17. Woolly (*lanatus*); covered with long, dense, curled, and matted hairs, resembling wool; as Verbascum Thapsus, Stachys germanica.
18. Hispid (*hispidus*); covered with long, rigid hairs; as the stem of Echium vulgare.
19. Floccose (*floccosus*); covered with dense hairs, which fall away in little tufts; as Verbascum floccosum, and pulverulentum.
20. Glandular (*glandulosus*); covered with hairs bearing glands upon their tips; as the fruit of roses, the pods of Adenocarpus.

21. Bearded (*barbatus, crinitus*); having tufts of long, weak hairs growing from different parts of the surface; as the leaves of Mesembryanthemum barbatum. It is also applied to bodies bearing very long, weak hairs in solitary tufts, or parcels; as the filaments of Anthericum, the pods of Adesmia.
22. Strigose (*strigosus*); covered with sharp, appressed, rigid hairs. *W.* Linnæus considers this word synonymous with hispid.
23. Silky (*sericeus*); covered with very fine, close-pressed hairs, silky to the touch; as the leaves of Protea argentea, Alchemilla alpina, &c.
24. †Peronate (†*peronatus*); laid thickly over with a woolly substance, ending in a sort of meal. *W.* This term is only applied to the stipes of Fungi.
25. Cobwebbed (*arachnoides*); covered with loose, white, entangled, thin hairs, resembling the web of a spider; as Calceolaria arachnoidea.
26. Ciliated (*ciliatus*); having fine hairs, resembling the eyelash, at the margin; as the leaves of Luzula pilosa, Erica tetralix, &c.
27. Fringed (*fimbriatus*); having the margin bordered by long filiform processes thicker than hairs; as the petals of Cucubalus fimbriatus.
28. Feathery (*plumosus*); consisting of long hairs, which are themselves hairy; as the pappus of Leontodon taraxacum, the beard of Stipa pinnata.
29. Stinging (*urens*); covered with rigid, sharp-pointed, bristly hairs, which emit an irritating fluid when touched; as the leaves of the Urtica urens.
30. Mealy (*farinosus*); covered with a sort of white scurfy substance; as the leaves of Primula farinosa, and of some poplars.
31. Leprous (*lepidotus, leprosus*); covered with minute peltate scales; as the foliage of Elæagnus.
32. Ramentaceous (*ramentaceus*); covered with weak, shrivelled, brown, scale-like processes; as the stems of many ferns.
33. Scaly (*squamosus*); covered with minute scales, fixed by one end; as the young shoots of the pine tribe.
34. Chaffy (*paleaceus*); covered with small, weak, erect, membranous scales, resembling the paleæ of grasses; **as the** receptacle of many compound plants.

C. *With respect to polish or texture.*

1. Shining (*nitidus*); having a smooth, even, polished surface; as many leaves.
2. Smooth (*glaber*, or *lævis*); being free from asperities or hairs, or any sort of unevenness.
3. Polished (*lævigatus*, †*politus*); having the appearance of a polished substance; as the testa of Abrus precatorius, and many seeds.
4. †Glittering (†*splendens*); the same as polished, but when the lustre is a little broken, from slight irregularity of surface.
5. Naked (*nudus, denudatus*); the reverse of hairy, downy, or any similar term: it is not materially different from *glaber*.
6. Opaque (*opacus*); the reverse of shining, dull.
7. Viscid (*viscidus, glutinosus*); covered with a glutinous exudation.
8. Mucous, or slimy (*mucosus*); covered with a slimy secretion; or with a coat that is readily soluble in water, and becomes slimy; as the fruit of Salvia Verbenaca.
9. †Greasy (†*unctuosus*); having a surface which, though not actually greasy, feels so.
10. Dewy (*roridus*); covered with little transparent elevations of the parenchyma, which have the appearance of fine drops of dew.
11. †Dusty (†*lentiginosus*); covered with minute dots, as if dusted; the calyx and corolla of Ardisia lentiginosa.
12. Frosted (*pruinosus*); nearly the same as roridus, but applied to surfaces in which the dewy appearance is more opaque, as if the drops were congealed; as the surface of the leaves of Rosa pruinosa and glutinosa.
13. Powdery (*pulverulentus*); covered with a fine bloom or powdery matter; as the leaves of Primula farinosa.
14. Glaucous (*glaucus*); covered with a fine bloom of the colour of a cabbage-leaf.
15. Cæsious (*cæsius*); like glaucous, but greener.
16. Whitened (*dealbatus*); covered with a very opaque white powder; as the leaves of many Cotyledons.

4. Of Texture, or Substance.

1. Membranaceous (*membranaceus*); thin and semitransparent, like a fine membrane; as the leaves of mosses.
2. Papery (*papyraceus, chartaceus*); having the consistence of writing paper, and quite opaque; as most leaves.

3. Leathery (*coriaceus*, †*alutaceus*); having the consistence of leather; as the leaves of Pothos acaulis, Prunus lauro-cerasus, and others.
4. Crustaceous (*crustaceus*); hard, thin, and brittle; as the testa of asparagus, or of Passiflora.
5. Cartilaginous (*cartilagineus*); hard and tough; as the testa of an apple-seed.
6. Loose (*laxus*); of a soft cellular texture, as the pith of most plants. The name is derived from the parts of the substance appearing as if not in a state of cohesion.
7. Scarious (*scariosus*); having a thin, dry, shrivelled appearance; as the involucral leaves of many species of Centaurea.
8. Corky (*suberosus*); having the texture of the substance called cork; as the bark of Ulmus suberosa.
9. Coated (*corticatus*); harder externally than internally.
10. Spongy (*spongiosus*); having the texture of a sponge; that is to say, very cellular, with the cellules filled with air; as the coats of many seeds.
11. Horny (*corneus*); hard, and very close in texture, but capable of being cut without difficulty, the parts cut off not being brittle; as the albumen of many plants.
12. Oleaginous (*oleaginosus*); fleshy in substance, but filled with oil.
13. Bony (*osseus*); hard, and very close in texture, not cut without difficulty, the parts cut off being brittle; as the stone of a peach.
14. Fleshy (*carnosus*); firm, juicy, easily cut.
15. Waxy (*ceraceus, cereus*); having the texture and colour of new wax; as the pollen masses of particular kinds of Orchis.
16. Woody (*lignosus, ligneus*); having the texture of wood.
17. Thick (*crassus*); something more thick than usual. Leaves, for instance, are generally papery in texture; the leaves of cotyledons, which are much more fleshy, are called *thick*.
18. Succulent (*succulentus*); very cellular and juicy; as the stems of Stapelias.
19. Gelatinous (*gelatinosus*); having the texture and appearance of jelly; as Ulvas, and similar things.
20. Fibrous (*fibrosus*); containing a great proportion of loose woody fibre; as the rind of a cocoa-nut.
21. †Medullary or pithy (†*medullosus*); filled with spongy pith.

22. Mealy (*farinaceus*); having the texture of flour in a mass; as the albumen of wheat.
23. Tartareous (*tartareus*); having a rough, crumbling surface; like the thallus of some Lichens.
24. Berried (*baccatus*); having a juicy, succulent texture; as the calyx of Blitum.
25. Herbaceous (*herbaceus*); thin, green, and cellular; as the tissue of membranous leaves.

5. Of Size.

Most of the terms which relate to this quality are the same as those in common use; and being employed in precisely the same sense, do not need explanation. But there are a few which have a particular meaning attached to them, and are not much known in common language. These are,—

1. Dwarf (*nanus, pumilus, pygmæus*); small, short, dense, as compared with other species of the same genus, or family. Thus, Myosotis nana is not more than half an inch high; while the other species are much taller.
2. Very small (*pusillus, perpusillus*); the same as the last, except that a general reduction of size is understood, as well as dwarfishness.
3. Low (*humilis*); when the stature of a plant is not particularly small, but much smaller than of other kindred species. Thus, a tree twenty feet high may be *low*, if the other species of its genus are forty or fifty feet high.
4. Depressed (*depressus*); broad and dwarf, as if, instead of growing perpendicularly, the growth had taken place horizontally; as some species of Cochlearia, Coronopus Ruellii, and many others.
5. Little (*exiguus*); this is generally used in opposition to large, and means small in all parts, but well proportioned.
6. Tall (*elatus, procerus*); this is said of plants which are taller than their parts would have led one to expect.
7. Lofty (*exaltatus*); the same as the last, but in a greater degree.
8. Gigantic (*giganteus*); tall, but stout and well proportioned.

To this class must also be referred words or syllables expressing the proportion which one part bears to another.

1. *Isos*, or equal, placed before the name of an organ, indicates

that it is equal in number to that of some other understood: thus, *isostemonous* is said of plants the stamens of which are equal in number to the petals. *De Cand.*

2. *Anisos,* or unequal, is the reverse of the latter: thus, *anisostemonous* would be said of a plant the stamens of which are not equal in number to the petals.

3. †*Meios,* or less, prefixed to the name of an organ, indicates that it is something less than some other organ understood: thus, †*meiostemonous* would be said of a plant the stamens of which are fewer in number than the petals.

4. *Duplo, triplo,* &c. or double, triple, &c. signify that the organs to the name of which they are prefixed are twice or thrice as numerous or large as those of some other.

The terms which express measures of length are the following:—

1. A hair's breadth (*capillus,* its adjective *capillaris*); the twelfth part of a line.
2. A line (*linea,* adj. *linealis*); the twelfth part of an inch.
3. A nail (*unguis*); half an inch, or the length of the nail of the little finger.
4. An inch (*pollex, uncia;* adj. *pollicaris, uncialis*); the length of the first joint of the thumb.
5. A small span (*spithama,* adj. *spithamæus*); seven inches, or the space between the thumb and the fore-finger separated as widely as possible.
6. A palm (*palmus,* adj. *palmaris*); three inches, or the breadth of the four fingers of the hand.
7. A span (*dodrans,* adj. *dodrantalis*); nine inches, or the space between the thumb and the little finger separated as widely as possible.
8. A foot (*pes,* adj. *pedalis*); twelve inches, or the length of a tall man's foot.
9. A cubit (*cubitus,* adj. *cubitalis*); seventeen inches, or the distance between the elbow and the tip of the fingers.
10. An ell (*ulna, brachium;* adj. *ulnaris, brachialis*); twenty-four inches, or the length of the arm.
11. A toise (*orgya,* adj. *orgyalis*); six feet or the ordinary height of man.
12. *Sesqui.* This term, prefixed to the Latin name of a measure, shows that such measure exceeds its due length by one half: thus, *sesquipedalis* means a foot and a half.

13. †A millimetre = $\frac{443}{1000}$ of a line.
14. †A centimetre = 4 lines and $\frac{432}{1000}$.
15. †A decimetre = 3 inches, eight lines, $\frac{320}{1000}$.
16. †A metre = 3 feet, 11 lines, $\frac{296}{1000}$.

Obs. The four last terms are French measures, which are rarely used, and for which no equivalent Latin terms are employed.

6. Of Duration.

The terms in ordinary use to express the absolute period of duration of a plant are sufficiently precise for common purposes, but are too inaccurate to be longer admitted within the pale of science. I have, therefore, adopted the phraseology of M. De Candolle, as far as relates to words expressive of the actual term of vegetable existence.

1. *Monocarpous;* bearing fruit but once, and dying after fructification; as wheat. Some live but one year, and are called annuals: the term of the existence of others is prolonged to two years; these are biennials: others live for many years before they flower, but die immediately afterwards; as the Agave americana. The latter have no English name. Annuals are indicated by the signs ⊙ or ①; biennials by ♂ or ②; and the others by ⚹.
2. *Polycarpous* (better *sychnocarpous*); having the power of bearing fruit many times without perishing. Of this there are two forms:—
    A. *Caulocarpous,* or those whose stem endures many years, constantly bearing flowers and fruits; as trees and shrubs. The sign of these is ♄.
    B. *Rhizocarpous,* or those whose root endures many years, but whose stems perish annually; as herbaceous plants. The sign of these is ♃.
3. *Hysteranthous;* when leaves appear after flowers; as the Almond, Tussilago fragrans, &c.
4. †*Synanthous;* when flowers and leaves appear at the same time.
5. †*Proteranthous;* when the leaves appear before the flowers.
6. *Double-bearing* (*biferus*); when any thing is produced twice in one season.
7. †*Often-bearing* (†*multiferus*); when any thing is produced several times in one season

Besides the foregoing, those that follow require explanation: —

1. Of an hour (*horarius*); which endures for an hour or two only; as the flowers of Talinum, Cistus, &c.
2. Of a day (*ephemerus*, †*diurnus*); which endures but a day, as the flower of Tigridia. *Biduus* is said of things that endure two days; and *triduus*, three days.
3. Of a night (*nocturnus*); which appears during the night, and perishes before morning; as the flowers of the night-blooming Cereus.
4. Of a month (*menstrualis*, †*menstruus*); which lasts for a month. *Bimestris* is said of things that exist for two months; *trimestris*, for three months.
5. Yearly (*annotinus*); that which has the growth of a year. Thus, *rami annotini* are branches a year old.
6. Of the same year (*hornus*); is said of any thing the produce of the year. Thus, *rami horni* would be branches not a year old.
7. Deciduous (*deciduus*); finally falling off, as the calyx and corolla of Cruciferæ.
8. Caducous (*caducus*); falling off very early, as the calyx of the poppy.
9. Persistent (*persistens*, †*restans*, Linn.); not falling off, but remaining green until the part which bears it is wholly matured; as the leaves of evergreen plants, the calyx of Labiatæ, and others.
10. Withering, or fading (*marcescens*); not falling off until the part which bears it is perfected, but withering long before that time; as the flowers of Orobanche.
11. Fugacious (*fugax*); falling off, or perishing very rapidly; as many minute Fungi, the petals of Cistus, &c.
12. Permanent (*perennans*); not different from persistent: it is generally applied to leaves.
13. Perennial (*perennis*); lasting for several years.

7. Of Colour.

The most useful books to consult for the distinctions of colours are Symes's *Book of Colours,* and the chromatic scale in the Duke of Bedford's publication upon Ericas.

The best practical arrangement of colours, as applied to

plants, is that of Dr. Bischoff, in his excellent *Terminology;* what follows is chiefly taken from that work.

There are eight principal colours, under which all the others may be arranged; viz. white, grey, black, brown, yellow, green, blue, and red.

I. White (*albus;* in words compounded of Greek, *leuco-*).

1. Snow-white (*niveus*); as the purest white; Camellia japonica.
2. Pure white (*candidus;* in Greek composition, *argo-*); very pure, but not so clear as the last; Lilium candidum.
3. Ivory-white (cream colour; *eburneus, eborinus*); white verging to yellow, with a little lustre; Convallaria majalis.
4. Milk-white (*lacteus;* in words compounded of Greek, *galacto-*); dull white verging to blue.
5. Chalk-white (*cretaceus, calcareus, gypseus*); very dull white with a little touch of grey.
6. Silvery (*argenteus*); a little changing to bluish grey, with something of a metallic lustre.
7. Whitish (*albidus*); any kind of white a little soiled.
8. Turning white (*albescens*); changing to a whitish cast from some other colour.
9. Whitened (*dealbatus*), slightly covered with white upon a darker ground.

II. Grey.

10. Ash-grey (*cinereus;* in words compounded of Greek, *tephro-* and *spodo-*); a mixture of pure white and pure black, so as to form an intermediate tint.
11. Ash-greyish (*cineraceus*); the same, but whiter.
12. Pearl-grey (*griseus*); pure grey a little verging to blue.
13. Slate-grey (*schistaceus*); grey bordering on blue.
14. Lead-coloured (*plumbeus*); the same with a little metallic lustre.
15. Smoky (*fumeus, fumosus*); grey changing to brown.
16. Mouse-coloured (*murinus*); grey with a touch of red.
17. Hoary (*canus,* or *incanus*); a greyish whiteness, caused by hairs overlying a green surface.
18. Rather hoary (*canescens*); a variety of the last.

III. Black.

19. Pure black (*ater;* in Greek composition, *mela-* or *melano-*); is black, without the mixture of any other colour.

Atratus and *nigritus;* when a portion only of something is black; as the point of the glumes of Carex.
20. Black (*niger*); a little tinged with grey. A variety is *nigrescens*.
21. Coal-black (*anthracinus*); a little verging upon blue.
22. Raven-black (*coracinus, pullus*); black with a strong lustre.
23. Pitch-black (*piceus*); black changing to brown. From this can scarcely be distinguished brown-black (*memnonius*).

IV. Brown.

24. Chesnut-brown (*badius*); dull brown, a little tinged with red.
25. Brown (*fuscus;* in Greek composition, *phæo-*); brown tinged with greyish or blackish.
26. Deep-brown (*brunneus*); a pure dull brown. Umber-brown (*umbrinus*) is nearly the same.
27. Bright brown (*spadiceus*); pure and very clear brown.
28. Rusty (*ferrugineus*); light brown with a little mixture of red.
29. Cinnamon (*cinnamomeus*); bright brown mixed with yellow and red.
30. Red-brown (*porphyreus*); brown mixed with red.
31. Rufous (*rufus, rufescens*); rather redder than the last.
32. † *Glandaceus*; like the last, but yellower.
33. Liver-coloured (*hepaticus*); dull brown with a little yellow.
34. *Fuligineus*, or *fuliginosus*; dirty brown, verging upon black.
35. Lurid (*luridus*); dirty brown a little clouded.

V. Yellow.

36. Lemon-coloured (*citreus* or *citrinus*); the purest yellow without any brightness.
37. Golden yellow (*aureus, auratus;* in Greek composition, *chryso-*); pure yellow, but duller than the last, and bright.
38. Yellow (*luteus;* in Greek composition, *xantho-*); such yellow as gamboge.
39. Pale yellow (*flavus, luteolus, lutescens, flavidus, flavescens*); a pure, but paler yellow than the preceding.
40. Sulphur-coloured (*sulphureus*); a pale, lively yellow, with a mixture of white.
41. Straw-coloured (*stramineus*); dull yellow mixed with white.

42. Leather yellow (*alutaceus*); whitish yellow.
43. Ochre-colour (*ochraceus*); yellow, imperceptibly changing to brown.
44. *Ochroleucus*; the same, but whiter.
45. Waxy yellow (*cerinus*); dull yellow, with a soft mixture of reddish brown.
46. Yolk of egg (*vitellinus*); dull yellow, just turning to red.
47. Apricot-colour (*armeniacus*); yellow, with a perceptible mixture of red.
48. Orange-colour (*aurantiacus, aurantius*); the same, but redder.
49. Saffron-coloured (*croceus*); the same, but deeper, and with a dash of brown.
50. *Helvolus*; greyish yellow with a little brown.
51. Isabella-yellow (*gilvus*); dull yellow with a mixture of grey and red.
52. Testaceous (*testaceus*); brownish yellow, like that of unglazed earthenware.
53. Tawny (*fulvus*); dull yellow with a mixture of grey and brown.
54. *Cervinus*; the same, darker.
55. Livid (*lividus*); clouded with greyish, brownish, and blueish.

VI. Green.

56. Grass-green (*smaragdinus, prasinus*); clear, lively green, without any mixture.
57. Green (*viridis*; in Greek composition, *chloro-*); clear green, but less bright than the last. *Virens, virescens, viridulus, viridescens*, are shades of this.
58. Verdigris-green (*æruginosus*); deep green with a mixture of blue.
59. Sea-green (*glaucus*, †*thalassicus, glaucescens*); dull green, passing into greyish blue.
60. Deep green (*atrovirens*); green, a little verging upon black.
61. Yellowish green (*flavovirens*); much stained with yellow.
62. Olive-green (*olivaceus*; in Greek composition, *elaio-*); a mixture of green and brown.

VII. Blue.

63. Prussian blue (*cyaneus*; in Greek composition, *cyano-*); a clear, bright blue.

64. Indigo (†*indigoticus*); the deepest blue.
65. Blue (*cæruleus*); something lighter and duller than the last.
66. Sky-blue (*azureus*); a light, pure, lively blue.
67. Lavender-colour (*cæsius*); pale blue with a slight mixture of grey.
68. Violet (*violaceus, ianthinus*); pure blue stained with red, so as to be intermediate between the two colours.
69. Lilac (*lilacinus*); pale, dull violet, mixed a little with white.

VIII. Red.

70. Carmine (*kermesinus, puniceus*); the purest red without any admixture.
71. Red (*ruber*; in Greek composition, *erythro-*); the common term for any pure red. *Rubescens, rubeus, rubellus, rubicundus*, belong to this.
72. Rosy (*roseus*; in Greek composition, *rhodo-*); pale, pure red.
   Flesh-coloured (*carneus, incarnatus*); paler than the last, with a slight mixture of red.
74. Purple (*purpureus*); dull red with a slight dash of blue.
75. Sanguine (*sanguineus*); dull red passing into brownish black.
76. Phæniceous (*phæniceus, puniceus*); pure, lively red, with a mixture of carmine and scarlet.
77. Scarlet (*coccineus*); pure carmine slightly tinged with yellow.
78. Flame-coloured (*flammeus, igneus*); very lively scarlet; fiery red.
79. Bright red (*rutilans, rutilus*); reddish with a metallic lustre.
80. Cinnabar (*cinnabarinus*); scarlet with a slight mixture of orange.
81. Vermilion (*miniatus*, †*vermiculatus*); scarlet with a decided mixture of yellow.
82. Brick-colour (*lateritius*); the same, but dull and mixed with grey.
83. Brown-red (*rubiginosus, hæmatiticus*); dull red with a slight mixture of brown.
84. *Xerampelinus*; dull red with a strong mixture of brown.
85. Coppery (*cupreus*); brownish red with a metallic lustre.
86. *Githagineus*; greenish red.

8. Of Variegation, or Marking.

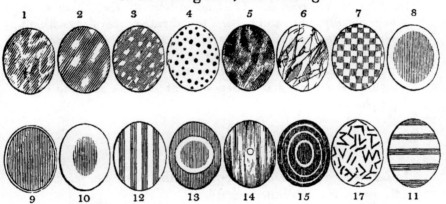

1. Variegated (*variegatus*); the colour disposed in various irregular, sinuous spaces.
2. Blotched (*maculatus*); the colour disposed in broad, irregular blotches.
3. Spotted (*guttatus*); the colour disposed in small spots.
4. Dotted (*punctatus*); the colour disposed in very small round spots.
5. Clouded (*nebulosus*); when colours are unequally blended together.
6. Marbled (*marmoratus*); when a surface is traversed by irregular veins of colour; as a block of marble often is.
7. Tessellated (*tessellatus*); when the colour is arranged in small squares, so as to have some resemblance to a tessellated pavement.
8. Bordered (*limbatus*); when one colour is surrounded by an edging of another.
9. Edged (*marginatus*); when one colour is surrounded by a very narrow rim of another.
10. Discoidal (*discoidalis*); when there is a single large spot of colour in the centre of some other.
11. Banded (*fasciatus*); when there are transverse stripes of one colour crossing another.
12. Striped (*vittatus*); when there are longitudinal stripes of one colour crossing another.
13. Ocellated (*ocellatus*); when a broad spot of some colour has another spot of a different colour within it.
14. Painted (*pictus*); when colours are disposed in streaks of unequal intensity.

15. Zoned (*zonatus*); the same as ocellated, but the concentric bands more numerous.
16. Blurred (*lituratus*). This, according to De Candolle, is occasionally, but rarely, used to indicate spots or rays which seem formed by the abrasion of the surface; but I know of no instance of such a character.
17. Lettered (*grammicus*); when the spots upon a surface assume the form and appearance of letters; as some Opegraphas.

9. Of Veining.

In terms expressive of this quality the word nerves is generally used, but very incorrectly.

1. Ribbed (*nervosus*, †*nervatus*); having several ribs; as Plantago lanceolata, &c.
2. One-ribbed (*uninervis*, †*uninervatus, costatus*); when there is only one rib; as in most leaves.
3. Three-ribbed (*trinervis*); when there are three ribs all proceeding from the base; as in Chironia Centaurium. *Quinquenervis*, when there are five; as in Gentiana lutea. *Septemnervis*, when there are seven; as in Alisma Plantago; and so on.
4. Triple-ribbed (*triplinervis*); when of three ribs the two lateral ones emerge from the middle one a little above its base; as in Melastoma multiflora. *Quintuplinervis*, &c. are used to express the obvious modifications of this.
5. †*Indirectè venosus;* when the lateral veins are combined within the margin, and emit other little veins. *Link.*
6. †*Evanescenti-venosus;* when the lateral veins disappear within the margin. *Ib.*
7. †*Combinatè venosus;* when the lateral veins unite before they reach the margin. *Ib.*
8. Straight-ribbed (†*rectinervis*, †*parallelenervis, directè venosus*, Link); when the lateral ribs are straight; as in Alnus glutinosa, Castanea vesca, &c. *Mirb.* When the ribs are straight and almost parallel, but united at the summit; as in grasses. *De Cand.*
9. †Curve-ribbed (†*curvinervis*, †*converginervis*); when the ribs describe a curve, and meet at the point; as in Plantago lanceolata.
10. †*Ruptinervis*, when a straight-ribbed leaf has its ribs interrupted at intervals. *De Cand.*

11. †*Penniformis;* when the ribs are disposed as in a pinnated leaf, but confluent at the point; as in the Date. *De Cand.*
12. †*Palmiformis;* when the ribs are arranged as in palmate leaves; as in the Chamærops. *Ib.*
13. †*Penninervis;* when the ribs are pinnated (*De Cand.*); as in Castanea vesca.
14. †*Pedatinervis;* when the ribs are pedate. *De Cand.*
15. †*Palminervis;* when they are palmated. *Ib.*
16. †*Peltinervis;* when they are peltate. *Ib.*
17. †*Vaginervis;* when the veins are arranged without any order; as in Ficoideæ. *Ib.*
18. †*Retinervis;* when the veins are reticulated, or like lace. *Ib.*
19. †*Nullinervis,* or *enervis;* when there are no ribs or veins whatever. *Ib.*
20. †*Falsinervis;* when the veins have no vascular tissue, but are formed of simple, elongated, cellular tissue; as in mosses, Fuci, &c.
21. †*Hinoideus;* when all the veins proceed from the midrib, and are parallel and undivided; as in Scitamineæ. *Link.* When they are connected by little cross veins, the term is †venuloso-hinoideus. *Ib.*
22. †*Venosus;* when the lateral veins are variously divided. *Ib.*

II. *Of Individual Relative Terms.*

These are arranged under the heads of *Estivation,* or the relation which organs bear to each other in the bud state; *Direction,* or the relation which organs bear to the surface of the earth, or to the stem of the plant which forms the axis, either real or imaginary, round which they are disposed; and *Insertion,* or the manner in which one part is inserted into, or adheres to, another.

1. Of Estivation.

The term *estivation,* or *profloration,* is applied to the parts of the flower when unexpanded; and *vernation* is expressive of the foliage in the same state. The ideas of their modifications are, however, essentially the same.

1. Involute (*involutiva, involuta*); when the edges are rolled inwards spirally on each side (*Link*); as the leaf of the apple.

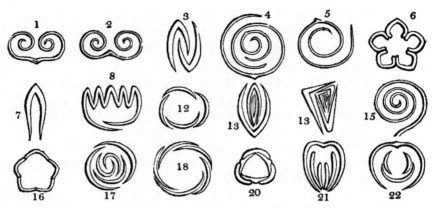

2. Revolute (*revolutiva, revoluta*); when the edges are rolled backwards spirally on each side (*Link*); as in the leaf of the rosemary.
3. Obvolute (*obvolutiva, obvoluta* (Link); *semi-amplexa*, De Cand.); when the margins of one alternately overlap those of that which is opposite to it.
4. Convolute (*convolutiva, convoluta*); when one is wholly rolled up in another, as in the petals of the wall-flower.
5. Supervolute (*supervolutiva*); when one edge is rolled inwards, and is enveloped by the opposite edge rolled in an opposite direction; as the leaves of the apricot.
6. Induplicate (*induplicativa*); having the margins bent abruptly inwards, and the external face of these edges applied to each other without any twisting; as in the flowers of some species of Clematis.
7. Conduplicate (*conduplicativa, conduplicata*); when the sides are applied parallelly to the faces of each other.
8. Plaited (*plicativa, plicata*); folded lengthwise, like the plaits of a closed fan; as the vine and many palms.
9. Replicate (*replicativa*); when the upper part is curved back and applied to the lower; as in the Aconite.
10. Curvative (*curvativa*); when the margins are slightly curved, either backwards or forwards, without any sensible twisting. *De Cand.*
11. Wrinkled (*corrugata, corrugativa*); when the parts are folded up irregularly in every direction; as the petals of the poppy.
12. Imbricated (*imbricativa, imbricata*); when they overlap each other parallelly at the margins without any involution. This is the true meaning of the term. M. De Candolle applies it in a different sense. (*Théorie*, ed. 1. p. 399.)

13. **Equitant** (*equitativa, equitans,* Link; *amplexa,* De Cand.); when they overlap each other parallelly and entirely, without any involution; as the leaves of Iris.
14. **Reclinate** (*reclinata*); when they are bent down upon their stalk.
15. **Circinate** (*circinatus*); when they are rolled spirally downwards.
16. **Valvate** (*valvata, valvaris*); applied to each other by the margins only; as the petals of Umbelliferæ, the valves of a capsule, &c.
17. **Quincunx** (*quincuncialis*); when the pieces are five in number, of which two are exterior, two interior, and the fifth covers the interior with one margin, and has its other margin covered by the exterior; as in Rosa.
18. **Twisted** (*torsiva, spiraliter contorta*); the same as contorted, except that there is no obliquity in the form or insertion of the pieces; as in the petals of Oxalis.
19. **Contorted** (*contorta*); each piece being oblique in figure, and overlapping its neighbour by one margin, its other margin being, in like manner, overlapped by that which stands next it; as Apocyneæ.
20. **Alternative** (*alternativa*); when, the pieces being in two rows, the inner is covered by the outer in such a way that each of the exterior rows overlaps half of two of the interior; as in Liliaceæ.
21. **Vexillary** (*vexillaris*); when one piece is much larger than the others, and is folded over them, they being arranged face to face; as in papilionaceous flowers.
22. **Cochlear** (*cochlearis*); when one piece, being larger than the others, and hollowed like a helmet or bowl, covers all the others; as in Aconitum, some species of personate plants, &c.

2. Of Direction.

1. **Erect** (*erectus, arrectus*); pointing towards the zenith.
2. **Straight** (*rectus*); not wavy or curved, or deviating from a straight direction in any way.
3. **Very straight** (*strictus*); the same as the last, but in excess.
4. **Swimming** (*natans*); floating under water; as Confervæ.
5. **Floating** (*fluitans*); floating upon the surface of water; as the leaves of Nuphar.

6. Submersed (*submersus, demersus*); buried beneath water.
7. Descending (*descendens*); having a direction gradually downwards.
8. Hanging down (*dependens*); having a downward direction, caused by its own weight.
9. Ascending (*ascendens, assurgens*); having a direction upwards, with an oblique base; as many seeds.

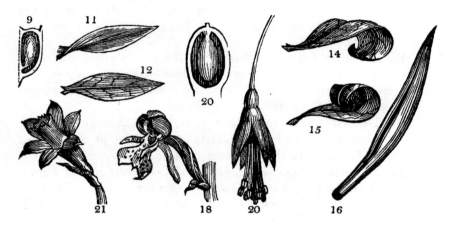

10. Perpendicular (*verticalis, perpendicularis*); being at right angles with some other body.
11. Oblique (*obliquus*); when the margin points to the heavens, the apex to the horizon; as the leaves of Protea and Fritillaria.
12. Horizontal (*horizontalis*); when the plane points to the heavens, the apex to the horizon; as most leaves.
13. Inverted (*inversus*); having the apex of one thing in an opposite direction to that of another; as many seeds.
14. Revolute (*revolutus*); rolled backwards from the direction ordinarily assumed by similar other bodies; as certain tendrils, and the ends of some leaves.
15. Involute (*involutus*); rolled inwards.
16. Convolute (*convolutus*); rolled up.
17. Reclining (*reclinatus*); falling gradually back from the perpendicular; as the branches of the banyan tree.
18. Resupinate (*resupinatus*); inverted in position by a twisting of the stalk; as the flowers of Orchis.
19. †Inclining (†*inclinatus, declinatus*); the same as reclining, but in a greater degree.
20. Pendulous (*pendulus*); hanging downwards, in consequence of the weakness of its support.

21. **Drooping** (*cernuus*); inclining a little from the perpendicular, so that the apex is directed towards the horizon.
22. **Nodding** (*nutans*); inclining very much from the perpendicular, so that the apex is directed downwards.

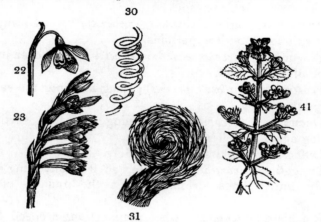

23. **One-sided** (*secundus*); having all the parts by twists in their stalks turned one way; as the flowers of Antholyza.
24. **Inflexed** (*inflexus, incurvus, introflexus, introcurvus, infractus*); suddenly bent inwards.
25. **Reflexed** (*reflexus, recurvus, retroflexus, retrocurvus, refractus*); suddenly bent backwards.
26. **Deflexed** (*deflexus, declinatus*); bent downwards.
27. **Flexuose** (*flexuosus*); having a gently bending direction, alternately inwards and outwards.
28. **Tortuous** (*tortuosus*); having an irregular, bending, and turning direction.
29. **Knee-jointed** (*geniculatus*); bent abruptly like a knee; as the stems of many grasses.
30. **Spiral** (*spiralis, anfractuosus*); resembling in direction the spires of a corkscrew, or other twisted thing.
31. **Circinate** (*circinatus, gyratus, circinalis*); bent like the head of a crosier; as the young shoots of ferns.
32. **Twining** (*volubilis*); having the property of twisting round some other body.
    a. To the right hand, or *dextrorsum;* when the twisting is from left to right, or in the direction of the sun's course; as the hop.
    b. To the left hand (*sinistrorsum*); when the twisting is from right to left, or opposite to the sun's course; as Convolvulus sepium.

414 GLOSSOLOGY. BOOK IV.

33. Turned backwards (*retrorsus*); turned in a direction opposite to that of the apex of the body to which the part turned appertains.
34. Turned inwards (*introrsus, anticus*); turned towards the axis to which it appertains.
35. Turned outwards (*extrorsus, posticus*); turned away from the axis to which it appertains.
36. Procumbent (*procumbens, humifusus*); spread over the surface of the ground.
37. Prostrate (*prostratus, pronus*); lying flat upon the earth, or any other thing.
38. Decumbent (*decumbens*); reclining upon the earth, and rising again from it at the apex.
39. Diffuse (*diffusus*); spreading widely.
40. Straggling (*divaricatus*); turning off from any thing irregularly, but at almost a right angle; as the branches of many things.
41. Brachiate (*brachiatus*); when ramifications proceed from a common axis nearly at regular right angles, alternately in opposite directions.
42. Spreading (*patens*); having a gradually outward direction; as petals from the ovarium.

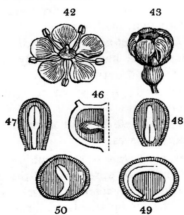

43. Converging (*connivens*); having a gradually inward direction; as many petals.
44. Opposite (*adversus,* †*oppositus*); pointing directly to a particular place; as the radicle to the hilum.
45. Uncertain (*vagus*); having no particular direction.
46. Peritropal (*peritropus*); directed from the axis to the horizon. This and the four following are only applied to the embryo of the seed.

CLASS I.   INDIVIDUAL RELATIVE TERMS.   415

47. Orthotropal (*orthotropus*;) straight, and having the same direction as the body to which it belongs.
48. Antitropal (*antitropus*); straight, and having a direction contrary to that of the body to which it belongs.
49. Amphitropal (*amphitropus*); curved round the body to which it belongs.
50. Homotropal (*homotropus*); having the same direction as the body to which it belongs, but not being straight.

3. Of Insertion.

A. *With respect to the mode of attachment, or of adhesion.*

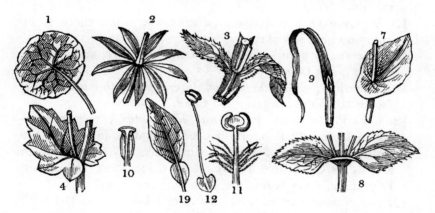

1. Peltate (*peltatus, umbilicatus*); fixed to the stalk by the centre, or by some point distinctly within the margin; as the leaf of Tropæolum.
2. Sessile (*sessilis*); sitting close upon the body that supports it without any sensible stalk.
3. Decurrent (*decurrens, decursivus*); prolonged below the point of insertion, as if running downwards.
4. Embracing (*amplectans*); clasping with the base.
5. Stem-clasping (*amplexicaulis*); the same as the last, but applied only to stems.
6. Half-stem-clasping (*semi-amplexicaulis*); the same as the last, but in a smaller degree.
7. Perfoliate (*perfoliatus*); when the two basal lobes of an amplexicaul leaf are united together, so that the stem appears to pass through the substance of the leaf.
8. Connate (*connatus*); when the bases of two opposite leaves are united together.

9. Sheathing (*vaginans*); surrounding a stem or other body by the convolute base: this chiefly occurs in the petioles of grasses.
10. Adnate (*adnatus, annexus*); adhering to the face of a thing.
11. Innate (*innatus*); adhering to the apex of a thing.
12. Versatile (*versatilis,* †*oscillatorius*); adhering slightly by the middle, so that the two halves are nearly equally balanced, and swing backwards and forwards.
13. Stipitate (*stipitatus*); elevated on a stalk which is neither a petiole nor a peduncle.
14. †Palaceous (†*palaceus*); when the foot-stalk adheres to the margin. *Willd.*
15. Separate (†*solutus, liber,* †*distinctus*); when there is no cohesion between parts.
16. Accrete (*accretus*); fastened to another body, and growing with it. *De Cand.*
17. Adhering (*adherens*); united laterally by the whole surface with another organ. *De Cand.*
18. Cohering (*coherens,* †*coadnatus, coadunatus,* †*coalitus,* †*connatus, confluens*): this term is used to express, in general, the fastening together of homogeneous parts. *De Cand.* Such are M. De Candolle's definitions of these three terms; but in practice there is no difference between them.
19. Articulated (*articulatus*); when one body is united with another by a manifest articulation.

B. *With respect to situation.*
1. Dorsal (*dorsalis*); fixed upon the back of any thing.
2. Lateral (*lateralis*); fixed near the side of any thing.
3. Marginal (*marginalis*); fixed upon the edge of any thing.
4. Basal (*basilaris*); fixed at the base of any thing.
5. Radical (*radicalis*); arising from the root.
6. Cauline (*caulinus*); arising from the stem.
7. Rameous (*rameus, ramealis*); of or belonging to the branches.
8. Axillary (*axillaris,* †*alaris*); arising out of the axilla.
9. Floral (*floralis*); of or belonging to the flower.
10. Epiphyllous (*foliaris, epiphyllus*); inserted upon the leaf.
11. Terminal (*terminalis*); proceeding from the end.
12. Of the leaf-stalk (*petiolaris*); inserted upon the petiole.
13. Crowning (*coronans*); situated on the top of any thing. Thus, the limbs of the calyx may crown the ovarium; a gland at the apex of the filament may crown the stamen; and so on.

14. Epigeous (*epigæus*); growing close upon the earth.
15. Subterranean (*hypogæus*, †*subterraneus*); growing under the earth.
16. Amphigenous (*amphigenus*).
17. Epigynous (*epigynus*); growing upon the summit of the ovarium.
18. Hypogynous (*hypogynus*); growing from below the base of the ovarium.
19. Perigynous (*perigynus*); growing upon some body that surrounds the ovarium.

Class II. Of Collective Terms.

It has already been explained, that collective terms are those which apply to plants, or their parts, considered in masses; by which is meant that they cannot be applied to any one single part or thing, without a reference to a larger number being either expressed or understood. Thus, when leaves are said to be *opposite*, that term is used with respect to several, and not to one; and when a panicle is said to be lax or loose, it means that the flowers of a panicle are loosely arranged; and so on.

1*. *Of Arrangement.*

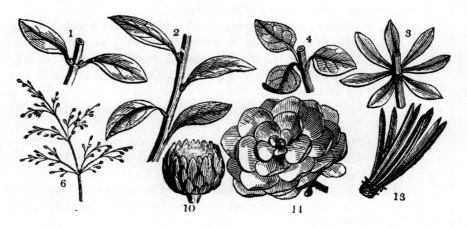

1. Opposite (*oppositus*); placed on opposite sides of some other body or thing on the same plane. Thus, when leaves are opposite, they are on opposite sides of the stem; when petals are opposite, they are on opposite sides of the ovarium; and so on.
2. Alternate (*alternus*); placed alternately one above the other on some common body, as leaves upon the stem.
3. Stellate (*stellatus, stelliformis, stellulatus*); the same as verticillate, No. 4., except that the parts are narrow and acute.
4. Whorled (*verticillatus*); when several things are in opposition round a common axis, as some leaves round their stem; sepals, petals, and stamens round the ovarium, &c.
5. Ternate (*ternus*); when three things are in opposition round a common axis.
6. Loose (*laxus*); when the parts are distant from each other, with an open, light kind of arrangement; as the panicle among the other kinds of inflorescence.
7. Scattered (*sparsus*); used in opposition to whorled, or opposite, or ternate, or other such terms.
8. Compound (*compositus*); when formed of several parts united in one common whole; as pinnated leaves, all kinds of inflorescence beyond that of the solitary flower.
9. Crowded (*confertus*); when the parts are pressed closely round about each other.
10. Imbricated (*imbricatus*); when parts lie over each other in regular order, like tiles upon the roof of a house; as the scales upon the cup of some acorns.
11. Rosulate (*rosulatus, rosularis*); when parts which are not opposite, nevertheless become apparently so by the contraction of the joints of the stem, and lie packed closely over each other, like the petals in a double rose; as in the offsets of houseleek.
12. Cæspitose (*cæspitosus*); forming dense patches, or turfs; as the young stems of many plants.
13. Fascicled (*fasciculatus*); when several similar things proceed from a common point; as the leaves of the larch, for example.
14. Distichous (*distichus, bifarius*); when things are arranged in two rows, the one opposite to the other; as the florets of many grasses.

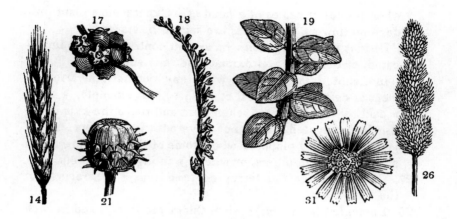

15. In rows (*serialis*); **arranged** in rows which are not necessarily opposite each other: *biserialis*, in two rows; *triserialis*, in three rows: but these are seldom used. In their stead, we generally add *fariam* to the end of a Latin numeral: thus, *bifariam* means in two rows; *trifariam*, in three rows; and so on.
16. One-sided (*unilateralis, secundus*); arranged on, or turned towards, one side only; as the flowers of Antholyza.
17. Clustered (*aggregatus, coacervatus, conglomeratus*); collected in parcels, each of which has a roundish figure; as the flowers of Cuscuta, Adoxa, Trientalis, &c.
18. Spiral (*spiralis*); arranged in a spiral manner round some common axis; as the flowers of Spiranthes.
19. Decussate (*decussatus*); arranged in pairs that alternately cross each other; as the leaves of many plants.
20. Fastigiate (*fastigiatus*); when all the parts are nearly parallel, with each pointing upwards to the sky; as the branches of Populus fastigiata, and many other trees.
21. Squarrose (*squarrosus*); when the parts spread out at right angles, or thereabouts, from a common axis; as the leaves of some mosses, the involucra of some Compositæ, &c.
22. Fasciated (*fasciatus*); when several contiguous parts grow unnaturally together into one; as the stems of some plants, the fruits of others, &c.
23. Scaly (*squamosus*); covered with small scales, like leaves.
24. Starved (*depauperatus*); when some part is less perfectly developed than is usual with plants of the same family. Thus,

when the lower scales of a head of a Cyperaceous plant produce no flowers, then scales are said to be starved.

25. Distant (*distans, remotus, rarus*); in contradiction to imbricated, or dense, or approximated, or any such words.
26. Interrupted (*interruptus*); when any symmetrical arrangement is destroyed by local causes; as, for example, a spike is said to be interrupted when here and there the axis is unusually elongated, and not covered with flowers; a leaf is interruptedly pinnated when some of the pinnæ are much smaller than the others, or wholly wanting; and so on.
27. Continuous, or uninterrupted (*continuus*); the reverse of the last.
28. Entangled (*intricatus*); when things are intermixed in such an irregular manner that they cannot be readily disentangled; as the hairs, roots, and branches of many plants.
29. Double, or twin (†*duplicatus, geminatus*); growing in pairs.
30. Rosaceous (*rosaceus*); having the same arrangement as the petals of a single rose.
31. Radiant (*radiatus*); diverging from a common centre, like rays; as the ligulate florets of any compound flower.

2*. *Of Number.*

1. None (*nullus*); absolutely wanting.
2. Numerous (*numerosus*); so many that they cannot be counted with accuracy; or several, but not of any definite number.
3. Solitary (*solitarius, unicus*); growing singly.
4. Many (in Greek compounds, *poly*); has the same meaning as numerous.
5. Few (in Greek compounds, *oligos*); means that the number is small, not indefinite. It is generally used in contrast with many (*poly*) when no specific number is employed; as in the definition of things, the number of which is definite, but variable.

Besides the above, M. De Candolle has the following Table of Numbers (*Théorie*, 502.) : —

Derived from the Latin.	Derived from the Greek.	
uni	mono	1.
bi	di	2.
tri	tri	3.
quadri	tetra	4.
quinque	penta	5.
sex	hexa	6.
septem	hepta	7.
octo	octo	8.
novem	ennea	9.
decem	deca	10.
undecim	endeca	11.
duodecim	dodeca	12, or from 11 to 19.
viginti	icos	20.
pauci	oligos	a small number.
pluri	- -	a middling number.
multi	poly	a great number.
bini, gemini	- -	2 together.
terni, ternati	- -	3 together.
quaterni, quaternati	- -	4 together.
quini, quinati	- -	5 together.
seni	- -	6 together.
septeni	- -	7 together.
octoni	- -	8 together.
noni, noveni	- -	9 together.
deni, †denarii	- -	10 together.
duodeni	- -	12 together.
viceni	- -	20 together.
simplici	- -	solitary, or simple.
duplici	- -	double.
triplici	- -	triple.
quadruplici	- -	quadruple.
quintuplici	- -	quintuple.
sextuplici	- -	sextuple.
multiplici	- -	multiple.
tripli	- -	triple, only applied to the ribs of leaves.

Class III. Of Terms of Qualification.

Terms of qualification are generally syllables prefixed to words of known signification, the value of which is altered by such addition. These syllables are often Latin prepositions.

1. *Ob*, prefixed to a word, indicates inversion : thus, *ob*ovate means inversely ovate; *ob*cordate, inversely cordate ; *ob*conical, inversely conical ; and so on. Hence it is evident that this prefix cannot be properly applied to any terms except such as indicate that one end of a body is wider than the other ; for if both ends are alike, there can be no apparent inversion : therefore when the word *ob*lanceolate is used, as by some French writers, it literally means nothing but lanceolate ; for that figure, being strictly regular, cannot be altered in figure by inversion.

2. *Sub*, prefixed to words, implies a slight modification, and may be Englished by somewhat : as, *sub*ovate means somewhat ovate ; *sub*viridis, somewhat green ; *sub*rotundus, somewhat round ; *sub*purpureus, somewhat purple ; and so on. The same effect is also given to a term by changing the termination into *ascens*, or *escens:* thus, virid*escens* signifies greenish ; rub*escens*, reddish ; and so on.

Signs.

In botany a variety of marks, or signs, are employed to express particular qualities or properties of plants. The principal writers who have invented these signs are Linnæus, Willdenow, De Candolle, Trattinnick, and Loudon : —

* *Linn., Willd., De Cand., Tratt.*, indicates that a good description will be found at the reference to which it is affixed.

† *Linn., Willd., De Cand., Tratt.*, indicates that some doubt or obscurity relates to the subject to which it is affixed.

! *De Cand.*, shows that an authentic specimen has been examined from the author to whose name or work it is annexed.

? The note of interrogation varies in its effect according to the place in which it is inserted. When found after a specific name, as *Papaver cambricum ?* it signifies that it is uncertain whether the plant so marked is that species, or some other of the genus; if after the generic name, as *Papaver? cambricum*, it shows an uncertainty whether the plant so marked belongs to the genus Papaver; when found affixed to the name of an author, as *Papaver cambricum* Linn., Smith, Lam.? it signifies that, while there is no doubt of the plant being the same as one described under that name by Linnæus and Smith, it is doubtful whether it is not different from that of Lamarck. It may be remarked, that when the interrogation has a general, and not a particular application, it should be placed at the commencement of the paragraph; as *?Papaver cambricum* Smith, &c., not *Papaver cambricum* Smith? &c. as is the usual practice.

♭ *Linn. Willd.* A tree or shrub.
⚴ *Loudon.* A deciduous tree.
⚵ *Loudon.* An evergreen tree.
♭ *Tratt.* A true tree; as the oak.
♭ *De Cand.* An under-shrub; as Laurustinus.
⚹ *Loudon.* A deciduous under-shrub.
⚺ *Loudon.* An evergreen under-shrub.
♭ *De Cand.* ⎫
☿ *Tratt.* ⎬ Caulocarpous.
5̄ *De Cand.* ⎫
♃ *Tratt.* ⎬ A shrub.
⚘ *Loudon.* A deciduous shrub.
⚛ *Loudon.* An evergreen shrub.
5 *De Cand.* A small tree.
5 *De Cand.* A tree more than twenty-five feet high.
♈ *Tratt.* ⎫
⚵ *Loudon.* ⎬ A simple-stemmed arborescent monocotyledonous tree; such as a palm.
♃ *Linn., Willd., De Cand., Tratt.* ⎫
△ *Loudon.* ⎬ A perennial.
⊙ *De Cand.* Monocarpous in general.
⊛ *De Cand.* ⎫
○ *Tratt.* ⎬ A monocarpous perennial.
⊙ *Linn., Willd., Tratt.* ⎫
① *De Cand.* ⎬ Annual.
○ *Loudon.* ⎭
♂ *Linn., Willd.* ⎫
② *De Cand.* ⎪
⊙⊙ *Tratt.* ⎬ Biennial.
☽ *Loudon.* ⎭

E E 4

♎	*Tratt.*	A plant that is propagated by new tubers, which perish as soon as they have borne a plant; as the potatoe.
♀	*Tratt.*	A plant that is propagated by suckers; as Poa pratensis.
♏	*Tratt.*	A plant that is propagated by runners; as the strawberry.
☿	*Tratt.*	A viviparous plant; or one that increases by buds which fall from it; as Lilium tigrinum.
●	*Tratt.*	A stemless plant; as Carduus acaulis.
♐	*Tratt.*	A plant whose flowers are borne upon a scape; as Hieracium Pilosella.
♓	*Tratt.*	A plant which bears its flowers and leaves upon two separate stems; as Curcuma Zedoaria. This sort of plant is called by Trattinnick *heterophytous*.
♍ *Tratt.* ♒ *Loudon.*		A Calamarious, or grassy, plant; as Bromus mollis.
— *De Cand.* ♑ *Tratt.*		A twining plant.
(	*De Cand.*	Which twines to the right.
)	*De Cand.*	Which twines to the left.
	Loudon.	A deciduous twining plant.
	Loudon.	An evergreen twining plant.
	Loudon.	A deciduous climbing plant.
	Loudon.	An evergreen climbing plant.
	Loudon.	A deciduous trailing plant.
	Loudon.	An evergreen trailing plant.
	Loudon.	A deciduous creeping plant.
	Loudon.	An evergreen creeping plant.
	Loudon.	A deciduous herbaceous plant.
	Loudon.	An evergreen herbaceous plant.
	Loudon.	A bulbous plant.
	Loudon.	A fusiform-rooted plant.
	Loudon.	A tuberous-rooted plant.
	Loudon.	An aquatic plant.
	Loudon.	A parasitical plant.
△	*De Cand., Tratt.*	An evergreen plant.
∞ *De Cand.* ⚬⚬ *Tratt.*		An indefinite number.
♂	*Willd., &c.*	The male sex.
♀	*Willd., &c.*	The female sex.
☿	*Willd., &c.*	The hermaphrodite sex.

♄	*Willd.*	The neuter sex.
♂—♀	*Willd.* }	Monœcious; or the male and female on one plant.
♒	*Tratt.* }	
♂:♀	*Willd.* }	Diœcious; or the male and female in different plants.
♋	*Tratt.* }	
☿|♀	*Willd.*	Hermaphrodite and female in one compound flower.
☿|♄	*Willd.*	Hermaphrodite and neuter in one compound flower.
☿—♂	*Willd.*	Hermaphrodite and male on one stem.
☿—♀	*Willd.*	Hermaphrodite and female on one stem.

Abbreviations.

These are only known in the botanical works which are written in Latin: they are of little importance, and, as will be seen by the mark † prefixed, are scarcely ever used. The following list is chiefly taken from Trattinnick. (*Synodus*, i. 16.) : —

† Æst.	*Æstate.*
Alb.	*Albumen.*
† Alp.	*Alpes, Alpinus.*
Anth.	*Anthera, Anthodium, Antheris.*
Apr.	*Aprilis, Apricus.*
† Arv.	*Arva, Arvensis.*
† Ar.	*Arena, Arenosus.*
† Art.	*Artificialis*
† Aug.	*Augustus.*
† Augm.	*Augmentum.*
Aut.	*Autumnus, Autumnalis.*
B.	*Beatus* or *Defunctus;* used in speaking of a person who is recently deceased, and is equivalent to our English word "late."
Br.	*Bractea*
Cal.	*Calyx.*
Cald.	*Caldarium.*
† Camp.	*Campus, Campestris.*
† Carpell.	*Carpellum.*
† Carpid.	*Carpidium.*
† Carpol.	*Carpologia.*
Cel.	*Celeberrimus.*
Char.	*Character, Characteristicus.*

Cl.	*Clarissimus, Classis.*
† Coll.	*Collis, Collinus, Collectanea.*
Cor.	*Corolla, Corollarium.*
† Cot.	*Cotyledon.*
Cult.	*Cultus, Cultura.*
Dec.	*December, Decas, Decandria.*
Descr.	*Descriptio.*
† Des.	*Desideratur.*
Diff.	*Differentia.*
† Diss.	*Dissepimentum, Dissertatio.*
† Dum.	*Dumetum.*
Ed.	*Editio, Editor, Edulis.*
† Excl.	*Exclusio.*
Embr.	*Embryo.*
Ess.	*Essentialis.*
Fam.	*Familia.*
Feb.	*Februarius.*
Fil.	*Filamentum.*
Fl.	*Flos, Flumen, Floret, Floralis.*
Fol.	*Folium.*
Fr.	*Fructus.*
Fructif.	*Fructificatio.*
† Fun.	*Funiculus umbilicalis.*
Gen.	*Genus, Genericus.*
Germ.	*Germen.*
† Glar.	*Glareosus.*
H.	*Herbarium, Habitat.*
Hab.	*Habitat, Habeo.*
Herb.	*Herbarium, Herba.*
† Hexap.	*Hexapodium.*
Hort.	*Hortus.*
† Hortul.	*Hortulanus, Hortulanarium, Hortulus.*
† Hosp.	*Hospes, Hospitatur.*
† Hum.	*Humidus, Humus.*
Ic.	*Icon,* b. *bona,* m. *mala.* p. *picta.* l. *lignea,* n. *nigra.*
Ill.	*Illustratio, Illustris.*
Infl.	*Inflorescentia.*
† Ind.	*Ineditus, Inedulis.*
Ind.	*Indicus, India,* a. *australis, or. orientalis, occ. occidentalis, Index.*
Inf.	*Inferus.*
† Inund.	*Inundatus.*

Jan.	*Januarius.*
Jul.	*Julius.*
Jun.	*Junius, Junior.*
† Juv.	*Juvenis, Juventus.*
Lat.	*Latus, Latitudo, Lateralis.*
† Lin.	*Linea, Linearis.*
Lit.	*Litera.*
† Litt.	*Littus, Litteralis.*
Long.	*Longus, Longitudo.*
L. c.	*Loco citato.*
† Loc.	*Loculamentum, Locusta.*
† Maj.	*Majus.*
† Mar.	*Mare, Marinus.*
† Mat.	*Matutinus, Maturus.*
Mart.	*Martius.*
Mont.	*Montes, Montanus.*
Mss.	*Manuscriptum.*
† Mus.	*Museum.*
† N.	*Numerus.*
Nat.	*Naturalis.*
† Nem.	*Nemus, Nemorosus.*
No.	*Numero.*
Nom.	*Nomen, gen.* genericum, *triv.* triviale, *s.* specificum, *barb.* barbarum, *leg.* legale, *syn.* synonymum.
Oct.	*Octobris.*
Obs.	*Observatio, Observandum.*
† Or.	*Origo, Originarium, Oriens, Orientale.*
Ord.	*Ordo, Ordinarium.*
Ov.	*Ovarium.*
P.	*Pagina, Pars.*
† Pal.	*Paludes, Paludosus.*
Ped.	*Pedunculus.*
Peric.	*Pericarpium.*
Perig.	*Perigonium.*
Pet.	*Petalum, Petiolus.*
† Phyll.	*Phyllum, Phyllodium.*
Pist.	*Pistillum.*
† Pom.	*Pomeridianum, Pomum.*
Plac.	*Placenta.*
Poll.	*Pollen, Pollicaris.*
† Pr. v.	*Primo vere.*
Rad.	*Radix, Radius, Radiatus.*
Ram.	*Ramus, Rameus, Ramosus.*

† Rec.	*Receptaculum, Recapitulatio.*
S.	*Seu, Sive.*
Sep.	*Sepalum, Sepes.*
† Salt.	*Saltus, Saltuarium.*
Sect.	*Sectio v. Divisio.*
† Segm.	*Segmentum.*
Sept.	*September, Septum.*
† Ser.	*Series.*
Sem.	*Semen, Semis.*
Sicc.	*Siccum.*
Stam.	*Stamen, Stamineum*
Stigm.	*Stigma.*
Stip.	*Stipes, Stipula, Stipularis.*
Sp.	*Species, Specificus.*
Spont.	*Spontaneus.*
† Spor.	*Sporula.*
† Sporang.	*Sporangium.*
Styl.	*Stylus.*
Subd.	*Subdivisio.*
Subv.	*Subvarietas.*
Sup.	*Superus.*
† Sylv.	*Sylvestris, Sylva.*
Syn.	*Synonymum, Synopsis, Synodus.*
T.	*Tabula, Tomus.*
† Tep.	*Tepidarium.*
† Temp.	*Tempestas, Temperatura.*
Trib.	*Tribus, Divisio.*
† Triv.	*Trivialis.*
† Turf.	*Turfosus.*
V.	*Volumen, Vide, Vel, Vulgo.*
Var.	*Varietas.*
† Vern.	*Vernalis, Vernaculum.*
† Vert.	*Vertex, Verticalis.*
V. s. c.	*Vidi siccam cultam.*
V. s. s.	*Vidi siccam spontaneam.*
V. v. c.	*Vidi vivam cultam.*
V. v. s.	*Vidi vivam spontaneam.*
Veg.	*Vegetabile, Vegetatio.*
† Vir.	*Viridarium, Vires, Viridis.*
† Vesc.	*Vesca, Vescarium.*
† Visc.	*Viscosus, Viscositas.*
† Volv.	*Volva, Volvaceus.*

The following excellent Table of Abbreviations was contrived by the late Mr. Ferdinand Bauer, to express all the subjects for which illustrations are required in botanical drawings. It is much to be regretted that these abbreviations, which are in every way unexceptionable, are not universally adopted for references to plates: they would not only form a common means of comparison between the figures of different authors, but would also keep continually within the view of artists the nature of the subjects they are employed to analyse. It may be added that the Table, if considered without reference to the abbreviations, is in itself an excellent sketch of the principal modes, degrees, and analogies of the regular morphosis, or developement, of fructification. When the letters used are capitals, they indicate that the object is magnified; when small, that it is of the natural size; when with a score (—) drawn beneath them, that it is less than the natural size: —

a.   A flower before expansion.
a 1. A flower expanded.
b.   The operculum of a flower; generally formed by the confluence of the calyx and corolla.
c.   The perianthium; the floral integument of monocotyledonous plants, and the generally simple one of dicotyledones. (Corolla of Linnæus; calyx of Jussieu.)
c 1. External leaflets of the perianthium; having generally the nature of a calyx. (Calyx of Linnæus.)
c 2. Internal leaflets of the perianthium, except c 3. and c 4.; having usually the texture of petals. (Corolla of Linnæus.)
c 3. The labellum, or its appendages. In Orchideæ.
c 4. The hypogynous scales of grasses. (Nectarium of Linnæus.)
c 4. Appendages of the perianthium.
d.   The calyx.
e.   A monopetalous corolla.
e 1. Petals.
e 2. Appendages of the corolla. (Nectarium of Linnæus; parapetala of Ehrhart.)
f.   The discus, whether hypogynous or epigynous.
f 1. Scales or glands, whether hypogynous or epigynous.

g.	Sexual organs combined in a column; in Orchideæ and Stylideæ.
g 1.	Sexual organs separate; the floral envelopes being removed.
h.	The stamens.
h 1.	An anther.
h 2.	Pollen.
h 3.	Pollen masses; in Orchideæ and Asclepiadeæ.
h 4.	Sterile stamens.
h 5.	The corona of a tube of stamens; in Asclepiadeæ. (Nectarium of Linnæus.)
i.	The pistil.
i 1.	The ovarium.
i 2.	The stigma.
i 3.	The indusium of the stigma; in Goodenoviæ and Brunoniaceæ.
i 4.	An ovulum.
l.	A compound fruit; common to several flowers.
l 1.	Several distinct pericarpia; belonging to a single flower.
m.	Induviæ; the remains of the flower, which either increase the fruit in size, or surmount it, or are adherent to it.
m 1.	Pappus.
m 2.	The calyptra of mosses.
n.	The pericarpium; comprehending all its species, from the simple caryopsis of grasses.
n 1.	Pericarpium open.
n 2.	A dissepiment.
n 3.	Valves.
n 4.	An operculum.
n 5.	The peristomum of mosses.
n 6.	The placenta. (Receptacle of the seeds of Gærtner.)
n 7.	Funiculus umbilicalis.
n 8.	The strophiola, or Caruncula umbilicalis.
n 9.	Arillus.
o.	The seed.
o 1.	Wing of the seed.
o 2.	Coma of the seed; in Asclepiadeæ and Epilobium.
o 3.	Integument of the seed.
o 4.	Albumen. (Perisperm of Jussieu; Endosperm of Richard.)
o 5.	Vitellus; in Scitamineæ and Nymphæa.

p.	The embryo.
p 1.	Cotyledon.
p 2.	Plumula.
p 3.	Radicle.
q.	A leaf.
q 1.	The petiole.
q 2.	A stipula.
r.	Portion of the stem or scape.
s.	Inflorescence; comprehending all the species except the two following, s 1. and 2.
s 1.	A compound flower.
s 2.	The locusta of a grass (either one-flowered or many-flowered).
t.	The involucrum of an umbel, or a head.
t 1.	The involucrum of a compound flower. (Calyx communis of Linnæus.)
t 2.	Glume of grasses. (Calyx of Linnæus.)
t 3.	Outer calyx of Malvaceæ, Dipsaceæ, Brunoniæ.
t 4.	Involucrum of ferns. (Indusium of Swartz.)
t 5.	Bracteæ.
t 6.	Scales of a catkin.
t 7.	Paleæ.
t 8.	The paraphyses of mosses.
t 9.	The calyptra, when formed of connate bracteæ.
u.	Receptacle of a single flower.
u 1.	Common receptacle either of a compound flower, a catkin, or a head.
*	Placed under one of the above (thus, $*^4$), shows that a part is expanded, or opened, by force.
†	Indicates a vertical section (used thus, $†^4$).
∴	Indicates a transverse section (used thus, $∴^4$).

BOOK V.

PHYTOGRAPHY; OR, OF THE RULES TO BE OBSERVED IN DESCRIBING AND NAMING PLANTS.

WE now proceed to investigate the principles upon which plants are described and named. It would be impossible for any person to recognise a plant discovered by another, unless such a description of it were put upon record as should express all its essential features; and unless it were, at the same time, furnished with a distinctive name, it could never be subsequently spoken of intelligibly. For these reasons, the mode of describing and naming plants is one of the most important practical subjects in the science.

It may appear, at first sight, extremely easy to describe a plant, and we constantly find travellers and others attempting to do so in vulgar language; but their accounts are always so vague, that no distinct idea can be formed of the subject of their descriptions, which remains an enigma until some botanist, following their steps, shall happen to be able to put its characters into scientific language.

The great object of descriptions in Natural History is to enable any person to recognise a known species after its station has been discovered in a classification; and also to put those who have not had the opportunity of examining a plant themselves into possession of all the facts necessary to acquire a just notion of its structure and affinities.

There are two means of effecting this object; the one by means of detailed descriptions, the other by the aid of briefer abstracts of the most essential characters only.

CHAPTER I.

OF DIAGNOSES; OR, OF GENERIC AND SPECIFIC CHARACTERS.

We have seen that plants are distinguished from each other by their characters: of the application of these characters we must now speak. Were each species to be characterised independently of other species, and to be described with all the minute circumstances of structure that belong to it, the progress of investigation would be too slow, and the length of time requisite to acquire information much too great:—for this reason, the process of enquiry has been simplified, by collecting in groups all those species which have certain great characters in common, and abstracting those characters, which then become the distinctions of classes: the species of a class are again collected into other groups, agreeing in some other common peculiarities which are, in like manner, abstracted, and form the characters of orders. Thus reduced in extent, the species of each order are submitted to the same process of combination; the characters by which they are combined become distinctive of genera; and the species are, finally, left shorn of the greatest part of their characters, which are thus reduced within a very narrow compass. Each plant has, therefore, four characters; or, if sub-classes, sub-orders, or other modes of division are adopted, as many separate characters as there may be divisions.

These characters are of two sorts; the one called essential, the other differential. The former are the most commonly employed for orders and genera; the latter are chiefly used in discriminating species: the former are the most valuable, and will probably, in time, supersede the others, which convey little information, and are only useful in aiding us in our analysis of large bodies of species; the latter are often called definitions; but, as no definite limits can be traced between living things, a strict definition in natural history becomes impracticable, for which reason the term differential must be admitted instead.

Differential characters express, in the least possible space, the distinctions between plants: they should contain nothing superfluous, nor any thing which can be considered implied by the contrasted characters of those with which they are to be compared. By this means the distinctions of species are brought into the least possible compass, and the analysis of heir characters becomes so effectual, that a botanist is ex-tpected to be able, without difficulty, to determine the exact station and name of any one of the 100,000 species supposed to exist. Nothing can sound better than this; but, unfortunately, the advantages of differential characters are not quite so great as would appear. In sacrificing every thing to brevity, it is found in practice that doubts and ambiguities are continually created; and for this especial reason, among others, that differential characters must necessarily be framed upon a consideration of what we know, and not with reference to what we do not know: on this account, a differential character, constructed in the most unexceptionable manner by one botanist, may be unintelligible to another, who possesses more knowledge, or a greater number of species. For example, when Linnæus framed the differential character of Rosa indica, " germinibus ovatis pedunculisque glabris, caule subinermi, petiolis aculeatis," it probably distinguished that species from all others that he knew: but our acquaintance with roses is so much more extensive than that of Linnæus, that we have many roses to which his character is equally applicable. A differential character, moreover, conveys no information beyond that of the differences between one thing and another, and can be viewed in no other light than as a convenient method of analysis. For this reason, the *essential* character is more generally adopted at the present day, either to the exclusion of the differential character, or in union with it.

The *essential* character of a plant expresses, as its name implies, those peculiarities which are known by experience to be most essential to it; but admits nothing unimportant or superfluous, or that is common to all the species of the same genus, or to all the genera of the same order, or to all the orders of the same class. It may be said to comprehend the chief differences and resemblances of bodies. In drawing up essential characters much discretion requires to be exercised: they may be over short, or over long; characters of importance may be omitted, and others of no importance introduced. Hence no better evidence need be desired of the merit of a botanist than his essential characters,—from which a practised eye will readily detect both how much the author knows, and what he does not know. As models of the manner in which these should be drawn up, no book can be consulted with more advantage than the *Genera Plantarum* of Jussieu, in which classical elegance of language, and as much rigid botanical precision as was supposed necessary at the time the work was written, are combined in a manner that has seldom been surpassed. The defects of that work were inseparable from the state of botany at the time it appeared; the characters of the genera and orders not embracing all those points of structure which are now known to be essential.

The following character, assigned by Mr. Brown to the order PROTEACEÆ (*Prodr. Fl. N. Holl.* p. 363.), may be taken as a specimen of the manner in which an essential character of the briefest kind ought to be constructed: —

"*Perianthium* tetraphyllum v. quadrifidum, æstivatione valvata. *Stamina* quatuor (altero nunc sterili), foliolis perianthii opposita. *Ovarium* unicum, liberum. *Stylus* simplex. *Stigma* subindivisum. *Semen* (pericarpii varii) exalbuminosum. *Embryo* dicotyledoneus (quandoque polycotyledoneus), rectus. *Radicula* infera."

In this character enough is expressed to distinguish the order from all others; and, at the same time, by a careful suppression of all superfluous terms, it is reduced within exceedingly narrow limits. Such a character as this leaves nothing to be desired, when the essence only of a mass of characters is the object in view.

The following, from the same author, is a specimen of an essential character of ACANTHACEÆ, of a more extended kind :—

"*Calyx* 5—4— divisus, partitus v. tubulosus, æqualis v. inæqualis; rarò multifidus v. integer et obsoletus: persistens. *Corolla* monopetala, hypogyna, staminifera, plerumque irregularis; limbo ringente v. bilabiato, rarò unilabiato; nunc subæqualis; decidua. *Stamina* sæpius duo, antherifera, modo 4 didynama, brevioribus quandoque effœtis. *Antheræ* v. biloculares, loculis insertione inæqualibus æqualibusve; v. uniloculares; longitudinaliter dehiscentes. *Ovarium* disco glanduloso basi cinctum, biloculare loculis 2 polyspermis. *Stylus* 1. *Stigma* bilobum, rarò indivisum. *Capsula* bilocularis, loculis 2 polyspermis, abortione quandoque monospermis, elasticè bivalvis. *Dissepimentum* contrarium, per axin (medio quandoque apertam) bipartibile, segmentis valvulis adnatis modò ab iisdem elasticè dissilientibus, integris v. rarò spontè bipartibilibus; margine interiore seminiferis. *Semina* processubus subulatis adscendentibus dissepimenti plerumque subtensa, subrotunda: *Testa* laxa. *Albumen* nullum. *Embryo* curvatus v. rectus; *Cotyledones* magnæ, suborbiculatæ: *Radicula* teres, descendens, et simul centripeta, curvata v. recta; *Plumula* inconspicua ———. *Herbæ* v. *Frutices*, intra tropicos præcipuè provenientes; pube, dum adsit, simplici, nunc capitatâ, rarissimè stellatâ. *Folia* opposita, rarò quaterna, exstipulata, simplicia, indivisa, integra v. serrata; rarò sinuata v. sublobata. *Inflorescentia* terminalis v. axillaris, spicata, racemosa, fasciculata, paniculata v. solitaria. *Flores* in spicis sæpius oppositi, nunc alterni, tribracteati, bracteis lateralibus rarò deficientibus, quandoque magnis foliaceis calycem nanum, interdum obsoletum, includentibus."

In this instance a much greater number of particulars is introduced than in the former; but still it comprehends nothing like all the characters that would be included in a general description.

The following is also a specimen of a generic and specific character, from the same author. It shows the plan upon which the essential characters of genera should be constructed:—

"VERONICA *L.*, *Juss.*
"Hebe *Juss.*

"*Calyx* 4-partitus, rarò 5-partitus. *Corolla* subrotata;

Tubus calyce brevior; *Limbus* 4-partitus, inæqualis, lobis indivisis. *Stamina* 2, antherifera, sterilia nulla. *Capsula* valvis medio septiferis v. bipartibilis ———. Herbæ v. Frutices. Folia *opposita, quandoque verticillata, v. alterna, sæpe dentata v. incisa.* Inflorescentia *varia.* Calyces *ebracteati.*

" § 1. *Capsula bipartibilis.*

" 1. V. *formosa,* fruticosa, foliis perennantibus decussatis lanceolatis integerrimis glaberrimis basi acutis, ramis bifariam pilosiusculis, corymbis axillaribus paucifloris." (*Prodr.* 434.)

In these characters it is difficult to say which is most to be admired, — the skill with which every thing superfluous is retrenched, or the ingenuity with which every thing essential is introduced. Nothing that is general to the order is introduced into the generic character; and nothing that the generic character comprehends is discoverable in the specific. By making the peculiarity of Capsula bipartibilis the distinction of a section, the necessity of introducing that circumstance into the specific characters of any of the species comprehended in the section is avoided.

Compare with this the following generic and specific characters taken from Labillardière's Sertum Austro-Caledonicum: —

" Microsemma; a genus of Ternströmiaceæ (?). *Calyx* 5-phyllus, rarè 6-phyllus, persistens, foliolis tribus interioribus. *Coronula* petaloidea, petalis 10—12 distinctis. *Stamina* numerosa (30 circiter), hypogina, filamentis inter se basi subconnatis, *antheris* bilocularibus reniformibus. *Germen* globulosum, superum, *stylo* simplici, *stigmate* 5—6-fido. *Capsula* ovata, 10—12—locularis, valvis medio septiferis, 10—12— valvis. *Semina* solitaria, in summo valvularum intùs affixa, *perispermo* carnoso, *radiculâ* superâ." (p. 58.)

Upon this character it may be observed, that the calyx is described awkwardly, and at a greater expense of words than is necessary: if he had said, calyx 5–6-phyllus imbricatus, the same idea would have been expressed: rarè should be rarò. In the next place, "coronula petaloidea" is a bad term, conveying no precise notion of the organ it is intended to designate. What is a coronula? If it is a row of petals, why call it otherwise? And it appears to be so, because it is immediately afterwards described as consisting of 10—12 distinct petals. In the next sentence, hypogina is misspelt; and the anthers are said to be bilocular and reniform, a character by no means essential; while their being covered with glandular dots, and the mode of their attachment to the filament, both of which should have been introduced, are omitted. Again, the germen, meaning the ovarium, is said to be globulose: what is globulose? Is it bullet-shaped, or round and small? If the former, the term is inapplicable; if the latter, the meaning is not expressed: it probably was intended for "subglobose." The capsule is said to be ovate, a quality of no consequence if it existed; but not true, inasmuch as it appears from the figure to be round. The construction of what follows is what we call in English putting the cart before the horse: instead of " valvis medio septiferis 10—12 valvis," it should have been, 10—12 valvis, valvis medio septiferis;" and all that is said about the attachment of the seeds might have been better expressed by two words, " semina pendula." It is said that they are attached to the top of the valves, in the inside: did any one ever hear of seeds being attached to the outside? Let the character be properly cut down, and see what remains of it.

MICROSEMMA.

"*Sepala* 5—6, imbricata, persistentia. *Petala* 10—12. *Stamina* numerosa, hypogyna, submonadelpha: *antheris* bilocularibus. *Ovarium* superum; *stylus* simplex; *stigmata* 5—6. *Capsula* 10-12-locularis, valvis totidem loculicidis; *semina* solitaria pendula; *albumen* carnosum; *radicula* supera."

But it is not in inaccuracy of language alone, or in the misplacing the members of a sentence, that an essential character may be defective: it may be expressed with a good selection of terms, and a due attention to arrangement; but the terms may be wrongly applied, or important characters may be omitted, or the author may not understand the structure of what he is describing. Take, as an instance, the following character of Carex, by the late Sir James Smith.

"*Barren flowers* numerous, aggregate, in one, or more, oblong, dense *catkins;* their *scales* imbricated every way. *Calyx* a single, lanceolate, undivided, permanent *scale* to each floret. *Corolla* none. *Filaments* 3, rarely fewer, capillary, erect, or drooping, longer than the scales. *Anthers* vertical, long, linear, of 2 cells.

"*Fertile flowers* numerous, in the same, or more usually in a different *catkin*, very rarely on a separate plant. *Calyx* as in the barren flower. *Corolla* a single, hollow, compressed, ribbed, often angular, permanent *glume* to each floret; contracted, mostly cloven, and often elongated at the extremity. *Germen* superior, roundish; with three, rarely but two, angles, very smooth. *Style* one, terminal, cylindrical, short. *Stigmas* three, more rarely two only, awl-shaped, long, tapering, downy, deciduous. *Seed* the shape of the germen, with unequal angles, loosely coated with the enlarged, either hardened or membranous, permanent corolla, both together constituting the fruit."

This character is carefully written, but full of inaccurate and confused applications of terms. The term catkin should be spike; for a catkin is deciduous, a spike persistent: and the inflorescence in Carex is of the latter kind. In the next place, what is called the calyx is a bractea. What is called the corolla of the fertile flowers is two confluent bracteæ; and, therefore, not a single glume, but a double one. Finally, what is called the seed is the pericarpium: in the young state it is called the germen, which is equivalent to ovarium; but, by the time the ovarium is ripe, it is metamorphosed into a seed.

Inaccuracies of this kind not only disfigure botanical writings, but very often lead the inexperienced botanist into errors and misconceptions.

In constructing essential and differential characters, it is customary to use the nominative case for genera and orders, and the ablative for species; but in English the nominative only is employed in both cases.

CHAPTER II.

OF DESCRIPTIONS.

We have now seen that the principal characters of a plant can be comprehended in the essential and differential characters. But as these contain only such peculiarities as are supposed to be most essential, a great number of circumstances are omitted from them which, in the view of the botanist drawing them up, may appear unessential, but which to another may seem of the first importance. On this account, a plant cannot be considered completely known until a full description of every part shall have been obtained. In this description every circumstance connected with the external or internal organisation should be included, and a full statement made of all the peculiarities of every part, however obscure or difficult to observe. It is upon descriptions of this kind that systematic botany is based: essential and differential characters are only relative to the degree of knowledge of the person who prepares them: a description is independent of all relative knowledge; it exhibits a plant as it actually is, without reference to its resemblances or differences. The former are adapted to the state of knowledge of a particular era; the latter, if complete, to that of all eras.

Notwithstanding their importance, descriptions of this kind are very rare: they occupy too much space in books to be inserted conveniently; they are difficult to draw up; and it seldom happens that an observer has the means of describing every part of a plant: the root, or the fruit, or the flower, or some other part, is probably not to be procured, and this renders a description, even in the best hands, necessarily imperfect.

In drawing up a description, care must be taken that every term is used in its strict sense; that all is perspicuous and free from ambiguity; and that the different parts are described in their just order, beginning with the root, and ending with the fruit. The following is the form in which a perfect description would be prepared: it shows the order in which the different parts are spoken of, and the points of structure to which it is desirable to advert. The student will do well to consult it carefully: he should take common plants, the descriptions of which he can find in books, and, for the sake of exercise, describe them himself according to this form; comparing them afterwards with the printed descriptions of botanists. A number of the points which I think it necessary to describe are usually overlooked by others, as unimportant, or as too difficult to ascertain: these I have marked with an asterisk; so that those points which are commonly adverted to may be distinguished from those that are usually omitted:—

Root. Its figure, quality, substance, duration, and *anatomical internal analysis.

Stem. Its figure, direction, duration, articulation, ramification, size, surface, and *internal analysis.

Leaves. Their *vernation, *internal structure, figure, articulation, insertion, margin, surface, venation, direction, colour, texture, and size.

Petiole. Its form, surface, and the proportion it bears to the leaf.

Stipulæ. Their position, texture, surface, insertion, duration, figure, and proportion to the petiole.

Inflorescence. Its nature, order of developement, ramification, position, and proportion to the leaves.

Bracteæ. Their numbers, figure, station, proportion to the adjacent parts, surface, texture, *venation.

Flowers. Their order, and time of expansion.

Calyx. Its structure, figure, station with respect to the ovarium and the axis of inflorescence, surface, æstivation, odour, size, proportion to the corolla, colour, and venation.

Corolla. Its structure, figure, station with respect to the ovarium and axis of inflorescence and adjacent parts, surface, æstivation, size, colour, proportion to the calyx and stamens, and venation.

Stamens. Their number, direction, æstivation, station with respect to the petals, insertion, proportion to the ovarium and corolla; whether separate, or combined in several parcels; whether in one series or several, of equal or unequal length. *Filaments*, their form, length, and surface. *Anthers*, their mode of insertion on the filament, dehiscence with respect to the axis, whether inwards or outwards, and, with respect to themselves, whether transversely or longitudinally, by pores, or otherwise; their form; *structure of the endothecium; surface, colour, size; the proportion they bear to the size of the filaments, the number of their valves, the nature of the connectivum.

Pollen. Its colour, form, size, surface; whether distinct or cohering; and *mode of bursting.

Disk. If present, its size, figure, texture, and station.

Ovarium. Its apparent, as well as theoretical, structure; the position of its carpella with respect to the organs around it; its surface; mode of division; number of ribs, if any; veins; cells. *Ovula*, their number; insertion upon the placenta; position with respect to the axis of the ovarium; *the situation of their foramen. *Styles*, their number, length, figure, surface, direction, and proportion. *Stigmata*, their number, form, and surface.

Fruit. Its texture, form; whether naked, or covered by the remains of the floral envelopes; whether sessile or stipitate; mode of dehiscence, if any; number of its valves and cells; situation of the placentæ; nature of its axis; number of its seeds.

Seed. Its position with respect to the axis of the fruit, mode of insertion, form, surface; the texture and nature of the testa, arillus, and other appendages, if any; *position of the raphe and chalaza. *Albumen*, its texture, if any. *Embryo*, its direction, position with respect to the axis of the fruit, to the hilum of the seed, and to the albumen; the proportion it bears to the mass of the latter; the form of its cotyledons and radicle; *its mode of germination.

The *medical* and *economical* qualities.

Its *distribution* on the surface of the earth.

The points in which it *agrees* or *disagrees* with other species.

Descriptions, it must be observed, are of two kinds, collective and specific: the former explaining minutely the characters common to several species, as in an order or a genus; the latter, the character of one species only. The difference between these two, and the manner of applying each of them, will be best understood by the following examples.

The following mode of fully describing an order is taken from M. Adolphe Brongniart's excellent memoir on RHAMNEÆ:—

"*Calyx* monophyllus 4–5-fidus, externè sæpiùs villosus. Tubus expansus subplanus, hemisphæricus, urceolatus, campanulatus vel subcylindricus, liber, vel inferiùs ovario adnatus, vel cum eo omninò cohærens; interiùs nudus, vel in pluribus, disco carnoso aut fauci limitato, aut in laciniis effuso, tectus. Laciniæ ovatæ, triangulares, rariùs subulatæ, acutæ, interiùs subcarnosæ, in pluribus in medio lineâ carnosâ prominente notatæ, et apice callosæ; in præfloratione valvatìm applicatæ.

"*Petala* cum calycis laciniis alternantia, ejusque fauci inserta, sæpiùs sub margine disci affixa, unguiculata, ungue plus minùsve longo. Laminæ rariùs patentes, planæ, superiùs integræ vel emarginatæ, in plerisque concavæ, convolutæ vel cucullatæ, stamina vel eorum filamenta involventes, in pluribus nullæ. Præfloratio complicata.

"*Stamina* petalis opposita. *Filamenta* calycis fauci vel margini disci inserta, et cum unguibus petalorum basi sæpiùs cohærentia, laciniis calycis breviora. *Antheræ* in petalis cucullatis reconditæ, vel è petalis convolutis exsertæ, parte mediâ vel inferiori dorsi ad apicem filamenti affixæ, versatiles, introrsæ (rarissimè extrorsæ); vel ovatæ, biloculares, loculis parallelis, aut basi divergentibus, rimâ longitudinali dehiscentibus; vel reniformes, uniloculares (loculis superiùs confluentibus), rimâ simplici arcuatâ bivalvìm hiantes. *Pollen* siccum ellipticum, sulco secundum longitudinem notatum; madefactum sphæricum, læve, vel trimamillosum.

Discus formâ maxime varians, in *Colletiâ* parvus, fundumque tubi calycis occupans; in plerisque tubum calycis strato plus minùsve crasso tegens ejusque formam accipiens (in *Zizypho, Paliuro, Ventilagine, Hoveniâ, Colubrinâ,* subplanus, pentagonus, angulis ad insertionem staminum emarginatis; in *Rhamno, Sageretiâ, Scutiâ,* urceolatus vel cupulæformis), et fauci margine distincto limitatus; in aliis (*Retanillâ, Cryp-*

tandrâ, Phylicâ, à plerisque auctoribus ut disco destitutis descriptis) super lacinias calycis etiam effusus, ejusque superficiem interiòrem à fundo usque ad apicem laciniarum substantiâ carnosâ incrustans; an in quibusdam nullus? (in *Pomaderri* et *Cryptandræ* speciebus;) margine petalis staminibusque insertionem præbens.

Ovarium liberum, disco plus minùsve immersum, vel calycis tubo semi-adhærens, seu omninò adhærens; ovatum vel subglobosum, bi-triloculare, rarissimè quadriloculare (in quibusdam *Rhamnis*); loculis monospermis.

Ovulum in quolibet loculo solitarium erectum, è fundo loculi natum, sessile vel podospermio brevi suffultum. *Podospermium*, dùm adest, ante evolutionem floris angustum, nec foramen testæ tegens, ad anthesin superiùs dilatatum, et ut cupula parva basin ovuli foramenque amplectens, celluloso-spongiosum, vasibus raphes percursum. *Testa* lævis vel dorso (in *Rhamnis*) sulco profundo notata, inferiùs prope hilum perforata. Foramen in ovulis sessilibus mammillæ albidæ endocarpii respondens, in pedicellatis cupulâ spongiosâ podospermi tectum, nec ei adhærens. Membrana testæ è stratis tribus formata, exterius cuticulatum tenuissimum, medium transversè fibrosum, testam seminis producturum, interius spongiosum, primùm maximam partem ovuli occupans, dehinc incremento nuclei evanescens, raphes vasa continens. *Membrana interior* albida, tenuis, primùm libera, deindè testæ plus minùsve adhærens (in *Pomaderri* semi-adnata, in *Phylicis, Rhamnis* aliisque pluribus omninò adnata), circum chalazam superiùs affixa, inferiùs tubulosa, perforata, tubulo in foramine testæ incluso. *Chalaza* superiùs notata, è duplici strato (ut in omnibus seminibus) formata; exterius vasculosum, vasorum raphes expansione productum, testæ insertum; interius spongiosum, in ovulo semi-evoluto fuscescens, nuclei membranâ continuum. *Nucleus* subcylindricus, liber, superiùs chalazæ affixus, pendulus, inferiùs in mammillâ brevi, foramine inclusâ, productus; interiùs laxè cellulosus, in medio sacculum amnii continens, è mammillâ usque ad chalazam extensum, in cujus cavitate granula parva natant; prope mammillam Embryo sub formâ globuli sphærici primùm visus est.

" *Fructus* subsphæricus, liber vel calyce adnato plùs minùsve tectus; pericarpium exteriùs carnosum, drupaceum, spongiosum vel siccum tenuissimum; interiùs (endocarpium) fibrosum, durum, plùs minùsve crassum; aut lignosum indehiscens, nucem 2-3-locularem (seu abortu unilocularem), seu nuculas 2—3 distinctas efformans; aut crustaceum dehiscens, capsulam tricoccam producens, coccis interiùs et inferiùs rimâ longitudinali dehiscentibus.

"*Semen* in quolibet loculo solitarium, erectum, sessile vel podospermio brevi cupulæformi suffultum. *Testa* lævissima, fusca, fibrosa, crustacea vel membranacea (in fructibus lignosis, ex. gr. *Zizyphis*), raphe laterali interiùs notata, vel raphe dorsali, sulco profundo exteriori inclusâ superiùsque testam perforante, prædita (in *Rhamnis*). *Chalaza*, ut in ovulo. *Nucleum* membranâ propriâ, liberâ, vel testæ subadhærente, inclusum. *Endospermium* carnosum, flavescens, cellulosum, lateribus embryonis applicatum. *Embryo* magnus, semini subconformis, sed magis compressus, flavescens vel virescens, cotyledonibus planis applicatis, carnosis; radiculâ brevi inferâ.

"*Arbores, frutices vel suffrutices, ramulis in pluribus spinescentibus.* Folia *simplicia, alterna, subopposita, vel rariùs exactè opposita (in* Colletiis), *penninervia vel triplinervia, sparsa vel subdisticha, basi sæpiùs bistipulata, stipulis parvis, caducis vel spinescentibus et persistentibus (in* Zizyphis, Paliuro). Flores *axillares, solitarii, fasciculati, umbellati, vel cymosi, rariùs spicati, in spicis simplicibus vel interruptis (ramulis nudis), glomeratìm dispositi (in* Sageretiâ, Gouaniâ, Ventilagine), *in quibusdam paniculas terminales efformantes (in* Ceanotho, Berchemiâ, Pomaderri), *vel glomerati seu capitati (in* Cryptandrâ, Phylicâ, &c.)"

As an instance of a somewhat different mode of describing an order, the following natural character of AMARANTACEÆ, by the learned Dr. Von Martius, may be studied with advantage: it exhibits the manner in which characters are valued by the philosophers of Germany: —

"Flores hermaphroditi, rarò diclines: dioici aut abortu polygamo-monoici, aut singuli aut nonnulli glomerati bracteati. *Perianthium* hypogynum, liberum, persistens, *duplex*, utrumque compagne simile, exterius (calyx) diphyllum, nunc deficiens (evanescens); interius (corolla) pentaphyllum, petalis distinctis aut rarò connatis; rarissimè triphyllum. *Stamina* hypogyna, quina, aut quinario numero dupla aut multipla, rarò pauciora, ultra quinque vix fertilia, uniserialia, nunc distincta, nunc monadelpha, in *cupulam* aut in *tubum* connata; *filamentis* fertilibus petalis oppositis; *antheris* medio dorso affixis, nunc didymis bilocularibus, nunc unilocularibus, longitudinaliter medio antice dehiscentibus, polline globoso, minuto, creberrimo. *Pistillum* unicum. *Ovarium* simplex, mono aut oligospermum, ovulis funiculo centrali libero appensis. *Stylus* unicus vel nullus, ex ovario transiens s. continuus, et in utriculo (plerumque) persistens. *Stigma* simplex vel multiplex. *Pericarpium*: *Utriculus* membranaceus, evalvis et irregulariter dehiscens, aut circumscissus mono aut oligospermus. *Semina* lentiformia, subglobosa v. elliptica, ad hilum emarginata, verticaliter appensa; testa crustacea, *membrana interna* tenui. *Albumen* centrale, farinaceum. *Embryo* periphericus, arcuatus; *cotyledonibus* plano-convexis incumbentibus; *plumula* inconspicua; *radicula* umbilicum spectante.

" *Herbæ* aut *suffrutices* ramosæ vel ramosissimæ, *caule* teretiusculo, rariùs angulato, humiles aut diffuso-incumbentes aliis vegetabilibus.

" *Folia* opposita vel alterna, simplicia, sæpè brevitèr petiolata, integra, subintegerrima uninervia, venis subparallelis combinatis, venulis creberrimè reticulatis, exstipulata.

" *Flores* pedicellis brevissimis subsessiles, sicciusculi, scariosi et quasi glumacei, glomerati, capitati vel spicati, colore vario. *Pubes* frequens, septata, articulata aut ganglionea, plerumque simplex, rarò stellata.

" *Evolutio.*

" *Cotyledones* epigææ, integerrimæ, glaberrimæ, subsuccosæ, in alternifoliis nonnunquam obliquè oppositæ, in oppositifoliis basi conjugatæ. *Radicula* subsimplex, fibrillosa et *Cauliculus* crassiusculi, internodio primario sæpè elongato. *Folia plumulæ* vernatione sursum complicata. *Gemmatio*: Gemmis nudis. *Æstivatio calycina* equitans. *Æstivatio corollina* interdum apice aperta (dum corolla calyce inclusa), quincuncialis: duobus tribusve petalis exterioribus, sibi lateraliter imbricatis et interiores subvalvulares, vel hinc imbricatas plus minusve tegentibus.

"*Æstivatio staminea* erecta, pistillo ante anthesin saepè stamina superante, postea incluso. *Prolepsis* florum composita et indeterminata (*Link*). *Anthesis* sursum peracta.

"*Propagatio.*

"*Antherarum* dehiscentia simultanea, completa, antheris effœtis explanatis vel tortis et versatilibus. *Stigma* pollen papillis pilisque affigens, dum divisum sensim sensimque expansum. *Disseminatio* aut floribus integris super pericarpium semenque clausis decidentibus, saepè ope lanae involventis volitantibus, seminibusve aut ex utriculo circumscisso libere aut una cum utriculo delabentibus.

"*Metamorphosis.*

"Folia sursum magnitudine decrescentia, floralia nunc reliquis minora, nunc omnino deficientia aut in squamas ad divisiones florescentiae mutata, bractearum sub specie contracta, sicca, scariosa atque calycis foliolis similia. Foliolorum calycinorum fabrica et species quasi repetita in petalis vix in orbem regularem dispositis, orbe non nisi in staminum monadelphorum perigynia absoluto.

"*Metamorphosis retrocedens* s. *negativa* in florum glomerulis, nonnullos flores in gemmulas spinosas coërcens.

"*Luxuries* caules rachesque florum fasciatos vel florum diclinorum hermaphroditismum incompletum sistens, aut semina in corpuscula vacua caudata extenuans.

"*Qualitas.*

"Herbae, praesertim junioris, *folia* textura laxiuscula molli, elementis mucilaginosis, saccharinis et fibrosis pollentia ideoque oleracea. *Semina* farinacea, amylo et muco pollentia. Virtus nutriens, emolliens, demulcens, in systema lymphaticum praevalens. Unicae hucusque speciei cognitae, *Gomphrenae officinalis* Mart., radix antidotalis, tonica, stimulans.

"*Statio et Habitatio.*

"Plantae et gregariae et solitariae; plures diffusae villosiores in siccis lapidosis arenosisve apricis regionibus, aliae erectae vel super alia vegetabilia decumbentes, pube rariore adspersæ, in sylvarum marginibus lucisque primævis vivunt; nonnullæ subsalsa maritimaque diligunt loca; in depressis, haud multum super oceanum elevatis, frequentiores ac in montanis. Obviam venit hæc plantarum familia in utroque hemisphærio; sub ipsa Æquatore rariùs, indè si versus Polos procedas, utrinque frequentior, ita ut ejus vis versus Tropicas augeri videatur. Cujusvis generis *Plaga* ampla, aliis Americæ, Asiæ, Novæ Hollandiæ peculiaribus, aliis paucissimis communibus, paucis hucusque Europæis et Africanis."

As an example of a full description of a species, the following account of Cephaelis Ipecacuanha is taken from Dr. Von Martius's *Materia Medica Brasiliensium*:—

"*Radix* perennis, simplex vel in ramos paucos divergentes divisa, oblique terram intrans, flexuosa, torta, 4—6 pollices longa, rarò longior, pennam anserinam circiter crassa, versus basin et apicem plerumque paulo attenuata, annulata, annulis ut plurimum ultra dimidiam radicis crassitiem latis inæqualibus; passim fibras agens tenues, flexuosas, simplices vel parum divisas in fibrillas patentes *epidermide* lævigata, glabra, in planta viva dilute fusca, in sicca umbrina et tandem umbrino-nigricante vel griseo-fusca obductas; cortice seu parenchymate, quod annulos exhibet, æquabili, primum molliusculo, subamylaceo, albo, tandem siccescente pallide rubente vel testaceo-roseo, resinoso-splendente, facilius a filo centrali lignoso tereti dilute flavido secedente, idque passim in conspectum dante.

"*Caulis* suffruticosus, 2—3 pedes longus, adscendens, interdum declinatus inque terra latitans, passim nodosus et e nodis radices agens reliquis similes, ut plurimum simplices, teres, crassitie pennæ anserinæ vel cygneæ, vel simplicissimus, vel adultior ramos paucos sarmentoso-emittens; *epidermide* crassiuscula lævigata vel longitudinaliter rimis aperta, in parte subterranea fusca, in parte extraterranea inferiore foliis destituta cinereo-alba glabra, in superiore viridi pubescente.

"*Folia* in apice caulis ramorumque 4—6, rarò plura, opposita, subhorizontaliter patentia, petiolata, oblongo-obovata, acuta, versus basin attenuata, margine integerrima vel obiter subrepanda, 3—4 pollices longa, 1—2 lata, uti pars suprema caulis et ramorum pilis brevibus adpressis scabriuscula, obscurè viridia, subtùs pallida, nervo medio venisque lateralibus ibidem prominentibus percursa.

"*Petioli* semiunguiculares, semiteretes, supra paulò caniculati, pubescentes.

"*Stipulæ* petiolos connectentes, erectæ adpressæ, basi membranaceæ supernè utrinque in lacinias setosas 4—6 fissæ, marcescentes et cum foliis deciduæ.

"*Pedunculi* solitarii, axillares, teretes, pubescentes, floriferi erectiusculi, fructiferi refracti, unciam et ultra longi.

"*Flores* in capitulum involucratum semiglobosum collecti, 8—12, rarò plures, in quovis involucro, singuli bracteati.

"*Involucrum commune* monophyllum, patens, profunde 4—, rarius 5—6—partitum in lacinias obovatas brevi acumine terminatas ciliatas.

"*Bracteæ (s. involucrum partiale)* pro singulo flore singulæ, ovato-oblongæ, acutæ, pubescentes.

"*Calyx* ovario adnatus, minutus, obovatus, albidus, extùs pubescens, supernè sectus in dentes 5 breves obtusiusculos erectos.

"*Corolla* alba, infundibuliformis, tubo cylindrico vix sursum dilatato extùs et in fauce tenuissimè pubescens, limbo quam tubus duplo breviore, in lacinias 5 ovatas acutiusculas patenti — reflexas diviso.

"*Stamina 5. Filamenta* filiformia, alba, glabra, in tubi parte superiore adnata. *Antheræ* lineares, quam filamenta paulò longiores, nonnihil exsertæ.

"*Ovarium* calyce inclusum, obovatum, in vertice disco carnoso medio umbilicato albido notatum. *Stylus* filiformis, longitudine tubi corollini, albus. *Stigmata* 2, linearia, obtusa, patentia.

"*Bacca* ovata, obtusa, magnitudine vix semen *Phaseoli multiflori* æquans, primum purpurea, dein violaceo-atra, carnosa, mollis, calyce parvo non ampliato coronata, bilocularis, dissepimento longitudinali carnoso, disperma.

"*Nuculæ* 2, hinc convexæ inde planæ ibidemque sulco tenui exaratæ, pallidæ, testaceæ, glabræ. *Nucleus* albus, albumine corneo, embryone erecto subclavato."

A briefer and comparative mode of describing species is,

however, more frequently employed; of which the following of Hypericum perforatum, from Sir James Smith's *English Flora*, is a good instance : —

"*Root* woody, somewhat creeping. *Stem* taller than the last (H. quadrangulum), and much more bushy, in consequence of the much greater length of its axillary leafy branches: its form round, with only two opposite ribs or angles, not so acute as those of *H. quadrangulum*. The whole herb is moreover of a darker green, with a more powerful scent when rubbed; staining the fingers with dark purple, from the greater quantity of coloured essential oil lodged in the herbage and even in the *petals*. *Leaves* very numerous, smaller than the last; elliptical, or ovate, obtuse, various in width. *Flowers* bright yellow, dotted and streaked with black or dark purple; numerous, in dense, forked, terminal panicles. *Calyx* narrow. *Styles* short, erect. *Capsule* large, ovate." (*English Flora*, iii. 325.)

In order to show the materials from which a plant is described, it has become customary to add, immediately after the indication of its native country, within a parenthesis, certain explanatory abbreviations; such as *v. s. sp.* (vidi siccam spontaneam), meaning that a wild specimen has been examined in a dried state; or *v. s. c.* (vidi siccam cultam), meaning that a cultivated specimen has been examined in a dried state; *v. v. sp.* (vidi vivam spontaneam), meaning that it has been seen wild in a living state; or *v. v. c.* (vidi vivam cultam), meaning that it has been seen cultivated in a living state; and the like. These are useful things to know, because it enables a reader to judge of the goodness of the materials from which an author has been describing. But they are capable of much improvement. It now appears, indeed, whether a plant has been seen alive or dried, wild or cultivated, but we have nothing to show what the nature of the examination has been to which it has been subjected in either case. A plant may have been seen alive, and not examined or analysed until it was dried: another may have been inspected in a dried state, without having been analysed; or if analysed, the analysis may have been very imperfect: no examination may have been made of the interior of the ovarium, of the fruit, or of the seed; all points upon which it is useful to possess information. It is, therefore, desirable that some alteration, or rather extension, of these abbreviations, should be contrived, something after the following manner: — *v. v. et ex. fl. ov. fr. s.*, seen alive; examined, flower, ovarium, fruit, and seed: if all these are named, they will all have been examined; if part only, then the other parts will be understood not to have been examined. The great necessity of making some such addition as this, will, I am sure, be felt by every one accustomed to consult botanical works. At all events, it is indispensable that it should be stated whether a plant has been examined sufficiently, as well as seen; because merely to inspect a plant in a herbarium will often enable the observer to form but a very imperfect idea of its organisation. For this reason I have introduced the abbreviation *exam.* (examinavi) into some of my own works, thus: —

"Habitat in *Mexico; Pavon. (exam. s. sp. in Herb. Lambert.)*"

Connected with this subject, is the mode of stating the native countries of plants, and of citing the authorities upon which the statement is made. For this purpose the two rules of M. De Candolle are unexceptionable.

1. If you have yourself seen a specimen collected in its native country, then the name of the collector, which is placed immediately after that of the country, is printed in italics: but, 2. If you have no other authority for the habitation than some printed book, or manuscripts, then the name of the author from whom you derive your information is printed in Roman characters; thus: —

Hab. in Mexico, *Graham ;* Caribæis, Jacquin; Florida, *Frazer ;* Louisiana, Rafinesque.

Here it is seen that you have examined Mexican specimens collected by Mr. Graham, and Florida ones from Frazer; but that you trust to the writings of Jacquin and Rafinesque for its being also found in the West Indies and in Louisiana.

CHAPTER III.

OF PUNCTUATION.

As the principle of composing and punctuating generic and specific characters, and descriptions, when written in Latin, differs from that employed in ordinary composition, a few rules upon the subject may with propriety be introduced here.

In the *characters of classes, orders,* or *genera,* the nominative case is employed, the ablative being only occasionally introduced: each adjective is separated by a comma; and the different members of a sentence by a semicolon, or a period; as, " *Perianthium* deciduum. *Ovarium* liberum, sessile, monospermum, ovulo erecto. *Stylus* brevissimus. *Stigma* sublobatum. *Semen* nucamentaceum, arillo multipartito. *Albumen* ruminatum, sebaceo-carnosum."

Or, —

" *Perianthium* deciduum. *Ovarium* liberum, sessile, monospermum, ovulo erecto; *stylus* brevissimus; *stigma* sublobatum. *Semen* nucamentaceum, arillo multipartito; *albumen* ruminatum, sebaceo-carnosum."

The latter is the better of the two, because the semicolons show that the parts connected by them all form a portion of the same organ; while, if the period is exclusively used, it would appear as if the parts divided by it were all so many distinct organs.

In *specific characters,* it is customary to employ the ablative case; not to separate the adjectives that belong to the same noun by any point; to use commas to divide the members of the sentence; to employ the colon to indicate when a new sentence forms a part of that which precedes; and to exclude the semicolon altogether, or to employ it to separate adjectives in the nominative case, when such are introduced, as is sometimes the case, from the ablative part of the character. Thus we write, —

" Stemodia *balsamea*, cauli procumbenti, ramis subhirsutis, foliis ovatis obtusis basi in petiolum brevem angustatis glabris: floralibus conformibus, floribus axillaribus sessilibus solitariis vel utrinque 2—3 glomeratis, calycibus 5-partitis : laciniis lanceolato-subulatis."

And not, —

" Stemodia *balsamea*, cauli procumbenti, ramis subhirsutis, foliis ovatis, obtusis, basi in petiolum brevem angustatis, glabris, floralibus conformibus, floribus axillaribus, sessilibus, solitariis, vel utrinque 2—3 glomeratis, calycibus 5-partitis, lanciniis lanceolato-subulatis."

If this character were punctuated in the latter manner, it would not be certain whether or not laciniis referred to calyx, or to any thing else; in the former case it is distinctly indicated.

If a semicolon is introduced into a specific character, it is when an adjective in the nominative case immediately follows the specific name, preceding all that part that is in the ablative: thus, —

" Gesneria *misera*, procumbens ; foliis obovatis villosis," &c.

In *detailed descriptions*, the mode of composing and punctuating is much the same as in the characters of genera ; the nominative case being chiefly used, and commas being placed between each adjective. The members of a sentence are divided by semicolons; and if colons are employed, it is in the same sense as in specific characters.

Although such are the most approved rules of punctuation, yet it must be confessed they are little attended to by many botanists; although it cannot be doubted that they tend very much to perspicuity and precision of language.

CHAPTER IV.

OF NOMENCLATURE AND TERMINOLOGY.

The following are the canons instituted by Linnæus, with reference to this subject. They are what guide the botanist in his doubts; and, although exceptionable in some points, as will hereafter appear, are, upon the whole, well deserving of attention and respect.

1. The names of plants are of two kinds; those of the class and order, which are *understood;* and of the genus and species, which are *expressed.* The name of the class and order never enter into the denominations of a plant.

2. All plants agreeing in genus, are to have the same generic name.

3. All plants differing in genus, are to have a distinct generic name.

4. Each generic name must be single.

5. Two different genera cannot be designated by the same name.

6. It is the business of those who distinguish new genera to name them.

7. Generic names derived from barbarous languages ought on no account to be admitted.

8. Generic names compounded of two entire words are improper, and ought to be excluded. Thus, Vitis Idæa must give way to Vaccinium, and Crista Galli to Rhinanthus.

9. Generic names formed of two Latin words, are scarcely tolerable: some of them have been admitted, such as Cornu-*copiæ*, Ros*marinus*, Semper*vivum*, &c., but this example is not to be imitated.

10. Generic names, formed half of Latin and half of Greek, are hybrid, and on no account to be admitted: such are Cardam*indum*, Chrysanthem*indum*, &c.

11. Generic names compounded of the entire generic name of one plant, and a portion of that of another, are unworthy of botany; such as Cann*acorus,* Lili*onarcissus,* Laur*ocerasus.*

12. A generic name, to which is prefixed one or more syllables, so as to alter its signification, and render it applicable to other plants, is not admissible. *Bulbo*castanum, *Cyno*crambe, *Chamæ*nerium, &c., are of this kind.

13. Generic names ending in *oides* are to be rejected; as, Agrimon*oides,* Aster*oides,* &c.

14. Generic names formed of other generic names, with the addition of some final syllable, are disagreeable, as Aceto*sella,* Balsam*ita,* Rap*istrum,* &c.

15. Generic names sounding alike, lead to confusion.

16. No generic names can be admitted, except such as are derived from either the Greek or Latin languages.

17. Generic names appertaining previously to Zoology, or other sciences, are to be cancelled, if subsequently applied in botany.

18. Generic names at variance with the characters of any of the species are bad.

19. Generic names the same as those of the class or order cannot be tolerated.

20. Adjective generic names are not so good as substantive ones, but may be admitted.

21. Generic names ought not to be misapplied, to gaining the goodwill or favour of saints, or persons celebrated in other sciences; they are the only reward that the botanist can expect, and are intended for him alone.

22. Nevertheless, ancient poetical names of deities, or of great promoters of the science, are worthy of being retained.

23. Generic names that express the essential character or habit of a plant are the best of all.

24. The ancient names of the classics are to be respected.

25. We have no right to alter an ancient generic name to one more modern, even although it may be for the better: this would, in the first place, be an endless labour; and, in the next place, would tend to inextricable confusion.

26. If new generic names are wanted, it must be first

ascertained whether no one among the existing synonyms is applicable.

27. If an old genus is divided into several new ones, the old name will remain with the species that is best known.

28. The termination and euphony of generic names are to be consulted, as far as practicable.

29. Long, awkward, disagreeable names are to be avoided, such as Calophyllodendron of Vaillant, Coriotragematodendros of Plukenet, and the like.

30. The names of classes and orders are subject to the same rules as those of genera. They ought always to express some essential and characteristic marks.

31. The names of both classes and orders must always consist of a single word, and not of sentences.

I have thought it right to give these Linnæan canons, firstly, because they are undoubtedly excellent in many respects; secondly, because we must attribute much of the greater perfection of natural history, since the time of Linnæus, to the adoption of them; and, thirdly, because they are constantly appealed to, by the school of Linnæus, as a standard of language, from which no departure whatever is allowable.

It is, however, necessary to remark, that, notwithstanding the undoubted excellence of many of these rules, yet there are others, adherence to which is often out of the question, and which have, indeed, fallen wholly into disuse. It seems to be an admitted principle, that it is of little real importance what name an object bears, provided it serves to distinguish that object from every thing else. This is the material point, to which all other considerations are secondary: thus, if A. or B. are universally known by the names of Thomas or John, it is quite as well as if they were called William or James. This being so, it will follow that Nos. 7, 9, 11, 12, 14, and 16, of the Linnæan canons, are either frivolous or unimportant; or, at least, that no person is bound, either in reason or by custom, to observe them. This is particularly apparent in considering the practice now universally adopted, although condemned by Linnæus, of converting the names by which plants are known in countries called barbarous, into scientific generic names, by adding a Latin termination to them. The advantage of this practice to travellers is known to be very great, as it puts them in possession of a certain part of the language of the country in which the plants are found. Such names are often not less euphonous than those admitted by the Linnæan school as unexceptionable: witness, Licaria and Eperua, rejected Caribean generic names; and Glossarrhena, Guldenstadtia, Schlechtendahlia; and similar admitted Linnæan names. Indeed, so impossible is it to construct generic names that will express the peculiarities of the species they represent, that I quite agree with those who think a good, well-sounding, *unmeaning* name by far the best that can be contrived. The great rule to follow is this: —

In *constructing* a generic name, take care that it is harmonious, and as unlike all other generic names as it can be. In *adopting* generic names, always take the most ancient, whether better or worse than those that have succeeded it. Attend as much as you will to the canons of Linnæus in forming a name of your own; but never allow them to induce you to commit the incivility of rejecting the names of other persons, because they do not think fit to acknowledge arbitrary rules which you are disposed to obey; and let the conduct of Schreber, a German botanist, who has been held up to universal scorn for having presumed, without authority or any sort of pretension to a knowledge of the plants of Aublet, to alter the whole nomenclature of that author, to the great confusion of science, be a warning to you, never to be induced to sanction any similar deviation from the rules of courtesy in science.

When species are named after individuals, the rule of construction is this: if the individual is the discoverer of the plant, or the describer of it, the specific name is then to be in the genitive singular; as Caprifolium Douglasii, Carex Menziesii; Messrs. Douglas and Menzies having been the discoverers of these species; and Planera Richardi, the species so called having been described by Richard: but if the name is merely given in compliment, without reference to either of these circumstances, the name should be rendered in an adjective form, with the termination *anus, a, um;* as Pinus

Lambertiana, in compliment to Mr. Lambert: and, for this reason, such names as Rosa Banksiæ and R. Brunonii are wrong; they should have been R. Banksiana and R. Brunoniana.

It is customary to name an order from the genus that most accurately represents its characters, adding the termination *aceæ* to such names as end in *a* or *as*, or even *us*. Rosaceæ from Rosa, Spondiaceæ from Spondias, Connaraceæ from Connarus; or by converting the terminations *us* or *um* into *eæ*; as Rhamneæ from Rhamnus, Menispermeæ from Menispermum. But this is not very strictly adhered to; many well-known old names, not constructed upon this principle, being still retained; such as Salicariæ, Leguminosæ, Caryophylleæ, Gramineæ, Palmæ, &c.

There is no rule for the construction of the names of the higher divisions in Botany.

In terminology, every name should have a distinct, positive meaning, which cannot be misunderstood; all terms that have two meanings being bad. For instance, the term nectarium, which is sometimes applied to glands secreting honey; sometimes to modifications of the petals or stamens; and even to the disk itself, is, in such an extended signification, unintelligible. Again, the term corolla, unless limited to the inner series of the floral envelopes, may be often applied to the calyx, and then ceases to have any precise signification. Capsule has been applied by various authors to a polyspermous dehiscent compound fruit, or to an indehiscent polyspermous fruit, or to an indehiscent monospermous fruit: so applied it has no distinct meaning. For this reason modern botanists have contrived a large number of new terms, which have contributed much to the perspicuity of botanical writings. But if this has been, in many cases, done advantageously, it has unfortunately happened that in others additional terms have been created uselessly, to the great confusion of the science. Thus, the old word albumen is perfectly well understood as the matter lying between the embryo and the seed coats when the seed is mature; nevertheless, we have the terms perisperm and endosperm contrived for the same part: testa is synonymous with episperm; putamen with endocarpium: for funiculus umbilicalis we have trophosperm and podosperm; and, unfortunately, numerous other instances might be adduced. The rule to be observed in terminology is evidently this; that as no word ought to have two applications or meanings, so no idea should be expressed by more than one term; and if a term, expressive of a distinct point of structure, already exists, no new term should, on any account, be created, from the fancy that it may be better, or more expressive, than the old one. To do so, is not only unwise, but absolutely mischievous.

CHAPTER V.

OF SYNONYMS.

The synonyms of plants are the names applied to particular species by different authors. Names are often unlike each other; in which case synonyms become indispensable to a right knowledge of a plant; but when one name only has been given by common consent, synonyms, in that case, are of less importance. The objects that they serve are these: they indicate—

1. The names of the authors who have described the species, and the place in their writings in which the description is to be found.
2. The chronology of the species, pointing out the period at which it was first made known to the world.
3. The works in which figures are to be found.
4. The various names under which it has, from time to time, been known.

Synonyms, therefore, if complete, present a brief but very instructive history of a plant. In monographs, or complete accounts of particular groups of plants, no synonyms of any importance whatever ought to be omitted: in more concise works, one or two of the principal are sufficient. The importance of a synonym depends upon its being that of some author who has written, in an original manner, upon a given plant. In proportion as originality decreases, the value of synonyms decreases also.

In arranging synonyms, a strict chronological order should be maintained, beginning with the most ancient name, and ending with the most recent. But, although the citation of the names must be strictly chronological, it does not therefore follow that the quotation of the works in which the names occur should be chronological also: this would lead to great confusion and inconvenience. It has been found practically better to arrange the names chronologically; and to arrange under each name, in chronological order, those authors who have spoken of the plant by each name.

This will be more apparent from the following example, from the *Systema Naturale* of De Candolle, in which the dates of the authors' works are introduced to demonstrate the chronological order of their quotations. In practice the dates are usually omitted: it would be, perhaps, an improvement if they were always added.

TROLLIUS ASIATICUS.

Helleborus aconiti folio flore globoso croceo. *Amm. ruth.* p. 76. n. 101. (1739.)

Trollius asiaticus. *Lin.! sp. pl.* 782 (*exclus. Buxb. et Tourn. syn.*) (1763.) — *Mill. dict.* n. 2. (1768.) — *Gmel. fl. sib.* 4. p. 190. n. 23. (1769.) — *Pall. itin.* 2. p. 528. (1793.) — *Curt. Bot. Mag.* t. 235. (1793). — *Willd. sp.* 2. p. 1334. (1799.) — *Poir.! dict.* 8. p. 122. (1808.)

T. europæus *Sobol. fl. petr.* p. 134. n. 376? (1799.)

T. sertiflorus *Salisb. in Lin. soc.* 8. p. 303. (1807.)

In order to show distinctly the different value of these synonyms, M. De Candolle marks with an asterisk (*) those in which good original descriptions are to be found; and to explain which have been ascertained by the actual inspection of authentic specimens, he marks such names with a note of admiration immediately succeeding the name of an author: thus, *Lin.! sp. pl.* 427. would mean that the original specimen from which the plant was described by Linnæus in the *Species Plantarum*, page 427., had been actually examined by himself; whereas, if the note of admiration had been omitted, it would have appeared that the only evidence, with respect to the plant described by Linnæus, was obtained from the book itself. This distinction is of great importance, as it shows upon which synonyms implicit reliance can be placed, and to which we can turn with less confidence.

In proportion to the importance of synonyms ought to be the care with which they are quoted. No synonyms ought to be adopted by a writer upon the credit of others; he should always judge for himself; or, if that should not be in his power, he should take care to show which have been ascertained by himself, and for which he trusts to others. It is especially important never to suppose that plants are the same whose names are the same. Upon this point it particularly behoves the botanist to be vigilant; for nothing is more common than for writers to mistake the plants intended by each other. Thus, R. pimpinellifolia of Linnæus, is R. spinosissima; R. pimpinellifolia of Pallas is a distinct variety, if not species, called altaica by Willdenow; R. pimpinellifolia of Villars is Rosa alpina; R. pimpinellifolia of Bieberstein is probably R. grandiflora. Care must also be taken not to suppose that the plants with different names are different species. It frequently happens that a known species, already described by one botanist, is described as new by another: this arises from a variety of causes; the original description is imperfect, or inaccurate, so that the species to which it refers cannot be recognised; or a species may have been described by one botanist, in a work unknown to another, who has therefore described it anew. This is an evil, for which there is no other remedy than vigilance on the part of those who take the lead in science; and who, from time to time, apply themselves to purify it from the errors that are daily accumulating. So difficult, however, is it to detect repetitions, that even in the publications of the most distinguished and skilful writers they occur in numberless instances: for instance, the Unonas uncinata, hamata, and esculenta of Dunal and De Candolle are identically the same.

CHAPTER VI.

OF HERBARIA.

To a botanist who studies the science with much attention, and with a view to becoming perfectly acquainted with it, neither books nor the most elaborate descriptions prove sufficient. He finds it indispensable to have continually within his reach some portion of as many species as he can procure. If he has access to a botanical garden, a great many species may thus be readily accessible; although, even in such a case, it is only at particular periods that he can study the flowers and fruit of any of them: a garden, too, seldom contains more than a fifteenth or a tenth of the number of known species; and far more frequently not a twentieth.

For these reasons, botanists have contrived a method of preserving, by drying and pressure, specimens of plants which represent all that it is most essential to recognise. A collection of such specimens was formerly known by the expressive name of *Hortus Siccus;* but is now universally called an *Herbarium.* If well prepared and arranged, such a collection is invaluable to any working botanist, because it enables him instantly, at all times, to compare plants themselves with each other, and with the accounts of other botanists; or to examine them with reference to points of structure not previously considered. It will, therefore, be useful to explain, shortly, the best modes of preparing, arranging, and preserving herbaria.

What is called the specimen of a plant, is a small shoot bearing flowers and fruit, either together or separately, pressed flat and dried, so that it may be conveniently fixed upon a sheet of paper. As a plant is, in all cases, an aggregation of individuals growing upon exactly the same plan, and producing the same kind of reproductive organs, it follows that a single shoot, comprehending leaves, flowers, and fruit, is a representation of the largest tree of the forest, and will give as distinct an idea of the individual as if a huge limb were before the botanist. It is this fact that enables us to form herbaria. Besides the dried twigs thus described, an herbarium should contain specimens of the wood of each species, and also a collection of fruits and seeds, which, being often large, hard, and incapable of compression, are not fit to be incorporated with the dried specimens themselves.

In selecting specimens for drying, care must be taken that they exhibit the usual character of the species; no imperfect or monstrous shoot should be made use of. If the leaves of different parts of the species vary, as is often the case in herbaceous plants, examples of both should be preserved. The twig should not be more woody than is unavoidable, because of its not lying compactly in the herbarium. If the flowers grow from a very large, woody part of the trunk, as is often the case, as in some Malpighias, Cynometra, &c., then they should be preserved with a piece of the bark only adhering to them. It is also very important that ripe fruit should accompany the specimen. When the fruit is small, or thin, or capable of compression without injury, a second dried specimen may be added to that exhibiting the flowers; but when it is large and woody, it must be preserved separately, in a manner I shall presently describe.

Next to a judicious selection of specimens, it is important to dry them in the best manner. For this purpose various methods have been proposed: some of the simplest and most practicable may be mentioned. If you are in a country where there is a great deal of sun-heat, it is an excellent plan to place your specimen between the leaves of a sheet of paper, and simply to pour as much sand or dried earth over it as will press every part flat, and then to leave it in the full sunshine. A few hours are often sufficient to dry a specimen thoroughly in this manner. But in travelling, when conveniences of this kind cannot be had, and in wild uninhabited regions, it is better to have two or more pasteboards of the size of the paper in which your specimens are dried, and some stout cord or leathern straps. Having gathered specimens until you are apprehensive of their shrivelling, fill each sheet of paper with as many as it will contain; and having thus formed a good stout bundle, place it between the pasteboards, and compress it with your cord or straps. In the evening, or at the first convenient opportunity, unstrap the package, take a fresh sheet of paper, and make it very dry and hot before a fire; into the sheet, so heated, transfer the specimens in the first sheet of paper in your package; then dry that sheet, and shift into it the specimens lying in the second sheet; and so go on, till all your specimens are shifted; then strap up the package anew, and repeat the operation at every convenient opportunity, till the plants are dry. They should then be transferred to fresh paper, tied up rather loosely, and laid by. Should the botanist be stationary, or in any civilised country, he may dry his paper in the sun; or, if the number of specimens he has to prepare is inconsiderable, he may simply put them between cushions in a press resembling a napkin-press, laying it in the sun, or before a hot fire. It is extremely important that specimens should be dried quickly, otherwise they are apt to become mouldy and rotten, or black, and to fall in pieces. Notwithstanding all the precautions that can be taken, some plants, such as Orchideæ, will fall in pieces in drying: when this is the case, the fragments are to be carefully preserved, in order that they may be put together when the specimen is finally glued down. In many cases, particularly those of Coniferæ, Ericæ, &c., the leaves may be prevented falling off by plunging the specimen, when newly gathered, for a minute into *boiling* water. The great object in drying a specimen is to preserve its colour, if possible, which is not often the case, and not to press it so flat as to crush any of the parts, because that renders it impossible subsequently to analyse them.

Specimens of wood should be truncheons, five or six inches long, and three or four inches in diameter, if the plant grows so much. They should be planed smooth at each extremity, but neither varnished nor polished.

Specimens of fruits simply require to be dried in the sun.

When specimens shall have been thoroughly dried, they should be fastened, by strong glue, not gum, nor paste, to half sheets of good stout white paper: the place where they were found, or person from whom they were obtained, should be written at the foot of each specimen, and the name at the lowest right-hand corner. If any of the flowers, or fruits, or seeds, are loose, they should be put into small paper cases, which may be glued, in some convenient place, to the paper. These cases are extremely useful; and fragments so preserved, being well adapted for subsequent analysis, will often prevent the specimen itself from being pulled in pieces.

The best size for the paper appears, by experience, to be $10\frac{3}{4}$ inches by $16\frac{1}{2}$. Linnæus used a size resembling our foolscap; but it is much too small: and a few employ paper $11\frac{1}{4}$ inches by $18\frac{1}{2}$; but that is larger than is necessary, and much too expensive.

In analysing dried specimens, the flowers or fruits should always be softened in boiling water: this renders all the parts pliable, and often restores them to their original position.

In arranging specimens when thus prepared, every species of the same genus should be put into a wrapper formed of a whole sheet of paper, and marked at the lower left corner with the name of the genus. The genera should then be put together according to their natural orders.

In large collections it is often found difficult to preserve that exact order which is indispensable to the utility of an herbarium; and, accordingly, we constantly find botanists embarrassed by multitudes of unarranged specimens. As this is a great evil, I trust that a few hints upon the subject may not be without their use; especially as, by attending to them myself, I have probably not 500 unarranged specimens in a collection of between 20,000 and 30,000 species.—Never suffer collections, however small, to accumulate; but the very day, if possible, that a parcel of dried plants arrives, put each in its place. For this purpose they should not be glued down; but each species, with a ticket explaining its origin, name, &c., should be laid loose upon a half sheet of waste paper, and then put into the cover of the genus to which it belongs: if the genus is not recognised, and there is no time for determining it, then take a cover, marked with the name of the order at its lower left-hand corner, and put them in it; or, if the order is not known, then put the specimens into covers marked with the names of countries instead of orders, after which you can examine them, from time to time, as opportunities may occur: in the herbarium above named there is about 300 species thus laid by for consideration. Afterwards, when leisure permits, those generic covers in which there appears to be the greatest accumulation of loose specimens should be examined, the species compared and sorted, new species glued upon fresh half sheets of paper, and duplicates taken out. The advantage of this plan is, that, under any circumstances, if it is wished to consult a particular order, all the materials you possess will be found, in some state or other, collected into one place. I am persuaded, that if this simple method were attended to, the confusion now so common in herbaria, and which renders so many of them almost useless, would never exist.

Fruits, if large, will be placed loose on shelves, in cases with glass fronts; or, if smaller, in little bottles, in which also seeds should be preserved; each fruit or bottle being labelled, and the whole arranged according to natural order. Specimens of wood may be conveniently combined with a carpological collection, and arranged on the same plan. When the sections of wood are very large, as is sometimes the case, it may be convenient to have an extra compartment at the base of the case, in which they can be placed.

The cases in which the specimens are arranged may be made of any well seasoned timber; mahogany is best; but pine wood will answer the purpose. They should consist of little closets, of a size convenient for moving from place to place; of which, two, placed one on the other, will form a tier. Each closet should have folding doors, and its shelves should be in two rows: the distance from shelf to shelf should be six inches. The sides and ends of the closets should be made of $\frac{3}{4}$ in. board; but for the shelves $\frac{3}{8}$ in. is sufficient.

To preserve plants against the depredations of insects, by which, especially the little Anobium castaneum, they are apt to be much infested, it has been recommended to wash each specimen with a solution of corrosive sublimate in camphorated spirits of wine; but, independently of this being a doubtful mode of preservation, it is expensive, and, in large collections, excessively troublesome. I have found that suspending little open paper bags, filled with camphor, in the inside of the doors of my cabinets, a far more simple and a most effectual protection. It is true that camphor will not drive away the larvæ that may be carried into the herbarium in fresh specimens; but the moment they become perfect insects they quit the cases, without leaving any eggs behind them.

In all large collections of specimens there must necessarily be a constant accumulation of duplicates; as they are of no utility to the possessor, he will, if he is a liberal man, and wish well to science, distribute them among his friends, or other men of science, in order that the means of observation and examination, upon which the progress of science depends, may be multiplied at the greatest possible number of points. He will not hoard them up till insects, dust, and decay destroy them; he will not plead want of leisure (meaning want of inclination) for looking them out, or, when applied to for them, invent some frivolous excuse for avoiding compliance with the request; on the contrary, he will be anxious to disembarrass himself of that which is superfluous, and it will be his greatest pleasure to find himself able to supply others with the same means of study as himself. Conduct with regard to the disposal of duplicate specimens is a sure sign of the real nature of a man's mind. We may be perfectly certain, for all experience proves it, that to be liberal in the distribution of duplicates, is a sign of a liberal, generous disposition, and of a man who studies science for its own sake; while, on the other hand, a contrary line of conduct is an equally certain indication of a contracted spirit, and of a man who studies science less for the sake of advancing it, than in the hope of being able to gain some little additional reputation by which his own fame may be extended. A private individual has, no doubt, a right to do as he likes with that which is his own, just as a miser has a right to hoard his money, if such is his taste; but, of the keepers of public collections, it is the bounden duty to take care that every thing in their charge be rendered, in every possible manner, available for the advancement of science. For acting to the contrary they are publicly answerable.

CHAPTER VII.

OF BOTANICAL DRAWINGS.

ANOTHER important method of indicating and preserving the characters of plants is by means of botanical drawings; which, if carefully executed, and accompanied by magnified analyses of the parts that are not visible upon external inspection, are the very best means of expressing the peculiarities of a species. But to render drawings really useful, there are many circumstances to be attended to.

In the first botanical works that were illustrated by figures, the drawings were rude, and ill calculated to convey any clear idea of the object they were intended to represent; but as a knowledge of the science advanced, great improvement took place in their execution, minute accuracy was introduced into the outline of the leaves, the form and position of the flowers were carefully expressed; and if the parts of fructification were neglected, it was because their importance was not understood. By degrees, the analysis of those parts began to be attended to; attempts were made, with various success, to represent the minute points in the organs of fructification. At last, the subject of carpology was taken up by the celebrated Gærtner, who published two quarto volumes, in which numerous plates represented, often in a magnified state, the internal structure of fruits, and especially of their seeds. From the appearance of this work, I think, it is that decided improvements in the drawings of the analysis of flowers may be dated. Since that period botanical drawings have been gradually improving, till, at last, many have been executed which seem to leave nothing to be desired.

A botanical drawing should represent a branch of the plant in flower, and also in fruit, of the natural size, in which all the characters of the leaves and ramifications, the direction and relative position of parts, the mode of expansion, the arrangement of the flowers, and, in short, all that can be seen by the naked eye should be accurately expressed. It should also contain analyses of all the parts of fructification, magnified so much that every character may be distinctly seen; and this analysis, to be complete, should express the state of the organs of fructification, not only at the period of the expansion of the flowers, but in the bud state, and when arrived at perfect maturity. If to this the germination, and vernation, and highly magnified anatomical representations of the tissue and internal structure of the stem and leaves be added, the drawing may be considered complete.

But as the expense of preparing and publishing such drawings would be enormous, botanists usually content themselves with a representation of those parts only that are supposed to be most essential; such as the structure of the flower when expanded, and of the fruit and seed when ripe; and this is found, for systematic purposes, sufficiently complete, provided such details as are introduced are perfectly clear and correct.

In order to enable the student, who is interested in this subject, to form a more distinct notion of the relative utility of botanical drawings, a reference to some of the most perfect that have yet been executed is subjoined.

As instances of the highest perfection of which botanical drawings are at present susceptible, the volume of illustrations of the structure of wheat, by Mr. Francis Bauer, preserved in the British Museum; the analysis of Rafflesia, published in the 12th volume of the Linnæan Transactions, and the microscopic drawings of the fructification of Orchideous plants, now in course of publication, both also by the same distinguished artist, may be justly said to be entitled to the highest place. Next to these come the drawings of New Holland plants in the Appendix to Captain Flinders's voyage to that country; and the three fascicles of figures of New Holland plants, by Mr. Ferdinand Bauer. A very high station is also claimed by Dr. Hooker's figures of British Jungermanniæ, in which great skill as an artist is combined with deep and accurate microscopic research. In all these works the details of analysis are carried to a great extent.

Among works in which fewer details are introduced, especial mention must be made of the drawings of Palms, and the figures that illustrate Dr. Von Martius's Nova Genera et Species Plantarum; Mr. Turpin's plates in Humboldt and Kunth's Nova Genera Plantarum, and Delessert's Icones Plantarum; and some excellent analyses of the parts of fructification of Rhamneæ and Bruniaceæ, in his memoirs upon those orders, by M. Adolphe Brongniart.

Almost every scientific work of reputation of the present day contains figures which are formed upon the models of those now enumerated; from which they differ in the quantity of analysis that is introduced, a circumstance generally regulated by the price at which they are published.

Of anatomical plates, the best are those of Kieser, in his Mémoire sur l'Organisation des Plantes; of M. Mirbel, in his Mémoire sur l'Ovule; of M. Francis Bauer, in his dissections of Orchideous plants; and of Mr. Adolphe Brongniart, in his various papers in the volumes of the Annales des Sciences.

I have mentioned these as instances of good drawings, because they are easily accessible, and incontestably are well adapted to improving the taste and execution of a student; but there are very many other modern works, in which the figures may be also studied with great advantage. Whatever bears the name of Francis or Ferdinand Bauer, Hooker, Greville, Mirbel, Poiteau fils, Redouté, Reichenbach, L. C. Richard, Sowerby, Sturm, or Turpin, may almost always be profitably studied.

BOOK VI.

GEOGRAPHY.

Under this head is to be considered the manner in which plants are affected by climate or station, and the conditions under which particular forms of vegetation are confined to certain zones of temperature; as the palms to the tropics, the true pines to extra-tropical regions.

This is one of the most curious and difficult subjects with which we can occupy ourselves. It embraces a consideration of the constitution of the atmosphere, and geological structure of all parts of the globe; and of the specific effects of particular conditions of climate and soil upon vegetation: all points upon which we can scarcely be said to know any thing. It involves the discussion of the plan upon which the world was originally clothed with verdure; and as Humboldt most truly observes, it is closely connected with " the physical condition of the world in general. Upon the predominance of certain families of plants in particular districts depend the character of the country, and the whole face of Nature. Abundance of grasses forming vast savannahs, or of palms or coniferæ, have produced most important effects upon the social state of the people, the nature of their manners, and the degree of developement of the arts of industry."

If we examine the surface of the globe, we shall find its vegetation varying according to its inequalities and its differences of soil; we shall see that the plants of the valleys are not those of the mountains, nor those of the marsh like the vegetables of the river or of dry grounds; it will also be seen that the vegetation of all valleys, all mountains, marshes, or rivers, has a similar character in the same latitudes. The flora of the granitic mountains of Spain and Portugal is very different from that of the calcareous mountains of the same kingdoms; in Switzerland, Teucrium montanum always indicates a calcareous soil; and the same may be said of certain Orchises, ustulata, and hircina, for instance, in our own country. Hence it is inferred, that the differences in the character of vegetation, depend upon circumstances connected with the soil or atmosphere in which they grow. A great deal of ingenious discussion upon this matter will be found in M. De Candolle's article on botanical geography, published in the 18th volume of the *Dictionnaire des Sciences Naturelles*.

But as I do not find much that can be called positive deductions from such facts as have been ascertained, I shall, without entering into speculations as to the causes why one description of plants grows in one situation, and others in another, confine myself to an exposition of the positive facts which appear to have been hitherto distinctly ascertained.

It has been found convenient to divide the surface of the earth into different stations, when treating of botanical geography. In this part of the subject I shall adopt the arrangement and distinctions of M. De Candolle; agreeing with him that they at least indicate the most remarkable differences of station, if they are not susceptible of any rigorous definition.

He admits the following classes:—

1. *Maritime*, or saline plants; that is to say, those which, without being plunged in salt water, and floating on its surface, are nevertheless constrained to live in the vicinity of salt water, for the sake of absorbing what may be required for their nourishment. Among these, it is requisite to distinguish those which, like the Salicornia, grow in salt marshes, where they absorb saline principles, both by their leaves and roots, from those which, like Roccella fuciformis, exist upon rocks exposed to the sea air, and appear to absorb by their leaves alone; and, finally, a third class, such as Eryngium campestre, which do not require salt water, but which live on the sea-coast, as well as elsewhere, because their constitution is so robust, that they are not affected by the action of salt.

2. *Marine plants*, also called *Thalassiophytes* by M. Lamouroux, which live either plunged in salt water or floating on its surface. These plants are distributed over the bottom of the sea, or of salt water, in proportion to the degree of saltness of the water, the usual degree of its agitation, the continuity or intermittence of their immersion, the tenacity of the soil, and perhaps also the intensity of the light.

3. *Aquatic* plants, living plunged in fresh water, either entirely immerged, as Confervæ; or floating on its surface, as Stratiotes; or fixed in the soil by their roots, with the foliage in the water, as several kinds of Potamogeton; or rooted in the soil, and either floating on the surface, as Nymphæa; or rising above it, as Alisma plantago. This last division is very near the following class.

4. Plants of *fresh water marshes*, and of very wet places, among which it is chiefly necessary to distinguish those of bogs, of marshy meadows, and of the banks of running streams; and, finally, those of places inundated in winter, but more or less dried up during the summer.

5. Plants of *meadows* and *pastures*, in the study of which it is requisite to distinguish those that by their natural or artificial association form the turf of the meadow, and those others which grow mixed together with the greatest facility.

6. Plants of *cultivated soil*. This class has been entirely produced by the agency of man: the plants which grow in cultivated land are those which, in a wild state, preferred light substantial soils: many have been transported from one country to another with the seeds of other cultivated plants. Those individuals of the same species, which are found in fields, vineyards, and gardens, are often different in some respects, according to the peculiar manner in which they have been cultivated.

7. *The plants of rocks;* these pass by insensible gradations to those of walls, rocky and stony places, and even of gravel; and the latter soil, as its fragments diminish in size, conduct us by degrees to the following class. Rock plants offer some remarkable singularities depending upon the nature of the rock.

8. *The plants of sands,* or of very *barren soil;* in the classification of which much difficulty is experienced: thus, plants of the sand of the sea-shore are confounded with saline plants; those of barren soil with the species of cultivated land, and those of coarse sand are not different from those of gravel.

9. Plants of *sterile places* that are very compact, as stiff clayey soil, or such as have their surface hardened by drought or heat, or those which are trodden hard by man or animals. This is an heterogeneous class, and contains plants of very uncertain characters.

10. *Plants which follow man.* These are few in number, and more fixed in their station, either in consequence of nitrous salts being necessary to their existence; or because, perhaps, azotized matter is required for their nutriment.

11. *Forest plants,* among which are to be distinguished, firstly, the trees that form the forest, and the herbs which grow beneath their shade. The latter are to be separated into two kinds, those which can support a considerable degree of shade during all the year, which are found in evergreen woods; or such as require light in the winter, like those which are found among deciduous trees.

12. *Bushes and hedge plants.* The shrubs which compose this division differ from the plants of the forest in their smaller size, and by the thinness of their leaves; the herbaceous kinds that grow among them are ordinarily climbing plants.

13. *Subterranean plants,* which live either in dark caverns as the byssus, or within the bosom of the earth, as the truffle. These can dispense altogether with light, and several cannot even endure it. Plants that grow in the hollows of old trees have great analogy with those of caverns.

14. *Mountain plants,* as subdivisions of which all the other stations may be taken. We generally class among mountain plants such as, in Europe, are not found lower than 500 yards; but this is quite an arbitrary limit. The most important division is between those which grow on mountains, the summit of which is covered with eternal snow, and those of mountains which lose their crest of snow in the summer. In

the former, the supply of water is not only continual, but more abundant and colder as the heats of summer advance; in the latter, on the contrary, the supply of water ceases when it becomes most requisite. The former are evidently much more robust than the latter.

15. *Parasitical plants;* that is to say, such as are either destitute of the power of pumping up their nourishment from the soil, or of elaborating it completely; or as cannot exist without absorbing the juices of other vegetables. These are found in all the preceding stations. They may be divided into, first, those which grow on the surface of others, as the Cuscuta and the Misletoe: and, secondly, intestinal parasites, which are developed in the interior of living plants, and pierce the epidermis to make their appearance outwardly, such as the Uredo and Æcidium.

16. *Epiphytes,* or *false parasites,* which grow upon either dead or living vegetables, without deriving any nourishment from them. This class, which has often been confounded with the preceding, has two distinctly characterised divisions. The first which approaches true parasites, comprehends cryptogamous plants, the germs of which, probably carried to their stations by the very act of vegetation, develope themselves at the period when the plant, or that part where they lie, begins to die, then feed upon the substance of the plant during its mortal throes, and fatten upon it after its decease; such are Nemasporas and many Sphærias: these are *spurious intestinal parasites.* The second comprehends those vegetables, whether cryptogamic, such as lichens and Musci, or phanerogamous, as Epidendrums, which live upon living plants, without deriving any nutriment from them, but absorbing moisture from the surrounding atmosphere; these are *superficial false parasites:* many of them will grow upon rocks, dead trees, or earth.

Thus we see that M. De Candolle has found it necessary to divide vegetation into sixteen stations. I do not attach much importance to several of them, because they are vague and uncertain of application, and frequently common to many plants; but it is, nevertheless, useful to bear in mind, that such distinctions do exist, and to point them out whenever they take any very decided peculiarity of character. This is, indeed, indispensable, in order to enable us hereafter to form any definite appreciation of the nature of the influence of the combined agency of soil, temperature, and atmosphere.

The next, and by far the most important head under which the geographical distribution of plants is to be considered, is with reference to temperature and light. These depend, firstly, upon latitude; and, secondly, upon elevation above the sea.

As we proceed from the pole towards the equator, we find the temperature gradually increasing: and, as we ascend from the surface of the ocean up into the atmosphere, we find the temperature gradually decreasing, until we reach a point at which perpetual frost holds his throne, and where vegetation ceases.

In like manner we find, as we recede from the equator to the pole, we quit the country of palms and other arborescent monocotyledonous plants, for the habitations of deciduous dicotyledonous trees, Coniferæ, and cryptogamic plants; and that as we rise into the atmosphere as considerable a change takes place. Thus, in Teneriffe, the foot of the mountain is occupied by Crithmum latifolium, succulent Euphorbias, Plocama pendula, and Prenanthes spinosa: to these succeed vines, corn, Canarina campanula, and Messerschmidia fruticosa: a third class, consisting of laurels, Ilex, Ardisias, heaths, and Viburnums, occupy the succeeding tract. These are surmounted by pines, Cytisus, and Spartium microphyllum; and, finally, the scenery is closed by Spartium nubigenum, Juniperus oxycedrus, Scrophularia, Viola, and Festuca. (See Humboldt's Travels.)

Therefore, in considering the matter of the vegetation of a given climate, it is necessary to take into account the temperature *peculiar to the latitude itself*, and the reduction *caused by elevation*.

The decrement of caloric, as we ascend into the air, will be understood by the following table, calculated by Mr. Daniell, from observations made by Mr. Green the aëronaut, in an aërial voyage performed in 1821. These are particularly instructive; because they were all made within the space of half an hour, under circumstances which varied as little as possible.

The temperature at the surface of the earth was	- 74°
at an elevation of 2,952 feet, was	- 70°
7,288 -	- 72°
9,993 -	- 69°
11,059 -	- 45°
11,293 -	- 38°

The difference between the temperature of the highest elevation and the earth's surface amounting to 36° in the space of twenty-seven minutes.

The amount of the decrement of heat, as compared with that of latitude, has been calculated to be, in France, equal to one degree of retrogressive latitude for every 540 feet of vertical elevation; that is to say, the temperature of a district of 3240 feet of elevation, in 45° N. lat., would be equal to the temperature of 51° N. lat. on a level with the sea. But, from Humboldt's computations, it appears that, nearer the equator, this proportion varies. He found, from careful and repeated observations, between 0 and 3000 feet of elevation, that in the middle of the temperate zone, the mean temperature of the year decreased in a degree equivalent to 2° of N. lat. for every 600 feet of elevation; the mean summer heat 1° 30'; the mean autumnal heat 1° 24'; or, on an average, the decrement of temperature was about 1° of latitude for every 396 feet of elevation. Temperature decreasing in this rapid ratio, it is evident that, if vegetation is affected by temperature, it will offer great differences in the ascent of a mountain. And accordingly it is found, as will be seen by the following tables, that the nature of the vegetation towards the upper limits at which plants grow, gradually changes from that of the base of the mountain, until plants entirely disappear at the limits of perpetual snow.

CHIMBORAZO (ANDES).

Lat. 2° 30′. S. — Height, 21,450 Feet.

Elevation in Feet.	Mean Temperature.	Vegetation.
0	Of the year - 80°	Palms.
3,250	Ditto - 71°	Palms cease to grow.
5,200	Ditto - 66°	Tree ferns cease.
9,750	Ditto - 60°	Cinchonas cease.
11,375	Ditto - 46°	Alstonias and Befarias cease.
13,325	- - -	Grasses cease.
14,300	- - -	Culcitium rufescens ceases.
15,600	Of the year - 29°	Limits of perpetual snow.

POPOCAYAN (MEXICO).

Lat. 19° 20′. N. — Height, 17,550 Feet.

Elevation in Feet.	Mean Temperature.	Vegetation.
10,400	Of the year - 53°	Oaks cease to grow.
11,375	- - -	Alnus mexicana ceases.
13,000	Of the year - 44°	Pinus occidentalis ceases.
15,275	- - -	Limits of perpetual snow.

ETNA (SICILY).

Lat. 38° 6′. N. — Height, 11,360 Feet.

Elevation in Feet.	Mean Temperature.	Vegetation.
0 to 100	Of the year - 64° / Of July and Aug. 76°	Palmæ, Musaceæ, Saccharum.
1,100	- - -	Oranges, olive, and rice cease to grow.
2,175	- - -	Vine, wheat, and maize cease.
4,350	- - -	Oaks and chestnuts cease.
6,500	- - -	Rye and Pinus sylvestris cease. Fagus sylvestris and Betula become shrubs.
8,125	- - -	Juniperus and Berberis cease.
9,750	- - -	Phænogamous plants disappear.
10,000	- - -	Lichens cease.

MONT BLANC (ALPS).

Lat. 44°. N. — Height, 15,600 Feet.

Elevation in Feet.	Mean Temperature.	Vegetation.
0	Of August - 69°	
0	Of the year - 53°	
1,950	- - -	The vine ceases.
2,925	- - -	Castanea vesca ceases.
3,900	- - -	Oaks cease.
4,650	- - -	Betula alba ceases.
5,850	- - -	Pinus Abies ceases.
6,695	Of the year - 32°	} Rhododendrons cease.
7,800	- - -	
8,190	- - -	Salix herbacea ceases.
8,780	- - -	Limits of perpetual snow.

MONT PERDU (PYRENEES).

Lat. 44°. N. — Height, 11,375 Feet.

Elevation in Feet.	Mean Temperature.	Vegetation.
3,250	Of the year - 42°	
5,280	- - -	Oaks cease to grow.
6,175	- - -	Pinus picea ceases.
7,800	- - -	Pinus rubra and uncinata cease.
8,780	{ Of August - 42° Of the year - 25°	} Limits of perpetual snow.

SULITELMA (LAPLAND).

Lat. 68° N. — Height, 6,175.

Elevation in Feet.	Mean Temperature.	Vegetation.
0	{ Of the year - 34° Of August - 60°	
975	Of the year - 31°	Pinus sylvestris ceases.
1,950	{ Ditto - 27° Of August - 54°	} Betula alba ceases.
2,925	- - -	Salix herbacea and lanceolata cease.
3,640	{ Of the year - 21° Of August - 49°	} Limits of perpetual snow.

The effect of elevation is not, in Europe, the same with all plants; there are many that grow indifferently upon the plains and upon mountains as high as perpetual snow. M. De Candolle speaks of 700 instances, with which he is acquainted, of the prevalence of this law. But, on the other hand, there are many plants, the limits of which are strictly circumscribed by elevation or equivalent temperature; as, for example, the chestnut does not rise higher in the Swiss Alps, in the parallel of 45°, than 2,400 feet; on Etna, in latitude 38°, it reaches no higher than 4000 feet. Many of the plants found on plains in the north of Europe occupy the mountains of the south. The olive, in 44° of latitude, its most northern range, will not grow at a greater elevation than 1200 feet, In general it is found that, as we approach the equator, vegetation becomes more and more affected by elevation; and that as we recede from it, the effects of elevation gradually cease.

The *cause* of the influence of elevation upon plants is ascribed, in the first place, to reduced temperature; secondly, to a greater intensity of solar light; and, thirdly, to a decrease in humidity. The rate at which temperature decreases as we ascend from the surface of the earth varies according to latitude: Humboldt has shown that, in the temperate and torrid zones, the decrement of heat is essentially different. In the equatorial zone, the temperature of the region lying at the height of between 3000 and 6000 feet, — on which the clouds repose that are visible to the natives of the plains, — decreases much more slowly than either above or below that elevation; but, in the temperate zone, the decrease is more gradual. In proof of this the following table has been formed by Humboldt: —

Elevation above the Sea in Feet.	Equatorial Zone, Lat. 0°—21°.		Temperate Zone, Lat. 45°—47°.	
	Mean Temperature of the Year.	Difference.	Mean Temperature of the Year.	Difference.
0	80°		53°	
		12°		12°
3,000	68°		41°	
		4°		9°
6,000	64°		32°	
		9°		9°
9,000	55°		23°	
		11°		
12,000	44°			
		10°		
15,000	34°			

The diminution of the density of the air as we ascend, produces a corresponding increase in the intensity of the light; a circumstance in which high elevation again corresponds with high latitudes.

It is said that the humidity of the atmosphere decreases as we ascend, and that to this may be ascribed much of the effect produced upon vegetation by great heights. That the humidity of the atmosphere does much affect vegetation is not to be doubted; and if it were certain that the air became gradually drier as we ascend, a second cause, as powerful as that of temperature, would be found for the effects of elevation upon vegetation. But it is certain that the humidity of the air does not change gradually, as we ascend, with the character of vegetation; on the contrary, it has been found that atmospheric humidity is either uniform or increased to heights far beyond uniformity of vegetation, and then suddenly diminishes to a large amount, vegetation not suddenly altering with it; so that it would seem as if the atmosphere were composed of deep beds of air, suddenly differing from each other in the elasticity of their aqueous vapour.

From observations made by Capt. Sabine, with a Daniell's hygrometer, at Ascension, it appears that, on that island, at seventeen feet above the sea, the amount of dryness was 5°; and, at 2237 feet higher, was 3° 5'; so that, in this case, the air became more humid as he ascended. At Trinidad the amount of dryness on a level with the sea was 5°; at 1060 feet higher the air was *saturated* with moisture; in this instance, also, humidity increased with elevation. At Jamaica it was found that, on a level with the sea, the degree of dryness was 7°; at 4080 feet higher the air was saturated with moisture; but at 4580 feet the dryness was 16°. Hence it is to be inferred that, in these observations, the lower bed of the atmosphere was not passed through, either at Ascension or in Trinidad; but that, in Jamaica, it had been left below at the time the third observation was taken; and that, in that island the lower stratum of air is something more than 4000 feet deep. In Mr. Green's voyage the degree of dryness of the air, at an elevation of 9893 feet, was 5°, nearly the same as it was observed to be on the surface of the earth below at the same time; but, at 11,059 feet, it was 13°; and at 11,293 feet, the highest point at which an observation was made, it was still 13°; so that it would seem that the humidity of the atmosphere, at that time, did not vary through a bed of air rising perhaps 2000 feet beyond the highest limits of vegetation in Europe.

It must be confessed that these observations are by no means sufficiently numerous to become the foundation of any thing connected with the effect of elevation upon the characters of plants; but they, at least, answer the purpose of showing that, in the present state of our information, the effects of humidity are not appreciable in investigating the subject.

Whether the increased rarity of the air, as we ascend, has any effect upon vegetation, is not determined. It is not easy to say in what way it can act, according to any yet known physiological laws, unless, as M. De Candolle remarks, in supplying an insufficient quantity of oxygen for absorption. But, as we find plants of the plains grow indifferently on the highest mountains, it does not seem that there is any such diminution of oxygen as interposes with the operations of vegetation. The diminution of atmospheric pressure, which of course takes place at high elevations, may facilitate evaporation; but we have yet to learn in what precise way that phenomenon influences vegetation.

From what has now been said, all that is apparent is, that,

as we ascend in the atmosphere, temperature diminishes and light increases in a proportion corresponding, to a certain degree, with the climate of higher latitudes; but even to this there are exceptions, depending upon particular circumstances, and especially upon the amount of summer heat, of which more will be said presently. Thus, at Enontekissi, in Lapland, in 68° 30′ N. lat., at an elevation of 1356 feet above the sea, a climate which, from its situation, should be scarcely clothed with herbage, Von Buch found corn, orchards, and a rich vegetation.

Having now seen what great differences are produced in the characters of vegetation by elevation above the sea, let us next take a view of the influence caused by latitude. In the countries lying near the equator, the vegetation consists of dense forests of leafy evergreen trees, palms, and arborescent ferns, among which are intermingled epiphytal herbs and rigid grasses: there are no rich verdant meadows, such as form the chief beauty of our northern climate; and the lower orders of vegetation, such as mosses, fungi, and confervæ, are very rare: Myrtaceæ, Melastomaceæ, Musaceæ, Piperaceæ, Scitamineæ, and frutescent Compositæ abound. As we recede from the equator these gradually give way to trees with deciduous leaves, to Coniferæ, Rosaceæ, and Amentaceæ; rich meadows appear, abounding with tender herbs; the epiphytal Orchideæ disappear, and are replaced by terrestrial fleshy-rooted species; mosses clothe the trunks of aged trees; decayed vegetables are covered with parasitical fungi; and the waters abound with Confervæ. Approaching the poles trees wholly disappear; dicotyledonous plants of all kinds become comparatively rare; and grasses and cryptogamic plants constitute the chief features of vegetation. To what cause, except that of temperature, and perhaps light, these effects are to be ascribed, is unknown. They are found to exist equally towards either pole; and it is evident, from the uniform manner in which the influence of the controlling cause, whatever it may be, is exercised, that the laws under which the geographical distribution of plants is determined, are as certain and immutable as any of those with the nature of which we are acquainted. It is probable that temperature is the principal cause, from the well-known fact that the vegetable productions of hot climates can be successfully cultivated in cold ones by the aid of heat; and that the plants of cold climates may be cultivated in hotter climates by an artificial reduction of temperature. But that other causes also operate, is apparent from the impossibility of cultivating the plants of any high latitudes in those considerably to the south. Thus, when living plants were brought to England from Melville Island, no means whatever could be discovered of keeping them alive, although the temperature at which they were maintained did not materially vary from that to which they must have been often exposed, in the summer season, in their own climate. Assuming, however, for the present, that temperature is the most efficient cause of variety in the distribution of plants, the first point to consider is, how far temperature and latitude are uniformly the same in either hemisphere. This has been discussed, with his habitual skill, by Baron Humboldt, of whose observations I must avail myself in nearly all that I can say upon the subject. According to this observer, the geographical parallels of latitude do not indicate corresponding temperature, either in the old and new world, or in the northern and southern hemispheres. In the new world the temperature decreases more rapidly as we recede from the equator than in the old world; and in the southern hemisphere, beyond the parallel of 34°, the summers are colder than in corresponding latitudes of the northern hemisphere; but the winters milder. On this account Baron Humboldt concludes that " the lines of equal mean annual heat, which may be called *isothermal*, are not parallel with the equator, but intersect the geographical parallels at a variable angle."

The following table shows the difference in the mean annual heat of the same latitudes in the old and new worlds: —

Latitude.	Mean Heat of the Year in the		Difference.
	Old World.	New World.	
0°	80°	80°	0°
20	77	77	0
30	70	67	3
40	63	54	9
50	50	38	12
60	40	24	16

Hence it appears that the old world is much warmer than the new, and that the temperature of America does not decrease, from Florida to the Gulf of St. Lawrence, in the same ratio as in Europe, from Egypt to Scandinavia. But although, in the temperate parts of North America, the mean annual heat of a given place is the same as that of Europe some degrees more to the northward, yet the temperature of particular seasons do not accord in the same degree; but the colder the winters the hotter the summers are found: thus,—

The summer of Philadelphia, lat. 39° 56' N. is the same as that of Rome - - - lat. 41° 53' N.
The winter of Philadelphia, lat. 39° 56' N. is the same as that of Vienna - - - lat. 48° 13' N.
The summer of Quebec, lat. 46° 47' N. is hotter than that of Paris - - - - lat. 48° 50' N.
The winter of Quebec, lat 46° 47' N. is colder than that of St. Petersburgh - - - lat. 59° 56' N.

In general, the summers of the temperate parts of North America, as far as 40° N. lat., are about 4° warmer than in Europe under the *same isothermal parallel;* whence it can be understood why magnolias and other equinoctial-looking trees extend so far to the north, since, in the parallel of 36°, the summer heat to which these trees are exposed scarcely differs from the mean annual heat of the equator. It is, therefore, extremely important in the study of botanical geography, to take into account, not only the mean temperature of the year, but also the mean summer heat.

According to Barton, the climate to the *west* of the Alleghany mountains is much warmer than that on the east, or Atlantic side, where the same plants exist 3° or 4° higher up on the west than on the east side of the range. It is probable, however, that this difference does not extend higher up than Lake Erie, in 42° N. lat.; for, both beyond Lake Superior and Hudson's Bay, the earth is said to be constantly frozen at three feet from the surface; a phenomenon which also occurs in Siberia, about the river Lena, in about 62° N. lat., near the town of Jakutsk; while, in Lapland, in 70° near Vadsoe, the temperature of the earth is found to be as much as 3° or 4° above the freezing point; whence it appears that the climate of the north of Europe is warmer than that of the same latitudes in Asia and America. We therefore shall not be far away, if we conclude that the isothermal lines bend towards the tropics in Europe, and towards the poles in Tartary and America.

As we approach the equator there appears to be little difference in the mean temperature of the year, either in the new or old world.

Of the Old World.

The mean temperature of Senegal is 79.7° in lat. 24° 30′ N.
of Madras is 80.4° in lat. 13° 5′ N.
of Batavia is 77.4° in lat. 6° 10′ S.
of Manilla is 78.0° in lat. 15° N.

Of the New World.

The mean temperature of Cumana is 81.6° in lat. 10° 27′ N.
of the Antilles is 81.6° in lat. 15° N.
of Vera Cruz is 78.0° in lat. 19° 12′ N.
of Havannah is 78.0° in lat. 23° 12′ N.

It is probable, however, that the summers of Asia are more fervid than those of America; for, according to Roxburgh, the mean temperature of Madras, in latitude 13° 5′ N., in the month of July, is 89.4°; while that of Cumana, in latitude 10° 27′, does not exceed 84.4°.

To the south of the equator, the temperature of the east seems to be higher than that of corresponding latitudes in the west: thus, the mean temperature of the Mauritius, in 20° 9′ S. lat., has been ascertained to be 80.4°; while that of

Rio Janeiro, in latitude 20° 59′ S., is as low as 74.3°; and at the Havannah, in nearly the same parallel in the northern hemisphere, it ranges between 77° and 77.9°. The whole of the western coast of South America, as far as the sands of Peru, in latitude 10° and 14° S., are affected so much by the continual prevalence of clouds and the low temperature (59.9°) of the currents setting round Cape Horn, that the mean temperature of the year in those parts does not exceed 68° or 69°. Hence the plants of Lower Peru live in a temperature not exceeding, by day, 68° or 72°, and by night 59° or 62°. Near the coast Humboldt observed the thermometer as low as even 55.4° in 12° 2′ S. lat. With this exception there is little difference in the temperature of the southern hemisphere as low as 34° S. lat., either in New Holland, Africa, or America. The mean temperature of Port Jackson, in 33° 51′ S. lat., has been ascertained to be 66.6°; of the Cape of Good Hope, in 33° 55′ S. lat., to be 66.8°; and of Buenos Ayres, in 34° 36′ S. lat., to be 67.6°. In the northern hemisphere the mean temperature, in latitude 34, is 67.8°. It is extremely probable that, as far as the parallel of 57° S. lat., the differences in the temperature of the two hemispheres are greater in the summer than the winter. The cold of the Falkland Istands, in latitude 51½° S., is less than that of London in the same latitude to the north. The arborescent ferns and epiphytal Orchideæ are often injured by the cold in Van Diemen's Island, latitude 42° S.; and in the southern part of New Zealand, latitude 46° S. Cook observed, in latitude 43°–44° S., in July in the middle of winter, that the thermometer at noon was usually between 46° and 51°. At Rome, latitude 41° 53′ N., the thermometer at noon in January rarely reaches 51°–53°: in Paris the mean noon-day temperature of January is, according to Arago, 38.7°. For this reason it is supposed that the climate of the southern hemisphere does not differ from that of the north so much in the greater coldness of the winters as of the summers. According to Humboldt, the greatest heat in the parallels of 48° and 58° of S. lat., does not exceed 43.7°–46.8°; while at St. Petersburgh and Umea, in 59° 66′ and 63° 50′ N. lat., it is 65.2° and 62.6°. In the Straits of Magellan, between 53° and 54° S. lat., snow falls almost daily in the middle of summer; and, in the same place, in the middle of December, the sun not setting for eighteen hours together, Krusenstern observed that the thermometer never rose higher than 52°; while, on the contrary, Von Buch remarked it as high as 79.4° in Lapland under the parallel of 70°. In 60° S. lat., which nearly answers to the position of St. Petersburgh in the northern hemisphere, Cook and Forster found the temperature at midsummer not higher than 36°; and icicles were continually forming on their ship. Even in the extreme points of Lapland, in 70° N. lat., the pines attain the height of sixty feet; while at the Straits of Magellan and in Station Island, near New Year's Harbour, in latitude 55° S., nothing like a tree is found, except scrubby birches and Wintereæ.

Viewing the distribution of plants with respect to longitude, we find that, while the great forms of vegetation are wholly controlled by circumstances attendant upon the parallels of latitude, there are wide differences, of a secondary nature, which correspond in some with the parallels of longitude; and that particular genera and species do not extend beyond the limits of particular districts, to which they give peculiar features. Thus, in North America, on the east of the Rocky Mountains, azaleas, rhododendrons, magnolias, vacciniums, actæas, and oaks, form the principal features of the landscape; while, on the western side of the dividing ridge, these genera almost entirely disappear, and no longer constitute a striking characteristic of the vegetation. The genera of Proteaceæ and the Ericeæ, at the Cape of Good Hope, are replaced in New Holland by different genera of Proteaceæ, and by Epacrideæ; while neither the one nor the other exist on the continent of South America, with the exception of some Rhopalas. The natural order of Bromeliaceæ is exclusively confined to America: Calathea, a genus of Marantaceæ, is only found on the same continent: cinnamon, cloves, and nutmegs are confined to the Indian Archipelago; and hundreds of other instances are to be named of similar exclusive stations. Whether these differences depend upon geological causes, or arise from some other circumstances, is entirely unknown.

Such are the most striking facts connected with the distribution of temperature with respect to vegetation. It will have been seen that little is known of the proportion of humidity in the atmosphere of different climates, and that the amount of light in various latitudes has scarcely been noticed. That the effect of both these agents upon vegetation is most important, cannot be doubted; especially of the latter, upon which the most material vital functions of vegetation mainly depend: but, unfortunately, there are no data from which the precise amount or action of light in different latitudes can be appreciated.

I shall now proceed to state what is known or conjectured of the distribution of the different orders or divisions of vegetables over the surface of the globe. In doing this, I shall merely translate a portion of the very valuable essay of Baron Humboldt upon the subject, as published in the *Dictionnaire des Sciences Naturelles*, vol. xviii. p. 422, in which is comprehended the sum of all that is known of the laws that are observed in the distribution of the various forms of vegetation. — " The numerical relations of the forms of vegetation are capable of being investigated in two very different modes. Supposing that the natural families of plants are studied without reference to their geographical distribution, the question will arise as to which type of organisation it is after which the greatest number of species have been created. Are there most Glumaceæ (Cyperaceæ, Gramineæ, and Junceæ are so called by M. De Humboldt,) or Compositæ in the world? Do these two tribes together constitute a fourth part of phænogamous vegetation? What proportion is borne by Monocotyledones to Dicotyledones? Questions of this kind refer rather to the science of vegetable organisation and of mutual affinities. But if, instead of studying natural groups of species in this abstract manner, we view them with reference to the relations they bear to climate or to the distribution over the surface of the globe, other questions of a much more varied nature will arise. Which families, for instance, are more predominant in the torrid zone than in the polar circle? Are Compositæ more numerous in the same parallel of latitude or in the same isothermal line in the old world or the new? Do those forms which are found to diminish in retreating from the equator to the pole follow a similar law of decrement in rising from the plains into the mountains of the equator? Do the proportions borne by one family to another vary on the same isothermal line; and are such proportions the same on either side of the equator? These are, properly speaking, questions of geographical botany: they are connected with the most important problems of meteorology, and of the physics of the globe in general.

" In studying the geographical distribution of particular forms, we can pause either at a consideration of particular species, genera, or natural families. It often happens that a particular species, especially of those kinds which I have called social, covers a vast extent of country: such, for instance are, in the north, the heaths and forests of pines; such are, in equinoctial America, the assemblages of multitudes of Cactus, Croton, Bambusa, and Brathys, of the same species. It is curious to examine such instances of multiplication and organic developement. We may enquire what species, in a given zone, produces the greatest number of *individuals?* and we may mark the families to which the predominant species belong in different climates.

" In a northern climate, where compositæ and ferns are to phænogamous plants in the relation of one to thirteen, and of one to twenty-five (that is to say, when these proportions are found by dividing the total number of phænogamous plants by the number of Compositæ and ferns), one single species of fern may occupy ten times as much land as all the Compositæ put together. In such a case, ferns would exceed Compositæ by their *mass*, by the number of individuals belonging to particular species of Pteris or Polypodium; but they would not exceed them if a comparison were instituted between the different forms exhibited by the two groups of Compositæ and ferns, and the sum total of phænogamous species. As the multiplication of all species does not follow a single law, and as they do not all produce an equal number of individuals, the *quotients* obtained by dividing the total number of phænogamous plants, by the number of species of different families, do not by themselves determine the aspect, or, it might almost be said, the nature, of the monotony of vegetation in different quarters of the world. A traveller is often surprised at the continual repetition of individuals of one species, and of the masses of such individuals which are continually occurring; but he has equal reason to wonder at the rarity of other species which are useful to mankind. Thus, in countries where whole forests are formed by Rubiaceæ (Cinchonaceæ), Leguminosæ, and Terebinthaceæ, the Cinchonas, logwood, and balsam trees are comparatively very rare.

" In the consideration of species, the subject may also be viewed *in an absolute manner* with reference to the number of species which prevail in particular zones. This interesting kind of comparison has been made in M. De Candolle's grand work, and Mr. Kunth has carried it into effect with more than 3500 Compositæ now known. It does not, indeed, indicate what families predominate, in a given degree, over other phænogamous plants, either with regard to the number of species, or the mass of individuals; but it determines the numerical relations of species of the same family in different latitudes. The most varied forms of ferns, for instance, are found in the tropics; it is in the mountainous, temperate, humid, and shady regions of those parts of the world that the family of ferns produces the greatest number of species. In the temperate zone there are fewer than in the tropics, and the total number continues to decrease as we approach the pole; but as a cold country, Lapland, for instance, produces species that have a greater power of resisting low temperature than the great mass of phænogamous plants, it happens that, in Lapland, the relative proportion borne by ferns to the rest of the flora is greater than in France or Germany. The *numerical relations* which appear in the tables that are now about to be produced, are entirely unlike the relations indicated *by an absolute comparison* of the species that vegetate under different parallels of latitude. The variation which is observable in proceeding from the equator to the poles is consequently different in those two methods. In that of fractions, which is adopted by Mr. Brown and myself, there are two causes of variation; that is to say, the total numbers of phænogamous plants do not vary in passing from one parallel of latitude, or rather from one isothermal zone to another, in the same proportions as the number of species of a given family.

" If from *species* or *individuals* of the same form, which reproduce themselves in conformity to certain fixed laws, we pass to those divisions of the natural system which are *abstractions* of different degrees of importance, we may either confine ourselves to genera, or orders, or sections of a still higher degree. There are certain genera and families which belong exclusively to certain zones, and a particular combination of the conditions of climate; but there is also a great number of genera and families, of which we find representatives under all zones and at all elevations. The earliest researches upon the geographical distribution of forms were those of M. Treviranus, published in his ingenious work on Biology (vol. ii. pp. 47. 63. 83. 129.), and the object of these was the stations of genera upon the globe. But it is more difficult to obtain general results from such a method than from that which compares the number of species of each family, or the great groups of a particular family to the whole mass of phænogamous plants. In the frozen zone, the variety of genuine forms does not diminish in any thing like the degree of decrement of species; a greater number of genera, in a given number of species, is always to be found in such countries: and so it also is with the summits of high mountains, which are colonised by a great number of genera supplied by the more abundant vegetation of the plains.

" It is very instructive to study the vegetation of the tropics and of the temperate zone between the parallels of 40° and 50°, in two different ways: firstly, in determining the numerical properties of the flora of a large extent of country, including both mountains and plains; and, secondly, in ascertaining those proportions for the plains only of the temperate and torrid zones. As in our herbaria we have indicated, by barometrical measurement, the elevation of each plant in more than 4000 cases above the level of the sea in equinoctial America, it will be easy, when the account of the species is completed (it is now completed), to separate those which grow at or above an elevation of 6000 feet from such as are inhabitants of a lower region. This operation will affect most sensibly those families that abound in alpine species; as, for instance, Gramineæ and Compositæ. At 6000 feet of elevation, the mean temperature of the air, on the back of the equatorial Andes, is 62° 6', which is equal to that of July at Paris. Although, upon the table-land of the Cordilleras, we find the same annual temperature as in high latitudes, yet it is not right to generalise too much such analogies between the temperate climates of equatorial mountains and low stations in the circumpolar zone. These analogies are not so great as is supposed; they are much influenced by the partial distribution of heat in different seasons of the year. The quotient does not regularly change, in rising from the plains into the mountains, in the same manner as it does in approaching the pole; as happens with Monocotyledones in general, ferns, and Compositæ.

"We may, moreover, remark, that the developement of the vegetation of different families depends neither upon geographical nor isothermal latitude alone; but that, on the contrary, the quotients are not in accordance on the same isothermal line of the temperate zone in the plains of America and of the old world. Under the tropics, there is a remarkable difference between America, India, and the western side of Africa. The distribution of organised beings over the surface of the globe depends not only upon very complicated conditions of climate, but also upon geological causes, the nature of which is wholly unknown, but which are connected with the original state of our planet. In the equinoctial zone of Africa palms are not very numerous, if compared with the much greater number in South America. Differences such as these, far from turning us from a search after the laws of nature, should, on the contrary, excite us to contemplate those laws in their most complicated forms. Lines of equal heat do not follow the parallel of the equator; they have convex and concave summits, which are distributed very regularly over the globe, and form different systems along the eastern and western sides of the two worlds, in the centre of continents, and in the vicinity of oceans. It is probable that, when the globe shall have been more correctly examined, it will be found that the lines of *maxima of grouping* (that is, lines drawn through those points where the fractions are reduced to their smallest denominator) will be isothermal lines. If we divide the globe into lines of longitude, and compare the numerical proportions of those lines under similar isothermal latitudes, the existence of different systems of grouping will at once be evident. From such systems can be distinguished, even in the present imperfect state of our knowledge, those of the new world, of western Africa, of India, and of New Holland. As we find that, notwithstanding the regular increase of heat from the equator to the poles, the maximum of heat is not always identical in different countries in different degrees of longitude; so there exist places where certain families attain a greater degree of developement than elsewhere; as is the case with Compositæ in the temperate region of North America, and especially at the southern extremity of Africa."

Now follow tables of the different numerical proportions of certain extensive families and divisions of plants, as far as they have been ascertained. I give them in Baron Humboldt's words, with a few interpolations, which are distinguished by being included within crotchets [   ].

" ACOTYLEDONES.

" Cryptogamic plants (fungi, lichens, mosses, and ferns); cellular and vascular Agamæ of M. De Candolle. Taking the plants of the plains along with those of the mountains, we have found, under the tropics, $\frac{1}{9}$; but their number ought to be much greater. Mr. Brown has shown that it is probable that, in the torrid zone, the proportion is $\frac{1}{15}$ for the plains, and $\frac{1}{5}$ for the mountains. In the temperate zone cryptogamous plants are generally to phænogamous as 1 to 2; in the frozen zone they maintain as large a proportion, and often much surpass it. [In Melville Island the numbers are 58 crypt. to 67 phænog., or nearly equal: in Sweden, according to the computation of Wahlenberg, they are something less than 4 to 1; and it is probable that this is a near approximation to the true proportions of Sweden, the cryptogamic flora of that country having been more accurately investigated than that of any other part of the world.]

" In separating cryptogamous plants into three groups, we observe that ferns are more numerous, the denominator of the fraction being smaller in the frozen than in the temperate zone. Lichens and mosses also increase towards the frozen zone. The geographical distribution of ferns depends upon the combination of local circumstances of shade, humidity, and moderate warmth. Their maximum (that is to say, the place where the denominator of the fraction of the group becomes the smallest possible,) is found to be in the mountainous parts of the tropics, especially in small islands, in which the proportion rises to $\frac{1}{3}$, and even higher. Not distinguishing the plains from the mountains, Mr. Brown finds the proportion of ferns in the torrid zone to be $\frac{1}{20}$: in Arabia, India, New Holland, and Western Africa (within the tropics) it is $\frac{1}{26}$: our American herbaria only indicate $\frac{1}{38}$: but ferns are rare in the wide valleys and arid table land of the Andes, where we were constrained to reside a long time. In the temperate zone ferns are $\frac{1}{70}$, in France $\frac{1}{73}$, in Germany, according to recent observations, $\frac{1}{71}$. The group of ferns is extremely rare on Atlas, and is almost entirely absent from Egypt. [In Sicily, Presl finds them $\frac{1}{80}$; in Sweden, according to Wahlenberg, they are about $\frac{1}{110}$.] In the frozen zone ferns appear to increase to $\frac{1}{25}$. [There are none in Melville Island.]

" MONOCOTYLEDONES.

" The denominator becomes progressively smaller in going from the equator to 62° N. lat.; it again increases in still more northern regions, on the coast of Greenland, where Gramineæ are very rare. [Mr. Brown remarks that, in the list of Greenland plants, Dicotyledones are to Monocotyledones as 4 to 1, or in nearly the equinoctial ratio; and in Spitzbergen, as well as can be judged, the proportion of Dicotyledones appears to be still further increased. This inversion was found to depend as much on the reduction of the proportion of Gramineæ as on the increase of certain dicotyledonous families, especially Saxifrageæ and Cruciferæ. The flora of Melville Island is, however, very different, Dicotyledones being to Monocotyledones as 5 to 2, or in as low a ratio as has any where been observed; while the proportion of grasses is nearly double that of any part of the world. *Parry's Appendix.*] The proportion varies from $\frac{1}{5}$ to $\frac{1}{6}$ in different parts of the tropics. Among 3880 phanerogamous plants found in equinoctial America by M. Bonpland and myself, there are 654 Monocotyledones and 3226 Dicotyledones; here, therefore, the great division of Monocotyledones forms $\frac{1}{6}$ of phænogamous plants. According to Mr. Brown, this proportion is in the old world (India, equinoctial Africa, and New Holland) $\frac{1}{5}$. Under the temperate zone it is found to be $\frac{1}{4}$; France 1 : $4\frac{2}{5}$; Germany 1 : $4\frac{1}{2}$; North America, according to Pursh, 1 : $4\frac{1}{2}$; kingdom of Naples 1 : $4\frac{1}{5}$; Switzerland 1 : $4\frac{1}{4}$; Great Britain 1 : $3\frac{3}{4}$; [Sweden 1 : $3\frac{6}{10}$; but in Sicily, according to Presl, it is 1 : $5\frac{3}{10}$, which is much too high]. In the frozen zone $\frac{1}{3}$.

"GLUMACEÆ (that is to say, the three families of Junceæ, Cyperaceæ, and Gramineæ united). — *Trop.* $\frac{1}{11}$; *Temp.* $\frac{1}{8}$; *Frozen* $\frac{1}{4}$. This increase towards the north is due to the greater prevalence of Junceæ and Cyperaceæ, which are much more rare, as compared with other phænogamous plants, in the temperate and torrid zones. Comparing the species of these three families, we find that Gramineæ, Cyperaceæ, and Junceæ are in the tropics as 25, 7, 1; in the temperate parts of the old world as 7, 5, 1; within the polar circle as $2\frac{2}{5}$, $2\frac{3}{5}$, and 1. In Lapland there are as many Gramineæ as Cyperaceæ; thence, towards the equator, Cyperaceæ and Junceæ diminish much more than Gramineæ. The form of Junceæ almost disappears in the tropics.

"JUNCEÆ alone. — *Trop.* $\frac{1}{400}$: *Temp.* $\frac{1}{90}$, (Germany $\frac{1}{94}$, France $\frac{1}{86}$), [Sicily $\frac{1}{303}$]; *Frozen* $\frac{1}{25}$, [Melville Island $\frac{1}{33}$].

"CYPERACEÆ alone. — *Trop.* America scarcely $\frac{1}{57}$, Western Africa $\frac{1}{18}$, India $\frac{1}{25}$, New Holland $\frac{1}{14}$; *Temp.* perhaps $\frac{1}{20}$, (Germany $\frac{1}{18}$, France, according to De Candolle, $\frac{1}{27}$, Denmark $\frac{1}{16}$,) [Sweden rather more than $\frac{1}{12}$, Sicily $\frac{1}{57}$]; *Frozen* $\frac{1}{9}$, in Lapland and Kamtchatka; [Melville Island $\frac{1}{17}$].

"GRAMINEÆ alone. — *Trop.* I have always supposed $\frac{1}{15}$; but Mr. Brown finds for western Africa $\frac{1}{12}$, for India $\frac{1}{12}$; and Mr. Horneman makes the proportion of Guinea $\frac{1}{10}$; *Temp.*

Germany $\frac{1}{13}$, France $\frac{1}{13}$, [Sweden not quite $\frac{1}{12}$, Sicily $\frac{1}{10}$]; *Frozen* $\frac{1}{10}$, [Melville Island nearly $\frac{1}{5}$].

"COMPOSITÆ. Not distinguishing plants of the plains from those of the mountains, we found them in equinoctial America $\frac{1}{8}$ and $\frac{1}{7}$; but of 534 compositæ of our herbaria, only 94 were found between the plains and 3000 feet of elevation; a height at which the mean temperature is 71° 3′, equalling that of Cairo, Algiers, and Madeira. From the plains to 6000 feet, where the mean temperature is that of Naples, we found 265 compositæ. Therefore the proportion of compositæ in the regions of equinoctial America, below 6000 feet, is from $\frac{1}{9}$ to $\frac{1}{10}$. This result is very remarkable, inasmuch as it proves that, within the tropics in the low and hot region of the new continent, there are fewer compositæ; and in the subalpine and temperate regions, more than under the same conditions in the old world. Mr. Brown finds for the Congo River and Sierra Leone $\frac{1}{23}$, for India and New Holland $\frac{1}{16}$. In the *temperate* zone compositæ are, in America, $\frac{1}{6}$; and this is probably the proportion borne by compositæ on the very high stations of equinoctial America, to the whole mass of phænogamous plants in the same places; at the Cape of Good Hope $\frac{1}{5}$, in France $\frac{1}{7}$, or more properly $\frac{2}{15}$, in Germany $\frac{1}{8}$, [in Sweden between $\frac{1}{10}$ and $\frac{1}{11}$, in Sicily rather less than $\frac{1}{8}$]. In the *frozen* zone compositæ are, in Lapland $\frac{1}{13}$, in Kamtchatka $\frac{1}{13}$, [in Melville Island $\frac{1}{13}$].

"LEGUMINOSÆ. — *Trop.* America $\frac{1}{12}$, India $\frac{1}{9}$, New Holland $\frac{1}{9}$, western Africa $\frac{1}{8}$; *Temp.* France $\frac{1}{16}$, Germany $\frac{1}{20}$, North America $\frac{1}{19}$, Siberia $\frac{1}{14}$, [Sweden $\frac{1}{22}$, Sicily $\frac{1}{7}$]; *Frozen* $\frac{1}{35}$, [Melville Island $\frac{1}{13}$].

"LABIATÆ. — *Trop.* $\frac{1}{40}$; *Temp.* North America $\frac{1}{40}$, Germany $\frac{1}{20}$, France $\frac{1}{24}$, [Sicily $\frac{1}{22}$, Sweden $\frac{1}{31}$]; *Frozen* $\frac{1}{70}$, [Melville Island 0]. The scarcity of Labiatæ and Cruciferæ, in the temperate zone of the new continent, is a very remarkable phenomenon.

"MALVACEÆ. — *Trop.* America $\frac{1}{47}$, India and Western Africa $\frac{1}{34}$, the coast of Guinea alone $\frac{1}{20}$; *Temp.* $\frac{1}{200}$; *Frozen* 0.

"CRUCIFERÆ. — *Trop.* Scarcely any, except in mountainous regions beyond from 7,000 to 10,000 ft. of elevation; France $\frac{1}{19}$, Germany $\frac{1}{18}$, [Sweden $\frac{1}{19}$, Sicily $\frac{1}{15}$, Balearic Islands according to Cambessedes $\frac{1}{21}$, Melville Island $\frac{1}{7}$] North America $\frac{1}{62}$.

RUBIACEÆ.—Without dividing the family into several sections we find for the *Tropics*, in America $\frac{1}{29}$, in Western Africa $\frac{1}{14}$; for the *Temperate* zone, in Germany $\frac{1}{70}$, in France $\frac{1}{73}$; for the *frozen* zone in Lapland $\frac{1}{80}$. Mr. Brown separates the great family of Rubiaceæ into two groups, distinguished by peculiar relations to climate. That of Stellatæ without stipulæ principally belongs to the temperate zone; it is almost wholly absent under the tropics, except on the summit of mountains. The group, with opposite stipulate leaves, (*Cinchonaceæ*, Lindl.) belongs exclusively to equatorial regions.

" EUPHORBIACEÆ. — *Trop*. America $\frac{1}{35}$, India and New Holland $\frac{1}{30}$, Western Africa $\frac{1}{28}$; *Temp*. France $\frac{1}{70}$, Germany $\frac{1}{100}$, [Sicily $\frac{1}{56}$, Sweden $\frac{1}{166}$, Balearic Islands $\frac{1}{43}$]; *Frozen*, Lapland $\frac{1}{500}$.

"ERICEÆ.— *Trop*. America $\frac{1}{130}$; *Temp*. France $\frac{1}{125}$, Germany $\frac{1}{90}$, North America $\frac{1}{36}$; *Frozen*, Lapland $\frac{1}{20}$.

"AMENTACEÆ. — *Trop*. America $\frac{1}{800}$; *Temp*. France $\frac{1}{50}$, Germany $\frac{1}{40}$, N. America $\frac{1}{25}$; *Frozen*, Lapland $\frac{1}{20}$.

" UMBELLIFERÆ. — Scarcely any in the *tropics* below 7000 ft., but taking together, in equinoctial America, both the plains and the high mountains, $\frac{1}{100}$; in the *Temp*. zone much more in the old than in the new world, France $\frac{1}{34}$, North America $\frac{1}{57}$; *Frozen*, Lapland $\frac{1}{60}$.

"In comparing the two worlds, we find in general in the new continent, under the equator, fewer Cyperaceæ and Cinchonaceæ, and more Compositæ; in the temperate zone fewer Labiatæ and Cruciferæ, and more Compositæ, Ericeæ, and Amentaceæ, than in the corresponding zones of the old world. The families that increase from the equator towards the poles (according to the method of fractions) are Glumaceæ, Ericeæ, and Amentaceæ; those which diminish from the equator to the pole are Leguminosæ, Rubiaceæ, Euphorbiaceæ, and Malvaceæ; the families that appear to attain their maximum in the temperate zone, are Compositæ, Labiatæ, Umbelliferæ, and Cruciferæ."

To these most instructive and interesting remarks Baron Humboldt has added the following table, with which this subject must terminate.

500   GEOGRAPHY.   BOOK VI.

Groups.	Proportion to the whole Mass of Phænogamous Plants.			Direction of Increase.
	Equatorial Zone, Lat. 0°—10°.	Temperate Zone, Lat. 45°—52°.	Frozen Zone, Lat. 67°—70°.	
Agamæ (*Ferns, Lichens, Mosses, Fungi*)	Plains, $\frac{1}{5}$; Mountains, $\frac{2}{3}$	$\frac{1}{2}$	$\frac{1}{1}$	↗
Ferns alone	Countries nearly flat, $\frac{1}{20}$; Countries very mountainous, $\frac{1}{3}$ to $\frac{1}{8}$	$\frac{1}{70}$	$\frac{1}{25}$	↓
Monocotyledones	Old Continent, $\frac{1}{6}$; New Continent, $\frac{1}{6}$	$\frac{1}{4}$	$\frac{1}{3}$	↗↗
Glumaceæ (*Junceæ, Cyperaceæ, Gramineæ*)	$\frac{11}{200}$	$\frac{1}{8}$	$\frac{4}{25}$	↗↗
Junceæ alone	$\frac{1}{22}$	$\frac{1}{90}$	$\frac{1}{5}$	↗↗
Cyperaceæ alone	Old Continent, $\frac{1}{25}$; New Continent, $\frac{1}{30}$	$\frac{1}{20}$	$\frac{1}{10}$	↗↗
Gramineæ alone	Old Continent, $\frac{1}{18}$; New Continent, $\frac{1}{12}$	$\frac{1}{12}$	$\frac{1}{15}$	↓
Compositæ	Old Continent, $\frac{1}{16}$; New Continent, $\frac{1}{10}$	$\frac{1}{18}$	$\frac{1}{35}$	↓
Leguminosæ	Old Continent, $\frac{1}{14}$; New Continent, $\frac{1}{25}$	$\frac{1}{60}$	$\frac{1}{80}$	↓
Rubiaceæ	$\frac{1}{32}$	—	$\frac{1}{500}$	↗
Euphorbiaceæ	$\frac{1}{40}$	America, $\frac{1}{40}$; Europe, $\frac{1}{25}$	$\frac{1}{70}$	↗
Labiatæ	$\frac{1}{135}$	$\frac{1}{500}$	0	↗
Malvaceæ	$\frac{1}{135}$	Europe, $\frac{1}{100}$; America, $\frac{1}{36}$	$\frac{1}{25}$	↗
Ericeæ	$\frac{1}{800}$	Europe, $\frac{1}{45}$; America, $\frac{1}{25}$	$\frac{1}{20}$	↓↑
Amentaceæ	$\frac{1}{300}$	$\frac{1}{40}$	$\frac{1}{60}$	↓↑
Umbelliferæ	$\frac{1}{800}$	Europe, $\frac{1}{18}$; America, $\frac{1}{60}$	$\frac{1}{24}$	
Cruciferæ				

EXPLANATION OF THE SIGNS : ⟶ , the denominator of the fraction diminishes from the equator towards the north pole ; ↗ , the denominator diminishes towards the equator ; ⟶ ⟵ , the denominator diminishes from the north pole and the equator towards the temperate zone ; ⟵ ⟶ , the denominator diminishes towards the equator and the north pole.

From what has now been said, it would seem that the forms assumed by vegetation in different latitudes are dependent upon particular conditions of climate and soil, and that it is to variations of these conditions that we are to ascribe the difference between the Flora of the equator and of the polar regions. And this is no doubt true: but there are, nevertheless, some plants which have a remarkable power of adapting themselves to all climates and circumstances; and there are others which readily naturalize themselves in climates similar to their own. Of the latter, examples present themselves at every step; all the hardy plants of our gardens may in some sort be considered of this nature; for although they do not grow spontaneously in the fields, they flourish almost without care in our gardens. The pine apple has gradually extended itself eastward from America, through Africa, into the Indian Archipelago; where it is now as common as if it were a plant indigenous to the soil; and in like manner the spices of the Indies have become naturalized on the coast of Africa and in the West Indian Islands. Of the former description the instances are not numerous, but they are very remarkable. In the woods of Georgia, in North America, grows the Rosa lævigata, which, while all the other species of rose of that country are entirely different from those of other regions, is identical with the R. sinica of China; to the Flora of which country, that of North America has no resemblance. Samolus Valerandi is found all over the world, from the frozen north to the burning south; associated here with Amentaceæ and similar northern forms, and there mixed with palms and the genuine denizens of the tropics. Above 350 species are said to be common to Europe and North America, and even among the peculiar features of the Flora of New Holland, Mr. Brown recognised 166 European species. The presence of many of such strangers may undoubtedly be referred to the agency of man, by whom they have been transported from climate to climate, along with corn and by other means; as, for example, at Pont Juvenal, near Montpellier, the vicinity of which abounds with Barbary plants; the seeds of which are known to have been brought across the Mediterranean along with the Barbary wool, which is disembarked at that station. In like manner the various kinds of corn have been carried about from country to country for the service of mankind, until their real home has become doubtful. Medicago sativa abounds in Chili, whither it has been transported by the Spaniards; and instances in abundance of similar cases could be produced. But it must not thence be inferred, that all cases of species growing in places far away from their kindred forms, are to be referred to migration: for this, the agency of man, of animals, of seas, of wind, and of torrents, will doubtless have done a great deal; but none of the causes, nor any other with which I am acquainted, will explain the identity of the Calypso borealis, Orchis viridis, and Betula nana of North America and of Europe; of the Potamogetons common to Europe and New Holland; of the Rosa, already adverted to, as common to North America and China; of the wide diffusion of Samolus valerandi, and, most especially, of the identity of the cryptogamic plants of various countries, plants incapable of cultivation, unconnected with the purposes of man, and of all others, the most difficult of transport under any form. To us it appears that such plants must have been originally created in the places where they now exist; the contingent circumstances under which they are found having been favourable to the particular mode of vegetable developement which was necessary for their formation. And this may, I think, be admitted, as a circumstance connected with the original creation of the world, without having recourse to the theory of some philosophers, that Nature exercises at this time the power of producing plants without parents; a subject upon which Professor Link remarks, that "we find buried in the earth the remains of plants which formerly existed, but which are now unknown. New forms have, therefore, been produced by nature different from the first. Wherever a salt spring breaks out at a distance from the sea, its vicinity immediately abounds with salt plants, although none grew there before. When lakes are drained a new kind of vegetation springs up: thus, when the Danish island of Zeland was drained, Vilny observed Carex cyperoides springing up, although that species is naturally not a native of Denmark, but native of the North of Germany. Hence it is easy to infer, that some plants have been produced at one time and others at another, some earlier and some later: and to this cause may be attributed the smaller number of species found upon islands than upon continents, the former having been produced the latest. Perhaps plants change from one to the other, as certain organic bodies when young belong to a less perfect class than when they are older. On the naked rocks we find Lichens, on the mud Confervæ, in ancient strata the remains of Monocotyledonous plants, in more recent strata those of Dicotyledons."

In concluding this important and very interesting subject, I must refer the reader who is desirous of further information to the writings of Mr. Brown in the appendix to Captain Flinders's *Voyagers*, and Captain Tuckey's *Expedition to the Congo;* to M. De Candolle's Essay upon the Geography of Plants, published in the 18th volume of the *Dictionnaire des Sciences Naturelles;* to the numerous writings of Baron Humboldt; and to the observations upon the subject by Professor Schouw, as translated in Brewster's *Edinburgh Journal.*

BOOK VII.

MORPHOLOGY; OR, OF THE METAMORPHOSIS OF ORGANS.

THAT part of botany which treats of the gradual transmutation of leaves into the various organs of a plant, which shows that bracteæ are leaves affected by the vicinity of the fructification, that the calyx and corolla are formed by the adhesion and verticillation of leaves, that the filament is a form of petiole, and the anther of lamina; and, finally, that the ovarium itself is a convolute leaf, with its costa elongated into a style, and the extremity of its vascular system denuded under the form of stigma, is called morphology.

This doctrine has already been treated of in this work, in connection with the different organs of which mention has been made; but it is so curious and important as to deserve especial mention. It seems to have originated with Linnæus, was deeply entered upon by the celebrated poet Göthe, has been universally adopted in Germany, and been partially received in France and England; in the former country by Du Petit Thouars, De Candolle, and others; and in this kingdom by most of the botanists of the present day.

The first idea of the subject appeared in the second volume of the tenth edition of the *Systema Naturæ*, published in 1759, in which Linnæus thus expresses himself: — " Leaves are the creation of the *present* year, bracteæ of the *second*, calyx of the *third*, petals of the *fourth*, stamens of the *fifth*, and the stamens are succeeded by the pistillum. This is apparent from ornithogalums, luxuriant plants, proliferous plants, double flowers, and cardui."

In December, 1760, these novel propositions were sustained by Linnæus in a thesis prepared in the name of his pupil Ullmark, called the Prolepsis Plantarum. The substance of this paper I shall endeavour to condense, in order that it may appear how far the discoveries or hypotheses of some modern writers are entitled to novelty; leaving out, however, all that relates to the physiological explanation given by Linnæus of his doctrine, which, being formed upon notions that, although entertained at that time, are now known to be inconsistent with facts, need not here be repeated.

Linnæus commences by remarking, that " as soon as leaves have expanded themselves in spring, a bud is observable in the axilla of each. This bud swells as the year advances, and in time becomes manifestly composed of little scales; in the autumn the leaves fall off, but the bud remains; and in the succeeding spring swells, disengages itself from its envelopes, and becomes lengthened: when its outer scales have dried up and fallen off, the inner ones are expanded into leaves (like the wings of a butterfly emerging from its pupa), which separate from each other by means of a gradual extension of the young branch, and presently each new leaf is found to contain in its bosom a little scaly bud, which, in the following season, will also be developed as a branch, with other leaves and other buds. Now when I see a tree adorned with leaves, and in the bosom of these leaves provided with its little buds, it is natural to enquire of what do these buds consist. Do they consist of the rudiments of leaves with their gemmules, the latter of other leaves and buds, and so on to infinity, or, at least, as far as the extension of the plant is likely to proceed? Nature organises living beings out of such minute particles, and even from fluids themselves, that the best eye may in vain seek to penetrate far into her mysteries. I shall, however, endeavour to show that the composition of buds does not extend further at one time than provision for six years; just as, among animals, we find the little *volvox globator* containing within the mother its children, grandchildren, great-grandchildren, and great-great-grandchildren down to the sixth generation." The substance of the subsequent observations is this.

"If a plant, which has flowered and fruited for many successive years in a pot, is transferred to a rich soil and warmer station, it breaks forth into branches instead of flowers. Hence it appears that branches and leaves can be produced from the provision made for flowers, provided circumstances are favourable to their development."

As to *bracteæ*, "the bulbs of hyacinths and ornithogalums afford good evidence of their nature. Both bulbs and buds are winter coverings of plants, with this difference, that bulbs are the bases of leaves of the previous year, while buds are the rudiments of leaves of a coming year. Wherever bulbs grow, there are formed the persistent bases of leaves, within which young leaves are to be developed: within these latter leaves repose the buds or rudiments of future plants, exactly the same as the buds of trees. These buds consist of the rudiments of leaves of the succeeding year; small indeed, but containing in their axillæ other rudiments of like nature. Now, if it happens that such a plant flowers, the bud, which would otherwise have produced leaves the year after, is converted into a scape a year earlier; whence it comes to pass that the rudimentary leaves lying in the bud lose a part of their nutriment, in consequence of the sap being drawn off to the fructification: in consequence of which those leaves continue small, assume a different structure, easily wither, and are called by botanists bracteæ. Thus bracteæ are nothing but leaves which would have been developed another year if the plant had not flowered."

As to the *calyx*. "That the calyx is only the approximated leaves of a plant, is apparent from several instances. The calyxes of Pyrus and Mespilus are often expanded into perfect leaves; the rose offers a similar instance of the change. The leaves of Mesembryanthemum barbatum are supplied with a most curious apparatus of hairs; and the calyx consists of five pieces, in all respects similar to the leaves of the stem. The calycine leaves, indeed, are often very small, juiceless, and different from those of the stem, as if scales of buds previous to their development; but that they still are nothing but leaves of the same nature as those of the stem, must be concluded from this, that when plants, roses and

Geum rivale for example, become, in consequence of excessive nutriment, proliferous, the calycine leaves, which before were small and dry, expand into leaves in size, colour, figure, texture, and substance, exactly like those of the stem. Hence it is not to be doubted, that the calyx and the leaves of the stem were in the beginning alike."

As to the *petals*, " It is often very difficult to distinguish them from the calyx. The white corolla of Helleborus niger, after flowering, assumes the green appearance of the calyx. In luxuriant flowers of Rosa and Geum, the corolla sometimes becomes wholly green, and assumes the foliaceous nature of the calyx. As the calyx is nothing but leaves, and as each leaf contains in its axilla the rudiment of a plant consisting of the rudiments of leaves of a future year, it follows that the petals are of necessity the rudiment immediately within the calycine leaves; the petals, therefore, would have been leaves another year, if flowers had not been produced."

As to the *stamens*. " From double flowers it is apparent that stamens do change into petals and petals into calyx. This is so well known that it need not be insisted on. Now, as from the axilla of every leaf arises the rudiment of a plant, and from the axilla of the calyx are produced the petals, which are nothing but mere tender leaves, and as these petals must have, like other leaves, the rudiments of leaves in their axilla, it follows that stamens are so, for they can be transmuted into petals, as the petals can into the leaves of the calyx."

As to the *pistillum*. The evidence of this being also reducible to leaves, is taken from a change observed in the flowers of Carduus heterophyllus and tataricus, in which the style was changed into two green leaves like bracteæ, and from the common conversion of the pistillum into leaves in proliferous individuals of the rose, the anemone, and others. I do not, however, find any clear evidence of Linnæus having entertained a distinct idea of the true origin and structure of the pistillum.

Such were the sentiments of this great naturalist. I adduce them, not only in justice to the memory of that highly gifted man, but also because I know that his opinion carries with it, especially in this country, just that degree of authority which is requisite to fix attention upon a theory apparently too mysterious or paradoxical to make its way without some such powerful assistance.

In the following remarks I shall endeavour to condense what has been said by writers on morphology under two heads: in the first, treating of regular metamorphosis, that is to say, of that which is connected with the structure of all vegetables; and in the second, of irregular metamorphosis, or of that which influences particular plants or parts of plants, and which occurs only in occasional instances.

CHAPTER I.

REGULAR METAMORPHOSIS.

If the structure of a perfect plant is attentively considered, it will be found to consist of a congeries of branches successively produced out of each other from one common stock, and each furnished with exactly the same organs or appendages as its predecessor. This continues until the fructification is produced, when an alteration takes place in the extremity of the fructifying branch, which is incapable, generally speaking, of further prolongation; but, as the branches, before they bore fruit, were repetitions the one of the other, so are the branches bearing fruit also repetitions of each other. If a thousand sterile or a thousand fertile branches from the same tree are compared together, they will be found to be formed upon the same uniform plan, and to accord in every essential particular. Each branch is also, under favourable circumstances, capable of itself becoming a separate individual, as is found by cuttings, budding, grafting, and other horticultural processes. This being the case, it follows, that what is proved of one branch is true of all other branches.

It is also known that the elementary organs used by nature in the construction of vegetables, are essentially the same; that the plan upon which these organs are combined, however various their modifications, is also uniform; that the fluids all move, the secretions all take place, the functions are all regulated, upon one simple plan; in short, that all the variations we see in the vegetable world are governed by a few simple laws, which, however obscurely they may be understood by us, evidently take effect with the most perfect uniformity.

Hence it is not only true, that what can be demonstrated of one branch is true of all other branches of a particular individual, but also, that whatever can be shown to be the principles that govern the structure of one individual, will also be true of all other individuals.

It is particularly requisite that this should be clearly understood, in order that a just estimate may be formed of the nature of the proofs to be now adduced in regard to the doctrines of morphology. Whatever can be demonstrated to be true, with regard to one single individual, is true of all other individuals: whatever is proved with reference to one organ, is proved by implication, as to the same organ, in all other individuals whatsoever.

Moreover, the fact of one organ being readily transformed into another organ, is in itself a strong presumption of the identity of their origin and nature; for it does not happen that one part assumes the appearance and functions of another if they are essentially different. Thus, while the functions of the hand may be performed by the feet, as we know they occasionally are in animals, nothing whatsoever leads the heart to perform the functions or assume the appearance of the liver, or the liver of any other organ. This is one of the arguments of Linnæus.

The first organ which requires consideration is the stipula. It is not present in all plants; but, when it does exist, is found at the base of the footstalk of the leaves. It generally is a membranous process, with a bundle of vessels passing up its centre; or it is entirely destitute of a vascular system. In the rose the former is the case; but nothing is more common than to find a leaflet accompanying the stipula; and in a specimen of Rosa bracteata which I once had in my possession there were no stipulæ, but, in their stead, two pinnated exstipulate leaves. Hence stipulæ are to be considered as rudimentary leaves.

The bracteæ are the organs intermediate between the leaves and the calyx. Their nature is extremely various, sometimes having a greater resemblance to the leaves, and sometimes to the calyx. In some roses, as R. canina, they are obviously dilated petioles, to which a leaflet now and then is attached: in other species, as R. spinosissima, they differ in no respect from the other leaves. In the tulip a bractea is occasionally present upon the scape, a little below the flower;

this is always of a nature partaking both of the leaf and the flower. In Pinus abies the purple scale-like bracteæ often become gradually narrower, and acquire a green colour like leaves. It has been stated by some botanists, that bracteæ are distinguishable from leaves by not producing buds in their axillæ; but the inaccuracy of such a distinction is apparent from a variety of cases. In Polygonum viviparum, and all viviparous plants, the flowers themselves are converted into buds within the bracteæ. There is a bud in the axillæ of every bractea of the rose. The common daisy often bears buds in the axillæ of the bracteæ of its involucrum; in which state it is commonly known in gardens by the name of hen and chickens. In the permanent monster called Muscari monstrosum, a small cluster of branches covered with minute imbricated coloured bracteæ is produced in lieu of each flower. Here all parts of the fructification, instead of remaining at rest to perform their functions, are attempting, but in vain, to become organs of vegetation; or, in other words, to assume that state from which, for the purposes of perpetuating the species, they had been metamorphosed by nature. Hence it is clear that bracteæ cannot be essentially distinguished from leaves.

Such being the case with the bracteæ, let us see if any positive line of demarcation exists between them and the calyx. With the calyx begins the flower properly so named; it forms what some morphologists call the outer whorl of the fructification, and with it commences a new order of leaves, —namely, those of the fructification,—said to be distinguished from the leaves of vegetation by their constantly verticillate arrangement, and by the want of buds in their axillæ. With the leaves of the fructification all power of further increase ceases; the energies of the plant being diverted, when they commence, from increasing the individual to multiplying the species. The general resemblance of the calyx to the ordinary leaves of vegetation is well known: its green colour, and tendency to develope itself into as many leaves as it consists of divisions, especially in double roses, is so notorious that it need not be insisted on. In the case of Mesembryanthemum barbatum, noticed by Linnæus, there is no difference whatsoever between the leaves of the calyx and those of the stem. In a specimen of a cowslip now before me, the calyx is formed of five perfect leaves, in no respect different from the others, except in being a little smaller. The resemblance, however, between the calyx and the stem leaves is often not apparent; but the identity of the calyx and bracteæ is usually more obvious. In Calycanthus, the transition from the one to the other is so gradual, that no one can say where the distinction lies; and in numberless Ericas the resemblance of the bracteæ and calyx is perfect. The divisions of the calyx are also occasionally gemmiferous. A case is mentioned by Röper, in which one of the sepals of Caltha palustris was separated from the rest, and furnished with a bud. And Du Petit Thouars speaks of a specimen of Brassica napus, in which branches were produced within the calyx.

I have myself a monster of Herreria parviflora (*fig.* A), of the same nature. From this it is apparent that the divisions of the calyx are not only not distinguishable from bracteæ, but that there is often a strong tendency in the former to assume the ordinary appearance of leaves. There is, however, another point to which it is necessary to advert, in order to complete the proof of the identity of calyx and leaves; this is the verticillate arrangement of the former.

Leaves are either opposite, alternate, or whorled; and it has been shown, in speaking of them, that these differences depend wholly upon their greater or less degree of approximation. If the leaves of a plant are rightly considered, they will be found to be inserted spirally round a common axis; that is to say, a line drawn from the base of the lower leaf to that of the one above it, thence continued to the next, and so on, would have a spiral direction. When leaves become approximated by pairs, the spire is interrupted, and the leaves are opposite: let the interruption be a little greater, and the leaves become ternate; and if the interruption be very considerable, what is called a whorl is produced, in which several leaves are placed opposite to each other round a common axis, as in Galium. Now a whorl of this nature is exactly of the nature of a calyx, only it surrounds the axis of the plant, instead of terminating it. As we know that such approximations often take place in the stem in the direct line of growth, when the propulsion of the matter of vegetation exists in its greatest activity, there is no difficulty in comprehending the possibility of such an approximation constantly existing at the end of the system of growth, where the propulsion of the matter of vegetation ceases. But the calyx and more inner whorls of the fructification do not always retain their verticillate position; on the contrary, they occasionally separate from each other, and assume the same position with regard to the axis of vegetation as is naturally proper to the leaves. This is particularly striking in a very common, permanent monster of Lilium album, known in the gardens by the name of the double white lily. In this plant the whole verticillation of the parts of fructification is destroyed; the axis is not stopped by a pistillum, but is elongated into a stem, around which the white leaves of the calyx are alternately imbricated; and in double tulips the outer whorl, representing the calyx, frequently loses its verticillate arrangement, and becomes imbricated like leaves of a stem. The same structure also occurs in the double white Fritillaria meleagris. Hence it cannot be doubted, that the calyx consists of leaves in a particular state.

The corolla forms the second line or whorl of the fructification. It consists of several divisions, usually not green, and always alternate with those of the calyx. It is a series of leaves arising within those of the calyx, from which it is sometimes, indeed, very easy to distinguish it; but from which it is so often impossible to discriminate it, that the difference between the calyx and corolla has been one of the most debatable subjects in botany. No limits can be found in Calycanthus; the same is true of Illicium, and several similar plants. In all Liliaceæ, Asphodeleæ, Orchideæ, and Scitamineæ, the only distinction that can be drawn between the calyx and corolla is, that the one is inserted within the other; they are alike in figure, colour, texture, odour, and function.

Whatever, therefore, has been proved to be true of the calyx is also true of the corolla. There are also cases in which the petals have actually reverted to the state of leaves. In a Campanula Rapunculus, seen by Röper, the corolla had become five green leaves like those of the calyx; the same was found in an individual of Verbascum pyramidatum, described by Du Petit Thouars; proliferous flowers of Geum and Rosa, in which the petals were converted into leaves, are adduced by Linnæus.

The third whorl, or series of the fructification, is occupied by the stamens. These often consist of a single row, equal in number to the divisions of the corolla, with which they are, in that case, alternate. The exceptions to this in flowers with a definite number of stamens are not numerous; and such as do occur are to be considered as wanting the outer row of stamens, and developing the second row instead. Thus in Primulaceæ, in which the stamens are opposite to the petals, and therefore belonging to a second whorl, the first makes its appearance in Schwenkia, which undoubtedly forms part of the order, in the form of clavate or subulate processes arising from the sinuses of the limb. These and similar processes, which are far from uncommon in plants, and which are known by a number of different names, such as scales of the orifice of the corolla, glands, nectary, cup, &c. are in most cases metamorphosed stamens. In Narcissus the cup is formed of three stamens of the first row, become petaloid and united at their margins; while the six, which form the second and third rows, are in their usual state, and within the tube. This is shown, firstly, by the frequent divisions of this cup into three lobes, which then alternate with the inner row of perianthium, or the petals; secondly, by a distinct tendency in double Narcissi, particularly N. poeticus, to produce abortive anthers on the margin of the lobes of the cup; and, thirdly, by the genus Brodiæa and its allies. In that genus the crown of the original species consists of three petaloid pieces, not united into a cup as in Narcissus, but wholly separate from each other: in Leucocoryne ixioides these pieces are not petaloid, but clavate; and in Leucocoryne odorata the pieces have the same figure as in L. ixioides, but almost constantly bear more or less perfect anthers. That the anthers are mere alterations of the margins of petals, there is no difficulty in demonstrating. In Nymphæa the passage from the one to the other may be distinctly traced. In double roses the precise nature of this metamorphosis is shown in a very instructive way: if any double rose is examined, it will be seen that those petals which are next the stamens contract their claw into the form of a filament; a distortion of the upper part, or lamina, also takes place; the two sides become membraneous, and put on the colour and texture of the anther; and sometimes the perfect lobe of an anther will be found on one side of a petal, and the half-formed, mis-shapen rudiment of another on the opposite side. In Aquilegia vulgaris this transformation is still more curious, but equally distinct: the petals of that plant consist of a long sessile purple horn or bag, with a spreading margin; while the stamens consist of a slender filament, bearing a small, oblong, two-celled yellow anther; in single and regularly formed flowers, nothing can be more unlike than the petals and stamens; but in double flowers the transition is complete: the petals, which first begin to change, provide themselves with slender ungues; the next contract their margin, and acquire a still longer unguis; in the next the purple margin disappears entirely; two yellow lobes like the cells of the anther take its place, and the horn, diminished in size, no longer proceeds from the base as in the genuine petal, but from the apex of the now filiform unguis: in the last transition the lobes of the anther are more fully formed, and the horn is almost contracted within the dimensions of the connectivum, retaining, however, its purple colour: the next stage is the perfect stamen. No further evidence can, I think, be required of the formation of stamens out of petals; if more is wished for, the first double flower that may present itself to the observer may be appealed to. The conversion of stamens into green leaves is far more uncommon; this, indeed, very rarely occurs. It was seen by Röper in the Campanula Rapunculus already referred to; and Du Petit Thouars found the stamens of Brassica napus converted into branches, bearing verticillate leaves. Thus it appears that the stamens, like the petals, calyx, and bracteæ, are merely modified leaves.

We now come to the consideration of a fourth series of the fructification, the discus: this is so frequently absent, and is of so obscure a nature, that few morphologists take it into their consideration. It is, however, necessary to understand it, if possible, especially as, when present, it occasionally presents itself under very singular and various forms. In many plants it consists of a mere annular fleshy ring, encompassing the base of the ovarium; in others it forms a sort of cup, in which the ovaria are inclosed as in certain Pæonies; and it very frequently makes its appearance as hypogynous glands or scales: it is almost always between the stamens and pistillum. That it is not an organ of a distinct nature may, I think, be safely inferred from its having no existence in a large number of flowers: but if it is not an organ of itself, it must be a modification of something else; and in that view, from its situation, it would be referable either to the stamens or pistillum. It has so little connection with the latter, from which it always separates at maturity, that it can scarcely be referred to it. With the stamens it has, perhaps, a stronger relation: it consists of the same cellular substance as the connectivum of the anthers; is very often of the same colour; whenever it separates into what are called hypogynous glands or scales, these always alternate with the innermost series of stamens. In the Pæony the discus may, in some measure, be compared to the inner row of scales which exists between the stamens and pistillum of the nearly related genus Aquilegia. M. Dunal has noticed half the disk of a Cistus bearing stamens; and a variety of instances may be adduced of an insensible gradation from the stamens to the most rudimentary state of this organ.

The fifth and last series of fructification is the pistillum. Let us first consider this organ in its simple state, and then advert to it in a state of composition. The simple pistillum, that of the pea for instance, consists of an ovarium, bearing its ovula on one side in two parallel contiguous rows, and at its upper extremity tapering into a style which terminates in a stigma. If this organ be further examined, it will be found that there is a suture running down each edge from the style to the base; it will be also seen that the ovula are attached to one of these sutures, and that the style is an elongation of the other; further it will be perceived, that the two sides of the ovarium are traversed by veins emanating from the suture that terminates in the style, and that these veins take a slightly ascending direction towards the suture that bears the ovula. Now, if when the pod of the pea is half grown, it be laid open through the suture that bears the ovula, all these circumstances will, at that time, be distinctly visible; and if it then be compared with one of the leaflets of the plant, it will be apparent that the suture bearing ovula answers to the two edges of the leaf, the suture without ovula to the costa, and the style to the mucro. Hence it might, almost without further evidence, be suspected that the ovarium was an alteration of the leaf; but if the enquiry be carried further in other plants, this suspicion becomes converted into certainty. In the first place, the suture without ovula, which has been said to be the costa, is always external with respect to the axis of fructification, as would be the case with the costa of a leaf folded up and terminating the fructification. In the next place, nothing is more common than to find the pistillum converted either into petals or into leaves: its change into petals is to be found in numerous double flowers; as, for example, double Narcissi, Hibiscus Rosa sinensis, wall-flowers, ranunculuses, saxifrages, and others. These, however, only show its tendency to revert to petals as the representatives of leaves. The cases of its reverting to other organs are much more instructive. In the double Ulex Europæus the ovarium is extremely like one of the segments of the calyx; its ovuliferous suture is not closed: in the room of ovula it sometimes bears little yellow processes, like miniature petals, and its back corresponds to what would be the back of the calyx; no style or stigma are visible; sometimes two of these metamorphosed ovaria are present: in that case the sutures which should bear ovula are opposite to each other, just as the inflexed margins of two opposite leaves would be. In Kerria Japonica, which is only known in our gardens in a double state, the ovaria are uniformly little miniature leaves, with serrated margins corresponding to the ovuliferous suture of the ovarium, and an elongated point representing the style; their interior is occupied by other smaller leaves. Nothing is more common among roses than to find the ovaria converted into perfect leaves; in such cases the margins uniformly occupy the place of the ovuliferous suture, and the costa that of the sterile suture. But the most instructive and satisfactory proof of the pistillum being merely a modified leaf, is to be found in the common double cherry of the gardens. In this plant the place of the ovarium is usually occupied by a leaf extremely similar to those of the branches, but much smaller; it is folded together; its margins are serrated, and, in consequence of the folding, placed so as to touch each other; and they occupy the place of the ovuliferous suture of a real pistillum. The costa of this leaf corresponds to the station of the sterile suture of the ovarium, and is not only lengthened into a process representing a style, but is actually terminated by a stigma. I think, therefore, that all doubt as to the foliaceous nature of the pistillum must cease. There is thus a greater identity of function between the pistillum and the other series of the fructification than would at first appear probable. We seldom, indeed, find it converted into stamens, but it often takes upon itself the form of petals, as has been shown above; and although cases are very rare of pistilla bearing pollen, yet several instances are known of ovula being borne by the stamens. This occurs continually in Sempervivum tectorum; I have shown it to happen in an Amaryllis, known in gardens as the double Barbadoes lily (see the *Horticultural Transactions*, vol. vi.); and it is constantly the case in a particular variety of the common wall-flower, cultivated in the Apothecaries' garden at Chelsea.

Thus we see that there is not only a continuous uninterrupted passage from the leaves to the bracteæ, from bracteæ to calyx, from calyx to corolla, from corolla to stamens, and from stamens to pistillum, from which circumstance alone the origin of all these organs might have been referred to the leaves; but that there is also a continual tendency on the part of every one of them to revert to the form of leaf. Some botanists say, that all this depends upon an alternation of expansion and contraction, which may be compared to the mechanical oscillation of the pendulum, and to the physical alternation observed in animated beings and the periods of life. The leaves, they say, are an expansion of vegetation, and the flower a contraction of it; and in the flower itself, while the calyx is contracted, the corolla is dilated, the stamens again are contracted, and the pistillum expanded; and with the ultimate contraction of the ovulum ceases the vegetable system.

All that has hitherto been said of the pistillum relates to it only in its simple state. In a state of composition it differs so much in appearance from its simple form, that all the old race of botanists was entirely ignorant of the true theory of its construction. No trace is to be found in the writings of Linnæus of any thing which can be construed into an acquaintance with this subject: nor can I find an indication of it in any writer before the appearance of Göthe's *Versuch die Metamorphose der Pflanzen zu Erklären*, in 1790. At section 78. of that work are the following remarkable words:— " Keeping in view the observations that have now been made, there will be no difficulty in discovering the leaf in the seed-vessel, notwithstanding the variable structure of that part, and its peculiar combinations. Thus, the pod is a leaf which is folded up, and grown together at its edges, and the capsule consists of several leaves grown together; and the compound fruit is composed of several leaves united round a common centre, their sides being opened so as to form a communication between them, and their edges adhering together. This is obvious from capsules, which, when ripe, split asunder; at which time each portion is a separate pod. It is also shown by different species of one genus, in which modifications exist of the principle on which their fruit is formed: for instance, the capsules of Nigella orientalis consist of pods assembled round a centre, and partially united; in Nigella Damascena their union is complete."

Having already spoken at length upon this subject, when considering the structure of the fruit, it is not necessary to repeat the arguments here. There is no doubt of the truth of the theory, which is now universally adopted by all philosophical botanists.

As it may thus be proved that all the parts of a flower are merely modified leaves, there can be no difficulty in admitting the following propositions as the basis of morphology: —

"Every flower, with its peduncle and bracteolæ, being the developement of a flower-bud, and flower-buds being altogether analogous to leaf-buds, it follows, as a corollary, that every flower, with its peduncle and bracteolæ, is a metamorphosed branch.

"And further, the flowers being abortive branches, whatever the laws are of the arrangement of branches with respect to each other, the same will be the laws of the arrangement of flowers with respect to each other.

"In consequence of a flower and its peduncle being a branch in a particular state, the rudimentary or metamorphosed leaves which constitute bracteæ, floral envelopes, and sexes, are subject to exactly the same laws of arrangement as regularly formed leaves." (*Outline of the First Principles of Botany*, edit. 2.)

Therefore all theories of structure inconsistent with these propositions must of necessity be vicious. For this reason there is no difficulty in rejecting the hypothesis of M. Dunal, that in a flower every organ consists of two parts, one standing *in front* of the other, and forming what he calls a *chorisie*; than which no doctrine more utterly unsupported by facts, or more entirely irreconcilable with whatever is known of structure, can easily be imagined. To admit it would be to overturn all the admitted rules of philosophy, by exalting a few seeming exceptions to great general laws above those laws themselves.

CHAPTER II.

IRREGULAR METAMORPHOSIS.

It is probable, that all plants have a particular range, in some cases more extended than in others, to which they are best suited in consequence of their constitutional peculiarities, which become visible in consequence of the effect produced by a change of situation, although not appreciable otherwise. The two great agents by which they are affected, that is to say, soil and atmosphere, will, in their natural situations, be nearly uniform. And so long as this uniformity of the conditions under which they exist continues, their structure will remain unchanged; but let an alteration take place, their atmosphere, for instance, change from that of the valley to that of the mountain; the soil from alluvial deposit to chalk or slate, and the mean temperature under which they are formed fall several degrees: or remove a plant from its native spot, and cultivate it in the rich soil of a garden for several generations; thus submitting it to what may be called the effect of domestication. Under such circumstances, an alteration will be produced in the structure of the plant, which will become manifest by external characters. This is what is called irregular metamorphosis; and may be considered the cause of the endless varieties of form into which garden plants are continually sporting. In a wild state these varieties are comparatively rare; while, on the contrary, new forms, miscalled species, are always starting up in every botanic garden. In the garden of Berlin, Link states, that Ziziphora dasyantha, after many years, changed to another form, which might be called Z. intermedia.

But although there is no reasonable doubt that irregular metamorphosis does take place in consequence of some change in the conditions under which plants are formed, the *cosmica momenta* of some writers, yet it is certain that we are entirely ignorant of the specific causes by which metamorphoses are effected. We know that the cellular tissue, and the secreted matter or proper juices, are what chiefly manifest their sensibility of change; but beyond this we know absolutely nothing whatsoever. In this want of information the simplest manner of treating this subject is to take the parts of vegetation in succession, and to state what is known of the irregular metamorphosis of each.

The roots and tubers undergo a vast variety of changes; some of which are the effects of domestication, and others produced in wild individuals. Some grasses, when growing in situations more dry than those to which they have been accustomed, acquire bulbs; as if laying by reservoirs of nourishment to meet the casual want of a sufficient supply of food. Other roots sport, when domesticated, into various forms and colours; as is familiarly exemplified in all those which supply our tables. In the turnip the form varies from spherical to depressed, oblong, and fusiform; the epidermis from white to yellow, purple and green: the same may be said of the radish. The celery, the root of which is fibrous when wild, produces under domestication a fleshy round root like that of a turnip, known in gardens by the name of celeriac. The common potato, the colour of which is usually yellow, produces a variety deeply stained, not on the epidermis only, but through its whole substance with purple. The parsnip varies from fusiform to spherical; and there are hundreds of similar cases of which every body must be aware.

Metamorphosis of the stem is much less frequent than those of the root. The stems of the common cabbage are naturally hard and stringy; but in a variety, called by the French *Chou moellier*, the stem is succulent and fusiform; and in the Kohl Rabi it forms a succulent tumour above the ground, in form and size resembling a turnip. In alpine situations the stem becomes shortened in proportion to the elevation at which it is produced, but it lengthens in low humid situations. Domestication has also rendered tall stems mere dwarf, and dwarf stems taller: the common Dahlia, the mean height of which may be estimated at six feet, has been reduced by cultivation to a stature not exceeding three. Cabbages and many culinary plants have undergone a similar change; while the common hemp has sported into a gigantic variety twice the usual size. The stem occasionally becomes fasciated; that is to say, assumes the appearance of a number of separate stems, glued together side by side, as in the common cockscomb Celosia. This was formerly believed to arise from the union of several stems; a manifest error, as an inspection of a dissected stem will prove: it is an extremely irregular formation, something analogous to that which constantly obtains in Bauhinia.

The leaves undergo a thousand metamorphoses, of which I shall only select a few remarkable cases. They become succulent and roll inwards, forming what gardeners call a heart; as in the cabbage and the lettuce. Their parenchyma extends more rapidly than the veins and margins; this produces puckering, as in curled leaves. If the parenchyma and margin are together produced in excess, we then have what gardeners call a curl, as in the plants known by the respective names of curled cress, curled savory, curled endive, &c. If this tendency to parenchymatous developement proceed much further, the surface is not merely puckered, but processes arise from it in every direction, and occasionally assume grotesque figures, or even the resemblance of other leaves: the Scotch kail of gardeners is an instance of this. The parenchyma is formed more slowly than the veins and margins; this produces what are called cut or pinnatifid leaves, as in many garden plants, such as the cut-leaved Fagus sylvatica, Alnus glutinosa, and others. Occasionally in compound leaves an unusual number of leaflets is produced, as seven in some trefoils in room of three; a doubly pinnate leaf in some roses in lieu of a simply pinnate one. In other plants the reverse occurs: there is a Dahlia, which constantly produces simple leaves in room of compound ones.

In flowers irregular metamorphoses are extremely common; they consist of a multiplication of the petals, of a transformation of petals into stamens, and of a change in colour or in smell. In roses the multiplication of petals is the nearly universal cause of the double state of their flowers; in the Rose Œillet, and many Anemones, impletion depends upon a conversion of petals into stamens. I have, in a paper read before the Horticultural Society, and published in the sixth volume of their Transactions, endeavoured to show that these changes always take place in the order of developement, or from circumference to centre; that is to say, that the calyx is transformed into petals, the petals into stamens, and the stamens into ovarium; but that the reverse does not occur. It is there observed, that alterations of another kind may happen, such as changes in the appearance of the stamens occasioned by abortion; but such metamorphoses are to be considered imperfect attempts on the part of particular organs to revert to their primitive forms. It is further remarked, that if metamorphosis took place from centre to circumference, or in a direction the inverse of developement, it would not be easy to show the cause of the greater beauty of double than of single flowers; because the inevitable consequence of a reversed order of transformation would be, that the rich or delicate colour of the petals, upon which all flowers depend for their beauty, would be converted into the uniform green of the calyx. Such a change, therefore, instead of producing a flower more beautiful than its original, would tend to destroy its beauty. But if the true order of alteration be from the circumference to the centre, and if the different organs of fructification are only susceptible of being converted into those which are next them, and the axis of inflorescence, and if no retrograde action takes place, the reason of the superior beauty of double flowers will be manifest. In the latter case the calyx may, indeed, throw off its dull green colour and assume the vivid hues of the petals, as in the Pæony and primrose, and the petals may dilate themselves, and in attempting to perform the functions of stamens may multiply and transform themselves into a hundred grotesque and curious appearances; but no diminution of beauty or loss of brilliant colours will take place. Such were the opinions I ventured to entertain in 1825; they concern a subject peculiarly exposed to doubt and difference of opinion: but I think that the weight of evidence is in favour of those opinions rather than the contrary.

With regard to colour, its infinite changes and metamorphoses in almost every cultivated flower can be compared to nothing but the alterations caused in the plumage of birds, or the hairs of animals by domestication. No cause has ever been assigned to these phenomena, neither has any attempt been made to determine the cause in plants. We are, however, in possession of the knowledge of some of the laws under which change of colour is effected. A blue flower will change to white or red, but not to bright yellow; a bright yellow flower will become white or red, but never blue. Thus the hyacinth, of which the primitive colour is blue, produces abundance of white and red varieties, but nothing that can be compared to bright yellow; the yellow hyacinths, as they are called, being a sort of pale yellow ochre colour verging to green. Again, the ranunculus, which is originally of an intense yellow, sports into scarlet, red, purple, and almost any colour but blue. White flowers, which have a tendency to produce red, will never sport to blue, although they will to yellow; the rose, for example, and Chrysanthemums. It is also probable that white flowers, with a tendency to produce blue, will not vary to yellow; but of this I have no instance at hand.

Smell varies in degree rather than in nature; some plants, which are but slightly perfumed, as the common China rose, acquire a powerful fragrance when converted to the variety called " the sweet-scented;" but I am not acquainted with any case among flowers in which a positive difference of smell exists in two varieties of the same species.

Metamorphoses of the fruit are very common, and administer largely to the wants of mankind. They consist of alteration in colour, size, flavour, smell, and structure. The wild blue sloe of our hedges has, in the course of ages, by successive domestication, been converted into the purple, white, and yellow plums of our desserts. The wild crab is the original from which have sprung the many coloured, Proteus-like variety of the apple; some of which are destitute of smell, others scented like the pine apple, and a few partaking of the perfume of the rose. In peas the parchment-like lining of the pod occasionally disappears, and the whole substance of the seed vessel consists of lax cellular tissue. In the orange a second fruit is sometimes produced in the inside, agreeing in all respects with the outer fruit, even in peel; this is doubtless due to an attempt at producing a second series of pistilla. In a variety of citrons called the fingered shaddock, well known in China, this tendency to form a second row of pistilla is not only in excess, but the cells of the fruit, in attempting to separate themselves into the simple individuals of which the fruit of the shaddock is ordinarily composed, divide it into distinct lobes irregularly arranged round a common axis.

Having thus passed in review the irregular metamorphoses of plants through all the different parts, there still remains a subject on which it is requisite to say a few words. This is the permanency of such metamorphoses, or their capability of being perpetuated by seeds. It is a general law of nature, that seeds will perpetuate a species but not a variety; and this is no doubt true, if rightly considered: and yet it may be urged, if this be so, how have the varieties, well known to gardeners and agriculturists, for many years been unceasingly carried on from generation to generation without change? The long, red, and round white radishes of the markets, for instance, have been known from time immemorial in the same state in which they now exist. The answer is this. A species will perpetuate itself from seed for ever under any circumstances, and left to the simple aid of nature: but accidental varieties cannot be so perpetuated; if suffered to become wild, they very soon revert to the form from which they originally sprung. It is necessary that they should be cultivated with the utmost care; that seed should be saved from those individuals only in which the marks of the variety are most distinctly traced; and all plants that indicate any disposition to cast off their peculiar characteristics should be rejected. If this is carefully done, the existence of any variety of annual or perennial plant may undoubtedly be prolonged through many generations; but in woody plants this scarcely happens, it being a rare occurrence to find any variety of tree or shrub producing its like when increased by seed.

Plate 1.

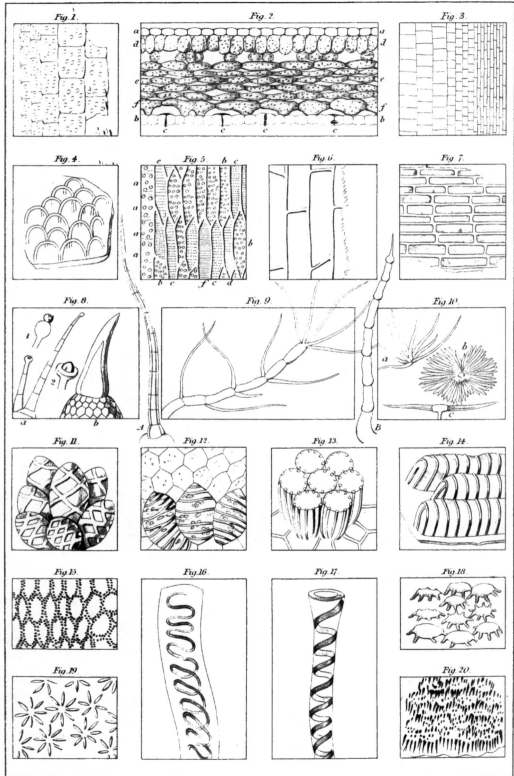

EXPLANATION OF THE PLATES.

N. B. All the figures in the plates, of which the following is an explanation, are more or less magnified: the drawings from which they have been prepared are in all cases original, except where it is stated to the contrary.

PLATE I.

Fig. 1. A small portion of a section of the cellular tissue of the pith of Calycanthus floridus, showing the pore-like spots upon the membrane.

Fig. 2. A section of the leaf of Lilium candidum, after A. Brongniart; *a*, cuticle of the upper surface; *b*, ditto of the lower surface; *c*, stomata cut through in different directions; these last are seen to open into cavities in the parenchyma; *d*, upper layer of parenchyma; *e*, intermediate ditto; *f*, lower ditto.

Fig. 3. Cubical cellular tissue, passing gradually into prismatical, from the stem of the gourd, cut vertically; after Kieser.

Fig. 4. Fibres forming arches in the endothecium of Linaria cymbalaria; after Purkinje.

Fig. 5. Fusiform cellules in the wood of a young branch of Viscum album; after Kieser; *a*, common hexagonal cells of the pith, with grains of amydon sticking to their sides; *b*, fusiform cellules, considered by Kieser to be pierced with holes; *c*, other cells of the same figure, with lines of dots spirally arranged on the membrane; *d*, others, in which the dots are run into lines; *e*, *f*, others, in which the cellules have all the appearance of short spiral vessels. Kieser considers these not as spiral vessels but as cellules of a peculiar kind, replacing spiral vessels in the viscum.

Fig. 6. A portion of the cuticle of Billbergia amæna, with the membrane torn on one side, showing that it does not tear with an even edge, but breaks into little teeth.

Fig. 7. Muriform cellular tissue, forming the medullary processes of Platanus occidentalis. Each cellule contains particles of brownish matter of very irregular size and form.

Fig. 8. *a*, Glandular hairs of the peduncle of Primula sinensis; 1, the glandular apex more highly magnified, with a particle of the viscid secretion of the species on its point; 2, the apex of another hair, showing that the end is open, a conical piece of the viscid secretion lying in the orifice: *b*, a hair of Dorstenia, showing the cellular base from which it arises, and that it consists of a single hollow conical curved cell.

Fig. 9. A branched hair from the cilia of the leaf of a species of Verbascum.

Fig. A. A simple coloured hair in Dichorizandra rufa.

Fig. B. A hair with tumid articulations from the leaf of Gesneria tuberosa.

Fig. 10. *a,* Stellate hairs from the leaf of a species of Hibiscus; *b,* a scale of the calyx of Elæagnus argentea; *c,* a hair of Chrysophyllum Cainito.

Fig. 11. Reticulated cellular tissue from the testa of Maurandya Barclayana.

Fig. 12. Spiral oblong cellules lying among the parenchyma of the leaf of Oncidium altissimum.

Fig. 13. Deep columnar cellules, with parallel fibres from the endothecium of Calla æthiopica; the top of each cell being flat; after Purkinje.

Fig. 14. Arched fibres connected by a membrane in the endothecium of Nymphæa alba; after Purkinje.

Fig. 15. Flat oval cellules, with marginal incisions in the endothecium of Phlomis fruticosa; after Purkinje.

Fig. 16. One of the elastic fibres upon the testa of Collomia linearis, unrolled spirally, and lying within its mucous sheath; magnified 500 times.

Fig. 17. A part of one of the elaters of a Jungermannia, showing a broad spiral fibre loosely twisted inside a transparent tubular membrane, with a dilated thickened mouth.

Fig. 18. Convex membranes, with lateral radiating fibres, forming together imperfect cells; in the endothecium of Veronica perfoliata; after Purkinje.

Fig. 19. Radiating fibres, in the place of cellules in the endothecium of Polygala chamæbuxus; after Purkinje.

Fig. 20. Prismatical depressed cells, with straight fibres on the walls; from the endothecium of Polygala speciosa; after Purkinje.

PLATE II.

Fig. 1. Common woody fibre; *a,* slightly magnified; *b,* very highly magnified, and shown as seen by transmitted light; the extremities only are seen: *c,* cellular tissue.

Fig. 2. Woody fibre from the leaf of Oncidium altissimum, from a preparation by Mr. Griffith. In this there are small tubercles growing from the surface of some of the fibres, irregularly, or arranged in a spiral direction; *a* is magnified 180 times; at *b,* which is magnified 350 times, the form of the tubercles is more distinctly shown; and it is seen that small granules are contained within the fibre.

Fig. 3. The dotted ducts of Zamia horrida. The little oval spaces that have been supposed to be holes, are shown to be opaque; most of them are oblique; but some of them are exactly transverse, rounder, and have a distinct line passing through their longer axis.

Fig. 4. Woody fibre from the stem of Calycanthus floridus: in this the sides of the tubes are marked with small oval dots, exactly as in Zamia.

Fig. 5. A minute portion of a section of the wood of a species of Gnetum from Tavoy; *a,* woody fibre, filled with loose and rather angular particles of greenish matter; *b,* glandular woody fibre, showing its large size in proportion to the other, and the appearance of its glands.

Fig. 6. A vertical radiant section of the wood of yew, magnified 250 times; after Kieser; showing what he calls the spiral porous cells. The spires vary from one to four in each cell; and the glands, when present, are always situated between the spires, as at *a, b, e,* and *g.* Some of the cellules have no glands, as *c, d, f, h, i.* The yew has true spiral vessels besides these.

Fig. 7. A small portion of a vessel in the wood of an Ephedra from Chili,

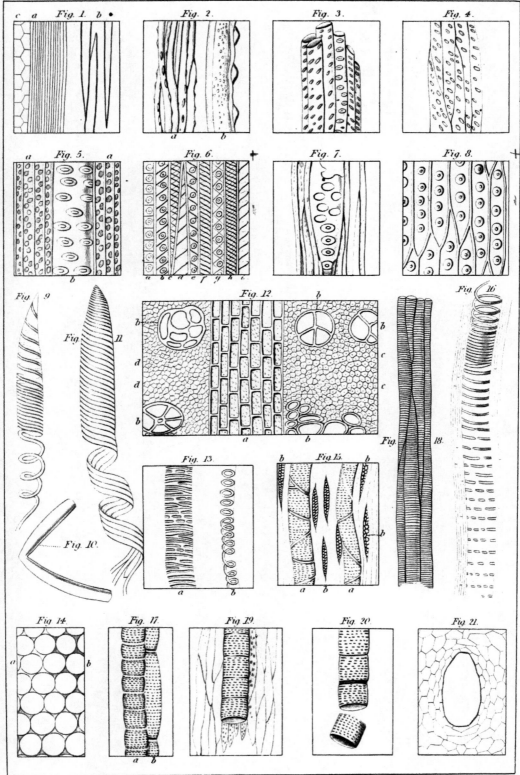

Plate II.

taken from the vicinity of the medullary sheath: its membrane is distinctly perforated at the upper part with oval holes: at the lower part the place of these holes is occupied by glands like those of Gnetum, in fig. 5.; or of Abies, in fig. 8. It would therefore seem, that the oval holes in the membrane of Ephedra are places from which the glands have fallen. Magnified 350 times.

Fig. 8. A vertical radiant section of the wood of Pinus Abies, showing the glands upon the walls of woody fibre, magnified 520 times; after Kieser; but corrected by showing that the glands are convex, and the supposed pores in their centre often opaque.

Fig. 9. The termination of a spiral vessel, extracted from the root of a hyacinth by Mr. Valentine: this shows distinctly the enveloping membrane.

Fig. 10. A fragment of a spiral vessel, bent abruptly to show that it is cylindrical and not flat; magnified 500 times.

Fig. 11. A spiral vessel, with each twist composed of four fibres, from the stem of Nepenthes distillatoria. This is the largest spiral vessel yet known.

Fig. 12. A minute portion of a transverse section of an Indian tree, showing a medullary ray, *a*; the mouths of several bundles of ducts, *b b b*; and the ends of the woody fibre, *c c, d d*, in which the latter are imbedded. The wood of this plant is exceedingly compact; evidently owing to the denseness and stoutness of the woody fibre.

Fig. 13. Two vessels from the stem of Impatiens balsamina; magnified 130 times; after Kieser: *a* is a duct, with the spires broken in some places, and inosculating in others, so as to form the reticulated vessel of this author; *b*, a spiral vessel, with the spires broken at the top into rings.

Fig. 14. A horizontal section of the stem of Tropæolum majus, magnified 130 times; after Kieser. This is to show the intercellular passages, which are unusually large; at *a* they are empty; at *b* filled.

Fig. 15. A tangental section of Sassafras wood, magnified 130 times; after Kieser; *a a*, two banded dotted vessels; or, as Kieser calls them, punctuated spiral vessels; *b b*, the mouths of the medullary rays, showing how they are connected with the bark.

Fig. 16. Dr. Bischoff's representation of the manner in which spiral vessels pass successively into annular and dotted ducts. This figure is imaginary; but Dr. Bischoff asserts that he has actually seen such a vessel in the garden spinage. (*Spinacia oleracea.*)

Fig. 17. Two sorts of dotted vessels from the wood of Phaseolus vulgaris, magnified 130 times; after Kieser: *a*, has the bands much more nearly approximated than *b*, in which the spaces *between* the bands is almost fusiform.

Fig. 18. A bundle of ducts from the stem of a Lycopodium; from a preparation by Mr. Griffith: this shows the manner in which such vessels are packed together when *in situ*, and their terminations.

Fig. 19. A dotted duct and short woody fibre from the stem of Phytocrene gigantea; from a drawing by Mr. Griffith, in Dr. Wallich's *Plantæ Asiaticæ*, t. 216.

Fig. 20. The same, from the same authority, showing that this sort of vessel is really composed of short cylindrical cellules, placed end to end, and opening into each other.

Fig. 21. A section of the cyst, or receptacle of oil in the rind of a lemon, showing that it is a mere cavity built up of cellular tissue.

PLATE III.

Fig. 1. A cluster of six-sided air-cells from the stem of Limnocharis Plumieri: they are formed entirely of prismatical cells; *a a*, partitions dividing the air cells in two.

Fig. 2. A partition or diaphragm of the last-mentioned plant, showing the open passages that exist at the angles of the cells. When dry the rims of the passages are dark, as at *a*: when immersed in water, the dark rim disappears, and the whole partition has the uniform appearance of *b*.

Fig. 3. A portion of the cuticle, and a stoma, of the leaf of Oncidium altissimum; *a*, the stoma, formed of two parallel glands or cells, which open by curving outwards. In this plant the stomata are very minute and few: on the membrane of each mesh of the cuticle are found sticking from four to six spherical semi-transparent green globules.

Fig. 4. Stomata of Strobilanthes Sabiniana. They are very large, and crowded together in an irregular manner.

Fig. 5. Ditto of Croton variegatum: this is an instance of a cuticle with sinuous lines. The orifice of each stoma is closed up with brownish matter.

Fig. 6. A stoma of Canna iridiflora.

Fig. 7. A cavity beneath the cuticle, in the parenchyma of Begonia sanguinea; seen from the inside, so that the cuticle is farthest from the eye. It is divided by sub-cylindrical cellules into five spaces, in each of which there lies a stoma.

Fig. 8. One of the stomata of the same, more magnified, and showing that the medial line does not touch either end, and that the cavity of the stoma is filled with granular matter.

Fig. 9. Stomata of the under side of the leaf of Caladium esculentum, with a portion of cuticle. These appear to be somewhat angular cellules, occupying the centre of every area of the cuticle. The stoma consists of an oval space, in the centre of which is a narrow cleft, with a border distinctly coloured orange or brownish, and having no communication with the circumference: the space between the cleft and the latter filled with a pale green granular substance. The cleft is sometimes seen closed, as at *a*, and then there is scarcely any appearance of a border.

Fig. 10. Cuticle and stomata of Yucca gloriosa: the latter lie in square areolæ, and consist of two parallelograms lying parallel with each other. Small spheroidal bodies, having a luminous appearance under the microscope, stick here and there to the inside of the cuticle.

Fig. 11. Stomata of Limnocharis Plumicri. These also lie in square areolæ, but they have the ordinary structure: they are found in different degrees of openness, or even quite closed, upon a small piece of the same specimen.

Fig. 12. Stamen of Lemna trisulca: anthers bursting vertically.

Fig. 13. Stamen of Polygonum Convolvulus; *a*, seen in front; *b*, from behind; *c*, the connectivum of the anther.

Fig. 14. Stamen of Correa alba; *a*, seen in front; *b*, from behind.

Fig. 15. Stamen of Stachys sylvatica; *a*, filament; *b*, connectivum; *c*, anther, its lobes separated at the base by the connectivum.

Fig. 16. Anther of Alchemilla arvensis; one-celled, and bursting transversely.

Fig. 17. Stamen of Scrophularia chrysanthemifolia; *a*, part of the filament, and

Plate III.

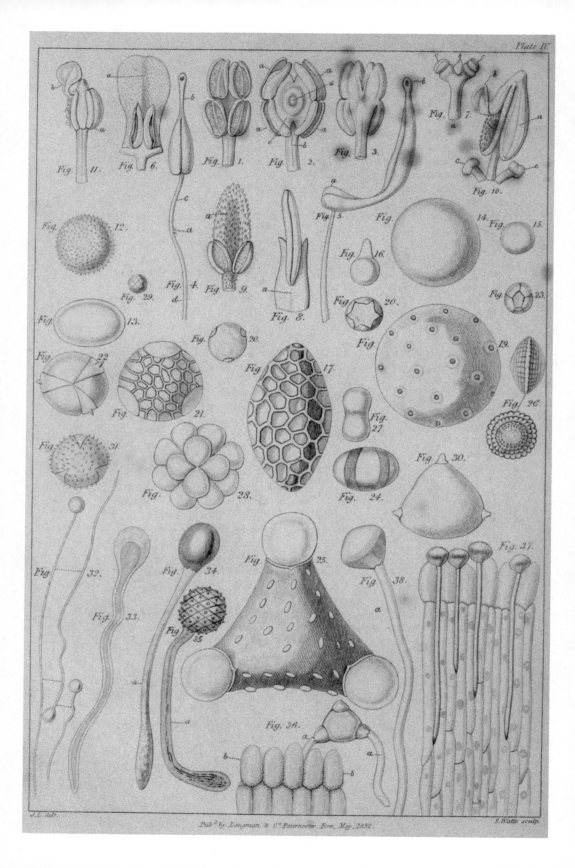

the anther, which is one-celled, after bursting; *b*, the same, before the dehiscence of the anther.

Fig. 18. Anther of Lamium album; its lobes, as in fig. 16., separated at their base by the large connectivum.

Fig. 19. Stamen of a species of Zygophyllum; *a*, the anther; *b*, the filament; *c*, the scale to which the filament adheres.

Fig. 20. The one-celled anther and filament of Callitriche.

Fig. 21. The stamen of Sparganium ramosum.

Fig. 22. The stamen of Vaccinium amænum; *a*, the pores by which the anther bursts.

Fig. 23. The anther of Begonia Evansiana; *a*, the oblique immersed cells; *b*, the connectivum.

Fig. 24. Anther of Cucumis sativa; *a*, seen from the front; *b*, from behind; *c*, the connectivum; *d*, the sinuous lobes of the anther.

Fig. 25. Stamen of Hermannia flammea; *a*, filament; *b*, scale to which the latter has grown.

Fig. 26. Halved stamen of Synaphea dilatata, after Ferdinand Bauer; *a*, filament; *b*, connectivum; *c*, single lobe of the anther after bursting.

Fig. 27. Stamen of Eupomatia laurina, after the same.

Fig. 28. Stamen of Cephalotus follicularis, after the same; *a*, a granular connectivum.

Fig. 29. Stamen of Pterospora Andromedea; *a*, an appendage of the anther.

Fig. 30. Stamen of Securinega nitida; the cells opening transversely.

Fig. 31. Stamen of Chloranthus monostachys; *a*, connectivum.

Fig. 32. Stamen of Eriodendron Samaüma, after Von Martius; anther sinuous and one-celled.

PLATE IV.

Figs. 1, 2, 3. Different views of the stamens and stigma of Stylidium violaceum, after Ferdinand Bauer; *a a*, anthers; *b*, a column formed by the union of their filaments; *c*, a cup-like disk, consisting of the flattened and united apices of the filaments; *d*, the stigma, the style of which is united with the column of filaments through its whole length. Fig. 1. The anthers when burst, seen in front; fig. 3. the same, from behind; fig. 2. the anthers pushed aside so as to show the stigma.

Fig. 4. Stamen of Rhynchanthera cordata, after Von Martius; *a*, a minute membrane that separates the filament *d* from the elongated connectivum *c*; *b*, the attenuated beak-like apex of the anther, opening by a single pore at the point.

Fig. 5. Stamen of Lasiandra Maximiliana, after Von Martius; *a*, dilated bases of the two cells of the anther; *b*, pore at the apex, through which the pollen is discharged.

Fig. 6. Stamen of Glossarrhen floribundus, after Von Martius; *a*, a dilated petaloid connectivum, to the face of which the lobes of the anther adhere; *b*, the filament.

Fig. 7. Stamen of Lacistema pubescens, after Von Martius; *a*, filament; *b*, forked connectivum; *c c*, separated lobes of the anther.

Fig. 8. Stamen of Gomphrena leucocephala, after Von Martius; *a*, broad dilated two-toothed filament, bearing a linear one-celled anther.

Fig. 9. Stamen of Humirium floribundum, after Von Martius; *a*, a large tuberculated petaloid connectivum.

Fig. 10. Stamen of a species of Cryptocarya, from Chili, in which the anther opens, as in other Laurineæ, by valves that roll back when they separate; *a*, one lobe of the anther, with the valve not separated; *b*, the other lobe, with the valve in the act of rolling back; *c c*, abortive stamens under the form of glands.

Fig. 11. Stamen of Berberis vulgaris, exhibiting the same phenomenon; *a*, valve closed; *b*, valve separated and recurved.

⁎⁎⁎ All the following figures of pollen are taken, with scarcely any alteration, from Purkinje, and are drawn to the same scale, so that their relative sizes are known.

Fig. 12. Pollen of Stratiotes aloides.
13.   Calla æthiopica.
14.   Elymus sabulosus.
15.   Avena latifolia.
16.   Scirpus romanus.
17.   Pancratium declinatum.
18.   Populus alba.
19.   Mirabilis Jalapa.
20.   Urtica dioica.
21.   Armeria fasciculata.
22.   Plumbago rosea.
23.   Cineraria maritima.
24.   Salvia interrupta.
25.   Stachytarpheta mutabilis.
26.   Polygala spinosa.
27.   Heracleum sibiricum.
28.   Acacia lophantha.
29.   Iresine diffusa.
30.   Fuchsia coccinea.
31.   Scorzonera radiata.

Fig. 32. Grains of pollen of Gesneria bulbosa emitting their tubes; magnified 180 times. The tube is of extreme tenuity, and may be withdrawn from the stigmatic tissue with great facility. Masses of granular matter may be seen descending the tubes at irregular intervals.

Fig. 33. A grain of pollen of the same plant, with its tube magnified 500 times; this shows that the tube is an extension of the outer membrane of the grain of pollen, if the latter was coated by more than one. The granular matter is seen passing down the tubes, and quitting the grain of pollen, which finally becomes a transparent empty vesicle.

Fig. 34. Grain of pollen of Datura stramonium, emitting its tube; after Brongniart; *a*, pollen-tube.

Fig. 35. Grain of pollen of Ipomæa hederacea, emitting its tube; after Brongniart; *a*, pollen-tube.

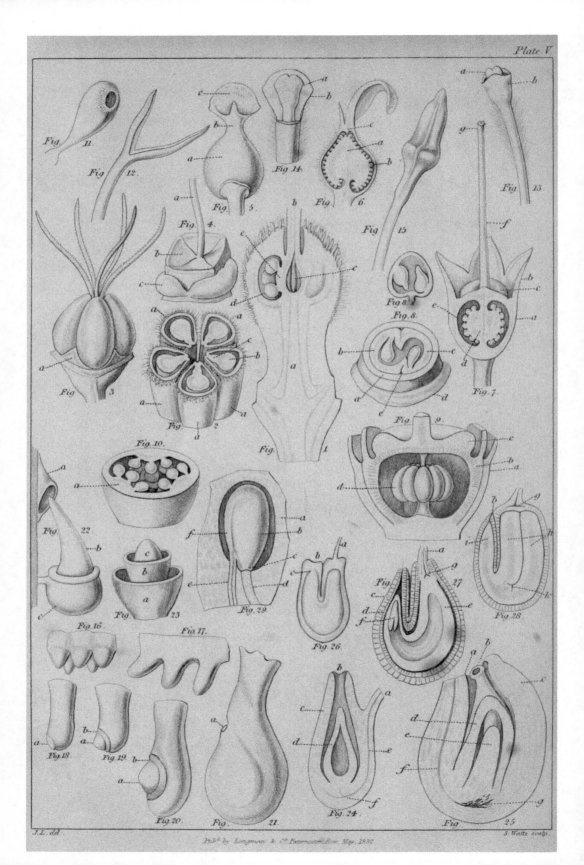

Plate V.

Fig. 36. Mode in which the pollen acts upon the stigma in Œnothera biennis: *a a*, pollen-tubes; *b b*, tissue of the stigma into which these tubes penetrate; after Brongniart.

Fig. 37. Mode in which the pollen acts upon the stigma in Antirrhinum majus; after Brongniart: the pollen sticks to the surface of the stigma, and the tubes plunge down between the utricles of cellular tissue, of which the stigma consists.

Fig. 38. A grain of pollen of the same plant with its tube, more highly magnified: *a*, the pollen-tube.

PLATE V.

Fig. 1. Vertical section of the ovarium of Dictamnus albus; *a*, gynophorus, or elongated base of the ovarium; *b*, base of the style; *c*, cavity where the carpella have not united; *d*, cell; *e*, placenta, with ovula attached to it.

Fig. 2. Transverse section of the same in a more advanced state, where the carpella are beginning to separate: *a a*, carpella; *b*, an ovulum cut through; *c*, placenta.

Fig. 3. Pistillum of Coriaria myrtifolia; consisting of five carpella, each bearing a single linear stigma, and collected round a common elevated axis, the base of which is seen at *a*.

Fig. 4. Ovarium of Lamium album; *a*, base of the style; *b*, carpella pressed together into a square concave body; *c*, fleshy lobed disk.

Fig. 5. Pistillum of Pinguicula vulgaris; *a*, ovarium; *b*, style; *c*, stigma consisting of two very unequal lobes.

Fig. 6. A vertical section of the same; *a*, the central free placenta; *b*, ovula; *c*, point where the placenta is connected, before fertilisation, with the stigmatic tissue.

Fig. 7. A perpendicular section of the pistillum of Vaccinium amænum; *a*, inferior ovarium combined with the tube of the calyx; *b*, limb of the calyx; *c*, epigynous disk; *d*, placenta; *e*, ovula; *f*, style; *g*, stigma.

Fig. 8. A transverse section of the ovarium of Hydrophyllum canadense, showing its remarkable placentation; *a*, wall of the ovarium; *b*, left placenta; *c*, right placenta; *e*, one of their points of union, the other is seen on the opposite side; *d*, a fleshy secreting annular disk. In this case, two placentæ grow up face to face from the base of the ovarium, and gradually unite at their edges, *e*, enclosing the ovula within the cavity they thus form; this is proved by Nemophila, in which the placentation is the same, except that the placentæ are always distinct from each other; one of these placentæ, the ovuliferous face turned towards the eye, is represented at fig. 8.

Fig. 9. A perpendicular section of the inferior ovarium of Thamnea uniflora, after A. Brongniart; *a*, tube of the calyx; *b*, wall of the ovarium; *c*, epigynous disk; *d*, ovula collected round a columnar placenta.

Fig. 10. Transverse section of the ovarium of Viola tricolor, showing its parietal placentation; *a*, one of the three placentas.

Fig. 11. Stigma of the same plant, which is inflated and hollow, with an orifice obliquely situated at its apex.

Fig. 12. Bifid stigma of Chloanthes stæchadis, after Ferdinand Bauer.

Fig. 13. Hairy apex of the style and stigma, with its indusium, of Brunonia australis, after Ferdinand Bauer; *a*, stigma; *b*, indusium.

Fig. 14. The same, divided perpendicularly; *a*, stigma; *b*, indusium.

Fig. 15. Stigma of Banksia coccinea, with a part of the style, after Ferdinand Bauer.

Fig. 16. The earliest state of the ovula of Cucumis anguria; this, and the succeeding figures, to 25 inclusive, are after Mirbel.

Fig. 17. Three of these ovules in a more advanced state.

Fig. 18. An ovulum at the period when the apex of the nucleus *a* is just appearing through the primine. The foramen has already become oblique with respect to the apex of the ovulum.

Fig. 19. An ovulum of the same, at the period when the secundine is appearing through the foramen; *a*, nucleus; *b*, border of secundine; the nucleus is now more oblique than before.

Fig. 20. An ovulum of the same, at a subsequent period, but still long before the expansion of the flower; the several parts are more developed; the nucleus, which at first was terminal, has now become lateral, and is evidently turning towards the base of the ovulum: *a*, nucleus; *b*, border of secundine.

Fig. 21. An ovulum of the same after fertilisation; in the interval between this state and the last, the primine has grown over the secundine and nucleus; the apex of the latter has turned completely to the base of the ovulum; and the foramen is contracted into the little perforation at *a*.

Fig. 22. Ovulum of Euphorbia lathyris, in a very young state, long before the expansion of the flower; *a*, kind of cap projecting from the wall of the ovarium, and into which the apex of the nucleus *b* is inserted; this hood finally closes over the foramen, into which it protrudes as the nucleus retreats; *c*, the primine; the secundine is a similar cap included within the primine.

Fig. 23. Very young ovulum of Ruta graveolens; *a*, the primine; *b*, the secundine; *c*, the nucleus: in the end the primine extends, contracts at its foramen, and closes over the secundine and nucleus.

Fig. 24. Vertical section of an ovulum of Alnus glutinosa; *a*, the umbilical cord; *b*, foramen; *c*, primine (and secundine perhaps united with it); *d*, nucleus; *e*, vessels of the raphe; *f*, place of the chalaza.

Fig. 25. An oblique vertical section of the fertilised ovulum of Tulipa gesneriana; *a*, foramen of the primine (or Exostome); *b*, foramen of the secundine (or Endostome); *c*, primine; *d*, secundine; *e*, nucleus, its apex concealed within that of the secundine; *f*, vessels of the raphe; *g*, place of the chalaza.

Fig. 26. Ovulum of Lepidium ruderale; after A. Brongniart: *a*, umbilical cord; *b*, foramen; *c*, point of the nucleus seen through the primine and secundine.

Fig. 27. Half ripe seed of the same, cut through perpendicularly; after Brongniart: *a*, the umbilical cord; *b*, foramen; *c*, primine; *d*, secundine; *e*, nucleus; *f*, embryo partially formed, its radicle pointing to the foramen; *g*, the point where the nourishing vessels of the placenta expand (the chalaza).

Fig. 28. A perpendicular section of the ripe seed of the same, after A. Brongniart; the primine and secundine are consolidated; and the nucleus is entirely absorbed by the embryo; *a*, umbilical cord; *b*, foramen, now become the micropyle; *g*, chalaza; *h*, cotyledons of the embryo; *i*, radicle; *k*, plumula.

Fig. 29. Mode of fertilisation in Cucurbita Pepo, after Adolphe Brongniart; *a*, portion of the placenta; *b*, ovulum; *c*, its foramen; *d*, the bundle of stig-

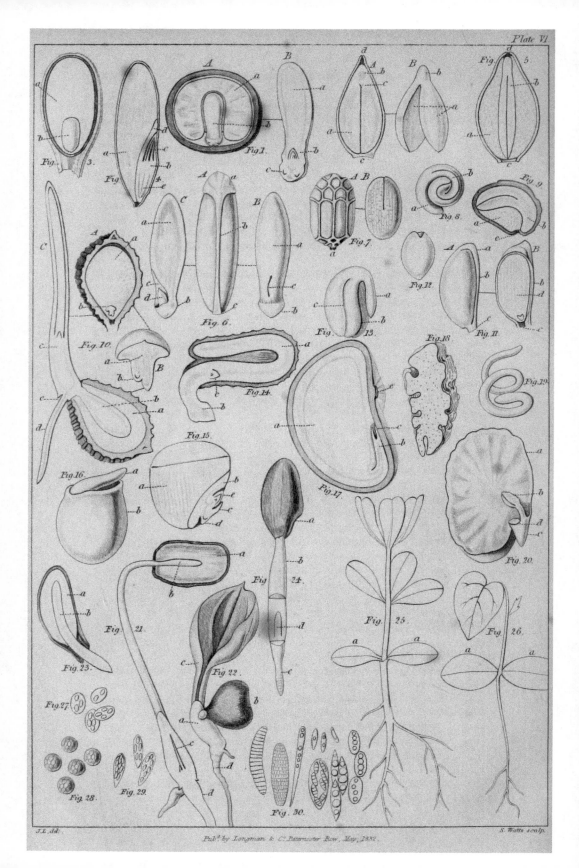

matic tissue through which the fertilising matter is conveyed, and to which the foramen is closely applied; *e*, the bundle of vessels that communicates with the umbilicus; *f*, the commencement of the raphe.

PLATE VI.

Fig. 1. A, Vertical section of the seed of Canna lutea; *a*, albumen; *b*, embryo. — B, Embryo extracted and divided vertically; *a*, cotyledon; *b*, plumula concealed within the embryo; *c*, radicle, with internal rudiments of roots.

Fig. 2. A, Vertical section of the seed of Myrica cerifera; *a*, cotyledons; *b*, radicle; *c*, plumula; *d*, remains of foramen; *e*, hilum. — B, Embryo extracted entire; *a*, cotyledons; *b*, radicle.

Fig. 3. Vertical section of the seed of Luzula campestris; *a*, albumen; *b*, embryo.

Fig. 4. Vertical section of the grain of Bromus mollis; *a*, albumen; *b*, embryo; *c*, its plumula; *d*, its cotyledon; *e*, its radicle, with internal rudiment of a root.

Fig. 5. Vertical section of the seed of Rheum rhaponticum; *a*, albumen; *b*, embryo; *c*, hilum; *d*, remains of foramen.

Fig. 6. A, Seed of Triglochin palustre; *a*, fungous chalaza; *b*, raphe; *c*, hilum. — B, Embryo of the same; *a*, cotyledon; *b*, radicle; *c*, fissure, within which the plumula lies. — C, The same halved vertically; *a*, cotyledon; *b*, radicle; *c*, fissure; *d*, plumula.

Fig. 7. A, Seed of a species of Begonia; *a*, hilum. — B, The dicotyledonous embryo.

Fig. 8. Coiled up embryo of Basella rubra; *a*, radicle; *b*, cotyledons.

Fig. 9. Vertical section of the seed of Mesembryanthemum crystallinum; *a*, albumen; *b*, radicle of the embryo.

Fig. 10. Anatomy of the grain, and germination of Scirpus supinus, after Richard. — A, A vertical section; *a*, albumen; *b*, embryo. — B, The embryo extracted, enlarged, and halved vertically; *a*, cotyledon; *b*, radicle. — C, The seed germinating and halved; *a*, albumen; *b*, cotyledon; *c*, plumula; *d*, young root; *e*, sheath of the latter.

Fig. 11. A, Seed of Ribes rubrum; *a*, chalaza; *b*, raphe; *c*, hilum. — B, The same, halved vertically, showing the minute embryo, with two spreading cotyledons lying at the base of the albumen; *b*, section of the raphe; *c*, hilum; *d*, albumen.

Fig. 12. Embryo of a species of Mammillaria; the cotyledons very small.

Fig. 13. Embryo of Geranium Robertianum; *a*, radicle; *b*, line of union of the two cotyledons; *c*, one of the plaits in the latter.

Fig. 14. Section of the seed of Alisma Damasonium, after Mirbel; *a*, cotyledon; *b*, radicle; *c*, plumula.

Fig. 15. Part of the seed of Olyra latifolia, after Richard; *a*, albumen; *b*, back cotyledon; *c*, front ditto; *d*, radicle; *e*, plumula.

Fig. 16. Embryo of Ruppia maritima, after Richard; *a*, plumula; *b*, cotyledon.

Fig. 17. Vertical section of the seed of Pekea tuberculosa, after Richard; *a*, radicle; *b*, collet; *c*, cotyledons.

Fig. 18. Embryo and ruminated albumen of Eupomatia laurina, after Ferdinand Bauer (a vertical section).

Fig. 19. Spiral twisted embryo of Cuscuta europæa.
Fig. 20. Half the embryo of a bean (Vicia Faba) after Mirbel; *a*, one of the cotyledons; *b*, plumula; *c*, radicle; *d*, scar from which the other cotyledon has been cut.
Fig. 21. Germination of Lachenalia serotina; *a*, albumen; *b*, cotyledon; *c*, plumula; *d*, radicle.
Fig. 22. Germination of Calla æthiopica, after Mirbel; *a*, exterior elongation of the cotyledon; *b*, seed; *c*, front leaf of the plumula; *d*, radicle.
Fig. 23. Germination of Allium Cepa; *a*, albumen; *b*, embryo elongated beyond the testa.
Fig. 24. The same further advanced; *a*, seed; *b*, base of the cotyledon; *c*, radicle or young root; *d*, plumula.
Fig. 25. Germination of Baptisia australis, after De Candolle; *a a*, cotyledons.
Fig. 26. Germination of Cercis siliquastrum, after De Candolle; *a a*, cotyledons.
Fig. 27. Thecæ, or Sporangiola of Erysiphe adunca, after Greville.
Fig. 28. Sporules of Phascum crassinervium, after Greville.
Fig. 29. Asci, or Thecæ of Sphæria tubæformis, after Greville.
Fig. 30. Thecæ of various Lichens, after Von Martius.

INDEX.

I. SUBSTANTIVES.

Abbreviations, 424.
Achenium, 170.
Acicula, 107.
Acid, 134.
Acinus, 166, 167.
Acrosarcum, 179.
Active molecules, 134.
Aculei, 44.
Additional membrane, 158.
Adductores, 201.
Adelphia, 124.
Adnascens, 52.
Adnascentia, 57.
Adnatum, 52. 57.
Æstivation, 106.
Affinity, 361.
Age of trees, 66.
Aiguillons, 42.
Ailes, 119.
Air-cells, 27.
Akena, 178.
Ala, 49.
Alæ, 119. 121.
Alabastrus, 106.
Albigo 299.
Albumen, 134. 187. 360.
Alburnitas, 298.
Alburnum, 65.
Amalthea, 175.
Ambitus, 381.
Amentum, 109.
Ammonia, 254.
Amphanthium, 108.
Amphigastria, 203.
Amphisarca, 177.
Amphispermium, 170.
Anabices, 55.
Analogy, 361.
Analytical method, 316.
Anasarca, 297.
Anatomical differences, 350.
Androcæum, 136.
Androphorum, 124.
Annulus, 196. 201. 208.
Anther, 126. 357.

Anthesis, 106.
Anthodium, 111.
Anthophorum, 152.
Anthozusia, 300.
Antrum, 179.
Apex, 87. 126. 163.
Apices, 122.
Apophysis, 202.
Apothecia, 205. 207.
Appendages of the axis, 78.
Appendages, 116.
Appendices, 57.
Appendix, 121.
Arbor, 48.
Arbrisseau, 48.
Arbusculus, 48.
Arbustum, 48.
Arcesthide, 180.
Arillus, 165. 186.
Arista, 103.
Arsenic, 288.
Artificial arrangements, 309.
Ascelli, 209.
Ascending axis, 44.
Asci, 204. 206.
Ascidium, 96.
Asimina, 176.
Asparagi, 47.
Aulæum, 117.
Awn, 103.
Axilla, 49. 79.
Axis, 107.
——— ascending, 44.
Azote, 1. 252.

Bacca, 166, 167, 168. 170. 178. 179.
Bacca sicca, 170.
Baccaularius, 176.
Bacilli, 53.
Back, 181.
Balausta, 179.
Bâle, 104.
Barbs, 39.
Bark, 61. 240.
Base, 87. 163.

Basigynium, 152.
Basis of vegetation, chemical, 1.
——————, organic, 1.
Baströhren, 12.
Beard, 103.
Blastéme, 188.
Blastus, 191.
Blatt, 79.
Bouquet, 110.
Bourgeon, 49.
Bourrelet, 95.
Bouton, 106.
Boyau, 133.
Brachium, 400.
Bracteæ, 101. 257. 354.
Bracteolæ, 101.
Branches, 47.
Branchlets, 47.
Brindilles, 47.
Bristles, 38.
Buisson, 48.
Bulb, 49. 51.
Bulbilli, 53.
Bulbo-tuber, 55.
Bulbus solidus, 52.
—————— squamosus, 53.

Cachrys, 166. 180.
Cænobio, 176.
Cætonium, 103.
Calamus, 56.
Calathidium, 111.
Calcar, 120.
Calybio, 178.
Calyculus, 102.
Calyptra, 201.
Calyx, 112. 257. 355.
—— communis, 102.
Cambium, 63. 246.
Canaux entrecellulaires, 26.
Canker, 298.
Cap, 208.
Capillamentum, 125.
Capillitium, 209.
Capillus, 400.
Capitulum, 110, 111. 126.
Capreolus, 96.
Capsella, 169.
Capsula, 104. 126. 166, 167, 168, 169. 177.
—————— circumscissa, 177.
—————— siliquiformis, 177.
—————— tricocca, 176.
Carbon, 1.
Carbonic acid, 251.
———— oxide, 254.
Carcerulus, 176.
Carcinoma, 298.
Caréne, 119.
Caries, 298.

Carina, 119.
Carpadelium, 179.
Carpella, 143.
Carpophorum, 152.
Carunculæ, 185.
Caryopsis, 176.
Catulus, 109.
Caudex, 76.
—————— ascendens, 45.
—————— intermedius, 45.
—————— repens, 58.
Caudicula, 131.
Cauliculi, 47.
Cauliculus, 188.
Caulis, 56.
Cayeu, 52.
Cells, 138.
Cellular integument, 61.
—————— tissue, 3. 221.
—————————— membranous, 8.
—————————— fibrous, 10.
—————————— reticulated, 10.
Cellule d'air, 28.
Centimétre, 401.
Cephalanthium, 111.
Cephalodium, 205.
Ceratium, 177.
Cerio, 176.
Cervix, 58.
Cicatricula, 80.
Ciliæ, 38.
Circumscription, 87. 381.
Cirrhus, 96.
Chaff of the receptacle, 102.
Chalaza, 156. 186.
Chalumeau, 56.
Characters, 433.
——————, value of, 349.
Chaton, 109.
Chaume, 56.
Chilblains, 300.
Chlorine, 252.
Chlorosis, 297.
Choking-up, 299.
Classes, 368.
Classification, 306.
Clavicula, 96.
Clavus, 299.
Claw, 118.
Clinanthium, 108.
Clinium, 152.
Clostres, 9.
Cloves, 52.
Clubbing, 300.
Coarcture, 45.
Coccum, 167.
Coleophyllum, 190.
Coleoptilum, 190.
Coleorbiza, 190
Collare, 95.

Collectors, 140.
Collet, 188.
Collum, 45.
Coloration, 300.
Colour, 274.
———, cause of, in tissue, 5.
Colum, 139.
Columella, 164.
Columna, 125.
Coma, 47. 102.
Concentric zones, 64.
Conceptacles, 196. *et seq.*
Conceptaculum, 166. 177.
Conidia, 205.
Coniocysta, 208.
Coniothecæ, 126.
Connectivum, 126. 128.
Contextus cellulosus, 3.
Conus, 180.
Copper, 288.
Corculum, 187.
Corolla, 117. 257. 355.
Corollula, 119.
Cormus, 54.
Cornua, 121.
Corona, 121.
——— staminea, 122.
Corpuscula vermiformia, 25.
Corrosive sublimate, 288.
Cortex, 83.
Cortina, 208.
Corymbus, 110.
Costa, 89.
Cotyledons, 80. 187. 193.
Coulant, 57.
Coussinet, 95.
Creeping stem, 55.
Cremocarpium, 179.
Crini, 38.
Crispatura, 300.
Crusta, 206.
Crypta, 27.
Cubit, 400.
Culmus, 45. 56.
Cupula, 102.
Curling, 300.
Cuticle, 31.
Cyanogen, 254.
Cyma, 47. 112.
Cynarrhodum, 176.
Cyphellæ, 205.
Cypsela, 178.
Cystella, 205.
Cystidium, 170.
Cystulæ, 204, 205.

Débris, 80.
Decimétre, 401.
Decoloration, 300.
Dehiscence, 164.

Descriptions, 440.
Diachyma, 83.
Diagnoses, 433.
Diclesium, 180.
Dieresilis, 176.
Diplöe, 83.
Diplotegia, 179.
Directions, 277.
Diseases, 297.
Disk, 137, 257. 357.
Dissepiments, 148.
———————, spurious, 151.
Dodrans, 400.
Dotting, 354.
Down, 38.
Double follicule, 177.
Dragon, 57.
Drawings, 469.
Dropsy, 297.
Drupa, 166, 167, 168. 170. 175.
Ducts, 23.
———, annular, 23.
———, dotted, 24.
———, reticulated, 23.
———, moniliform, 25.
———, necklace-shaped, 25.
———, strangulated, 25.
Ductus intercellulares, 26.
Dumus, 48.
Duramen, 65.

Elateres, 203.
Elaterium, 176.
Elementary organs, 221.
——————————, spurious, 26.
Ell, 400.
Elytriculi, 106.
Embryo, 187. 360.
———, fixed, 49.
Embryotega, 185.
Endocarpium, 163.
Endogenous stems, 72.
Endopleura, 183.
Endoptiles, 192.
Endorhizæ, 190.
Endosmose, 237.
Endospermium, 187.
Endostome, 155.
Endothecium, 128.
Entrenœud, 46.
Epi, 108.
Epicarpium, 163.
Epiblastus, 191.
Epidermis, 61.
Epillet, 109.
Episperme, 182.
Ergot, 299.
Erythrostomum, 175.
Estivation, 409.
Etærio, 175.

Etendard, 119.
Etiolation, 297.
Excipulus, 206.
Exogenous stem, 59.
Exoptiles, 192.
Exorhizæ, 190.
Exosmose, 237.
Exostome, 155.
Exostosis, 300.
Exothecium, 126.
Extravasation, 298.

Face, 181.
Fasciculus, 110.
Fasergefässe, 12.
Fausses trachées, 23.
Faux, 118.
Ferrugo, 299.
Fibre, 1, 2.
Fibrillæ, 76, 77. 206.
Fibrous cellular tissue, 10.
Filament, 125.
Fistulæ spirales, 17.
Fixed embryo, 49.
Flagella, 47.
Flagellum, 57.
Fleuron, 106.
Flocci, 209.
Florets, 106.
Florets of the ray, 111.
——————— disk, 111.
Flosculi, 106.
Flowerless plants, 305.
Foliation, 53.
Folliculus, 166, 167, 168. 175. 197.
Foliolum, 94.
Foot, 400.
Foramen, 155.
Fornix, 122.
Fovilla, 134.
Fries's dogmata, 336.
Frond, 195.
Fructus inferus, 163.
——————— superus, 163.
Fruit, 160. 266. 359.
Frutex, 48.
Fruticulus, 48.
Fulcra, 96.
Fundus plantæ, 45.
Funiculus, 154.
——————— umbilicalis, 139.

Galbulus, 180.
Galea, 118.
Galls, 298.
Gangrene, 297.
Gemma, 49.
Gemmæ, 204.
Gemmula, 188.
Generic characters, 433.

Geniculum, 46.
Genus, 367.
Geography, 472.
Germen, 138.
Germination, 270.
Gills, 208.
Glands, 42.
Glandes à godet, 43.
——————— cellulaires, 43.
——————— vasculaires, 43.
——————— corticales, 33.
——————— miliaires, 33.
——————— epidermoidales, 33.
——————— vesiculaires, 27.
Glandulæ cutaneæ, 33.
——————— hypogynæ, 137.
——————— impressæ, 27.
——————— lenticulares, 44.
——————— squamosæ, 196.
——————— urceolares, 43.
——————— utriculares, 43.
Glandular hairs, 39.
Glans, 178.
Globules, 207.
Globuli, 205.
Globuline, 6.
Globulus, 205.
Glochis, 39.
Glomeruli, 205.
Glomerulus, 111.
Glomus, 111.
Gluma, 103.
Glumella, 104.
Glumellula, 104.
Gluten, 134.
Goldbach's speculations, 330.
Gonophorum, 152.
Gongyli, 206, 207.
Gousse, 175.
Grands pores, 33.
Granula, 207.
Grappe, 108.
Groh's speculations, 333.
Grossification, 139.
Grossus, 166.
Growth, rapidity of, 7.
Gum, 134.
Gumming, 298.
Gynæceum, 138.
Gynobasis, 137.
Gynophore, 139.
Gynostemium, 126.
Gyroma, 196. 205.
Gyrus, 196.

Hairiness, 38.
Hairs, 38. 352.
———, lymphatic, 40.
———, secreting, 40.
Hairsbreadth, 400.

INDEX. 541

Hami, 39.
Hampe, 107.
Heat in vegetables, 260.
Heart-wood, 65.
Helmet, 118.
Hemigyrus, 175.
Herbaria, 463.
Hesperidium, 178.
Hilofère, 183.
Hilum, 185.
Hirsuties, 38.
Hooks, 39.
Horns, 121.
Hybernaculum, 49.
Hybridism, 301.
Hydrogen, 1. 255.
Hymenium, 208.
Hypanthodium, 108.
Hypha, 207.
Hypoblastus, 191.
Hypophyllium, 95.

Inch, 400.
Indusium, 140. 196.
Induviæ, 80.
Inflammability, 275.
Inflorescence, 106. 354.
Inflorescences, 354.
Inner bark, 61.
Innovations, 47.
Insertion, 353.
Insolation, 297.
Intercellular passages, 26.
Internodium, 46.
Intervenium, 91.
Involucellum, 102.
Involucrum, 102. 196, 197.
Irritability, 288.
Iulus, 109.

Jaundice, 300.
Jet, 57.
Jeterus, 300.
Juba, 112.

Keel, 119.

Labellum, 118.
Lacinula, 120.
Lacuna, 28. 205.
Lamarck's system, 316.
Lame, 118.
Lamella, 121. 208.
Lamina, 79. 86. 118.
———— proligera, 206.
Languette, 95.
Latex, 86.
Laub, 79.
Laying, 57.
Lead, 288.

Leaf, 79. 247. 352.
Leaf-buds, 49. 352.
Leaflet, 94.
Lecus, 54.
Legumen, 166. 168. 170. 175.
———— lomentaceum, 175.
Lenticelles, 44.
Lenticular glands, 44.
Lepals, 136.
Lepicena, 103.
Lepis, 41.
Lepisma, 137.
Liber, 61.
Ligneæ fistulæ, 12.
Ligula, 95. 100. 121.
Limbus, 86. 118.
Lime, 134.
Limes communis, 45.
Line, 400.
Linea, 400.
Linnean system, 310.
Lip, 118.
Lirella, 205.
Loculi, 126.
Locusta, 103. 109.
Lodicula, 104.
Lomentum, 169. 175.
Lorica, 182.
Lorulum, 206.
Loss of colour, 300.
Lousiness, 300.
Luftbehälter, 28.
Lymphæducts, 23.

Mace, 165.
Malate of potash, 134.
Malic acid, 134.
Malleolus, 57.
Marcor, 297.
Margin, 87.
Maturation, 267.
Meatus intercellulares, 26.
Medulla seminis, 187.
Medullary rays, 63. 240.
———— sheath, 64. 239.
Melligo, 299.
Melonidium, 179.
Membrane, 1, 2.
Membranous cellular tissue, 8.
Membranula, 196.
Mercury, 288.
Merithallus, 46.
Mesophyllum, 83.
Metre, 401.
Microbasis, 176.
Micropyle, 185.
Milk-vessels, 27.
Millimétre, 401.
Moria, 106.
Morphology, 504.

Muriate of barytes, 288.
Muscarium, 110.
Mycelia, 209.

Nacelle, 119.
Nail, 400.
Naked seeds, 161.
Natural system, 308.
Nauca, 185.
Neck, 45. 188.
Necrosis, 299.
Nectar, 121.
Nectarilyma, 122.
Nectarium, 104. 121, 122.
Nectarostigma, 122.
Nectarotheca, 120.
Nees's speculations, 324.
Nitrogen, 1. 252.
Nitrous acid, 252.
Nodus, 46.
Nœud, 46.
——— vital, 45.
Nomenclature, 454.
Nucamentum, 109.
Nucelle, 155.
Nucleus, 52. 155. 187. 200. 206.
——— proligerus, 206.
Nucula, 170. 178. 207.
Nuculanium, 178.
Number, 420.
Nux, 167, 168. 170. 174.
——— baccata, 180.

Ochrea, 99, 100.
Oeffnungen, 33.
Offsett, 58.
Oken's speculations, 328.
Olefiant gas, 255.
Omphalodium, 185.
Onglet, 118.
Operculum, 96. 201.
Oplarium, 205.
Orbilla, 205.
Orbiculus, 121. 209.
Order, 368.
Organs, elementary, 1.
——— of vegetation, 100.
——— of fructification, 100.
Orgya, 400.
Ostiolum, 209.
Outline, 381.
Outre, 96.
Ovarium, 138.
Ovula, 138. 152. 359.
Oxalate of lime, 29.
Oxygen, 1. 251.

Pagina, 87.
Paillette, 102.
Palate, 118.

Paleæ, 102.
Paleolæ, 104.
Palm, 400.
Panicle, 111.
Papillæ, 43.
Pappus, 114.
Papulæ, 44.
Paquet, 109.
Paracorolla, 122.
Parapetalum, 122.
Paraphyllia, 99.
Paraphyses, 201. 122.
Parastades, 122.
Parastemon, 122.
Parenchyma, 3. 9.
Paries, 150.
Parties élémentaires, 1.
——— similaires, 1.
Patellula, 205.
Pecten, 209.
Pedicel, 106.
Pediculus, 125.
Pedunculus, 106.
Pelta, 205.
Pentakenium, 179.
Pepo, 167, 168. 179.
Peponida, 179.
Perapetalum, 122.
Peraphyllum, 115.
Perianthium, 113.
Pericarpium, 162.
Pericladium, 95.
Periclinium, 102.
Peridiola, 207.
Peridiolum, 209.
Peridium, 209.
Peridroma, 195.
Perigonium, 114.
Perigynandra communis, 102.
——————— interior, 117.
Perigynium, 104. 137.
Periphoranthium, 102.
Perisperm, 182. 187.
Perisporum, 105.
Peristachyum, 103.
Peristomium, 201.
Perithecium, 209.
Pernio, 300.
Pes, 400.
Petals, 117.
Petiole, 79. 86. 94.
Petiolule, 94.
Petits tubes, 12.
Phorantium, 108.
Phosphate of lime, 134.
——————— of magnesia 4.
Phthiriasis, 300.
Phycomater, 207.
Phycostemones, 137.
Phyllodia, 95.

INDEX.

Phyllum, 114.
Physiological differences, 351.
Pileus, 208.
Pili, 38.
—— capitati, 39.
—— Malpighiacei, 40.
—— pseudo-Malpighiacei, 40.
—— biacuminati, 40.
—— subulati, 38.
—— scutati, 41.
Pilidium, 205.
Pilula, 166. 180.
Pistillidia, 201.
Pistillum, 138. 262. 358.
Pith, 60. 239.
Pithy part, 3.
Placenta, 139. 144.
————, free central, 151.
Plateau, 54. 139.
Plopocarpium, 175.
Plumula, 188.
Podetia, 205.
Podogynium, 152.
Podospermium, 139.
Poils en écusson, 41.
—— en goupillon, 40.
—————— navette, 40.
—————— fausse navette, 40.
Poinçon, 110.
Poisons, 288. 293.
Polakenium, 179.
Polarity, 325.
Polexostylus, 176.
Pollen, 129. 357.
—— tubes, 133.
Pollex, 400.
Polychorion, 175.
Polyphorum, 152.
Polysecus, 175.
Pomum, 167, 168. 170. 179.
Pores of the epidermis, 33.
—— corticaux, 33.
—— allongés, 33.
—— évaporatoires, 33.
—— in tissue, nonexistent, 2. 4.
Potash, 288.
Prefloration, 409.
Prickles, 44. 352.
Primine, 155.
Propaculum, 58.
Propagines, 53.
Propago, 57.
Propagula, 205.
Proper juice, 230.
Proportion, 360.
Prosenchyma, 9.
Prosphyses, 201.
Protoxide of nitrogen, 255.
Prussic acid, 295.

Pseudobulb, 58.
Pseudohymenium, 207.
Pseudoperidium, 207.
Pseudoperithecium, 207.
Pseudo-tuber, 77.
Pteridium, 176.
Pterodium, 176.
Pubescence, 38.
Pulpa, 3.
Pulvinuli, 205.
Pulvinus, 95.
Punctuation, 452.
Putamen, 163.
Pyrenarium, 179.
Pyridium, 179.
Pyxidium, 177.

Quartine, 158.
Quintine, 158.

Raceme, 108.
Rachis, 107. 195.
Rachitis, 300.
Radicle, 187.
Radiculæ, 76.
Radiculoda, 191.
Radii, 110.
Ramastra, 94.
Rameau, 47.
Ramenta, 41.
Rami, 47.
Ramilles, 47.
Ramuli, 47.
Raphe, 126. 156. 186.
Raphides, 29.
Receptacle 152. 196.
—————— of the seeds, 139.
—————— of the flower, 107.
Receptacles of secretion, 27.
—————— of oil, 27.
Receptacula succi, 27.
Regma, 176.
Reliquiæ, 80.
Respiration, 251.
Réservoir d'air, 28.
Réservoirs accidentels, 27.
—————— en cœcum, 27.
—————— vésiculaires, 27.
—————— du suc propre, 27.
Reticulum, 99.
Rhizina, 76.
Rhizoma, 58. 187.
Rhizula, 76.
Rictus, 118.
Rootstock, 58.
Root, 76. 227. 351.
Rostella, 39.
Rostellum, 187.
Rostrum, 121.

INDEX.

Runner, 57.
Rupturing, 165.
Russia mats, 62.

Sac of the embryo, 158.
Saccus, 121.
Saftröhren, 23.
Salsugo, 299.
Samara, 167, 168. 176.
Sap, 230.
Sap-vessels, 23.
Sapwood, 65.
Sarcobasis, 176.
Sarcocarpium, 163.
Sarcodermis, 183.
Sarcoma, 137.
Sarmentum, 57.
Sautilles, 53.
Scabrities, 43.
Scale, 102.
Scales, 41.
Scaliness, 300.
Scape, 107. 188.
Scapellus, 188.
Scaphium, 119.
Scar of leaves, 80.
Schelling's speculations, 327.
Schraubengefässe, 17.
Scion, 47.
Scobina, 107.
Scorching, 297.
Scutella, 204.
Scutellum, 191.
Scutum, 122.
Scyphus, 121. 205.
Secondine, 155.
Secundinæ internæ, 187.
Seed, 162. 181. 270. 360.
Seeds, naked, 161. 194.
Sepals, 114.
Septa, 148.
——— in woody fibre, 14.
Septum, 126.
Sertulum, 110.
Seta, 201.
———, of grasses, 104.
Setæ, 38. 42.
———, hypogynous, 105.
Sexual system, 310.
Sheath, 95.
Shields, 204.
Shrub, 48.
Signs, 422.
Silica, 134.
Silicula, 177.
Siliqua, 166. 168. 170. 177.
Silver grain, 63.
Similary parts, 1.
Small span, 400.
Smell, 274.

Soboles, 55. 57.
Solubility, 165.
Soredia, 204, 205.
Sori, 196.
Sorosis, 180.
Souche, 54.
Sous-arbrisseau, 48.
Spadix, 110.
Span, 400.
Spatha, 103.
Spathella, 104.
Species, 365.
Specific characters, 433.
Speculative opinions, 324.
Spermaphorum, 139.
Spermatocystidium, 126. 200. 203.
Spermidium, 174.
Spermodermis, 182.
Sphærula, 209.
Sphalerocarpum, 180.
Spicula, 103. 109.
Spike, 108.
Spilus, 185.
Spines, 48.
——— of the leaves, 98.
Spiral vessels, 17. 223.
Spiralgefässe, 17.
Spithama, 400.
Spongiolæ seminales, 43.
Spongiole, 77.
Sporangiola, 209.
Sporangium, 209.
Sporidia, 207. 209.
Sporidiola, 209.
Sporule, 196, *et seq.*
Spotting, 299.
Spurious bacca, 169.
——— drupe, 169.
——— capsule, 169.
——— nut, 169.
Squama, 102.
Squamatio, 300.
Squamelles, 102.
Squamulæ, 104.
Staining, 300.
Stamens, 122. 262. 356.
Staminidia, 203.
Standard, 119.
Starch, 134.
Steffens's speculations, 328.
Stem, 44. 56. 361.
Stephanöum, 178.
Sterigmum, 176.
Stigma, 141.
Stimuli, 38.
Stings, 38.
Stipellæ, 99.
Stipes, 45. 195. 208.
Stipulæ, 99. 203. 354.
Stole, 57.

INDEX.

Stolo, 57.
Stomata, 33. 84. 354.
Stomatia, 33.
Stragulum, 103.
Straw, 56.
Striga, 42.
Strobilus, 167. 169. 180.
Stroma, 209.
Strophiolæ, 185.
Struma, 95. 202.
Style, 139.
Stylopodium, 137.
Stylotegium, 121.
Sucker, 57.
Suffocatio, 299.
Suffrutex, 48.
Sugar, 134.
Sulphate of magnesia, 288.
——— of potash, 134.
Sulphuretted hydrogen, 252.
Sulphuric acid, 288.
Sulphurous acid gas, 252.
Surculus, 57.
Surface, 87.
Surgeon, 57.
Suture, 144.
Syconus, 180.
Syncarpium, 176.
Synochorion, 176.
Synonyms, 460.
System, Lamarck's, 316.
———, analytical, 316.
———, natural, 318.
———, Jussieu's, 321.
———, Linnean, 310.
———, artificial, 339.
———, natural, 339.
———, factitious, 339.
———, mathematical, 340.
———, physiological, 340.
———, philosophical, 340.
———, speculative, 325.

Tabes, 297.
Talaræ, 119.
Taste, 274.
Tegmen, 103. 183.
Tegmenta, 51.
Tela cellulosa, 3.
Tendril, 96.
Tercine, 155.
Terminology, 454.
Testa, 182.
Testiculus, 126.
Testis, 126.
Thalamus, 108. 152. 209.
Thallus, 202. 209. 267.
Theca, 126.
Thecæ, 126. 196, *et seq.*
Thecaphore, 139. 152.

Thecidium, 174.
Throat, 118.
Thyrsula, 112.
Thyrsus, 112.
Tigelle, 188.
Tin, 288.
Tissue, cellular, 3.
———, vascular, 17.
Tissu cellulaire allongé, 12.
——————— ligneux, 12.
——— organique, 1.
Toise, 400.
Tomentum, 38.
Torus, 152.
Tracheæ, 17.
Transportation of seed, 272.
Tree, 48.
Trica, 205.
Trichidium, 209.
Tronc, 47.
Trophopollen, 126.
Trophospermium, 139.
Truncus, 45.
———, ascendens, 45.
Tryma, 178.
Tube, 118. 139.
Tuber, 55.
Tuberculum, 55. 205.
Tubes corpusculifères, 23.
Tumid excrescences, 298.
Tunic, 53.
Tunica externa, 182.
——— interna, 183.
Turio, 47.
Turpentine vessels, 27.
Tympanum, 201.

Ulna, 400.
Umbel, 110.
Umbilical cord, 189.
Umbilicus, 185.
Unci, 39.
Uncia, 400.
Undershrub, 48.
Unguis, 118. 400.
Urceolus, 104.
Uredo, 299.
Utriculus, 167, 168. 170.

Vaisseaux en chapelet, 25.
——— étranglés, 25.
——— lymphatiques, 23.
——— pneumatiques, 23.
——— fendu, 24.
——— propres, 27.
——— propres fasciculaires, 12.
Vagina, 95.
Vaginellæ, 41.
Value of characters, 349.
Valves, 104. 126.

Valvulæ, 104.
Vasa fibrosa, 12.
—— moniliformia, 25.
—— propria, 27.
—— spiralia, 17.
Vascular system, 223.
—————— tissue, 17.
Vasculum, 96.
Vegetable tissue, 1.
Veil, 208.
Veins, 88.
Velvet, 38.
Velum, 208.
Velumen, 38.
Venæ, 90.
Venulæ, 90.
Verminatio, 300.
Vernation, 53. 409.
Verrucæ, 43.
Verticillaster, 112.
Verticillus, 80.
Vesicula, 95. 207.
———— amnios, 158. 185.
———— colliquamenti, 185.
Vessels of the latex, 86.
Vexillum, 119.
Villus, 38.

Vimen, 47.
Vine, 58.
Virgultum, 47.
Vital vessels, 13.
Vitellus, 185.
Viticula, 58.
Vittæ, 27.
Volva, 208.
Vrille, 96.

Wagner's speculations, 330.
Warts, 43.
Welting, 297.
Whorl, 80.
Wilbrand's speculations, 326.
Wings, 119.
Wood, 241.
Woody fibre, 12. 222.
—————————, glandular, 15.
—————————, granular, 15.

Xylodium, 174.

Yellow resin, 134.

Zellengewebe, 3.

II. ADJECTIVES.

Abdominal, 381.
Abnormal, 381.
Abruptè pinnatus, 390.
Acaulis, 58.
Accisus, 386.
Accretus, 416.
Acerosus, 382.
Acetabuleus, 379.
Aciculated, 394.
Acinaciformis, 375.
Acormous, 58.
Accumbent, 193.
Aculeatus, 394.
Acuminatus, 385.
Acuminose, 385.
Acute, 385.
Adhering, 416.
Adnate, 128. 208. 416.
Adventitious, 50.
Adversus, 414.
Æqualis, 384.
Æqualivenium, 91.
Æquilaterus, 384.
Æruginosus, 405.
Aggregatus, 419.
Alaris, 416.
Alatus, 377.
Albescens, 403.
Albidus, 403.

Albus, 403.
Alsinaceous, 119.
Alternately pinnate, 390.
Alternative, 411.
Alternus, 418.
Alutaceus, 398. 405.
Alveolatus, 393.
Amphigenus, 417.
Amphitropal, 193. 415.
Amplexus, 411.
Amplexicaulis, 415.
Amplectans, 415.
Anastomozing, 393.
Anatropous, 156.
Anceps, 375.
Androus, 124.
Anfractuosus, 413.
Angiocarpien, 171.
Angular, 375. 387.
Anisodynamous, 192.
Anisobrious, 192.
Anisos, 400.
Anisostemonous, 400.
Annexus, 416.
Annotinus, 402.
Annual, 249.
Annular, 23.
Annulatus, 394.
Anthracinus, 404.

INDEX.

Anticus, 128. 414.
Antitropal, 193. 415.
Appendiculate, 116.
Apetalous, 119.
Apice circinatus, 385.
Apiculatus, 385.
Apocarpus, 144.
Arachnoid, 39. 396.
Arcuatus, 374.
Areolate, 394.
Argenteus, 403.
Argo, 403.
Aristatus, 385.
Armeniacus, 405.
Arrectus, 411.
Arrow-headed, 383.
Articulatus, 392. 416.
Artiphyllous, 46.
Ascending, 159. 412.
Ascens, 422.
Asper, 39. 395.
Aspergilliformis, 392.
Assurgens, 412.
Ater, 403.
Atratus, 404.
Atrovirens, 405.
Attenuatus, 384.
Aurantiacus, 405.
Aurantius, 405.
Auratus, 404.
Aureus, 404.
Auriculatus, 383.
Autocarpien, 163.
Avenium, 91.
Awl-shaped, 382.
Awned, 385.
Axe-shaped, 375.
Axillary, 79. 416.
Azureus, 406.

Baccatus, 385.
Baccien, 171.
Badius, 404.
Banded, 407.
Band-shaped, 381.
Barbatus, 39. 396.
Basal, 416.
Basilaris, 416.
Basinervia, 93.
Beaked, 376. 385.
Bearded, 39. 125. 396.
Bell-shaped, 378.
Bellying, 381.
Berried, 399.
Biconjugatus, 391.
Biconjugato-pinnatus, 390.
Bicornis, 376.
Bidentate, 387.
Bidigitato-pinnatus, 390.
Biduus, 402.

Bifarius, 418.
Biferus, 401.
Bifidus, 388.
Bifoliolate, 390.
Bifurcate, 125.
Bigeminate, 391.
Bijugus, 391.
Bilobus, 388.
Bimestris, 402.
Bimorious, 106.
Binate, 390.
Bipartitus, 388.
Bipinnate, 391.
Biserrate, 387.
Bitten, 386.
Bitten off, 77.
Bladdery, 380.
Blotched, 407.
Blunt, 385, 386.
Blunt with a point, 385.
Blurred, 408.
Boat-shaped, 376.
Bony, 398.
Bordered, 407.
Botuliformis, 380.
Brachialis, 400.
Brachiate, 47. 414.
Branched, 39. 391.
Bristle-pointed, 385.
Bristly, 394.
Brunneus, 404.
Brush-shaped, 379. 392.
Buckler-shaped, 374.
Butterfly-shaped, 119.
Byssaceous, 392.

Caducous, 249. 402.
Cæspitose, 418.
Calcareus, 403.
Calyptrate, 114.
Campaniformis, 378.
Campanaceus, 378.
Campanulatus, 118. 378.
Campulitropous, 156.
Canaliculatus, 376.
Cancellate, 393.
Candidus, 403.
Canescens, 403.
Canus, 403.
Capillaris, 376. 400.
Capitatus, 386.
Capsulaire, 171.
Carcerulaire, 171.
Carinatus, 376.
Carneus, 406.
Carnosus, 398.
Cartilaginous, 398.
Caryophyllaceous, 119.
Cassideous, 119.
Catapetalous, 118.

Caudatus, 385.
Caulinus, 416.
Caulocarpous, 401.
Cenobionnaire, 171.
Centrifugal (inflorescence), 110.
——————— (embryo), 193.
Centripetal (embryo), 193.
——————— (inflorescence), 110.
Ceraceus, 398.
Cereus, 398.
Cerinus, 405.
Cernuous, 413.
Cervinus, 405.
Chaffy, 396.
Chalk-white, 403.
Channelled, 376.
Chartaceus, 397.
Chloro, 405.
Chryso, 404.
Cicatrisatus, 394.
Ciliated, 396.
Cineraceus, 403.
Cinereus, 403.
Cinnabarinus, 406.
Cinnamomeus, 404.
Circinate, 53. 411. 413.
Circumscissile, 164.
Cirrhous, 385.
Citreus, 404.
Citrinus, 404.
Clavate, 125. 373.
Claviformis, 373.
Closed, 47.
Clouded, 407.
Club-shaped, 573.
Clustered, 419.
Clypeatus, 374.
Coacervatus, 419.
Coadnatus, 416.
Coalitus, 416.
Coated, 398.
Cobwebbed, 396.
Cochlearis, 411.
Cochleate, 373.
Coccineus, 406.
Cœruleus, 406.
Cæsious, 397. 406.
Cohering, 416.
Comb-shaped, 389.
Combinatè venosus, 92. 408.
Compositus, 389. 418.
Compound, 21. 46. 389.
Compound (leaf), 87.
Compressed, 9. 182. 374.
Conduplicate, 53. 410.
Confertus, 418.
Confluens, 416.
Conglomeratus, 419.
Conical, 372.
Conjugatus, 391.

Connate, 82. 415.
Connivens, 414.
Conoidal, 372.
Continuus, 420.
Contortus, 411.
Convergenti-nervosus, 91.
Converginervis, 408.
Converging, 414.
Convolute, 53. 410. 412.
Coracinus, 404.
Cordatus, 383.
Cordiformis, 383.
Coriaceus, 398.
Corky, 398.
Corneus, 393.
Corniculatus, 376.
Cornutus, 376.
Corollaris, 96.
Coronans, 416.
Corrugatus, 410.
Corticatus, 398.
Corymbose, 110.
Costatus, 92. 408.
Cotyliform, 379.
Crassus, 398.
Crateriformis, 379.
Cream colour, 403.
Creeping (stem), 55.
Crenated, 386.
Crescent-shaped, 383.
Crested, 377.
Cretaceus, 403.
Crinitus, 396.
Crispus, 387.
Cristatus, 377.
Croceus, 405.
Crowded, 418.
Cruciate, 119.
Crustaceous, 398.
Cubical, 373.
Cubitalis, 400.
Cucullatus, 380.
Cunearius, 382.
Cuneatus, 382.
Cuneiformis, 382.
Cupola-shaped, 379.
Cupreus, 406.
Cup-shaped, 378.
Cupuliformis, 379.
Curbed, 387.
Curvative, 410.
Curve-ribbed, 408.
Curved, 374.
Curve-veined, 92.
Curvinervis, 408.
Curvinervium, 92.
Curvivenium, 92.
Cushioned, 374.
Cuspidate, 385.
Cut, 388.

INDEX. 549

Cyaneus, 406.
Cyano, 406.
Cyathiformis, 378.
Cyclical, 192.
Cylindrical, 9. 372.
Cymbiformis, 376.

Dædaleous, 386.
Dealbatus, 397. 403.
Decandrous, 125.
Deciduous, 249. 402.
Declinate, 124. 412, 413.
Decompound, 389.
Decreasingly pinnate, 390.
Decumbens, 414.
Decurrent, 81. 208. 415.
Decursively pinnate, 390.
Decursivus, 415.
Decussatus, 419.
Deflexus, 413.
Dehiscent, 163.
Deliquescens, 392.
——————— (caulis), 48.
Deliquescent (panicle), 111.
Deltoid, 383.
Demersus, 412.
Dendroides, 392.
Dentatus, 387.
Denudatus, 397.
Depauperatus, 419.
Dependens, 412.
Depressed, 182. 374. 399.
Descending, 412.
Determinatus (caulis), 48.
Dewy, 397.
Diadelphous, 124.
Diandrous, 124.
Dichotomous, 392.
Didynamous, 124.
Didymus, 392.
Dieresilien, 171.
Diffuse, 414.
Digitato-pinnate, 390.
Digitaliformis, 380.
Digitatus, 388.
Digitinervius, 93.
Dimidiate, 201. 377. 384.
Dipterus, 377.
Directè venosus, 93. 408.
Disappearing, 392.
Discoidal, 374. 407.
Distans, 420.
Distichous, 418.
Distinctus, 416.
Distractile, 126.
Diurnus, 402.
Divaricating, 40. 94. 414.
Diverging, 94.
Divided, 46.
Dodecandrous, 125.

Dodrantalis, 400.
Dolabriformis, 375.
Dorsal, 144. 416.
Dotted, 24. 394. 407.
Double-bearing, 401.
Downy, 395.
Drooping, 413.
Dumosus, 48.
Duplicato-dentate, 387.
——————-pinnate, 391.
——————-serrate, 387.
Duplicatus, 420.
Duplo, 400.
Dusty, 397.
Dwarf, 399.

Eared, 383.
Eborinus, 403.
Eburneus, 403.
Echinatus, 394.
Edged, 407.
Egg-shaped, 382.
Elatus, 399.
Ellipsoidal, 374.
Ellipticus, 382.
Emarginate, 386.
Embracing, 415.
Empty, 101.
Endophyllous, 190.
Endogenous, 59.
Enervis, 409.
Enneandrous, 125.
Ensiformis, 382.
Entangled, 420.
Entire, 46. 386.
Ephemerus, 402.
Epigæus, 417.
Epigynous, 123. 417.
Epiphyllus, 416.
Epipterus, 377.
Equal, 120. 384.
Equal-veined, 91.
Equatus, 394.
Equitant, 411.
Equally pinnate, 390.
Equal-sided, 384.
Erect, 159. 411.
Erosus, 387.
Erythro, 406.
Escens, 422.
Etairionnaire, 171.
Evanescentè venosus, 92. 408.
Even, 394.
Evergreen, 249.
Exaltatus, 399.
Exaspiratus, 395.
Excurrens (caulis), 47.
Excurrent, 392.
Exiguus, 399.
Exogenous, 59.

Exophyllous, 190.
Exserted, 124.
Extrorsus, 128. 414.

Fading, 402.
Falcate, 375.
Falsely-ribbed, 93.
Falsinervis, 409.
Falsinervius, 91.
Fan-shaped, 377.
Farinaceus, 398.
Farinosus, 396.
Fasciarius, 381.
Fasciatus, 407. 419.
Fasciculated, 78. 418.
Fastigiate, 419.
Favosus, 393.
Feather-veined, 93.
Feathery, 396.
Ferrugineous, 404.
Fibrous, 77. 398.
Fiddle-shaped, 384.
Filiformis, 376.
Fimbriatus, 396.
Fingered, 388.
Fissus, 388.
Fistulous, 372.
Five-ribbed, 93.
Flabelliformis, 377.
Flagelliformis, 376.
Flammeus, 406.
Flavidus, 404.
Flavescens, 404.
Flavovirens, 405.
Flavus, 404.
Fleshy, 398.
Flexuosus, 413.
Floating, 411.
Floralis, 416.
Floccose, 395.
Fluitans, 411.
Foliaceus, 53. 377.
Foliaris, 96. 416.
Foliiformis, 377.
Forked, 392.
Foxglove-shaped, 380.
Free, 208.
Fringed, 396.
Frosted, 397.
Fugacious, 249. 402.
Fulcraceus, 53.
Fuligineus, 404.
Fuliginosus, 404.
Fulvus, 405.
Fumeus, 403.
Fumosus, 403.
Funalis, 376.
Fungilliformis, 380.
Fungiformis, 380.
Funiliformis, 376.

Funnel-shaped, 378.
Furcatus, 392.
Furrowed, 394.
Fuscus, 404.
Fusiform, 9. 77. 374.
Fusinus, 374.

Galacto, 403.
Gamopetalous, 118.
Gamosepalous, 116.
Ganglioneous, 40.
Gelatinous, 398.
Geminatus, 420.
Geniculate, 125. 413.
Gibbous, 374.
Giganteus, 399.
Gigantic, 399.
Gilvus, 405.
Githagineus, 406.
Glaber, 397.
Gladiatus, 382.
Glandaceus, 404.
Glandular, 395.
Glaucous, 397. 405.
Glaucus, 405.
Glaucescens, 405.
Glittering, 397.
Globose, 372.
Globulosus, 372.
Glochidatus, 39.
Glutinosus, 397.
Gnawed, 387.
Goblet-shaped, 379.
Gongylodes, 377.
Grammicus, 408.
Granular, 392.
Granulatus, 392.
Greasy, 397.
Griseus, 403.
Gromonical, 192.
Grumous, 377.
Guttatus, 407.
Gymnocarpien, 171.
Gypseus, 403.
Gyratus, 413.

Hæmatiticus, 406.
Hair-pointed, 385.
Hair-shaped, 376.
Hairy, 395.
Halberd-headed, 383.
Half-netted, 393.
Half-terete, 374.
Halved, 377. 384.
Hastatus, 383.
Headed, 386.
Heart-shaped, 383.
Hebetatus, 386.
Helvolus, 405.
Hepaticus, 404.

Heptandrous, 125.
Herbaceous, 399.
Heterocarpien, 163.
Heterotropal, 193.
Hexandrous, 125.
Hidden-veined, 93.
Hinoideus, 92. 409.
Hirtus, 395.
Hispid, 395.
Hoary, 395.
Homotropal, 415.
Honeycombed, 393.
Hooded, 380.
Hooked, 305.
Hooked-back, 384.
Horarius, 402.
Horizontal, 412.
Horned, 376.
Horny, 398.
Hornus, 402.
Humifusus, 414.
Humilis, 399.
Hypocrateriform, 118. 378.
Hypogæous, 417.
Hypogynous, 123. 417.
Hysteranthus, 401.

Ianthinus, 406.
Icosandrous, 125.
Igneus, 406.
Imbricated, 410. 418.
Imparipinnatus, 390.
Inæqualis, 384.
Inæquilaterus, 384.
Incanus, 395. 403.
Incarnatus, 406.
Incisus, 388.
Inclinate, 412.
Included, 124.
Incumbent, 193.
Incurvus, 413.
Indefinite, 125.
Indehiscent, 163.
Indeterminatus (caulis), 48.
Indirectè venosus, 92. 408.
Indigoticus, 406.
Induplicate, 410.
Induviatus, 80.
Inermis, 394.
Inferior (calyx), 115.
Inflatus, 380.
Inflexed, 413.
Infra-axillary, 79.
Infractus, 413.
Infundibuliform, 118. 378.
Innate, 128. 416.
Integer, 386.
Integerrimus, 386.
Interruptedly pinnate, 390.
Interruptus, 420.

Intricatus, 420.
Introcurvus, 413.
Introflexus, 413.
Introrsus, 128. 414.
Introvenium, 93.
Inverted, 412.
Involute, 409. 412.
Irregular, 9. 120. 381.
Isobrious, 192.
Isodynamous, 192.
Isos, 399.
Isostemonous, 400.
Ivory-white, 403.

Jointed, 392.

Keeled, 376.
Kermesinus, 406.
Kidney-shaped, 383.
Knee-jointed, 413.
Kneepan-shaped, 379.
Knoblike, 377.
Knotted, 376.

Labiate, 118. 378.
Labiose, 119.
Lachrymæformis, 373.
Lacerus, 388.
Laciniatus, 388.
Lacteus, 403.
Lacunose, 393.
Lævigatus, 397.
Lævis, 397.
Lamellar, 386.
Lamellatus, 386.
Lanatus, 395.
Lanceolate, 38.
Latent, 50.
Lateral, 416.
Laterinervius, 92.
Lateritius, 406.
Laxus, 398. 418.
Leaf-like, 377.
Leathery, 398.
Lens-shaped, 374.
Lenticular, 374.
Lentiformis, 374.
Lentiginosus, 397.
Lepidotus, 41. 396.
Leprous, 396.
Leprosus, 396.
Lettered, 408.
Leuco, 403.
Liber, 416.
Ligneus, 398.
Lignosus, 398.
Ligulatus, 381.
Lilacinus, 406.
Liliaceous, 119.
Limbatus, 407.

INDEX.

Linealis, 400.
Linear, 381.
Lineatus, 394.
Lined, 394.
Linguiformis, 375.
Little, 399.
Lituratus, 408.
Lividus, 405.
Lobatus, 388.
Lobed, 9. 388.
Loculicidal, 164.
Loculosus, 393.
Lofty, 399.
Loose, 398. 418.
Loratus, 381.
Low, 399.
Lunatus, 383.
Lunulatus, 383.
Luridus, 404.
Luteolus, 404.
Lutescens, 404.
Luteus, 404.
Lymphatic, 40.

Macrocephalous, 183.
Macropodous, 191.
Maculatus, 407.
Manicate, 39.
Many-headed, 77.
Marbled, 407.
Marcescens, 402.
Marginal, 416.
Marginatus, 407.
Marmoratus, 407.
Masked, 118.
Mealy, 396. 398.
Medullary, 398.
Meios, 400.
Meiostemonous, 400.
Mela, 403.
Melano, 403.
Melon-shaped, 374.
Membranous, 86.
Membranaceous, 397.
Memnonius, 404.
Meniscoideus, 380.
Menstrualis, 402.
Menstruus, 402.
Milk-white, 403.
Millsail-shaped, 377.
Miniatus, 406.
Mitriform, 201.
Mixtinervius, 92.
Modioliformis, 380.
Molendinaceus, 377.
Monadelphous, 124.
Monandrous, 124.
Monocarpous, 401.
Moniliform, 25. 376.
Monopetalous, 118.

Monophyllus, 116.
Morious, 106.
Much-branched, 392.
Mucous, 397.
Mucronate, 385.
Multiceps, 77.
Multiferus, 401.
Multifidus, 388.
Multijugus, 391.
Multidigitato-pinnatus, 391.
Muricated, 395.
Muriform, 9.
Murinus, 403.
Muscariformis, 379.
Mushroom-headed, 380.
Muticus, 386.

Naked, 52. 397.
Naked (seeds), 161. 194.
Nanus, 399.
Napiformis, 373.
Natans, 411.
Nave-shaped, 380.
Navicularis, 376.
Nebulosus, 407.
Necklace-shaped, 25. 376.
Needle-shaped, 382.
Nervatus, 93. 408.
Nerved, 88.
Nervosus, 408.
Netted, 92. 393.
Niger, 404.
Nigritus, 404.
Nitidus, 397.
Niveus, 403.
Nocturnus, 402.
Nodding, 413.
Nodosus, 376.
Nodulose, 77. 400.
Normal, 381.
Nudus, 397.
Nullinervis, 409.
Nullinervius, 91.
Nutans, 413.

Ob, 422.
Oblique, 94. 384. 412.
Oblong, 8. 382.
Obtusus, 385.
——— cum acumine, 385.
Obvolute, 410.
Ocellated, 407.
Ochraceus, 405.
Ochroleucus, 405.
Octandrous, 125.
Often-bearing, 401.
Oleaginous, 398.
Oligos, 420.
Olivaceus, 405.
One-ribbed, 408.

One-sided, 40. 413. 419.
Opaque, 397.
Operculate, 114.
Opposite, 414. 418.
Orbicular, 382.
Orgyalis, 400.
Orthotropous, 156. 193. 415.
Oscillatorius, 416.
Osseous, 398.
Oval, 382.
Ovate, 382.
Ovoidal, 374.

Painted, 407.
Paired, 391.
Palaceus, 416.
Paleaceous, 114. 396.
Palmaris, 400.
Palmate, 388.
Palmatifidus, 389.
Palmatilobatus, 389.
Palmatipartitus, 389.
Palmatisectus, 389.
Palmiformis, 409.
Palminervis, 93. 409.
Panduratus, 384.
Panduriformis, 384.
Papery, 397.
Papilionaceous, 119.
Papillosus, 395.
Papulosus, 395.
Papyraceus, 397.
Parabolical, 382.
Parallelinervis, 408.
Paralleli-nervosus, 91.
Parietal (pistilla), 139.
Paripinnatus, 390.
Parted, 388.
Partial (umbel), 110.
Partiale, 208.
Partitioned, 393.
Partitus, 388.
Patelliformis, 379.
Patens, 414.
Pear-shaped, 373.
Pectinatus, 389.
Pedalinervius, 93.
Pedalis, 400.
Pedate, 388.
Pedatifidus, 389.
Pedatilobatus, 389.
Pedatinervis, 409.
Pedatipartitus, 389.
Pedatisectus, 389.
Peduncularis, 96.
———————— (cirrhus), 112.
Peltate, 415.
Peltinervis, 409.
Peltinervius, 93.
Pendulous, 159. 412

Pennatipartitus, 388.
Penniformis, 409.
Penninervis, 92. 409.
Pennivenius, 93.
Pentandrous, 125.
Perennans, 402.
Perennial, 249. 402.
Perfoliate, 82. 415.
Perforated, 393.
Perichætial, 201.
Perigynous, 123. 417.
Peripterus, 377.
Peritropal, 414.
Permanent, 402.
Peronate, 396.
Persistent, 249. 402.
Personate, 118. 378.
Perpendicular, 412.
Perpusillus, 399.
Pertusus, 393.
Pervious, 47.
Petal-like, 377.
Petaloideous, 377.
Petiolaceus, 53.
Petiolaris, 96. 416.
Phragmiger, 393.
Phœniceus, 406.
Phylloideus, 377.
Piceus, 404.
Pictus, 407.
Piliferus, 385.
Pilose, 114. 395.
Pimpled, 395.
Pinnate with an odd one, 390.
Pinnated, 39. 390.
Pinnaticisus, 388.
Pinnatifid, 388.
Pinnatilobatus, 389.
Pinnatisectus, 389.
Pitcher-shaped, 378.
Pithy, 398.
Pitted, 393.
Pivotante, 77.
Placenta-shaped, 373.
Plaited, 410.
Plane, 374.
Pleiophyllous, 46.
Plicativus, 410.
Plumbeus, 403.
Plumosus, 39. 114. 396.
Poculiform, 379.
Pointless, 386.
Pointletted, 385.
Polished, 397.
Politus, 397.
Pollicaris, 400.
Poly, 420.
Polyadelphous, 124.
Polyandrous, 125.
Polycarpous, 401

Polymorious, 106.
Polypetalous, 117.
Porphyreus, 404.
Posticus, 128. 414.
Pouch-shaped, 379.
Powdery, 397.
Præmorse, 77. 386.
Prasinus, 405.
Prickly, 394.
Prism-shaped, 372.
Prismatical, 9.
Proboscideus, 376.
Procerus, 399.
Procumbens, 414.
Pronus, 414.
Prostratus, 414.
Proteranthous, 401.
Pruinosus, 397.
Pseudo-costatus, 93.
Pterus, 377.
Pubens, 395.
Pubescens, 395.
Pullus, 404.
Pulley-shaped, 379.
Pulverulentus, 397.
Pulvinatus, 374.
Pumilus, 399.
Punctatus, 394. 407.
Puniceus, 406.
Pungent, 385.
Pure white, 403.
Purpureus, 406.
Pusillus, 399.
Pygmæus, 399.
Pyramidalis, 372.
Pyriformis, 373.

Quadridigitato-pinnatus, 391.
Quinate, 390.
Quincunx, 411.
Quintuplinervius, 93.

Radiating, 93.
Radiatus, 420.
Radicalis, 107. 416.
Ramealis, 416.
Rameous, 416.
Ramentaceous, 86. 396.
Ramosus, 391.
Ramosissimus, 392.
Rarus, 420.
Reclinate, 411, 412.
Rectinervis, 408.
Rectivenius, 91.
Rectus, 411.
Reflexed, 94.
Regular, 50. 120. 381.
Remotus, 420.
Renarius, 383.
Reniformis, 383.

Repand, 387.
Replicate, 410.
Restans, 402.
Resupinate, 412.
Reticulated, 23. 88. 92. 393.
Retinervis, 92. 409.
Retrorsus, 414.
Retuse, 385.
Revolute, 410. 412.
Rhizocarpous, 401.
Rhodo, 406.
Rhomboid, 383.
Ribbed, 92. 408.
Right-angled, 94.
Ringed, 394.
Ringent, 118. 378.
Rope-shaped, 376.
Roridus, 397.
Rosaceous, 119. 420.
Roseus, 406.
Rostellatus, 385.
Rostratus, 385.
Rosulate, 418.
Rotate, 378.
Rotundatus, 382.
Rotundus, 382.
Rough, 39. 395.
Roughish, 395.
Roundish, 382.
Rubellus, 406.
Rubens, 406.
Ruber, 406.
Rubescens, 406.
Rubicundus, 406.
Rubiginosus, 406.
Rufescens, 404.
Rufus, 404.
Rugose, 393.
Ruminated, 187. 393.
Runcinate, 384.
Ruptinervius, 92. 408.
Rutilans, 406.
Rutilus, 406.

Saddle-shaped, 380.
Sagittatus, 383.
Salver-shaped, 378.
Sanguineus, 406.
Sausage-shaped, 380.
Sawed, 387.
Scaber, 395.
Scabridus, 395.
Scabrous, 43.
Scaly, 396. 419.
Scarious, 398.
Scarred, 394.
Schistaceus, 403.
Scimitar-shaped, 375.
Scrobiculatus, 393.
Scrotiformis, 379.

Scutatus, 374.
Scutelliform, 379.
Scutiformis, 374.
Secreting, 40.
Secundatus, 40.
Secundus, 413. 419.
Sellæformis, 380.
Semi-amplexa, 410.
Semi-amplexicaulis, 415.
Semilunatus, 383.
Semireticulatus, 393.
Semiteres, 374.
Sepalous, 116.
Septatus, 393.
Septicidal, 164.
Septifragal, 164.
Serialis, 419.
Sericeus, 39. 396.
Serratus, 387.
Sesqui, 400.
Sesquipedalis, 400.
Sessile, 94. 415.
Setiger, 385.
Setosus, 38. 114. 385.
Shaggy, 395.
Sharp-pointed, 385.
Sheathing, 95. 416.
Shield-shaped, 374.
Shining, 397.
Sigmoid, 192.
Silky, 39. 396.
Silvery, 403.
Simple, 21. 46. 111. 389.
——— (leaf), 87.
Simplicissimus, 389.
Sinuate, 387.
Sinuolatus, 387.
Sinuous, 9.
Slashed, 388.
Slimy, 397.
Smaragdinus, 405.
Smooth, 397.
Snow-white, 403.
Solutus, 416.
Spadiceus, 404.
Sphæricus, 372.
Spheroidal, 374.
Spatulate, 382.
Sparsus, 418.
Spiculate, 395.
Spindle-shaped, 374.
Spiny, 394.
Spiral, 125. 373. 412. 419.
Spiraliter contortus, 411.
Spithamæus, 400.
Splendens, 397.
Split, 388.
Spodo, 403.
Spongy, 398.
Spotted, 407.

Spreading, 94. 414.
Squamosus, 41. 396. 419.
Square, 9.
Squarrose, 419.
Squarrose-slashed, 388.
Squarroso-laciniatus, 388.
Stalked, 42.
Starved, 419.
Stellate, 40. 392. 418.
Stelliformis, 418.
Stellulatus, 418.
Stemless, 58.
Stemclasping, 415.
Sterile (stamens), 136.
Stinging, 396.
Stipitatus, 416.
Stipulaceus, 53.
Straggling, 414.
Straight, 411.
Straight-ribbed, 408.
——— veined, 91, 92.
Stramineus, 404.
Strangulated, 25.
Strapshaped, 381.
Striated, 394.
Strictus, 411.
Strigose, 42. 396.
Striped, 407.
Strombuliformis, 373.
Strombus-shaped, 373.
Stupose, 125.
Sub, 422.
Submersed, 412.
Sub-parallel, 94.
Subramosus, 391.
Suberosus, 398.
Subrotundus, 382.
Subulatus, 382.
Succisus, 386.
Succulent, 86. 398.
Sulcatus, 394.
Sulphureus, 404.
Superior (calyx), 115.
Supervolute, 410.
Supra-axillary, 79.
Supradecompound, 390.
Suspended, 159.
Sutural, 164.
Swimming, 411.
Sword-shaped, 382.
Sychnocarpous, 401.
Synanthous, 401.
Syncarpous, 144.

Tænianus, 380.
Tail-pointed, 385.
Tall, 399.
Taper, 374.
Taper-pointed, 385.
Tapering, 384.

Tap-rooted, 77.
Tapeworm-shaped, 380.
Tartareous, 399.
Tear-shaped, 373.
Tephro, 403.
Terete, 374.
Tergeminate, 391.
Terminal, 416.
Ternate, 390. 418.
Ternato-pinnatus, 391.
Testaceus, 405.
Tessellated, 407.
Testiculatus, 377.
Tetradynamous, 124.
Tetrandrous, 125.
Thalassicus, 405.
Thallodes, 206.
Thick, 398.
Threadshaped, 375.
Three-cornered, 375.
Three-edged, 376.
Three-ribbed, 93.
Thrice digitato-pinnate, 391.
Tomentose, 395.
Tongue-shaped, 375.
Toothed, 39. 387.
Tooth-letted, 39.
Top-shaped, 373.
Torn, 388.
Torsivus, 411.
Tortuous, 413.
Torulose, 40. 376.
Trapeziform, 383.
Treelike, 392.
Triadelphous, 124.
Triandrous, 124.
Triangular, 375. 383.
Tricornis, 376.
Tricostatus, 93.
Tridentatus, 386.
Tridigitato-pinnatus, 391.
Triduus, 402.
Trifariam, 419.
Trifidus, 388.
Trifoliolate, 390.
Trigonus, 375.
Trijugus, 391.
Trilobus, 388.
Trimestris, 402.
Trinervis, 93.
Tripartitus, 388.
Tripinnate, 391.
Triple-ribbed, 93. 408.
Triplinervius, 93. 408.
Triplicostatus, 93.
Triplo, 400.
Triqueter 375.
Triternate, 391.
Trochlearis, 379.
Trumpet-shaped, 376.

Truncate, 386.
Tubæformis, 376.
Tubatus, 376.
Tubercled, 395.
Tubular, 372.
Tunicated, 52.
Turbinate, 373.
Turgid, 380.
Turnip-shaped, 77. 373.
Twin, 392.
—— digitato-pinnate, 390.
Twining, 413.
Twisted, 411.
Two-edged, 375.

Ulnaris, 400.
Umbilicate, 415.
Umbonatus, 374.
Umbraculiformis, 380.
Umbrella-shaped, 380.
Umbrinus, 404.
Unarmed, 394.
Uncatus, 385.
Uncertain, 414.
Uncialis, 400.
Uncinatus, 385.
Unctuosus, 397.
Undulatus, 385.
Unequal, 120. 384.
Unequal-sided, 384.
Unguiculate, 119.
Unicus, 420.
Unijugatus, 391.
Unijugus, 391.
Unilateralis, 419.
Uninervis, 408.
Universal (number), 110.
Universale, 208.
Urceolatus, 378.
Urens, 396.

Vacuus, 101.
Vaginans, 416.
Vaginervis, 93. 409.
Vagus, 193. 414.
Valvaris, 411.
Valvate, 411.
Variegatus, 407.
Vascularis, 380.
Vase-shaped, 380.
Veinless, 91.
Velutinus, 38. 395.
Velvety, 395.
Venous, 88.
Venosus, 92. 409.
Ventral, 144.
Ventricosus, 381.
Venuloso-hinoideus, 92.
Venuloso-nervosus, 91.
Vermicularis, 376

Vermiculatus, 406.
Verrucosus, 395.
Versatile, 128. 416.
Vertebrate, 390.
Verticalis, 412.
Verticillatus, 418.
Vexillary, 411.
Virens, 405.
Virescens, 405.
Virgate, 47.
Viridis, 405.
Villosus, 395.
Vimineous, 47.
Violaceus, 406.
Viscid, 397.
Viridescens, 405.
Viridulus, 405.
Vitellinus, 405.
Vittatus, 407.
Volubilis, 413.

Wandering, 193.

Wavy, 384.
Waxy, 398.
Wedge-shaped, 382.
Wheel-shaped, 378.
Whip-shaped, 376.
White, 403.
Whitened, 397.
Whitish, 403.
Whorled, 418.
Winged, 377.
Withering, 402.
Woody, 398.
Woolly, 395.
Worm-shaped, 376.
Wrinkled, 410.

Xantho, 404.
Xerampelinus, 406.

Yearly, 402.

Zoned, 408.

THE END.

LONDON:
Printed by A. & R. Spottiswoode,
New-Street-Square.

[London, April 1834.]

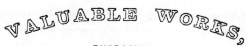

EMBRACING

HISTORY,	GEOGRAPHY,	NATURAL HISTORY,
THEOLOGY,	TOPOGRAPHY,	POETRY,
BIOGRAPHY,	COMMERCE,	VOYAGES AND TRAVELS,

&c. &c.

PRINTED FOR

LONGMAN, REES, ORME, BROWN, GREEN, & LONGMAN.

ENCYCLOPÆDIA of GEOGRAPHY: comprising a complete Description of the Earth; exhibiting its Relation to the Heavenly Bodies, its Physical Structure, the Natural History of each Country, and the Industry, Commerce, Political Institutions, and Civil and Social State of all Nations. By HUGH MURRAY, F.R.S.E.

Assisted in

ASTRONOMY, &c. by PROFESSOR WALLACE, | BOTANY, &c. by PROFESSOR HOOKER,
GEOLOGY, &c. by PROFESSOR JAMESON, | ZOOLOGY, &c. by W. SWAINSON, Esq.

With 82 Maps, drawn by Sidney Hall; and upwards of 1000 other Engravings on Wood, from Drawings by Swainson, T. Landseer, Sowerby, Strutt, &c., representing the most remarkable objects of Nature and Art in every Region of the Globe. To form 12 Monthly Parts, 5s. each. Parts I. and II. are published.

HISTORY of the REVOLUTION in ENGLAND in 1688: comprising a View of the Reign of James II., from his Accession, to the Enterprize of the Prince of Orange. By the late Right Honourable SIR JAMES MACKINTOSH. And completed to the Settlement of the Crown, by the Editor. To which is prefixed, a Notice of the LIFE, WRITINGS, and SPEECHES of SIR JAMES MACKINTOSH. 4to. with a Portrait, engraved by C. Turner. 3l. 3s. in cloth.

"Highly gifted by nature, deeply read, and singularly accomplished, the view of one of the most remarkable epochs in English history could not have been undertaken by any man of a capacity to do it justice, in every respect, superior to this eminent individual."—*Literary Gazette.*

"In every page we perceive the anxiety of the historian to hold the balance of justice with unfaltering hand, and to watch its slightest vibrations."—*Athenæum.*

THE LIFE and ADVENTURES of JOHN MARSTON HALL. By the Author of "Darnley," "Mary of Burgundy," &c. 3 vols. post 8vo. 31s. 6d. bds.

EGYPT and MAHOMMED ALI; or, Travels in the Valley of the Nile: containing a Description of all the remarkable Ruins, and other Monuments of Antiquity, in Egypt and Nubia, from the Mediterranean to the Second Cataract; with a Comparison between the Greek and Egyptian Schools of Art; together with an Account of the Government and Personal Character of the Pasha, his Harems, Palaces, Gardens, Baths, &c.; Sketches of Native Manners, Schools, Colleges, Manufactories, &c.—Excursion to the beautiful Nome of Arsinoë, and Lake Moeris—History of the War in Syria, &c. &c. By JAMES AUGUSTUS ST. JOHN. 2 vols. 8vo. 30s. bds.

A DICTIONARY, PRACTICAL, THEORETICAL, and HISTORICAL, of COMMERCE and COMMERCIAL NAVIGATION. By J. R. M'CULLOCH, Esq. 1 large vol. 8vo. with Maps and Plans, Second Edit. greatly enlarged and amended. TO APPEAR IN MAY.

This edition contains a full account of the late arrangements with respect to the BANK OF ENGLAND, the EAST INDIA COMPANY, and the TRADE to INDIA and CHINA. Copious abstracts are given of the new CUSTOMS ACTS, the ACT abolishing SLAVERY, &c.; and the modifications of the Law that bear upon Commercial Transactions are carefully specified. Much valuable information as to the Trade, Commercial Regulations, Shipping Charges, &c. in FOREIGN PORTS, has been obtained from the British Consuls, and from other sources. The GAZETTEER department has been greatly augmented, and, it is hoped, materially improved; two new Maps are added (exhibiting the Canals, Rail-roads, and Lighthouses of Britain, the Mouths of the Mersey and Dee, with the country from Liverpool to Manchester), and Plans of several of the principal British and Foreign Sea-Ports are given.

Modern Publications, and New Editions of Valuable Standard Works,

DR. LARDNER'S CABINET CYCLOPÆDIA, published in Monthly Volumes, small 8vo. price 6s. each, in cloth.

Volumes published.

51. HISTORY OF NATURAL PHILOSOPHY, 1 Vol. By Professor Powell, Oxford.
50. HISTORY OF ROME, 2 Vols. Vol. I.
47. TREATISE ON THE ARTS, MANUFACTURES, MANNERS, AND INSTITUTIONS OF THE GREEKS AND ROMANS, 2 Vols. Vol. I.
46. LIVES OF THE MOST EMINENT FOREIGN STATESMEN. By E. E. Crowe. Vol. I.
45, 49, 53. HISTORY OF EUROPE DURING THE MIDDLE AGES, 4 Vols. By the Author of "History of Spain and Portugal." Vols. I. to III.
44. CHRONOLOGY OF HISTORY, 1 Vol. By Sir Harris Nicolas.
43. TREATISE ON ASTRONOMY, 1 Vol. By Sir John Herschel.
41, 52. HISTORY OF THE CHURCH, 2 Vols. By the Rev. Henry Stebbing, M.A.
40, 48. NAVAL HISTORY OF ENGLAND. By Robert Southey, Esq. Vols. I. and II.
39. TREATISE ON HEAT, and its Applications, 1 Vol. By Dr. Lardner.
34. TREATISE ON CHEMISTRY, 1 Vol. By M. Donovan, Esq.
31. HISTORY OF SWITZERLAND, 1 Vol.
29, 30, 32, 35, 38. HISTORY OF SPAIN AND PORTUGAL, 5 Vols.
27. HISTORY OF THE ITALIAN REPUBLICS, 1 Vol. By J. C. L. de Sismondi.
26. HISTORY OF THE MANUFACTURES OF PORCELAIN AND GLASS, 1 Vol.
25, 28, 36. LIVES OF BRITISH GENERALS, 3 Vols. By the Rev. G. R. Gleig.
24, 42. TREATISE ON THE MANUFACTURES IN METAL. Iron and Steel, Vols. I. & II.
22. TREATISE ON THE SILK MANUFACTURE, 1 Vol.
21. LIVES OF BRITISH STATESMEN, 3 Vols. Vol. I.
20. HISTORY OF POLAND, 1 Vol.
19. TREATISE ON OPTICS, 1 Vol. By Sir D. Brewster.
17. HYDROSTATICS AND PNEUMATICS, 1 Vol. By Dr. Lardner.
14. PRELIMINARY DISCOURSE ON NATURAL PHILOSOPHY. By Sir J. Herschel.
13, 33. THE WESTERN WORLD, 4 Vols. Vols. I. and II.—UNITED STATES, 2 Vols.
12, 15, 23. HISTORY OF FRANCE, 3 Vols. By E. E. Crowe.
2, 11, 16. HISTORY OF MARITIME AND INLAND DISCOVERY, 3 Vols.
10. HISTORY OF THE NETHERLANDS, 1 Vol. By T. C. Grattan, Esq.
9. OUTLINES OF HISTORY, 1 Vol.
8, 18, 37. SIR JAMES MACKINTOSH'S HISTORY OF ENGLAND. Vols. I. to III.
7. CITIES AND TOWNS OF THE WORLD, 2 Vols. With Woodcuts. Vol. I.
6. LIVES OF EMINENT BRITISH LAWYERS, 1 Vol. By H. Roscoe, Esq.
5. MECHANICS, 1 Vol. By Capt. H. Kater and Dr. Lardner.
1, 4. HISTORY OF SCOTLAND, 2 Vols. By Sir Walter Scott, Bart.
3. DOMESTIC ECONOMY. By M. Donovan, Esq. Vol. I.

In immediate preparation.

MANUFACTURES IN METALS, Vol. III.—Tin, Lead, Copper, &c.
TREATISE ON GENERAL GEOGRAPHY, 5 Vols. By W. D. Cooley, Esq. Vol. I.
HISTORY OF THE DECLINE OF CIVILIZATION IN EUROPE. By J. C. L. de Sismondi.

THE DOCTOR, &c. 2 vols. post 8vo. 21s. in cloth.

"This work has excited more attention than any one belonging to, or approaching, the class of *novels*, which has appeared in England for a considerable number of years; and we are not at all disposed to wonder that such should have been the case."—*Quarterly Review.*

PRINCIPLES of POLITICAL ECONOMY, deduced from the Natural Laws of Social Welfare, and applied to the Present State of Britain. By G. POULETT SCROPE, M.P. F.R.S. &c. Small 8vo. 7s. in cloth.

"As a writer, Mr. Scrope is distinguished for a broad common-sense view of his subject, a great knowledge of facts, and a pleasant mode of stating them. The present volume is intended for the *masses*. One of its objects is to instruct them as to the causes of their present condition, and to suggest the true remedies for its evils. It is clearly and plainly written."—*Spectator.*

LECTURES on the HISTORY and PRINCIPLES of PAINTING. By THOMAS PHILLIPS, Esq. R.A. F.R.S. and F.S.A. late Professor of Painting in the Royal Academy. 8vo. 13s. in cloth.

"There is not a passage in them with which a liberally-educated gentleman should not be acquainted."—*Lit. Gazette.*

MEMOIR of the COURT and CHARACTER of CHARLES the FIRST. By LUCY AIKIN. 2 vols. 8vo. with Portrait, 2d edit. 28s. bds.

"Miss Aikin's present work, and her previous Memoirs of the Courts of Elizabeth and of James I., are very acceptable additions to our literature."—*Edinburgh Review.*

Printed for Longman, Rees, Orme, Brown, Green, and Longman.

TREATISE on ROADS; in which the Right Principles of Road-making are explained and illustrated by the Plans, Specifications, &c. used by Sir T. Telford, on the Holyhead Road. By the Right Hon. Sir HENRY PARNELL, Bart. 8vo. Plates. 21s. in cloth.

BIOGRAPHICAL MEMOIRS of BARON CUVIER. By Mrs. R. LEE, formerly Mrs. T. ED. BOWDICH. 8vo. with Portrait, 12s. bds.
" We recommend to our readers Mrs. Lee's interesting Biography of Cuvier."
Jameson's Edinburgh Philosophical Journal.

JOHN HOPKINS'S NOTIONS on POLITICAL ECONOMY. By the Author of "CONVERSATIONS ON CHEMISTRY," "POLITICAL ECONOMY," &c. 12mo. 4s. 6d. cloth; smaller edition, 1s. 6d.
" Admirably adapted, by plain straightforward sense, for the improvement of the labouring classes."
Edinburgh Review.

TRADITIONARY STORIES of OLD FAMILIES, and LEGENDARY ILLUSTRATIONS of FAMILY HISTORY. With Notes, Historical and Biographical. By A. PICKEN, Author of "The Dominie's Legacy," &c. 2 vols. post 8vo. 21s.
" Of these volumes we feel disposed to speak in unmeasured terms of approbation."—*Monthly Review.*

BOOK of NATURE: a Popular Illustration of the general Laws and Phœnomena of Creation, in its Unorganized and Organized, its Corporeal and Mental Departments. By J. MASON GOOD, M.D. & F.R.S. 3d Edit. 3 Vols. fcap. 8vo. 24s. in cloth.
" The best philosophical digest of the kind which we have seen."—*Mon. Rev.*

NATURALIST'S LIBRARY, conducted by Sir WM. JARDINE, Bart. F.R.S.E. F.L.S. &c. With numerous coloured Plates, descriptions, woodcuts, and Portraits and Lives of celebrated Naturalists. Fcap. 8vo. 6s. each vol. in cloth.
Vol. I.—HUMMING BIRDS, Vol. I.: 35 col'd Plates, and Portrait and Life of LINNÆUS.
II.—MONKEYS, Vol. I.: 31 col'd Plates, and Portrait and Life of BUFFON.
III.—HUMMING BIRDS, Vol. II.: 31 col'd Plates, and Portrait and Life of PENNANT.
IV.—LIONS, TIGERS, &c.: 37 col'd Plates, and Portrait and Memoir of CUVIER.
V.—GAME BIRDS: col'd Plates, and Life of ARISTOTLE. *Just ready.*
" The most interesting, most beautiful, and cheapest series, yet offered to the public."—*Athenæum.*
⁎⁎* The two vols. of Humming Birds may be had in one, in rich silk binding, price 14s.

SELECTIONS from the EDINBURGH REVIEW; comprising the best Articles in that Journal, from its commencement to the present time: consisting of Characters of Eminent Poets, Painters, Divines, Philosophers, Statesmen, Orators, Historians, Novelists, and Critics; Dissertations on Poetry and the Drama; Miscellaneous Literature; Education; Political History; Metaphysics; Foreign and Domestic Politics; Political Economy; Law and Jurisprudence: Parliamentary Reform; Church Reform; the Liberty of the Press; the State of Ireland; and West India Slavery. With a Preliminary Dissertation, and Explanatory Notes. Edited by MAURICE CROSS, Esq. Secretary of the Belfast Historic Society. 4 large vols. 8vo. 3l. 3s. bds.

LECTURES on POETRY and GENERAL LITERATURE, delivered at the Royal Institution, in 1830 and 1831. By JAMES MONTGOMERY, Author of "The World before the Flood," &c. &c. Post 8vo. 10s. 6d. bds.
" A fine specimen of pure English composition; the style is simple—just what prose ought to be; and yet every sentence breathes of poetry."—*New Monthly Mag.*

STEEL'S SHIP-MASTER'S ASSISTANT, and OWNER'S MANUAL: containing General and Legal Information necessary for Owners and Masters of Ships, Ship-Brokers, Pilots, and other persons connected with the Merchant Service. 21st Edition, newly arranged, and corrected to 1833-4 (containing the New Customs Laws, &c.), by J. STIKEMAN, Custom-House Agent. With Tables of Weights, Measures, Monies, &c. by Dr. KELLY, Author of the "Universal Cambist." 1 large and closely-printed vol. 21s. bds.; 22s. 6d. bd.

AMERICA and the AMERICANS. By a Citizen of the World.
8vo. 12s. bds.

"We hail with pleasure the performance before us, as, by the sound judgment and impartiality which it displays, we at once place confidence in the statements and opinions it contains."—*Monthly Rev.*

OUTLINE of the SMALLER BRITISH BIRDS; for the Use of
Ladies and Young Persons. By R. A. Slaney, Esq. M.P. Fcap. 8vo. with Cuts, 4s. 6d. cloth.

"A delightful little book, which in its spirit and manner emulates that of the beloved White."
Magazine of Natural History.

PHILOSOPHICAL CONVERSATIONS; in which are familiarly explained the Effects and Causes of many Daily Occurrences in Natural Phenomena.
By Frederick C. Bakewell. 12mo. 5s. 6d. bds.

"Amply explanatory of the scientific principles upon which the Phenomena are founded."—*Monthly Review.*

MEMOIRS and CORRESPONDENCE of the late SIR JAMES
EDWARD SMITH, M.D. F.R.S. President of the Linnæan Society, &c. &c. Edited by Lady Smith. 2 large Vols. 8vo. with Portrait and Plates, 31s. 6d. bds.

"This work is among the books which, from their moral beauty, are to be regarded as the most precious treasures of literature."—*Tait's Magazine.*

GREEK TESTAMENT, with English Notes, Critical, Philological, and Exegetical.
By the Rev. S. T. Bloomfield, D.D. F.S.A. of Sidney College, Cambridge; Vicar of Bisbrooke, Rutland; Author of "Recensio Synoptica Annotationis Sacræ;" &c. 2 large and closely-printed vols. 8vo. 36s. bds. Adapted to the use of Academical Students, Candidates for Holy Orders, and Theological Readers generally.

By the same Author,

HISTORY of THUCYDIDES; newly translated into English, and illustrated with very copious Annotations, &c. 3 vols. 8vo. with Maps, 2l. 8s. bds.

"A version as literal and as perspicuous as erudition and industry combined can render it."—*Eclectic Rev.*

MEMOIR on SUSPENSION BRIDGES, comprising their History; with Descriptions.
Also, Experiments on Iron Bars, Wires, and Rules for facilitating Computations, &c. By Charles Stewart Drewry, Civil Engineer. 8vo. with numerous plates and woodcuts, 12s. bds.

"A complete manual on Suspension Bridges, worthy the subject, and worthy the high character for practical knowledge which our engineers enjoy."—*Monthly Review.*

INTRODUCTION to BOTANY. By John Lindley, LL.D.
F.R.S. L.S. and G.S. Professor of Botany in the University of London; &c. 8vo. with numerous Plates and Wood Cuts, 18s. cloth.

By the same Author,

INTRODUCTION to the NATURAL SYSTEM of BOTANY; or a Systematic View of the Organization, Natural Affinities, and Geographical Distribution of the Vegetable Kingdom; with the Uses of the most important Species. 8vo. 12s. cloth.

SYNOPSIS of the BRITISH FLORA, arranged according to the Natural Orders; containing Vasculares, or Flowering Plants. 12mo. 10s. 6d. bds.

OUTLINE of the FIRST PRINCIPLES of BOTANY. Plates. 18mo. 3s.

OUTLINE of the FIRST PRINCIPLES of HORTICULTURE. 18mo. 2s.

GUIDE to the ORCHARD and KITCHEN-GARDEN; or,
an Account of the most valuable Fruit and Vegetables cultivated in Great Britain: with Calendars of the Work required in the Orchard and Kitchen-Garden during every Month in the Year. By George Lindley, C.M.H.S. Edited by John Lindley, LL.D. &c. Assistant Secretary to the Horticultural Society of London. 1 large volume, 8vo. 16s. bds.

Printed for Longman, Rees, Orme, Brown, Green, and Longman. 5

GENERAL INDEX to the EDINBURGH REVIEW, from Vols. XXI. toL. inclusive (forming Nos. 113 and 114 of the work). 12s.

GENERAL INDEX to Vols. I. to XX. inclusive. 15s. bds.

LEGENDS of the LIBRARY at LILIES. By Lord and Lady Nugent. 2 vols. post 8vo. 21s. bds.

"Two delightful volumes."—*Literary Gazette.*——"The 'Legends' will be eagerly read, and valued for their intrinsic power of imparting pleasure."—*Tait's Magazine.*

TREATISE on HAPPINESS; consisting of Observations on Health, Property, the Mind, and the Passions; with the Virtues and Vices, the Defects and Excellencies, of Human Life. 2 vols. post 8vo. 21s. bds.

"Unexceptionable and amusing."—*Lit. Gazette.*——"Overflowing with entertainment."—*Spectator.*

SIR EDWARD SEAWARD's NARRATIVE of his SHIPWRECK, and Consequent Discovery of certain Islands in the Caribbean Sea. With a Detail of many Extraordinary Events in his Life, from 1733 to 1749, as written in his own Diary. Edited by Miss Jane Porter. Second Edition. 3 vols. small 8vo. 21s. cloth.

"The most curious and instructive work that has appeared since the first dawn of discovery, and in the history of navigation."—*Spectator.*

TALES and CONVERSATIONS; or, the New Children's Friend. By Mrs. Markham, Author of Histories of England and France. 2 vols. 12mo. 10s. 6d. cloth.

"We cannot praise too much this excellent work, in which good sense and good feeling are as conspicuous as good taste."—*La Belle Assemblée.*

HINTS on PICTURESQUE DOMESTIC ARCHITECTURE; In a Series of Designs for Gate Lodges, Gamekeepers' Cottages, and other Rural Residences. By T. F. Hunt, Architect. 4to. New Edit. with Additions, and a new set of Plates, 21s. bds.; India Proofs, 31s. 6d.

By the same Author,

DESIGNS for LODGES, GARDENERS' HOUSES, and other BUILDINGS, in the Modern or Italian Style; in a Series of 12 Plates, with Letterpress. Royal 4to. 21s. bds.; India Proofs, 31s. 6d.

DESIGNS for PARSONAGE-HOUSES, ALMS-HOUSES, &c. &c.; in a Series of 21 Plates, with Letterpress. Royal 4to. 21s. bds.; India Proofs, 31s. 6d.

EXEMPLARS of TUDOR ARCHITECTURE, adapted to Modern Habitations: with illustrative Details, selected from Ancient Edifices; and Observations on the Furniture of the Tudor Period. Royal 4to. with 37 Plates, 2l. 2s.; India Proofs, 3l. 3s.

PLAIN INSTRUCTIONS to EXECUTORS and ADMINISTRATORS, shewing the Duties and Responsibilities incident to the due Performance of their Trusts; with Directions respecting the Probate of Wills, and taking out Letters of Administration, &c.; with a Supplement, containing an elaborate Fictitious Will, comprising every description of Legacy provided for by the Legacy Acts, with the Forms properly filled up for every Bequest. By John H. Brady, late of the Legacy Duty Office, Somerset House. 8vo. 4th Edit. enlarged. 8s. bds.

GUIDE to all the WATERING and SEA-BATHING PLACES, including the SCOTCH WATERING-PLACES; containing full Descriptions of each Place, and of striking Objects in its Environs; forming an agreeable Companion during a Residence at any of the Places, or during a Summer Tour. With a Description of the Lakes, and a Tour through Wales. 1 portable vol. with 94 Views and Maps, 15s. bd.

GEOLOGY of SUSSEX, and of the adjacent parts of **Hampshire**, Surrey, and Kent. By GIDEON MANTELL, Esq. F.R.S. F.G.S. &c. &c. 8vo. with 75 Plates, Maps, and Cuts, 21s. cloth.

"A work of great interest to all geologists; and its eloquence, together with the wonders it tells of, render it likely to be a favourite with all."—*Athenæum*.

SELECT VIEWS in GREECE: 104 Plates, engraved in the best Line-Manner, from Drawings by H. W. WILLIAMS, Esq. Edinburgh. With Descriptions. 2 vols. imperial 8vo. 7l. 10s. in cloth; royal 4to. with India Proofs, 12l. 12s.

Any of the Numbers may be had separately, to complete sets, price each—in imp. 8vo. 12s.; royal 4to. (India Proofs) 21s.; India Proofs before letters (a few impressions) 31s. 6d.

ARRANGEMENT of BRITISH PLANTS, according to the latest Improvements of the Linnæan System; with an Easy Introduction to the Study of Botany. By WILLIAM WITHERING, M.D. &c. 7th Edition, with Additions, by WILLIAM WITHERING, Esq. LL.D. &c. 4 Vols. 8vo. 2l. 16s. bds.

ENCYCLOPÆDIA of PLANTS; comprising the Description, Specific Character, Culture, History, Application in the Arts, and every other desirable particular, respecting all the Plants indigenous, cultivated in, or introduced to Britain: combining the advantages of a Linnean and Jussieuean Species Plantarum, an Historia Plantarum, a Grammar of Botany, and a Dictionary of Botany and Vegetable Culture. Edited by J. C. LOUDON, F.L.S. &c. 1 large Vol. 8vo. with nearly 10,000 Engravings on Wood, 4l. 14s. 6d. bds.

By the same Author,

ENCYCLOPÆDIA of AGRICULTURE; comprising the Theory and Practice of the Valuation, Improvement, and Management of Landed Property; the Cultivation and Economy of the Animal and Vegetable Productions of Agriculture; a general History of Agriculture in all Countries; &c. New Edition, with considerable Improvements, in 1 large vol. 8vo. with upwards of 1100 Engravings on Wood. 2l. 10s. bds.

ENCYCLOPÆDIA of GARDENING; comprising the Theory and Practice of Horticulture, Floriculture, Arboriculture, and Landscape Gardening. 8vo. New Edit. (of which 5 Parts have appeared), with upwards of 1200 Engravings on Wood. To be completed in 20 Monthly Parts, 2s. 6d. each.

ENCYCLOPÆDIA of COTTAGE, FARM, and VILLA ARCHITECTURE; illustrated by upwards of 2000 Engravings on Wood and nearly 100 Lithographic Plates, embracing Designs of Cottages, Farm-Houses, Farmeries, Villas, Country Inns, Public Houses, Parochial Schools, &c. with their interior Finishing and Furniture; accompanied by Critical Remarks, &c. Complete in 12 Parts, 5s. each; or in 1 handsome volume, £3 bds.

HORTUS BRITANNICUS: a Catalogue of all the Plants Indigenous, Cultivated in, or Introduced to Britain.—PART I. The Linnean Arrangement, in which nearly 30,000 Species are enumerated, &c. &c.: preceded by an Introduction to the Linnean System.—PART II. The Jussieuean Arrangement of nearly 4000 Genera; with an Introduction to the Natural System, and a Description of each Order. 8vo., with the First additional Supplement, 23s. 6d. cloth; the Supplement separately, 2s. 6d.

MAGAZINE of NATURAL HISTORY, and JOURNAL of ZOOLOGY, BOTANY, MINERALOGY, GEOLOGY, and METEOROLOGY. 8vo. with Wood Engravings. Nos. 1 to 38, 3s. 6d. each. Continued every Two Months. Vols. I. to VI. may be had in boards, price 6l. 7s.

GARDENER'S MAGAZINE, and Register of Rural and Domestic Improvement. 8vo. with Wood Engravings. Nos. 1 to 49. Vols. I. to IX. may be had in boards, price 8l. 4s. 6d.

The GARDENER'S MAGAZINE will be continued Monthly, at 1s. 6d., commencing May 1.

ARCHITECTURAL MAGAZINE; or, Popular Journal of Improvements in ARCHITECTURE, BUILDING, and FURNISHING, and the various Arts and Trades more immediately connected therewith. With numerous illustrative woodcuts. Nos. I. and II. price 1s. 6d. each. To be continued Monthly.

Printed for Longman, Rees, Orme, Brown, Green, and Longman. 7

NEW GENERAL ATLAS of FIFTY-THREE MAPS, with the Divisions and Boundaries carefully COLOURED; constructed entirely from New Drawings, and engraved by SIDNEY HALL. Corrected to 1834. (Complete in Seventeen Parts, any of which may be had separately, price 10s. 6d. each.)

Folded in half, and bound in canvas	£8	18	6
Ditto, half-bound in Russia	9	9	0
In the full extended size of the Maps, half-bound in Russia	10	0	0
Ditto, with Proofs on India paper, half-bound in Russia	14	5	0

For favourable opinions of this Atlas, see the Literary Gazette, Gentleman's Magazine, the Sphynx (conducted by J. S. Buckingham, Esq.), New Monthly Magazine, Globe, &c.

ALPHABETICAL INDEX of all the NAMES contained in the above ATLAS, with References to the Number of the Maps, and to the Latitude and Longitude in which the Places are to be found. Royal 8vo. 21s. cloth.

SUNDAY LIBRARY: a Selection of SERMONS from Eminent Divines of the Church of England, chiefly within the last Half Century. With Notes, &c. by the Rev. T. F. DIBDIN, D.D. Complete in 6 vols. small 8vo. with Six Portraits of Distinguished Prelates. 30s. cloth.

"*A little library for a churchman; and a treasure for the pious among the laity.*"—*Literary Gazette.*

ORIGINAL PICTURE of LONDON, corrected to the present time; with a Description of its Environs. Re-edited, and mostly written, by J. BRITTON, F.S.A. &c. 18mo. 27th Edition, with upwards of 100 Views of Public Buildings, Plan of the Streets, and 2 Maps, 9s. neatly bound; with the Maps only, 6s.

THE NEW GIL BLAS; or, PEDRO of PENAFLOR. By H. D. INGLIS, Author of "Spain in 1830," &c. New and cheaper edit. in 2 vols. post 8vo. 16s. bds.

"*We have read these volumes with great delight.*"—*Metropolitan.* "*A very vivid picture of Spanish habits, customs, and manners.*"—*Monthly Mag.*

POETICAL WORKS of L. E. L. With uniform Titles and Vignettes. 4 Vols. foolscap 8vo. 2l. 2s. bds.

The above may also be had in separate Portions, viz.—

THE VENETIAN BRACELET; and other POEMS. 10s. 6d. bds.

THE GOLDEN VIOLET; and other POEMS. 10s. 6d. bds.

THE TROUBADOUR. 4th Edition. 10s. 6d. bds.

THE IMPROVISATRICE. 6th Edition. 10s. 6d. bds.

MEMOIR of the LIFE, WRITINGS, and CORRESPONDENCE of JAMES CURRIE, M.D. F.R.S. of Liverpool, Fellow of the Royal College of Physicians of Edinburgh, &c. Edited by his Son. W. W. CURRIE. 2 Vols. 8vo. with a Portrait. 28s. bds.

"*There is so much sterling value in*" these volumes, "*and they address so extensive a class of intelligent men, that we cannot doubt of their ample success.*"—*Lit. Gazette.*

TRADITIONS of LANCASHIRE. By J. ROBY, Esq. M.R.S.L. with highly-finished Plates by Finden, and numerous Woodcuts by Williams, &c.

FIRST SERIES.—2 vols. demy 8vo. 2l. 2s. in cloth; royal 8vo. with India Proofs, 4l. 4s.; and with India Proofs and Etchings, 4l. 14s. 6d.

SECOND SERIES.—2 vols. demy 8vo. 2l. 2s. in cloth; royal 8vo. with India Proofs, 3l. 3s.; and with India Proofs and Etchings, 4l. 4s.

"*A work which must be seen to be estimated as it ought.*"—*Literary Gazette.*

LITERARY RECOLLECTIONS. By the Rev. R. WARNER, F.A.S. Rector of Great Chalfield, Wilts, &c. &c. In 2 vols. 8vo. price 26s. bds.

"*We have seldom seen so much good sense and good humour, united with a greater abundance of charitable feeling and innocence of purpose. The style is remarkably forcible, chaste, and elegant.*"—*Monthly Rev.*

DR. ARNOTT's ELEMENTS of PHYSICS, or NATURAL PHILOSOPHY; written for universal use, in plain or non-technical language. 8vo. 5th Edition, enlarged.

Vol. I. (price 21s.) has Treatises on Mechanics, Hydrostatics (with an account of the FLOATING BED, lately contrived by Dr. Arnott for the relief of the bed-ridden,) Pneumatics, Acoustics, Animal Mechanics, &c.; Vol. II. Part I. (price 10s. 6d.) on Heat, Optics, &c.; and Vol. II. Part 2 (to complete the work) on Electricity, Magnetism, and Astronomy.

"A useful and excellent work."—*Sir J. Herschel.*

BIBLIOTHECA CLASSICA; or a CLASSICAL DICTIONARY, on a plan entirely new, containing a minute and accurate account of the Proper Names which occur in Greek and Latin Authors, relating to HISTORY, BIOGRAPHY, MYTHOLOGY, GEOGRAPHY, and ANTIQUITIES. By JOHN DYMOCK, LL.D. and THOMAS DYMOCK, M.A. 1 large vol. 8vo. 16s. cloth.—*The quantities of the proper names are marked throughout the work, the inflexions and genders are pointed out, and the adjectives and other derivatives subjoined—advantages which no other classical dictionary possesses.*

CONRAD BLESSINGTON: a TALE. By a LADY. Post 8vo. 7s. boards.

"A very graceful and pleasing volume. The story is interesting, the language refined, and the sentiments those of an accomplished and amiable woman."—*Lit. Gazette.*

NEW SYSTEM of GEOLOGY, in which the great Revolutions of the Earth and Animated Nature are reconciled at once to Modern Science and to Sacred History. By ANDREW URE, M.D. F.R.S. Member of the Geol. and Astron. Societies of London, &c. In 1 Vol. 8vo. with 7 Plates and 51 Woodcuts. Price 1l. 1s. bds.

.... "We regard this New System of Geology as one of the most valuable accessions lately made to the Scientific Literature of our country."—*Brande's Journal of Science.*

LIFE of FREDERIC the SECOND, KING of PRUSSIA. By LORD DOVER. 2 vols. 8vo. with Portrait, Second Edition, 28s. bds.

"A most delightful and comprehensive work.—Judicious in selection, intelligent in arrangement, and graceful in style."—*Lit. Gazette.*

SERMONS. By RALPH WARDLAW, D.D. Glasgow. In 8vo. 12s. bds.

By the same Author,

DISCOURSES on the Principal Parts of the SOCINIAN CONTROVERSY. 8vo. 15s. bds. 4th Edit. much enlarged.

MEDICAL GUIDE: for the Use of the Clergy, Heads of Families and Seminaries, and Junior Practitioners in Medicine: embracing the Discoveries of the most eminent Continental, American, and British Practitioners, which are entitled to the attention of the public, or of the medical profession of this country. By RICHARD REECE, M.D. &c. 8vo. 16th Edition, with considerable additions, 12s. Bds.

ANNUAL BIOGRAPHY and OBITUARY, for 1834; forming Vol. XVIII.: containing Memoirs of Lord Exmouth, Sir Geo. Dallas, Bart., Sir John Malcolm, Earl Fitzwilliam, Lord Dover, Sir Henry Blackwood, W. Wilberforce, Esq., Sir E. G. Colpoys, Capt. Lyon, R.N., Rajah Rammohun Roy, Admiral Boys, J. Heriot, Esq., Mrs. Hannah More, Sir Christopher Robinson, Rev. Rowland Hill, Edmund Kean, Esq., Sir Thomas Foley, Sir John A. Stevenson, Lord Gambier, Sir Banastre Tarleton, &c. &c. &c. 8vo. 15s. bds.—*⁎⁎* A few complete sets of the work can be had.

WORKS of WILLIAM PALEY, D.D. with additional Sermons, &c. and a Life of the Author. By the Rev. EDMUND PALEY, M.A. Vicar of Easingwold. A New Edition. 6 Vols. 8vo. 2l. 14s. bds.

By the same Author,

SERMONS on SEVERAL SUBJECTS. 8th Edition. 10s. 6d. bds.

Printed for Longman, Rees, Orme, Brown, Green, and Longman.

TREATISE on the STEAM ENGINE; Historical, Practical, and Descriptive. By JOHN FAREY, Engineer. In 4to. illustrated by numerous Woodcuts, and 25 Copperplates, engraved by Wilson Lowry, from Drawings by Messrs. Farey. 5l. 5s. bds.—*Vol. II. is in the Press.*

EDINBURGH GAZETTEER; containing a Description of the various Countries, States, Cities, Towns, Mountains, Seas, Rivers, Harbours, &c. of the World. Abridged from the larger work. 2d Edit. in 1 large Vol. 8vo. with an Abstract of the Population Return of 1831, and Maps, 18s. bds.

As an Accompaniment to the above Work,

NEW GENERAL ATLAS, constructed by A. ARROWSMITH, Hydrographer to the King; comprehended in Fifty-four Maps, including Two New Maps, with recent Discoveries. Royal quarto, half-bound, 1l. 16s. plain; coloured, 2l. 12s. 6d.

FRANCE in the LIVES of her GREAT MEN. By G. P. R. JAMES, Esq. Vol. 1—HISTORY of CHARLEMAGNE. 8vo. with Portrait, 16s. bds.

" Our author has executed the present work with so much ability, that we shall look forward to those which are to succeed it with much interest."—*Quarterly Review.*

INTRODUCTION to ENTOMOLOGY; or, Elements of the Natural History of Insects. By WM. KIRBY, M.A. F.R.S. and L. S. and WM. SPENCE, Esq. F.L.S. 4 thick Vols. 8vo. with Plates, and Portraits of the Authors, price 4l.—A SCIENTIFIC INDEX may be had, price 2s.

POETICAL WORKS of W. WORDSWORTH, Esq. New Edition, including the contents of the former edition in 5 vols., and some additional Pieces. 4 vols. fcap. 8vo. 24s. bds.

The EXCURSION may be had separately. 7s. bds.

By the same Author,

DESCRIPTION of the SCENERY of the LAKES in the North of England. With a Map. 5s. 6d.

MARY of BURGUNDY; or, the Revolt of Ghent. By the Author of "RICHELIEU," "DARNLEY," &c. 3 vols. post 8vo. 31s. 6d. bds.

" Decidedly the very best romance that Mr. James has produced."—*Lit. Gazette.*

PRINCIPLES of CHRISTIAN PHILOSOPHY; containing the Doctrines, Duties, Admonitions, and Consolations of the Christian Religion. By JOHN BURNS, M.D. Regius Professor of Surgery in the University of Glasgow, &c. 12mo. 4th Edit. 7s. bds.

HISTORY of ENGLAND, from the earliest Period to the Death of Elizabeth. By SHARON TURNER, F.A.S. R.A.S.L. 12 Vols. 8vo. 8l. 3s. bds.

The same may also be had in the following separate portions:—

REIGNS of EDWARD VI. MARY, and ELIZABETH; being the Second Part of the Modern History England. 2 Vols. 8vo. 2d Edit. 32s. bds.

HISTORY of the ANGLO-SAXONS; comprising the History of England from the earliest Period to the Norman Conquest. 3 Vols. 8vo. 5th Edit. 2l. 5s. bds.

HISTORY of ENGLAND, during the MIDDLE AGES; comprising the Reigns from William the Conqueror to the Accession of Henry VIII., and also the HISTORY of the LITERATURE, POETRY, RELIGION, the PROGRESS to the REFORMATION, and of the LANGUAGE of ENGLAND during that Period. 5 Vols. 8vo. 3d Edition, 3l. bds.

HISTORY of the REIGN of HENRY VIII.; comprising the Political History of the Commencement of the English Reformation: being the First Part of the Modern History of England. 2 Vols. 8vo. 3d Edit. 26s. bds.

Modern Publications, and New Editions of Valuable Standard Works,

MILITARY MEMOIRS of FIELD-MARSHAL the DUKE of WELLINGTON. By Major MOYLE SHERER. 2 vols. fcap. 8vo. 10s. cloth.

" Major Sherer has ably completed his difficult task of compressing the history of the Duke of Wellington and of the wars of our times into this excellent epitome. We rejoice to see the work accomplished so creditably to the writer."—*United Service Journal.*

Uniform with the above,
LIFE AND REIGN OF GEORGE IV. By W. WALLACE, Esq. Barrister-at-Law. 3 Vols. 15s.
ANNUAL RETROSPECT OF PUBLIC AFFAIRS for 1831. 2 vols. 10s.
HISTORICAL MEMOIRS OF THE HOUSE OF BOURBON. 2 vols. 10s.

BRITISH FLORA; comprising the PHENOGAMOUS or FLOWERING PLANTS, and the FERNS. By WILLIAM JACKSON HOOKER, LL.D. Regius Professor of Botany in the University of Glasgow, &c. &c. 12s. cloth.

By the same Author,
THE MOSSES, and the rest of the CRYPTOGAMIA: forming Vol. V. of Smith's " English Flora," or Vol. II. of the Author's " British Flora." Part I. (MUSCI, HEPATICÆ, LICHENS, CHARACEÆ, and ALGÆ), 12s. bds.; the concluding Part (FUNGI) is in progress.

MUSCOLOGIA BRITANNICA; containing the Mosses of Great Britain and Ireland, systematically arranged and described; with Plates. By W. J. HOOKER, LL.D. F.L.S. &c. and T. TAYLOR, M.D. F.L.S. &c. 8vo. 2d Edit. enlarged. 3l. 6s. plain; 3l. 3s. col'd.

PRACTICAL TREATISE on RAIL-ROADS, and INTERIOR COMMUNICATION IN GENERAL: containing an Account of the Performances of the Locomotive Engines at and subsequent to the Liverpool Contest; upwards of 260 Experiments; with Tables of the Value of Canals and Rail-roads; &c. By N. WOOD, Civil Engineer, &c. 8vo. New Edition, numerous Plates, 18s. bds.

LALLA ROOKH. An Oriental Romance. By T. MOORE, Esq. New Edition, with 4 Engravings, from Paintings by R. Westall, R.A. Fcp. 8vo. 14s. bds.

Another Edition of this Work, in 8vo. 14s.; Illustrations of the same, by Westall, 12s.

By the same Author,
LIFE and DEATH of LORD EDWARD FITZGERALD. 2 vols. small 8vo. with a Portrait, Third Edition, 21s. bds.
THE LOVES of the ANGELS. 8vo. 5th Edition, 9s. bds.; WESTALL'S ILLUSTRATIONS of the same, 5s.
THE EPICUREAN. A Tale. Fcp. 8vo. 5th Edit. 9s. bds.

MEMOIRS of the COURT of QUEEN ELIZABETH. By LUCY AIKIN. 6th Edition. 2 Vols. 8vo. 1l. 5s. bds.

Also may be had,
MEMOIRS of the COURT of KING JAMES I. By LUCY AIKIN. 2 Vols. 8vo. 3d Edit. 24s. bds.
ANNALS of the REIGN of GEORGE III. By JOHN AIKIN, M.D. 2 Vols. 8vo. 3d Edition, 25s. bds.

LETTERS to a YOUNG NATURALIST on the STUDY of NATURE and NATURAL THEOLOGY. By JAMES L. DRUMMOND, M.D. Professor of Anatomy and Physiology, Belfast. 12mo. with Cuts, 2d Edition, 7s. 6d. bds.

" Happily calculated to generate in a young mind, to sustain in the matured, and to renovate in the old, an ardent love of nature under all her forms."—*Monthly Review.*

By the same Author,
FIRST STEPS to BOTANY, intended as popular Illustrations of the Science, leading to its Study as a Branch of Education. 12mo. numerous Woodcuts. 3d Edit. 9s.

" This answers more completely to the proper notion of an Introduction to Botany than any work we have seen."—*Eclectic Review.*

Printed for Longman, Rees, Orme, Brown, Green, and Longman. 11

NARRATIVE of a NINE MONTHS' RESIDENCE in NEW ZEALAND, in 1827; with a Journal of a Residence in Tristan d'Acunha. By A. EARLE, Draughtsman to his Majesty's Discovery Ship "The Beagle." 8vo. with Engravings, 13s. bds.

"One of the most extraordinary narratives of personal adventure which have fallen within our observation for some time."—*Monthly Review.*

ELEMENTS of PLANE GEOMETRY. By THOMAS KEITH. 8vo. 3d Edit. 10s. 6d. bds.

By the same Author,

INTRODUCTION to the Theory and Practice of PLANE and SPHERICAL TRIGONOMETRY. 8vo. 6th Edition, improved, 14s. bds.

NEW TREATISE on the USE of the GLOBES. Designed for the Instruction of Youth. 12mo. with Plates, New Edit. 6s. 6d. bd.

SYSTEM of GEOGRAPHY, for the Use of Schools. 12mo. illustrated by Maps and Plates, 6s. bd.

ILLUSTRATED INTRODUCTION to LAMARCK'S CONCHOLOGY, contained in his "Histoire Naturelle des Animaux sans Vertèbres;" being a Literal Translation of the Descriptions of the recent and Fossil Genera, accompanied by Twenty-two highly-finished Lithographic Plates, in which are given Instructive Views of the various Genera and their Divisions, drawn from Nature, from characteristic and generally well-known Species. By EDMUND A. CROUCH, F.L.S. Royal 4to. 1l. 11s. 6d. plain, or 3l. 3s. coloured.

Also,

EPITOME of LAMARCK'S ARRANGEMENT of TESTACEA: with Illustrative Observations and comparative Tables of the Systems of Linnæus and Lamarck. By C. DUBOIS, F.L.S. and F.H.S. 8vo. 14s. bds.

RODERICK, the LAST of the GOTHS: a Poem. By ROBERT SOUTHEY, LL. D. &c. &c. 2 Vols. fcap. 8vo. 16s. bds.

By the same Author,

THALABA, 2 Vols. 16s.; Madoc, 2 Vols. 16s.; Curse of Kehama, 2 Vols. 14s.; Minor Poems, 3 Vols. 18s.; Pilgrimage to Waterloo, 10s. 6d.; Tale of Paraguay. 10s. 6d.; Carmen Triumphale; and Carmen Aulica, for 1814, 5s.

A VISION of JUDGMENT, a Poem. 4to. 15s. bds.

ELEMENTS of MUSICAL COMPOSITION; comprehending the Rules of Thorough Bass, and the Theory of Tuning. By WILLIAM CROTCH, Mus. Doc. Professor of Music in the University of Oxford. 2d edit. small 4to. with Plates, 12s. in cloth.

By the same Author,

SUBSTANCE of several COURSES of LECTURES on MUSIC. 7s. 6d.

FAMILY SHAKSPEARE; in which nothing is added to the Original Text; but those Words and Expressions are omitted which cannot with propriety be read aloud in a Family. By THOMAS BOWDLER, Esq. F.R.S. &c. New Edition. In 1 large vol. 8vo. with Illustrations by Smirke, Howard, &c. engraved on wood by Thomson. 30s. in cloth; or 31s. 6d. with gilt edges.

The same work, in 8 vols. 8vo. 4l. 14s. 6d. bds.

"We are of opinion, that it requires nothing more than a notice to bring this very meritorious publication into general circulation."—*Edin. Rev.*

By the same Editor,

GIBBON'S HISTORY of the ROMAN EMPIRE: for the Use of Families and Young Persons. Reprinted from the original Text, with the careful Omission of all Passages of an irreligious or immoral Tendency. 5 Vols. 8vo. 3l. 3s. bds.

CONVERSATIONS on the **EVIDENCES** of CHRISTIANITY; in which the leading Arguments of the best Authors are arranged, developed, and connected with each other: for the Use of Young Persons and Students. 12mo. 8s. bds.

SKETCH of **ANCIENT** and **MODERN GEOGRAPHY**, for the Use of Schools. By SAMUEL BUTLER, D.D. F.R.S. &c. Head Master of Shrewsbury Royal Free Grammar School. 8vo. New Edition, with important Additions, 9s. bds.

By the same Author,

ATLAS of **MODERN GEOGRAPHY**, consisting of 22 Coloured Maps, from a New Set of Plates, with an Index of all the Names. 8vo. 12s. half-bound.

ATLAS of **ANCIENT GEOGRAPHY**, consisting of 21 Coloured Maps, with a complete accentuated Index. 8vo. 12s. half-bound.

GENERAL ATLAS of **ANCIENT** and **MODERN GEOGRAPHY**, 43 Coloured Maps, and two Indexes. 4to. 24s. half-bound.

*** The latitude and longitude are given in the Indexes to these Atlases.

OUTLINE GEOGRAPHICAL COPY-BOOKS, with the Lines of Latitude and Longitude only; adapted to Dr. Butler's Atlases. 4s. each; or 7s. 6d. together.

OUTLINE MAPS of **ANCIENT GEOGRAPHY**; being a Selection, by Dr. Butler, from D'Anville's Ancient Atlas. Folio, 10s. 6d.

PRAXIS on the **LATIN PREPOSITIONS**; being an Attempt to illustrate their Origin, Power, and Signification, in the way of Exercise. 8vo. 4th Edit. 6s. 6d. bds.

KEY to the same. 8vo. 6s. bds.

MANUAL of the **LAND** and **FRESH-WATER SHELLS** of the BRITISH ISLANDS; with an Index of English Names. By W. TURTON, M.D. Fcap. 8vo. with 10 coloured Plates, comprising Figures of 150 Specimens, 10s. 6d. in cloth.

LIVES of **ENGLISH FEMALE WORTHIES.** By Mrs. JOHN SANDFORD. Vol. I., containing LADY JANE GREY and MRS. COLONEL HUTCHINSON. Foolscap 8vo. 6s. 6d. cloth.

"A most interesting work, with a high tone of moral and religious feeling."—*Lit. Gazette.*

By the same Author,

WOMAN, in her Social and Domestic Character. Fcap. 8vo. 2d Edit. 6s. cloth.

LINNEAN SYSTEM of **CONCHOLOGY.** By J. MAWE. 8vo. with a Plate to each Genus (37); plain, 1l. 1s.; coloured after Nature, 2l. 12s. 6d.

By the same Author,

SHELL-COLLECTOR'S PILOT, with Coloured Frontispiece; also, the best Methods of preserving Insects, Birds, &c. 4th Edit. 5s. bds.

NEW DESCRIPTIVE CATALOGUE of **MINERALS.** 7th Edit. 6s. bds.

HISTORICAL MEMOIRS of the **HOUSE** of **RUSSELL**; from the time of the Norman Conquest. By J. H. WIFFEN, M.R.S.L. &c. With much curious unpublished Correspondence. 2 vols. 8vo. with Plates and Portraits, 2l. 2s. in cloth; royal 8vo. (India Proofs), with the FIRST RACE OF ANCESTRY, &c. 3l. 13s. 6d.

Also, separately,

HISTORICAL MEMOIRS of the **FIRST RACE** of **ANCESTRY** whence the House of Russell had its origin. Royal 8vo. 7s.

Printed for Longman, Rees, Orme, Brown, Green, and Longman. 13

DOMESTIC DUTIES; or, Instructions to Young Married Ladies, on the Management of their Households, and the Regulation of their Conduct. By Mrs. WILLIAM PARKES. 12mo. 3d Edit. 10s. 6d. bds.

"The volume before us is a perfect *vade mecum* for the young married lady, who may resort to it on all questions of household economy and etiquette."—*New Monthly Mag.*

COLLECTIONS from the GREEK ANTHOLOGY. By the late Rev. ROBERT BLAND, and others. New Edition; comprising the fragments of early Lyric Poetry, with Specimens of all the Poets included in Meleager's Garland. By J. H. MERIVALE, Esq. F.S.A. Post 8vo. 14s. cloth. (A Second Volume is in progress.)

"A very delightful volume."—*Blackwood's Magazine.*

PICTURESQUE ANTIQUITIES of the **ENGLISH CITIES**; containing 60 Engravings by Le Keux, &c. and 24 Woodcuts: with Historical and Descriptive Accounts of the Subjects, and of the Characteristic Features of each City. By JOHN BRITTON, F.S.A. &c. Med. 4to. elegantly hf.-bd. 7l. 4s.; imp. 4to. with Proofs, 12l.

By the same Author,

DICTIONARY of the ARCHITECTURE and ARCHÆOLOGY of the MIDDLE AGES; including the Words used by Old and Modern Authors. Part I. with 12 Engravings by J. Le Keux; and Part II. with 10 Engravings. The Volume will contain at least 40 Engravings, and be completed in 4 Parts Royal 8vo. 12s. each; med. 4to. 21s.; imp. 4to. 31s. 6d.

CATHEDRAL ANTIQUITIES of ENGLAND; or, an Historical, Architectural, and Graphical Illustration of the English Cathedral Churches. 12s. per Number in Med. 4to.; and 20s. Imp. 4to.: 52 Numbers are published. Each Size classes with the *Architectural Antiquities of Great Britain*. The following are complete, and may be had separately, viz.

Salisbury Cathedral, with 31 Engravings, med. 4to. 3l. 3s.; imp. 4to. 5l. 5s.; cr. fol. 8l.; sup.-roy. fol. 11l. Bds.

Norwich, with 25 Plates, med. 4to. 2l. 10s.; imp. 4to. 4l. 4s.; cr. fol. 6l. 10s.; sup.-roy. fol. 8l. 16s. Bds.

Lichfield, with 16 Engravings, med. 4to. 1l. 18s.; imperial 4to. 3l. 3s.; sup-roy. fol. 6l. 6s. Bds.

York, with 35 Engravings, med. 4to. 3l. 15s.; imp. 4to. 6l. 6s.; cr. fol. 9l. 9s.; sup.-roy. fol. 12l. 12s. Bds.

Bristol, 14 Engravings, med. 4to. 1l. 4s.; imp. 4to. 2l. 2s.

Oxford, with 11 Engravings, med. 4to. 1l. 4s.; imp. 4to. 2l. 2s.; sup.-roy. fol. 4l. 4s. Bds.

Canterbury, 26 Engravings, med. 4to. 3l. 3s. imp. 5l. 5s.; sup.-roy. fol. 11l. Bds.

Exeter, with 22 Engravings, med. 2l. 10s.; imp. 4to. 4l. 4s.; sup.-roy. fol. 8l. 16s. Bds.

Wells, with 24 Engravings, med. 4to. 2l. 10s.; imp. 4l. 4s.; sup.-roy fol. 8l. 16s.; or with proofs and etchings, 16l. 16s.

Peterborough, 17 Plates, med. 4to. 1l. 18s.; imp. 3l. 3s, with proofs & etchings, 6l. 6s.; sup.-roy. fol. 6l. 6s, with pr. & etch. 12l. 12s.

Gloucester, 22 Engravings and 2 Woodcuts, med. 4to. 2l. 10s.; imp. 4l. 4s.

Winchester, 30 Engravings, med. 4to. 3l. 3s.; imp. 4to. 5l. 5s.; cr. fol. 8l.; sup.-roy. fol. 11l. Bds.

Hereford, with 16 Engravings, med. 1l. 18s.; imp. 3l. 3s.; sup.-roy. fol. 6l. 6s.

Worcester Cathedral will be comprised in 3 Nos. Nos. I. and II. are published.

HISTORY and ANTIQUITIES of BATH ABBEY CHURCH; with Ten Engravings, by J. and H. LE KEUX, from Drawings by MACKENZIE, &c. Royal 8vo. 12s.; Med. 4to. 21s.; Imp. 4to. 31s. 6d.

HISTORY and ILLUSTRATION of REDCLIFFE CHURCH, Bristol. Royal 8vo. 16s; imp. 4to. 31s. 6d.

ARCHITECTURAL ANTIQUITIES of GREAT BRITAIN. 4 Vols. Med. 4to. 21l.; or Imp. 4to. 32l. half-bd.

CHRONOLOGICAL and HISTORICAL ILLUSTRATIONS of the ANCIENT ARCHITECTURE of GREAT BRITAIN. 4to. 6l. 12s.; large paper, 11l.

⁎⁎ To correspond with the "Architectural Antiquities," of which this work forms the 5th Volume.

The ARCHITECTURAL ANTIQUITIES may be purchased in Ten Separate Parts, boards, at Two Guineas each. Parts XI. and XII. at 2l. 8s. each; and Part XIII. which completes the Fifth Volume, or *Chronological Series*, 1l. 16s.

THE ANALYSIS of INORGANIC SUBSTANCES. By J. J. BERZELIUS. From the French, by G. O. REES. 12mo. with a Plate, 5s. bds.

"A valuable acquisition to the student of practical chemistry."—*Medical Gazette.*
"The notes of Mr. Rees are highly valuable."—*Monthly Review.*

TRAVELS of an IRISH GENTLEMAN in search of a RELIGION. With Notes and Illustrations by the Editor of "Captain Rock's Memoirs." 2 vols. fcap. 8vo. 2d edition, 18s. bds.

"These volumes are amongst the most interesting records of which the operations of the human mind ever formed the theme."—*Monthly Review.*

ENGLISH FLORA. By SIR JAMES E. SMITH, M.D. F.R.S. President of the Linnæan Society, &c. 4 Vols. 8vo. New edition, 2l. 8s. bds.

Part I. of Vol. V. (CRYPTOGAMIA, by Dr. HOOKER,) 12s.; Part II. *in progress.*

Also by Sir J. E. SMITH,

COMPENDIUM of the ENGLISH FLORA. 12mo. 7s. 6d. cloth.

COMPENDIUM FLORÆ BRITANNICÆ. 12mo. 5th Edit. 7s. 6d. bds.

INTRODUCTION to the STUDY of PHYSIOLOGICAL and SYSTEMATICAL BOTANY. 8vo. New Edition, with Illustrations of the Natural Orders (combining the object of Sir J. Smith's "Grammar" with that of his "Introduction,") by W. J. HOOKER, LL.D. &c. 36 Plates, 16s. cloth.

PEN TAMAR; or, the History of an Old Maid. By the late Mrs. H. M. BOWDLER. Second Edition. Post 8vo. 10s. 6d. bds.

"Written with great simplicity, and in the most engaging spirit of benevolence."—*Monthly Review.*

HISTORY of the REIGN of GEORGE the THIRD. By ROBERT BISSET, LL.D. Author of the Life of Burke, &c. &c. A new Edition, completed to the Death of the King, in 6 Vols. 8vo. Price 3l. 3s. bds.

INSTRUCTIONS to YOUNG SPORTSMEN in all that relates to Guns and Shooting: Difference between the Flint and Percussion System; PRESERVATION of GAME; Getting Access to all Kinds of Birds; Specific Directions, with new Apparatus, for WILD FOWL SHOOTING, both on the Coast and in Fresh Water; New Directions for TROUT FISHING; and Advice to the Young Sportsman on other Subjects. By Lieut. Col. P. HAWKER. 8vo. 7th Edit. enlarged and improved, with an ABRIDGMENT of the OLD and NEW GAME LAWS, and 30 Plates and Woodcuts, 18s. cloth.

"Col. Hawker is one of the best shots in England, and his 'Instructions to Sportsmen' the very best book we have on the subject."—*Blackwood's Magazine.*

REMAINS of HENRY KIRKE WHITE, selected, with prefatory Remarks, by ROBERT SOUTHEY, Esq. The only complete Editions. 2 Vols. 8vo. 24s. bds.; and 1 Vol. 24mo. with engraved title and vignettes, 5s. bds.

N. B. The property of the Family having been invaded, it is necessary to state that these are the *only Editions* which contain the Life by Mr. Southey, and the whole of the contents of the Third Volume.

CONVERSATIONS on ALGEBRA; being an Introduction to the First Principles of that Science, designed for those who have not the Advantage of a Tutor, as well as for the Use of Students in Schools. By W. COLE. 12mo. 7s. bds.

EXPLANATORY PRONOUNCING DICTIONARY of the FRENCH LANGUAGE, in French and English. By L'ABBE TARDY, late Master of Arts in the University at Paris. In 12mo. new Edition, revised, 6s. bd.

Printed for Longman, Rees, Orme, Brown, Green, and Longman. 15

SELECT WORKS of the BRITISH POETS, from JONSON to BEATTIE. With Biographical and Critical Prefaces. By Dr. Aikin. Complete in 1 Vol. 8vo. for Schools, &c. 18s. in cloth; or neatly done up, gilt edges, 20s. Also, in 10 Vols. post 18mo. price 2l.; in royal 18mo. to match the British Essayists and Novelists, 3l.

SELECT WORKS of the BRITISH POETS, from CHAUCER to WITHERS. With Biographical Sketches. By Robert Southey, LL.D. Poet Laureate. 1 vol. 8vo. uniform with " Aikin's Poets," 30s. in cloth; or neatly done up, with gilt edges, 1l. 11s. 6d.

HISTORY of ROMAN LITERATURE, from its earliest Period to the end of the Augustan Age. By J. Dunlop, Esq. 3 Vols. 8vo. 2l. 7s. 6d. bds.

By the same Author,

HISTORY of FICTION. 3 Vols. post 8vo. 2l. 2s. bds.

CONVERSATIONS on VEGETABLE PHYSIOLOGY; comprehending the Elements of Botany, with their Application to Agriculture. 2 vols. 12mo. with Plates, 2d Edition, 12s.

" These instructive little volumes are composed by an author (Mrs. Marcet) already well known by similar works on other branches of science, all of which have been received with great and merited favor."—*Edin. Rev.*

By the same Author,

CONVERSATIONS on CHEMISTRY, in which the Elements of that Science are familiarly explained and illustrated by Experiments; with a Conversation on the Steam Engine. 2 Vols. 12mo. 12th Edit. with Plates by Lowry, 14s. bds.

CONVERSATIONS on NATURAL PHILOSOPHY. 7th Edition, 10s. 6d. bds. With 22 Engravings by Lowry.

CONVERSATIONS on POLITICAL ECONOMY. 12mo. 6th Edit. 9s. bds.

MALTE BRUN's SYSTEM of GEOGRAPHY; with an Index of 44,000 Names. Complete in 9 vols. 8vo. price £7, bds.

In the translation now offered to the public, many important corrections and additions have been introduced. The additions to the Description of Great Britain and Ireland are more especially extensive: in fact, this portion of the translation has been entirely re-written, and rather merits the title of an original work. The Geographical Index will be found to be the most accurate and comprehensive work of the kind in our language: it is so constructed as to be a complete Table of Reference to the whole work, while it forms an extensive and useful Gazetteer.

*** Subscribers are requested to complete their Sets forthwith, as several of the Parts are nearly out of print.

" The translators of M. Malte Brun's Geography have done great service to the public, by rendering so valuable a work accessible to the English reader."—*Edinburgh Review.*
" Infinitely superior to any thing of its class which has ever appeared."—*Literary Gazette.*

CONVERSATIONS on the ANIMAL ECONOMY of MAN. By a Physician. 2 Vols. 12mo. illustrated by Plates, &c. 16s. bds.

" The Author, in our opinion, has succeeded in producing an accurate, interesting, and highly amusing account of the Animal Economy."—*Jameson's Philosophical Journal.*

LIFE and PONTIFICATE of GREGORY the SEVENTH. By Sir Roger Greisley, Bart. F.A.S. 8vo. 12s. bds.

CONVERSATIONS on MINERALOGY; with upwards of 400 (including 12 beautifully coloured) Figures. 2 Vols. 12mo. New Edition preparing.

" One of the most desirable Text Books that have issued from the British Press."—*Monthly Review.*

Modern Publications, &c. printed for Longman and Co.

TREATISE on the VALUATION of PROPERTY for the POOR'S RATE; shewing the Method of rating Lands, Tithes, Woods, &c. By J. S. BAYLDON, Land Agent and Appraiser. 8vo. new Edition, 7s. 6d. bds.

By the same Author,

ART of VALUING RENTS and TILLAGES, and the Tenant's Right on entering and quitting Farms explained. For the Use of Landlords, Land-Agents, Appraisers, Farmers, and Tenants. 8vo. 4th Edit. corrected and enlarged, 7s. bds.

CONVERSATIONS on BOTANY, with 21 Engravings. 12mo. 7th Edition, enlarged. 7s. 6d. plain; 12s. col'd.

SACRED HISTORY of the WORLD, from the Creation to the Deluge: attempted to be philosophically considered, in a Series of Letters to a Son. By SHARON TURNER, F.S.A. and R.A.S.L. 8vo. 4th Edition. 14s. bds.

INTRODUCTION to GEOLOGY; intended to convey a Practical Knowledge of the Science, and comprising the most important recent Discoveries. By ROBERT BAKEWELL. 8vo. 4th Edition, with considerable Additions (including 5 entirely new chapters), new Plates, and numerous Cuts, 21s. bds.

"We consider that the present is by far the best introduction extant."—*Lit. Gazette.*

LACON; or, MANY THINGS in FEW WORDS. By the Rev. C. C. COLTON, late Fellow of King's Coll. Cambridge. New Edit. in 1 vol. 8vo. 12s. cloth.

THE PELICAN ISLAND, in 9 Cantos; and other Poems. By JAMES MONTGOMERY. Foolscap 8vo. 3d Edit. 8s. bds.

By the same Author,

THE WANDERER of SWITZERLAND. Fcap. 8vo. 10th Edit. 6s. bds.

THE WORLD before the FLOOD. Fcap. 8vo. 8th Edit. 9s. bds.

THE WEST INDIES, and other POEMS. Fcap. 8vo. 7th Edit. 6s. bds.

GREENLAND, and other POEMS. Fcap. 8vo. 3d Edit. 8s. bds.

SONGS of ZION, being IMITATIONS of PSALMS. Fcap. 8vo. 3d Edit. 5s. bds.

VERSES to the MEMORY of R. REYNOLDS. 2s.

MEMOIRS of JOHN, DUKE of MARLBOROUGH; with his Original Correspondence. By the Rev. Archdeacon COXE. 6 Vols. 8vo. with an Atlas, 5l. 5s.

By the same Author,

MEMOIRS of the ADMINISTRATION of the Right Hon. H. PELHAM. 2 vols. 4to. with Portraits, 5l. 5s. bds.; large paper, 10l. 10s.

MEMOIRS of HORATIO LORD WALPOLE. 2 Vols. 8vo. 3d Edition. 26s. bds.

MEMOIRS of the KINGS of SPAIN of the HOUSE of BOURBON, from 1700 to 1788. 5 Vols. 8vo. 2d Edition. 3l. bds.

HISTORY of the HOUSE of AUSTRIA, from 1218 to 1792. 5 vols. 8vo. 3l. 13s. 6d. bds.

ORIENTAL CUSTOMS applied to the ILLUSTRATION of the SACRED SCRIPTURES. By SAMUEL BURDER, A.M. &c. 12mo. 8s. 6d. bds.